University of
Waterloo

The UW Library
gratefully acknowledges
the support of
the Yu family
for the Kresge Challenge,
a fundraising effort to
help create the Next
Generation Library.

Parents IN PARTNERSHIP

Limnological Analyses

Third Edition

Robert G. Wetzel
Gene E. Likens

Limnological Analyses

Third Edition

With 89 Illustrations

 Springer

Robert G. Wetzel, Bishop Professor of Biology, Department of Biology, University of Alabama, College of Arts and Sciences, Tuscaloosa, AL 35487, USA

Gene E. Likens, Director, Institute of Ecosystem Studies, The New York Botanical Garden, Cary Arboretum, Millbrook, NY 12545, USA

Cover illustration: Cover art supplied by Gene E. Likens.

Library of Congress Cataloging-in-Publication Data
Wetzel, Robert G.
    Limnological analyses/Robert G. Wetzel, Gene E. Likens.—3rd ed.
        p.   cm.
    Includes bibliographical references.
    ISBN 0-387-98928-5 (hc.: alk. paper)
        1. Limnology—Field work.   2. Limnology—Laboratory manuals.
    I. Likens, Gene E.   II. Title.
    QH96.57.A1W48   2000
    577.6—dc21   99-042459

The first edition of this volume was published by W.B. Saunders Co. © 1979 W.B. Saunders Co.

ISBN 0-387-98928-5   Printed on acid-free paper.

Printed in the United States of America.   (BS/HAM)

9 8 7 6 5 4 3 2

springeronline.com

We acknowledge with appreciation the early limnological training and inspiration provided by David C. Chandler and Arthur D. Hasler

Preface to the Third Edition

The continuing positive responses by students and instructors to the first and second editions of *Limnological Analyses* over the past two decades are most gratifying. The exercises in this book were set forth initially because of our frustrations in teaching such a complex, multifaceted discipline as limnology. As with any ecological subject, understanding requires experiencing the techniques used for analysis of properties of the ecosystems being studied and the responses of organisms within them. Our intention is to provide a series of coupled field and laboratory exercises on the basic subjects of limnology. The exercises examine, as before, a broad range of topics within both standing and running fresh waters. Methods evolve over time, however, and here we present a revised, updated, and expanded version of the second edition.

The previous style of presentation of each exercise worked effectively in our teaching and was very well received by others. Indeed, a recent book on methods in analyses of streams imitated our style closely. Each exercise begins with a brief introduction to the topic and its relevance within the science or to human activities. Throughout, we attempt to present the details of the methods in relation to important questions and problem solving. Methods presented are generally of contemporary research quality. Although the methods are presented for instruction and problem solving, and not as a rigorous manual on limnological methods, many sections could be and have been used for research purposes. As previously, some methods presented are not the best available for the subject but were selected to introduce the subjects to students within the constraints of time and facilities of the classroom.

The importance of problem solving is reflected in the structure of the book and its exercises. Initial exercises address physical components of lakes and streams and major chemical constituents within them. These topics are then followed by quantitative methods for analyzing biota and metabolic processes. The later exercises often build on experience gained in the earlier ones. The previous provision of several options for doing the exercise continues in expanded form here in an attempt to accommodate various logistic, time, and equipment constraints. Simultaneously, these options offer suggestions for more ambitious students to extend their inquiries beyond what was undertaken in regular class periods.

The questions set forth at the end of each exercise are important to the learning experience. We have modified many of these, again based on continuing experience in the classroom. We urge their use to enhance and hone rationale underlying the methods, limitations of techniques, and interpretation of the information obtained.

We do not have a solution to the perplexing problems of needing to identify taxonomically the dominant aquatic flora and fauna. Students in initial courses in limnology usually come with diverse backgrounds and experience in taxonomy. We include only a few rudimentary pictorial keys to get the students with little experience started. Thereafter it is essential for instructors to assist with general and regional guides to identification.

In addition to the many people cited in the Preface to the Second Edition who have offered guidance and constructive criticism of our previous editions, we would particularly like to acknowledge the inputs of the following persons to the present revision: Jean-Christian Auclair, Arthur C. Benke, Arthur S. Brooks, Paul R. Burten, Jonathan J. Cole, Gary N. Ervin, Steven N. Francoeur, Stephen K. Hamilton, Richard Hauer, Stuart H. Hurlbert, Mark D. Johnson, Klaus Jürgens, Winfried Lampert, Måns Lindell, Joseph C. Makarewicz, Ian D. McKelvie, Michael L. Pace, Vincent Resh, Richard D. Robarts, David L. Strayer, Keller Suberkropp, Karen Wiltshire, and J. Wulfhorst. R. Wetzel acknowledges the stimulating atmosphere of the Max-Planck-Institut für Limnologie, Plön, Germany, that served as a refuge where much of his writing was accomplished. Inputs from student users have been most helpful. We appreciate greatly the assistance of all those mentioned and other persons who have encouraged us to prevail in this endeavor. The final responsibilities for the contents, however, rest with us.

Tuscaloosa, Alabama   Robert G. Wetzel
Millbrook, New York   Gene E. Likens

Preface to the Second Edition

The response by both instructors and students to the first edition of *Limnological Analyses*, which appeared slightly over a decade ago, has been most gratifying. The intent then was to provide a series of interrelated field and laboratory exercises on basic subjects of limnology. Those objectives have not changed. Here we present a revised, updated, and expanded version.

The 29 exercises examine a broad range of topics concerned with both standing and running fresh waters. The complex mechanisms controlling the metabolism and dynamics of biotic populations and communities in aquatic ecosystems require an understanding of physical and chemical variables. About one-third of the exercises address the major physical components of lakes and streams, important mineral nutrients, and organic matter. The remainder of the exercises provide rationale and methods for quantitative analyses of the biota, as well as some integrated analyses of whole ecosystems. A few exercises address the effects of human activities on aquatic ecosystems.

The later exercises usually build on experience gained in the earlier ones. Although most of the exercises can be performed largely in an intensive afternoon, sometimes additional time beyond the scheduled classroom hours may be needed. Certain subjects, such as decomposition, do not lend themselves to simple, short-term analyses. We urge that exercises requiring more time not be avoided, because the subjects are of major importance, and we have found them to be particularly informative in our limnological classes. Often these longer exercises provide an opportunity for more independent individual or small-team projects, with results that may be reported to the entire class in effective learning experiences.

Each exercise consists generally of a brief introduction to the topic with a statement of its limnological relevance or importance. We selected methods to illustrate a variety of problems and how they might be solved. Although the methods presented are nearly always of contemporary research quality, our intentions were not to provide a rigorous manual of limnological methods, even though many sections could be and have been used for research purposes. Certain techniques presented in the book are not the best available, but they provide an introduction to the subject within the constraints of time, facilities, and experience of the classroom. Where better methods are available, we have attempted to provide references to these techniques.

As previously, each exercise provides three or more options in an attempt to accommodate various time, logistic, and equipment constraints or to allow some students to pursue a more ambitious course of study. We have been gratified that students often have pursued aggressively many of these suggested extensions of the exercises. Wherever possible, we have attempted to combine and integrate field and laboratory analyses. However, some options can be performed completely in the classroom laboratory. Although we consider these options to be the least desirable, some topics are more appropriately conducted in the laboratory at this level of instruction. For example, the important and complex subject of hydrodynamics can be introduced in a most illuminating and enjoyable manner by means of lake model experiments.

We have expanded the questions at the end of each exercise. We encourage students to answer as many of them as possible, for they often generate critical evaluations of the problems being addressed and the efficacy of the techniques being used. The questions and references provided for each exercise can lead students to explore further the myriad complexities of freshwater ecosystems.

The taxonomy of dominant aquatic flora and fauna is consistently a source of frustration both in the teaching of courses in limnology and in the preparation of this book. The taxonomic backgrounds and experience of students entering an initial course in limnology are usually diverse. Even superficial keys to common organisms from bacteria to fish would exceed the length of the present text. In the first edition, we included a few basic keys to orient students with little taxonomical experience to major separations. These keys were not received well by systematists and were not used extensively by students. Therefore, we include only a few rudimentary "starter" taxonomic keys in this edition, and we recommend that the instructors use general and regional guides to the taxonomy of the flora and fauna of particular regions. We cite example references of useful general taxonomic works.

We recognize that it would be difficult to accomplish all of the exercises in this book in an intensive course in limnology during a single semester. Hopefully our efforts, however, will assist in reducing the labors in such courses and enhance insights into the operational integrity of aquatic ecosystems. We found the experiences of others important to the improvements to this book. We welcome further suggestions on how the exercises might be improved or expanded.

Our ideas for effective instruction of certain topics had a long and diffuse developmental history. As a result, we are no longer certain of their exact origins. Effective teachers in our background certainly contributed at least indirectly to aspects that appear in some exercises, particularly A.D. Hasler, W.T. Edmondson, G.H. Lauff, D.C. Chandler, and J.R. Vallentyne. Gordon L. Godshalk read critically the entire work and offered many helpful suggestions to both the first edition and parts of the second edition. Constructive criticism and good counsel have been received from numerous persons, including students as they grappled with the techniques in classes. Discussions about and suggested revision to portions of the first and/or second editions were provided by H.L. Allen, J.R. Barnes, R. Bilby, J. Cole, W.G. Crumpton, H.W. Cunningham, F. deNoyelles, J.A. Dickerman, J. Eaton, R. Edwards, P. Godfrey, J.B. Grace, C.A.S. Hall, D.J. Hall, R. Hall, G. Hendrey, M. Mattson, W. McDowell, J. Meyer, S. Nodvin, D.K. Nordstrom, P.H. Rich, B. Riemann, D.W. Schindler, J. Sloane, A. Stewart, D. Strayer, R. Walter, A.K. Ward, G.M. Ward, and W. Youngs, Jr. Special insights were received on specific topics by several individuals, particularly toward improvements in the second edition. Arthur C. Benke offered great refinements and improvements to the treatments of benthos. Michael F. Coveney, Clifford Ochs, and Michael Pace counseled us on our new treatment of bacterial productivity. The new exercise on predator-prey analyses was guided by the efforts of Kenneth Wagner and W. Charles Kerfoot. Mark D. Mattson contributed significantly to the revision of the exercise on the inorganic carbon complex. The exercises on manipulation of model ecosystems, diurnal changes in stream ecosystems, and effect of sewage outfall on stream ecosystems contain input from F.J. deNoyelles, P. Godfrey, C.A.S. Hall, J. Sloane-Richey, and Raymond Barrett. William D. Taylor helped with facets of phytoplankton biomass analyses, Kathleen Weathers with inorganic nutrients, Michael Pace with zooplankton feeding, Nina Caraco with whole ecosystem analyses, Jonathon Cole with decomposition, and Robert E. Moeller with paleolimnological analyses. Anita J. Johnson prepared many of the final figures. Phyllis Likens provided invaluable assistance with proofreading. We gratefully acknowledge the assistance of these and other persons that have encouraged us with this effort. The final responsibilities for the contents, however, rest with us.

Tuscaloosa, Alabama   Robert G. Wetzel
Millbrook, New York   Gene E. Likens

Contents

Preface to the Third Edition vii

Preface to the Second Edition ix

EXERCISE 1
Lake Basin Characteristics and Morphometry   1

EXERCISE 2
Light and Temperature   15

EXERCISE 3
Physical Characteristics: Lake Models   33

EXERCISE 4
The Heat Budget of Lakes   45

EXERCISE 5
Morphology and Flow in Streams   57

EXERCISE 6
Dissolved Oxygen   73

EXERCISE 7
Inorganic Nutrients: Nitrogen, Phosphorus, and Other Nutrients   85

EXERCISE 8
The Inorganic Carbon Complex: Alkalinity, Acidity, CO_2, pH,
Total Inorganic Carbon, Hardness, Aluminum.   113

EXERCISE 9
Organic Matter   137

EXERCISE 10

Composition and Biomass of Phytoplankton   147

EXERCISE 11

Collection, Enumeration, and Biomass of Zooplankton   175

EXERCISE 12

Benthic Fauna of Lakes   189

EXERCISE 13

Benthic Fauna of Streams   209

EXERCISE 14

Primary Productivity of Phytoplankton   219

EXERCISE 15

Feeding Rates by Protists and Larger Zooplankton   241

EXERCISE 16

Zooplankton Production   251

EXERCISE 17

Predator–Prey Interactions   257

EXERCISE 18

Enumeration of Fish or Other Aquatic Animals   263

EXERCISE 19

Bacterial Growth and Productivity   271

EXERCISE 20

Decomposition: Relative Bacterial Heterotrophic Activity
on Soluble Organic Matter   289

EXERCISE 21

Decomposition: Particulate Organic Matter   301

EXERCISE 22

The Littoral Zone   313

EXERCISE 23

Experimental Manipulation of Model Ecosystems   325

EXERCISE 24

Diurnal Changes in a Stream Ecosystem:
An Energy and Nutrient Budget Approach 339

EXERCISE 25

Diurnal Changes in Lake Systems 349

EXERCISE 26

Special Lake Types 355

EXERCISE 27

Historical Records of Changes in the Productivity of Lakes 361

EXERCISE 28

Effect of Sewage Outfall on a Stream Ecosystem 369

EXERCISE 29

Estimates of Whole Lake Metabolism:
Hypolimnetic Oxygen Deficits
and Carbon Dioxide Accumulation 373

APPENDIX 1

General Chemical Relationships 383

APPENDIX 2

Basic Definitions Used in Community Analyses 389

APPENDIX 3

Useful Relationships Relative to the Use of Colorimeters
and Spectrophotometers 397

APPENDIX 4

Characteristics and Taxonomic Sources of
Common Freshwater Organisms 399

APPENDIX 5

SI Conversion Factors 421

Index 425

Lake Basin Characteristics and Morphometry

Limnological analyses of a lake or stream very often require a detailed knowledge of morphometry, particularly of the volume characteristics of the body of fresh water. Depth analyses, including measurement of areas of sediments and of water strata at various depths, volumes of strata, and shoreline characteristics, are often critical to detailed analyses of biological, chemical, and physical properties of fresh waters. Morphometric parameters are needed, for example, to evaluate erosion, nutrient loading rates, chemical mass, heat content and thermal stability, biological productivity and effectiveness of growth, and many other structural and functional components of the ecosystem. Management techniques, such as the loading capacity for effluents and the selective removal of undesirable components of the biota, are also heavily dependent on a detailed knowledge of the morphometry and water retention times in freshwater ecosystems.

Accurate hydrographic maps of lakes and streams are rarely available in sufficient detail for the limnologist. It is a characteristic feature that the morphometry of lakes and streams changes with time, so even if bathymetric (i.e., depth contour) maps are available from governmental or other sources, their accuracy should be checked carefully. It is essential, therefore, that the rudiments of the construction of bathymetric maps and the computation of morphometric parameters be understood.

LOCATION WITHIN PROJECTIONS: GEOGRAPHICAL COORDINATES

Latitude, Longitude, and Bearings

A level surface, such as the surface of a lake, is parallel everywhere to the mean surface of the earth and so is essentially spherical. A level line is the arc of a circle whose radius varies according to the elevation above or below mean sea level. In the survey of small areas, or when sight lines are short, level surfaces are assumed to be planes, and level lines are assumed to be straight lines.

A great circle is the trace of a level line on a vertical plane and may be illustrated by imagining a vertical plane passing through the polar axis of the earth. The trace of the earth's surface on this vertical plane is considered to be a great circle.

Geographic meridians and parallels of latitude are a system of spherical coordinates by which it is possible to locate points or to describe precisely the locations of points on the surface of the earth. Geographic meridians (longitudes) are of great circles, since they are imaginary planes that bisect the earth, and, in this special case, these planes also bisect the polar axis. The equatorial plane bisects the earth normal (i.e., perpendicular) to the polar axis; hence, the equator is a great circle. Parallels of latitude other than the equator are not great circles but may be considered to be a series of equally spaced planes parallel to the equatorial plane. The center of the earth is the point of origin for the angular designations given to both latitudes and longitudes.

Longitudes are described as west or east of any selected meridian, which is called a primary meridian. The primary meridian most commonly used is the one passing through the observatory at Greenwich, England. Longitudes never exceed 180° east or 180° west of the primary meridian. Longitudes are true north–south lines; thus, no two north–south lines are exactly parallel.

The direction of lines may be expressed as bearings and azimuths. Both are measured in terms of degrees, minutes, and seconds of arc from a selected line of reference or meridian. Azimuths are measured clockwise through 360°, starting from the reference meridian. They may be referred to as either the north or south point, but never both on the same survey. Bearings are measured either clockwise or counterclockwise in quadrants and never exceed 90°.

Bearings are true, magnetic, or assumed, depending on the status of the reference line used. When the reference is a geographic meridian, the bearings or azimuths are true. When the reference line is arbitrarily selected for the immediate purpose of the survey, the bearings or azimuths are assumed. When the bearings or azimuths are referred to a magnetic north (compass) line they are magnetic. The relationships between true and magnetic bearings and azimuths are illustrated in Fig. 1.1.

When the magnetic poles of the earth do not coincide with the geographic poles, the compass needle does not point to true north; rather, it points along a line called a magnetic meridian, and the angular deviation of the magnetic meridian from a true, or geographic, meridian is called a magnetic declination. Magnetic declinations are east if the compass needle points east of true north and west if the needle points west of true north. The amount and direction of declination depend on both the

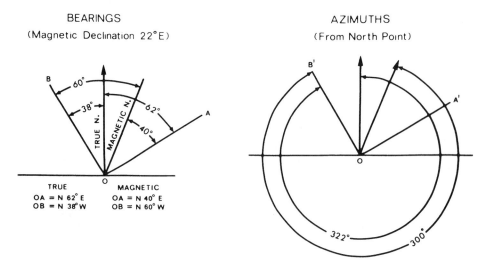

Figure 1.1. Relationships of magnetic and true bearings (*left*) and azimuths (*right*).

geographical position of the observer and the year and day on which the observations are made. The time is related to the shifting of the earth's magnetic poles. Important souces of error are annual variations, irregular variations of unknown cause, and local attractions caused by magnetic substances in the immediate vicinity of measurements.

A small-scale diagram, called an isogonic chart, is included on detailed maps to indicate the extent of declination of magnetic north from true north. An isogonic chart consists of a set of roughly parallel solid lines representing lines of equal magnetic declination. Another set of lines, which are dashed and cut across the solid lines, indicate the annual rate of change in the amount of declination. Those lines east of the zero (agonic) line show west declination; those west of it show east declination. Isogonic charts are dated so that declinations may be determined for dates after the chart was prepared, with the assistance of data from governmental or other sources.

Location within Public Lands

In 1785, a law was enacted for the purpose of subdividing public lands into townships and sections, the townships to be $6\,mi^2$ and the sections $1\,mi^2$. In 1976, the original system was amended so that sections were to be numbered in east–west rows, alternately west and east, starting with 1 in the northeast corner and ending with 36 in the southeast. This system remains in use today for all surveys throughout a majority of the United States (McEntyre, 1978). Exceptions include the states of New England, New York, Pennsylvania, and the Atlantic coastal states [cf., Hunt (1974)].

Because of the convergence of all true north–south lines (meridians), it is not possible to lay out a truly rectangular system of coordinates over large areas and maintain true east–west and north–south boundaries. Therefore, all townships are not $6\,mi^2$, nor are all sections $1\,mi^2$, even though field measurements have been done without error. An attempt is made, therefore, to lay out the townships and sections in such a manner as to place the necessary corrections for convergence in predetermined rows of sections within the townships (Fig. 1.2).

The starting point for the numbering of townships (the initial point) is a selected intersection of a geographic meridian, called a principal meridian, and a geographic parallel, called a base line. Secondary meridians are established east and west of the principal meridian at intervals calculated to contain a number of full-sized townships, usually four townships or 24 mi. Such secondary meridians are designated first, second, and so on, "guide meridians east" or "guide meridians west," depending on whether they lie east or west, respectively, of the principal meridian. At intervals of 24 mi, or other multiples of 6 mi, secondary parallels of latitude are established north and south of the base line. These lines are designated first, second, and so on, "standard parallels north" or "standard parallels south" of the base line.

The principal meridian and all base lines, standard parallels, and north and south boundaries of townships are continuous true north–south and east–west lines. Guide meridians are true north–south lines but are continuous only for the distance between two standard parallels. Along the base lines and standard parallels, guide meridians (and range lines) are offset in order to adjust for convergence.

Theoretically, the east and west sides of townships are each exactly 6 mi long, any discrepancies being due entirely to error in measurement. Along the north and south boundaries, only the south sides of those townships bordering on, and immediately

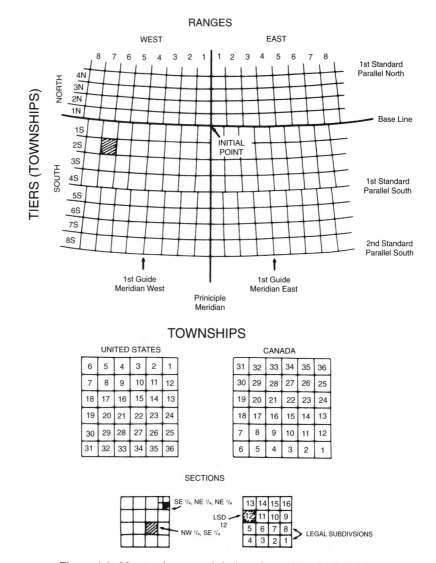

Figure 1.2. Nomenclature and designations of land subdivision.

north of, the base line and standard parallels are fully 6 mi wide; all others are 6 mi less the amount of convergence. The east and west boundary lines of townships are called range lines, or ranges; the north and south boundaries are tiers, or, more commonly, township lines.

Specific townships are referenced by tier and range. For example, the township represented by hatch marks in Fig. 1.2 is located at Tier 2 South, Range 7 West (abbreviated T2S, R7W).

Subdivision of townships into sections progresses westward from the east range line. All sections should be as parallel as possible to the east range line. Since the range lines converge to the north in the northern latitudes, it is obvious that the section in the westernmost row will not be rectangular, nor will they be a full square mile in area. If all field measurements were without error, the east and west sides of sections would be exactly 1 mi in length; but since errors of varying magnitudes are usually present, it is common practice, wherever feasible, to combine these errors into the northern half of the westernmost row of sections. Reference to a certain section is accomplished by citing its number, as illustrated in Fig. 1.2. A small area is designated by a system of fractions and of directional location of subdivisions of a section, as illustrated in Fig. 1.2. The designation of the 10-acre (4.05-hectare) plot in the upper right of the section would be as follows:

$SE\frac{1}{4}$, $NE\frac{1}{4}$, $NE\frac{1}{4}$, Sec,____, T____ (N or S), R____ (E or W) of the ____ th principal meridian.

Because of the necessity of making new surveys in isolated regions opened for settlement, particularly in the west, it was not always possible or expedient to conduct the work in a systematic manner from a few well-chosen initial points. Large errors in geographic position and in the bearings of lines thus were allowed to develop so that, when the various isolated surveys were finally joined to more reliable work, the errors were found to be too large to adjust by approved means. There are, likewise, a number of areas where small groups of townships do not fit into the regional system of numbering, and it is difficult, if not impossible, to determine from maps the base lines and meridians used in the original surveys.

In Canada, the scheme of land subdivision is very similar to that used in the United States, except that there are fewer irregularities in the Canadian surveys. Townships are subdivided into section $1\,mi^2$, but the numbering starts in the southeast corner and ends in the northeast. Sections are subdivided into 16 parts, called legal subdivisions, which are numbered from 1 to 16, starting in the southeast corner of the section and ending in the northeast. Each legal subdivision in a regular section contains 40 acres. Townships are numbered from south and north from the international boundary in western Canada, and ranges are designated east and west from the principal meridian.

Geographic Information System

A Geographic Information System (GIS) is a relational database that serves as a platform for mapping and spatially distributed modeling. Data layers (also called themes or coverages), such as topography, water features, roads, and vegetation type, can be stored in two forms. Linear features, such as roads or streams, are often stored in vector form as a series of azimuths and distances tracing the path of the landscape feature. Other attributes, such as topography, soil type, land use, or vegetation type, may be represented with vectors enclosing a polygon(s) or as matrices of position and attribute data in *primary* layers. For example, topography is mathematically represented with a Digital Elevation Model (DEM) with x and y as Longitude and Latitude, Easting and Northing, or UTM grid coordinates, and z as elevation above mean sea level. For other layers, identification numbers or attribute (integer) codes to differentiate soil, vegetation, or land use types are the z-value at a given (x,y) location. *Secondary* layers are formed with attribute data (as real numbers) pertaining to a primary layer (e.g., soil: thickness, infiltration capacity, and permeability or vegetation: density, biomass, and condition). GIS layers also may be comprised of point data, such as building, wells, septic systems, or other features of limited or discrete size.

Point, line, and area (polygon) data can be combined to map and model interrelationships, calculate areas, or create new layers. For example, the DEM is routinely used to generate a slope layer (by calculating the change in elevation between adjacent grid cells). The slope layer can be used to estimate flow path (cell-to-cell linkages from the watershed divide to stream valleys) and contributing area (upslope of any given grid cell). These *derivative* layers are valuable for watershed modeling and management. In sum, a GIS provides analytical and operational capabilities that were once limited to small experimental watersheds (e.g., the Hubbard Brook Ecosystem Study) for a wide range of needs.

Shore Lines of Lakes and Streams

A reasonably accurate hydrographic survey is essential for the evaluation of basic morphometric characteristics of lakes and sections of streams. Such surveys can be performed with a minimum of equipment, although the work can be facilitated and made more accurate when simple surveying equipment is available.

Often the general outline of the lake or stream in question can be obtained from aerial photographs. The shore line then is measured and formulated, section by section, until the entire area is circumscribed. It is sometimes easier to make shore line surveys during the winter, when the lakes or streams are frozen, than during the open-water season, especially if the margins are swampy or heavily wooded.

Most shoreline surveys use the traverse method, in which a series of points are connected by straight lines of known lengths and angles from each other. Computerized total station surveying instruments are also available with automatic data collection capability (e.g., Lietz Co., Overland Park, KA). The use of transit or alidade surveying equipment is necessary for the construction of bathymetric maps of research investigations. Sophisticated methods of point location and survey are possible by use of the Global Positioning System, where coordinates are determined from triangulation from satellites. The principles, however, of the morphometric characteristics can be obtained with a simple compass and measuring tape, but at considerable sacrifice of accuracy. The rudiments of such surveying techniques are elaborated on in a number of surveying manuals; application to lakes and streams is detailed in Welch (1948) and Håkanson (1981). Several methods are outlined in the exercises proposed later in the text.

Sounding for Depth Contours

The bathymetry of a lake basin is determined most accurately from a continuous record of the basin contours with an accurate sonar device (depth sounder or fathometer). The depth sounder is moved across the lake at a constant speed along straight lines from known shoreline points. The continous record shows small irregularities in bottom contours that easily could be missed by manual sounding methods. The disadvantage of depth sounders is that most are quite inaccurate in water less than 2 m deep.

Manual soundings of depth are made along transects between shoreline points with weighted, calibrated lines or, in shallow waters, with metered sounding poles. Where sediments are soft, the weights or ends of the poles must consist of a plate (30 to 40 cm in diameter) to prevent appreciable penetration into the sediments.

The depth measurements along transects are then plotted precisely to scale within the map of the shore line. Points of equal depth at convenient intervals (e.g., 1 m, 2 m, or 5 m) then are interpolated between the points of known depth in several directions. These points of equal depth then are connected by contour lines (isopleths).

EXERCISES

OPTION 1. FIELD TRIP

In this exercise you will construct a simple map of the shore line of a small lake or a section of a stream and establish the depth contours. From this map, a number of important morphometric parameters can be calculated and used in subsequent physical, chemical, and biological analyses. Select a small lake or section of a stream of reasonable size that permits measurements to be made in the time allotted.

Triangulation Method

1. Establish a base line along a relatively straight portion of the shore line and stake this line at each end. The base line should be as long (e.g., 30 to 60 m) as the configuration of the shore line will permit and should be located so that nearly all points on the entire shore line are visible.
2. Establish a plane table (or other sturdy table) at each end of the base line (points A and B on Fig. 1.3). Measure the length of the base line accurately. Securely attach to the surface of each plane table some high-quality transparent or translucent paper (linen cloth or vellum is good, especially on a rainy day!). The plane table should be level and stable. With a compass, establish the north–south direction and draw a magnetic north line.
3. One or more persons then proceed to position a stadia rod (or brightly painted pole) at numerous positions around the shore line of the lake. The distance between the positions is governed by the configuration of the shore line—the closer the interval, the more accurate will be the representation of the true shore line.
4. At each position, a sighting is made from each plane table. With a pin set firmly in the plane table at point A (or B) of the base line, sight along a straight edge (a sighting alidade

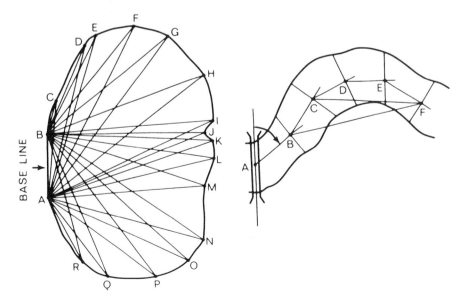

Figure 1.3. Diagrams of a small lake and section of a stream showing the locations of shore-line stations as determined by lines of sight of an alidade on a plane table.

is preferable, but homemade cross-hairs and a ruler will do) and draw a line along the straight edge. Repeat for each position.

5. Remove the paper with the sightings from each table and superimpose the base lines. Adjust and measure the distance between points A and B on this base line to establish a convenient scale of the map of the lake. The outline of the lake is determined simply by connecting the points where sighting lines intersect on the superimposed drawings.

6. In winter it is convenient to establish the base line near the center of an ice-covered lake and proceed as above.

7. Time can be saved and accuracy improved if several boat crews (or persons making soundings through ice) determine depths along the transects while the shore line is being established. The boat crews call out depths at points of distance along the transect, and each is recorded by the workers at the plane table.

8. En route, the flagpersons should note marked deviations in physical features of the shore line from a general circular configuration, such as inlets, outlets, and bays.

Alidade Method

1. As with the triangulation method, establish a base line along a relatively straight portion of the shore line and stake this line at each end. The base line should be reasonably long (e.g., 30 to 60 m) and located so that nearly all points on the whole shore line are visible.

2. Set and level a plane table (or other sturdy table) over the stake of one end (A) of the base line. Securely attach to the surface of the plane table some high-quality paper (linen cloth or vellum). With a compass, establish the north–south direction and draw a magnetic north line.

3. Place stakes at intervals along the shore line, numbering each and coding them periodically with red and white cloth flags, until the entire shoreline area is circumscribed (Fig. 1.3).

4. Estimate the total size of the lake or section of stream being mapped in relation to the length of the base line. The scale of your map should be as large as possible but obviously must fit entirely on the paper of the board.

5. With a pin set firmly at point A of the base line, sight the alidade in line with the flagperson at the other end of the base line. Draw the base line. This distance (AB) must be measured accurately with a tape measure.

6. Rotate the alidade against the pivotal pin until the line of sight aligns with the next shore line flagperson at position C. Draw the line of sight; record distance if a telescopic alidade is used or estimate length if an extendable open-sight alidade is used. Repeat the procedure for each stake at shore line positions until the area is circumscribed.

7. Move the plane table to the other end of the base line (B) and repeat all angular and distance sightings.

8. Construct an outline map as discussed earlier.

9. Accuracy of the positioning of depth soundings is improved when persons making the soundings (from boats or through ice) determine depths along the transects as the shore line is being established. The depths are noted against distance sightings by the alidade along the transects.

10. Deviations in physical features of the shore line from a general circular configuration, such as inlets, outlets, and bays, should be noted by the flagpersons as the periphery is traversed.

Alternative Methods

1. When the body of water is very small, such as a pond or a section of a stream, measurements often can be made directly with tape measures. Simply establish a continuous series of triangles, the lengths of whose sides are known, and the apices of which are located at various points on the shore line.

2. Small bodies of water also may be mapped by direct measurements made perpendicularly from a base line extending along the entire length of the water body. The distance from the base line to the intersection of the parallel transverse lines from points on the shore line on both sides permits an accurate evaluation of the periphery.
3. A rather crude map can be constructed by using only a surveying compass and a tape measure. Angular measurements from a base line oriented north–south are made by compass bearings at points of measured distance around the shore line.

In the construction of the map of a body of water, the map should include the shore line, depth contours at meter intervals, a scale in metric units, the name of the body of water, the exact geographical location including section, date of the survey, and names of the surveying crew.

Complete the calculations and questions following Option 2.

OPTION 2. LABORATORY EXERCISE

1. From the field data given in Table 1.1, construct a map of the shore line of the exemplary lake. Label completely.
2. From the data given in Fig. 1.4, interpolate points of uniform depth at 2- or 5-m intervals. Draw in bathymetric contour lines.
3. Complete the following calculations and questions.

Morphometric Calculations

Maximum Length (L). The maximum distance on the lake surface between any two points on the shore line. This length is potentially the maximum fetch or effective length for wind to act on the surface of the lake without land interruption.

Table 1.1. Exemplary data obtained from field measurements of a plane table—alidade survey of the shore line of a small lake.

Lake:	Omega		Location:	Sec. 14, T1N, R3W,
Date:	August 18, 1976			Theta County, North Dakota
Workers:	J. Blown		Reference:	3.5 m below USGS
	S. Scatter			Bench Mark at 146.3 m
	W. Cloud			above mean sea level
			Base line Bearing:	N57°30′W

Station lines	Distance (m)	Angle between AB and line	Station lines	Distance (m)	Angle between BA and line
AB (base line)	43.2	—	BA (base line)	43.2	—
AC	58.5	13°	BC	19.0	136°
AD	81.5	26°	BD	47.5	131°
AE	92.6	38.5°	BE	65.0	117°
AF	109.0	59°	BF	94.5	97.5°
AG	111.5	73°	BG	107.0	84°
AH	83.0	88°	BH	93.0	64.5°
AI	80.0	106°	BI	101.5	50°
AJ	73.5	119.5°	BJ	102.0	39.5°
AK	70.0	132°	BK	104.5	30.5°
AL	62.0	140.5°	BL	99.0	24°
AM	28.5	154°	BM	70.0	10.5°

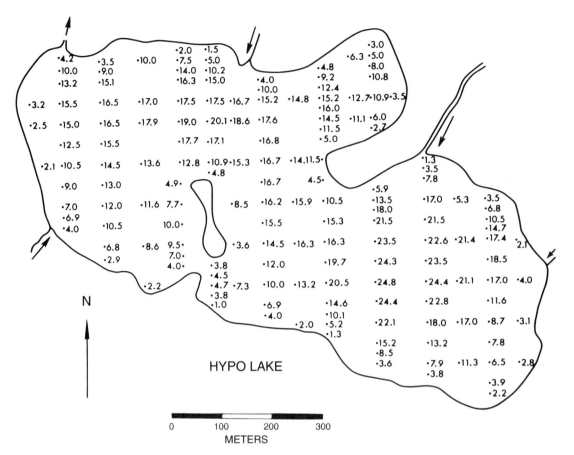

Figure 1.4. Depth soundings (in meters) in exemplary Hypo Lake from which depth contours and morphometric calculations can be made.

Maximum Width or Breadth (b). The maximum distance between the shores perpendicular to the line of maximum length. The mean width ($\bar{b}$) is equal to the surface area (A_0) divided by the maximum length (L):

$$\bar{b} = A_0/L.$$

Area (A). The area of the surface and of each contour interval at depth z may be determined by a digitizer or planimetry of a detailed bathymetric map. Computers with appropriate software and other electronic devices are available to digitize and measure two-dimensional distributions of data such as line graphs, maps, etc. (e.g., Sigma Scan Pro, SPSS Science, Sausalito, CA). A less precise but effective method is by a grid enumeration analysis [cf., Olson (1960)].

A polar planimeter is an ingenious but relatively inexpensive instrument that permits an integration of the area of any plane surface regardless of its shape. A planimeter consists of two units: the pole arm unit and the tracer arm-carriage unit. The tracer arm, containing the tracer point and magnifying lens, connects to the carriage unit. The carriage unit consists of a dial, a vernier, a measuring wheel, a support wheel, and a zero-setting wheel.

In practice, the operator traces the periphery of a plane surface with the tracer point in a clockwise direction (Fig. 1.5). The planimeter units of the enclosed irregular area can then be read directly from the dial, measuring wheel, and vernier. Then, using the same pole and tracer arm settings as used for the measurement of the unknown area, one can measure a region of known area. On a map, a square of side dimensions equal to 5 to 10 cm, made equiv-

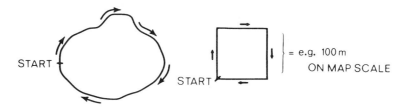

Figure 1.5. Directional movements of the tracing point of a planimeter in evaluating the areas of an unknown irregular plane surface (*left*) and of a known surface of proportional area (*right*).

alent to some number of meters from the scale of the map, makes a convenient regular plane surface of known area. The area in question can then be determined by a direct proportional equation:

$$\frac{A}{PU} = \frac{A'}{PU'}$$

where A = area of known plane surface, PU = planimeter units of known plane surface, A' = area of unknown plane surface, and PU' = planimeter units of unknown plane surface.

In using the planimeter, gently place the carriage unit, with the tracer arm, in the center of the area to be measured. Place the pole weight or pin at some distance from the ball-and-socket joint of the carriage unit and connect both with the pole arm. The angle formed between the pole arm and the tracer arm should be approximately 90°. *Never* allow this to approach 180°, or a straight line, or less than 20°, as the planimeter is measuring to the limit of its capability at these angles.

Before measuring, always record the tracer and pole arm settings. Be sure the vernier scale is adjusted precisely to 0.000 before starting. Make a rough preliminary circuit of the area before the actual measurement. The paper to be traced should be absolutely flat, smooth, and preferably in the middle of the table. The paper should also be clean and free of particles. When measuring, do not lift or press the tracer head or rotate the grip, as this action will allow the guide wheel to slip from the paper. When this happens accidentally during the measurement, the measurement must be repeated. On sharp corners of the curve, where the guide wheel tends to move rapidly across the paper, move slowly or else the guide wheel will slip. *Never* touch the guide wheel with your fingers or the matte surface will clog, reducing its friction on the surface of the paper. Measure the area contained by each contour at least twice. The variance between readings should be limited to 0.003 units or less, depending on the size of the figure and the accuracy. The standard area should be measured at least five times to obtain a mean value. When the unknown area is too large to be accommodated by the working area of the planimeter, it should be divided into a number of fractional areas; the areas of each then are determined and summed. Areas of islands or submerged lake mounts must be subtracted; areas of planes at equal depth but separated (e.g., multiple depressions in a lake basin) must be added.

Volume (V). The volume of the basin is the sum of the volumes of the strata at successive depths from the surface to the point of maximum depth. The volume is approximated closely by plotting the areas of contours, as closely spaced as possible, against depth and integrating the area of this curve by planimetry. Alternatively, the volume can be estimated by summation of a series of truncated irregular cones of the strata:

$$V = \frac{h}{3}(A_1 + A_2 + \sqrt{A_1 A_2})$$

where h = vertical depth (thickness) of the stratum, in meters; A_1 = area of the upper surface, in m², and; A_2 = area of the lower surface of the stratum whose volume is to be determined.

It has been assumed commonly that the increase in benthic area below an area of lake surface resulting from the slope of the lake or stream bottom is negligible. This assumption can be valid for very shallow lakes or shallow areas of a littoral zone where the slope is slight. The error increases significantly, however, with increasing degree of slope. Estimates of benthic area can be made for smooth or rocky substrata [cf., Loeb et al. (1983)]. For example, benthic areas below $1\,m^2$ of lake surface for different bottom slopes ($\theta°$) are

Area of bottom below $1.00\,m^2$ of lake surface

$\theta°$	Smooth slope	Rocky slope
15°	1.04	1.27
25°	1.10	1.47
35°	1.22	1.70
45°	1.44	2.00

Maximum Depth (z_m). The greatest depth of the lake.

Mean Depth ($\bar{z}$). The volume divided by its surface area at zero depth:

$$\bar{z} = \frac{V}{A_0}$$

Relative Depth (z_r). The maximum depth as a percentage of the mean diameter [cf., Hutchinson (1957)]:

$$z_r = \frac{50 z_m \sqrt{\pi}}{\sqrt{A_0}}$$

Most lakes have a z_r of less than 2%, whereas deep lakes with small surface areas exhibit greater resistance to mixing and usually have a $z_r > 4\%$.

Shore Line (SL). The intersection of the land with permanent water is of nearly constant length in most natural lakes. The length of the shore line can, however, fluctuate widely in ephemeral lakes and in reservoirs in response to variations in amounts of precipitation and discharge.

The length of the shore line can be determined directly by measurement or from maps with map measures (chartometer, rotometer). These inexpensive devices measure the lengths of lines by means of a tracing wheel whose revolutions are recorded directly on a graduated dial. The shore line is traced from a starting point through the periphery of the area. This distance is then converted to real units by a simple proportional comparison to the distance measured on the scale of the map with the same map measurer.

Shore Line Development (D_L). The ratio of the length of the shore line (SL) to the length of the circumference of a circle whose area is equal to that of the lake:

$$D_L = \frac{SL}{2\sqrt{\pi A_0}}$$

As the length of the shore line becomes more irregular, the shoreline development deviates more and more from the minimum D_L-value of one, i.e., that of a circle. Only a few lakes, such as Crater Lake in Oregon and a few kettle lakes, approach this circular shape. Many subcircular and elliptical lakes have D_L-values of about 2; lakes of flooded river valleys have much larger D_L-values. Shoreline development is of interest because it reflects the potential for development of littoral communities, which are usually of high biological productivity.

Hypsographic Curves. A hypsographic curve, or depth-area curve, is a graphic representation of the relationship between the surface area of a lake basin and its depth. This relationship may be expressed in absolute units of area (m^2, hectares, or km^2), which are at a

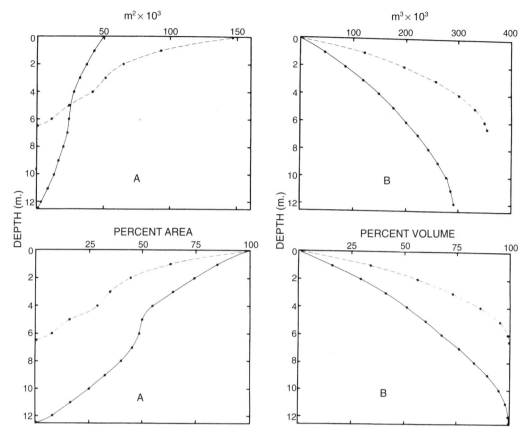

Figure 1.6. Hypsographic (depth-area) curve (A) and depth-volume curve (B) of mesotrophic Lawrence Lake (——) and hypereutrophic Wintergreen Lake (-----), southwestern Michigan. (From Wetzel, 1983.)

given depth, or in terms of percent of lake area which is at a given depth (Fig. 1.6). Plot the area of your lake at each depth and similarly plot the area at each depth as a percentage of the surface area.

Depth-Volume Curves. The depth-volume curve is closely related to hypsographic curves and represents the relationship of lake volume to depth. Similarly, the units may be expressed in volume units against depth or in percent of lake volume above a specific depth (Fig. 1.6). That depth above which 50% of the lake volume occurs approximates the mean depth of the lake. Plot the accumulated volume of your lake from the surface for each stratum and, for each value, calculate and plot the volume as an accumulated percent from the surface of the lake (see Fig. 1.6).

Questions

1. How would multiple depressions influence (a) your calculations of area and volume of different depth strata, and (b) hypolimnetic structure within the basin as a whole?
2. How would islands and subsurface lakemounts influence (a) your calculations of area and volume of different depth strata, (b) fetch (effective length) in relation to wind exposure and circulation within the lake basin, and (c) the amount of sediment surface available within the photic zone for colonization by attached algae and larger aquatic plants?

3. Using data from your lake or from Hypo Lake (Fig. 1.4), approximately how much sediment surface area would be available to benthic invertebrate fauna requiring:
   (a) Large sediment particles? Fine, organic-rich sediment particles?
   (b) High oxygen concentrations at all times of the year?
4. How might these parameters change in a lake of greater relative depth (z_r)?
5. Assume that the free-floating (planktonic) algae in your lake require at depth at least 0.1% of solar irradiance impinging on the surface of the lake to maintain a net positive growth. In your lake or the example provided, what would be the volume of the stratum in which photosynthesis could occur (the euphotic, photic, or trophogenic zone) if:
   (a) Light were attenuated under very productive conditions to the 0.1% level at a depth of 2 m?
   (b) Light at the 0.1% level were not attenuated until reaching a depth of 5.5 m?
   (c) Continual turbidity from silt caused the light to be attenuated to the 0.1% level at a depth of 25 cm?
6. If a mysid crustacean were adapted to cold temperatures and optically to blue light in deep water (below 10 m), what volume of water would be available to this organism?

Apparatus and Supplies

1. Plane table and tripod, heavy paper, pins, pencils
2. Alidade (telescopic or open-sight) or transit
3. Rulers
4. Mental measuring tapes (ca. 50 m)
5. Stakes (30 to 100), with numbering and flags
6. Surveyor's or military compass
7. Depth sounding lines, graduated in 0.5-m intervals, with platelike weights
8. Graduated sounding poles and endplates
9. Boats and motors
10. Sonar-type depth sounder, if available
11. Ice augers, if work is done in winter
12. Planimeter(s) or electronic digitizer
13. Protractors
14. Graph paper

References

Håkanson, L. 1981. A Manual of Lake Morphometry. Springer-Verlag, Berlin. 78 pp.

Hunt, C.B. 1974. Natural Regions of the United States and Canada. 2nd Ed. Freeman, San Francisco, 725 pp.

Hutchinson, G.E. 1957. A Treatise on Limnology. Vol. 1. Geography, Physics, and Chemistry. Wiley, New York. 1015 pp.

Loeb, S.L., J.E. Reuter, and C.R. Goldman. 1983. Littoral zone production of oligotrophic lakes. pp. 161–167. *In*: R.G. Wetzel, Editor. Periphyton of Freshwater Ecosystems. Dr. W. Junk Publ. The Hague.

McEntyre, J.G. 1978. Land Survey Systems. Wiley, New York. 537 pp.

Olson, F.C.W. 1960. A system of morphometry. Int. Hydrogr. Rev. *37*:147–155.

Welch, P.S. 1948. Limnological Methods. Blakiston Philadelphia. 381 pp.

Wetzel, R.G. 1999. Limnology: Lake and River Ecosystems. 3rd Ed. Academic Press, San Diego (in press).

Light and Temperature

Solar radiation is vital to the metabolism, indeed to the very existence, of freshwater ecosystems. Nearly all energy that drives and controls the metabolism of lakes and streams is derived from solar energy, which is converted biochemically via photosynthesis to potential chemical energy. The photosynthetic synthesis of organic matter occurs within the lake or river (autochthonous) or within the terrestrial drainage basin (allochthonous) and is transported to the aquatic ecosystem in various forms of dissolved and particulate organic matter by "vehicles" for movement [e.g., air, water, animals; cf., Likens and Bormann (1972)]. In addition to direct biological utilization, the absorption of solar energy and its dissipation as heat markedly affect the thermal structure and stratification of water masses and circulation patterns of lakes, reservoirs, and streams. These characteristics in turn have profound effects on nutrient cycling and the distribution of dissolved gases and the biota. The optical properties of fresh waters, therefore, exert important regulatory controls on the physiology and behavior of aquatic organisms.

Light impinging on the surface of water does not penetrate completely—a significant portion is reflected and backscattered [cf., Wetzel (1983) and Exercise 4, p. 45]. Within the water, light is rapidly attenuated with increasing depth by both absorption and scattering mechanisms. Absorption is defined as diminution of light energy with increasing depth by transformation to heat [cf., Westlake (1965)]. Absorption is influenced by the molecular structure of water itself, by particles suspended in the water, and particularly by dissolved organic compounds. The result is a selective absorption and attenuation of light energy with increase in depth, influenced by an array of physical, chemical, and, under certain conditions, biotic properties of the water. These optical properties are dynamic, changing seasonally and over geological time for individual freshwater ecosystems.

Over half of the solar radiation that penetrates into water is absorbed and dissipated as heat. As we will see in subsequent analyses (Exercises 3 and 4), the distribution of this heat is influenced greatly by wind energy. In this exercise, we will evaluate methods for the measurement and description of the distribution of light and temperature in water.

LIGHT

Units of Measurement

Solar radiant energy that reaches the surface of the earth has a spectral range from about 300 nm (ultraviolet) to about 3000 nm (infrared). Light is that portion, about

Table 2.1. Interrelationships of selected units of irradiance and illuminance.

	μmol quanta/m^{-2}-sec	J/m^2-sec	erg/cm^2-sec	gcal/cm^2-min	lux[a] (illuminance unit)
μmol quanta/m^{-2}-sec	1	1.20×10^{-1}	1.20×10^2	1.72×10^{-4}	~5.12×10^1
J/m^2-sec (= W/m^2)	5.03	1	10^3	1.43×10^{-3}	~2.5×10^2
gcal/cm^2-min	5.83×10^3	6.98×10^2	6.98×10^5	1	~1.8×10^5
lux[a] or m-candle (= 1 lumen/m^2 or 0.0929 ft-candle)	~1.953×10^{-2}	~4.0×10^{-3}	~4.0	~5.7×10^{-6}	1

[a] Energy equivalents are given in terms of visible range in daylight (380–720 nm).
1 W/m^2 ≈ 4.6 μmol quanta/m^2-sec; 1 lux ≈ 0.01953 μmol quanta/m^2-sec; 1 lux = 4.1×10^{-7} W/cm^2 = 6.0×10^{-6} gcal/cm^2-min; 1 lumen = 4.17×10^{-3} W (1 W = 240 lumens); 1 ft-candle = 4.6×10^{-6} W/cm^2 = 6.5×10^{-5} gcal/cm^2-min; 1 W/cm^2 = 2.4×10^6 lux = 2.2×10^5 ft-candles; and 1 gcal/cm^2-min = 1.6×10^5 lux = 1.5×10^4 ft-candles.

half, of the total radiant energy that can be detected by the human eye (380 to 780 nm). Photosynthetically active radiation (PAR) occurs between approximately 390 and 710 nm and forms about 46 to 48% of the total energy impinging on the earth's surface (Strickland, 1958; Talling, 1957; Westlake, 1965; Kirk, 1994b). Little of the ultraviolet radiation (UV, 280–400 nm) is utilized directly in photosynthesis. However, UV can potentially damage aquatic biota and modify the organic chemistry of aquatic ecosystems (e.g., Karentz et al., 1994; Wetzel et al., 1995). Reductions of stratospheric ozone have increased substantially fluxes of the most energetic radiation (UV-B, 280–320 nm), particularly at high latitudes (>35°) (e.g., Kerr and McElroy, 1993).

Radiant flux is the quantity of electromagnetic energy flow over time, expressed as quanta (= photons) per second (or as watts or joules/sec). *Irradiance* (= intensity or flux density) is the radiant flux per unit area of a surface, expressed as mol quanta per second per square meter. This rate of incident energy has been expressed in a number of ways: (a) in the meter-kilogram-second system as joules/area-time, J/m^2-sec; (b) watts per area, W/m^2; or (c) as gram-calories per area-time, gcal/cm^2-min (= langley/min). Selected interconversions are given in Table 2.1 and in Wetzel (1983, pp. 756–757). Irradiance is now internationally expressed in moles of photons per area per time as μmol quanta m^{-2}s^{-1} (formerly microeinsteins/m^{2}s^{-1}, where 1 μEinst = 6.02×10^{17} photons or quanta). Modern instruments measure in situ irradiance flux densities directly in μmol quanta m^{-2}s^{-1} (= μEinst m^{-2}s^{-1}).

Photosynthetic irradiance is the radiant energy flux density of PAR and is expressed as the radiant energy (400 to 700 nm) incident on a unit of surface per unit time. Conversion of quantum measurements (μmol quanta/m^{-2}, 400 to 700 nm) to radiometric illuminance units (W/m^2, 400 to 700 nm) is complicated. Conversion factors differ with different light sources and their spectral distribution curves (Table 2.1).

Measurement of Surface Irradiance

The irradiance impinging on the surface of a lake or stream is most commonly measured with a pyrheliometer or solarimeter. The most accurate pyrheliometers contain 10 to 50 thermocouples (thermopile), which measure any difference in temperature between a polished surface that theoretically reflects away all incoming radiation and a black body that theoretically absorbs all incoming solar energy per unit area. The electromotive force output from the thermopile is processed by a millivoltmeter for direct recording, calibrated in μmol quanta/m^{-2}-sec, and integrated over a given time. Less expensive pyrheliometers mechanically record the reflection–absorption difference upon a calibrated clock-driven drum. These instruments (e.g., Belfort of Baltimore, MD) should be calibrated periodically against more accurate pyrheliometer systems (Eppley Laboratories, Newport, RI, Li-Cor, Inc., Lincoln, NE, or Kipp and Zonen, Amsterdam, The Netherlands), commonly employed at governmental or other meteorological stations.

Local cloud cover is often so variable that extrapolation of solar radiation measurements from a standard weather station should not be made when the station is more than about 50-km distant from the study area. The irradiance impinging on a woodland stream can be highly variable, depending on the adjacent forest canopy, so direct measurements of irradiance are mandatory. Portable pyrheliometers have distinct advantages in these situations.

Pyrheliometers are sensitive in the range of 300 to 5000 nm. Slightly less than half of this measured total irradiance is available for photosynthetic activity. Under a clear sky and unobscured sunlight, with the sky light contributing about one-third of the total radiation, the percentages of the total irradiance within various ranges of wavelengths are as follows [after Strickland (1958)]:

380–720 nm	50.0%
380–490 nm	17.0%
490–560 nm	12.0%
560–620 nm	8.5%
620–720 nm	12.0%

The amount and spectral composition of solar radiation impinging on the surface of a lake or stream are influenced by an array of dynamic environmental factors. Direct solar radiation reaching the water surface varies with the angular height of the radiation and, therefore, with time of day, season, and latitude (Wetzel, 1983, 1999). The quantity and quality of light also vary with the molecular transparency of the atmosphere and the distance the light must travel through it; therefore, it varies with altitude and meteorological conditions.

In addition, light is scattered as it passes through the atmosphere, producing indirect solar radiation, some of which eventually reaches the surface of the water. The height of the sun and the distance which light must pass through the atmosphere influence the amount and spectral composition of indirect irradiance. For example, at a sun elevation of 10° from the perpendicular, 20 to 40% of light received may be from indirect irradiance, whereas at a sun elevation of 40° this amounts to 8 to 20%.

A significant portion of the light is reflected by the surface of the water. Such light is unavailable to the aquatic system but may be backscattered to the lake again from the atmosphere or surrounding topography. The extent of reflection of direct and indirect irradiance varies widely with the angle of incidence of incoming energy, the surface characteristics of the water, the surrounding topography, and meteorological conditions (Wetzel, 1983, 1999). Reflection of direct solar radiation increases with increasing angle of the sun from the perpendicular. Reflection also increases with disturbance of the surface by wave action, when the angle of the sun is low. Ice, and especially snow, increase reflection of light from the surfaces of fresh waters.

Of the total irradiance impinging on the surface of a lake, the average amount reflected on a clear summer day is about 5 to 6% and increases to about 10% in winter when ice covered. At least 70% of incident light striking fresh, dry snow is reflected (see Exercise 4).

Measurement of Underwater Irradiance

As solar radiation penetrates water, portions are absorbed both by water itself and by dissolved and suspended materials contained in it. A significant portion of this light is also scattered, that is, deflected by the molecular components of water, its solutes, and particulate materials suspended in the water. The scattering of light energy can be viewed as a composite of reflection from a massive array of angles existing internally within a lake or stream (Wetzel, 1983, 1999). The extent of scattering in a specific volume of water varies greatly with the composition, quantity, and relative transparency of suspended materials. Therefore, one would anticipate variations in scattering of light in relation to, for example, distribution of inorganic and organic suspended matter and proximity to sediments.

Solar radiation in water is absorbed and dispersed by the processes discussed above. Light attenuation is the reduction of radiant energy with depth by both scattering and absorption mechanisms. The measurement of transmission or absorption of light in water can be made in several ways. The percentage absorption through a given depth of water may be expressed in percent according to the relationship developed by Birge (1915, 1916):

$$\frac{100(I_0 - I_z)}{I_0}$$

where I_0 = irradiance at the surface of the lake or some discrete layer within the lake and I_z = irradiance at depth z, usually taken at 1-m intervals below I_0.

The percentile absorption of pure water is very high in the infrared portion of the spectrum and results in rapid heating of water by incident light. About 53% of total light energy is transformed into heat in the first meter of water. Absorption by pure water decreases markedly in the shorter wavelengths to a minimum absorption in the blue and increases again in the violet and ultraviolet spectral wavelengths (Fig. 2.1).

Solar irradiance I_z at depth z is a function of the intensity at the surface, I_0, multiplied by the antilog of the negative extinction coefficient (η) at depth z in meters:

$$I_z = I_0 e^{-\eta z}$$

or

$$\ln I_0 - \ln I_z = \eta z$$

Although theoretically the extinction coefficient (η) is constant for a given wavelength in water, in nature underwater irradiance is a composite of many wavelengths. Therefore the relationship is imperfect under natural conditions and represents a composite for multichromatic light and the various characteristics of water that affect absorption and scattering.

The in situ or total extinction coefficient (η_t) is a composite of absorption by the water itself (η_w), by particles suspended in the water (η_p), and especially by dissolved organic compounds or "color" (η_c) (Åberg and Rodhe, 1942):

Figure 2.1. Transmission of light by distilled water at six wavelengths (R = red, 720 nm; O = orange, 620 nm; Y = yellow, 510 nm; G = green, 510 nm; B = blue, 460 nm; V = violet, 390 nm). Percentage of incident light that would remain after passing through the indicated depths of water expressed on linear (*upper*) and logarithmic (*lower*) scales. (Wetzel, 1999, after Clarke.)

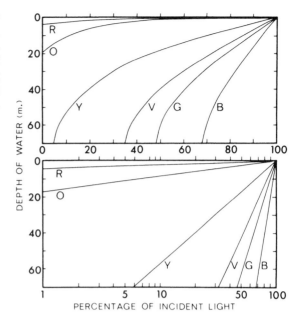

$$\eta_t = \eta_w + \eta_p + \eta_c$$

Variation in transparency of inland waters to ultraviolet (UV) radiation is great. Attenuation depths ($z_{1\%}$) for UV-B radiation extend from several centimeters to >10 m (Morris et al. 1995). Much (>90%) of the among-lake variation in diffuse attenuation for downward irradiance can be explained by differences in dissolved organic carbon (DOC) concentrations, even though absorption coefficients for different classes of DOC compounds vary by at least one order of magnitude.

Of the ultraviolet radiation band (100–400 nm), nearly all of UV-C (<280 nm) is absorbed by the stratospheric gases and by water of aquatic ecosystems. Although relatively little UV-B (280–320 nm) passes through the stratosphere, UV-B is highly energetic and an important photoactivating agent in fresh waters. UV-A (320–400 nm) is less energetic than UV-B but is absorbed less readily in water and penetrates more deeply.

Suspensions of particulate matter have relatively little effect on absorption at low concentrations. Particulate matter can be removed by centrifugation or filtration, although caution is needed as some dissolved or colloidal humic substances, which afford color to water, can absorb to filter materials.

The attenuation of total irradiance from direct and indirect insolation within a lake can be measured with a variety of instruments. Ideally, the total underwater irradiance would be measured with a receptor system at a specific depth. It is this amount of energy, whether from direct incident light or from scattered light sources, that is collectively important to organisms for photosynthesis or as behavioral stimuli. This amount of irradiance is the photon scalar irradiance, defined as the total number of photons from all directions about the point of measurement when all directions are weighted uniformly (Smith and Wilson, 1972). Instruments for measuring photon scalar irradiance are available commercially (e.g., Li-Cor, Lincoln, NE). Specifically designed radiometers are required to analyze the underwater distribution and quantification of UV irradiance (cf., Kirk, 1994a; Morris et al., 1995).

Approximations of underwater irradiance commonly are made with less sophisticated instrumentation. Many older underwater photometers contain selenium photocells; more recent instruments employ silicon photovoltaic cells. These photocells respond to light of wavelengths between approximately 300 and 750 nm and usually have a pronounced peak in response between 500 and 600 nm (Strickland, 1958; Sauberer, 1962; Westlake, 1965).

The physical structure of the photoreception unit is important. Most underwater photometers are flat (a 2π system) and respond largely to incident light from above. Little or none of the laterally scattered light, as could be utilized by an organism, is measured. Various hemispherical designs with diffusing materials permit a more realistic measurement of incident as well as scattered light [e.g., Rich and Wetzel (1969)]. Combined with a downward hemisphere, the photometer approaches a 4π design that permits an estimation of both incident and scattered light from above and reflected and scattered light from below.

Some photometers currently in use give relative measurements. That is, underwater values are compared, as a percentage, to values measured immediately below the surface of the water (e.g., 10-cm depth, to avoid wave-mediated reflection and scattering). While such data are useful, more information is gained if measurements are in absolute units. Recent photometers have quantum sensors that permit measurement of photosynthetically active radiation within the 400- to 700-nm portion of the spectrum directly in μmol quanta/m^{-2}-sec. Other sensors measure directly in watts/m^2 or in lux (see Table 2.1).

None of these instruments measures the spectral distribution of the total underwater irradiance. However, when filters with a specific and narrow range of wavelengths are positioned over the photocells, the spectral attenuation of irradiance at depth can be evaluated. Radiometers that scan the spectrum and measure the amount of energy at all wavelengths within the visual spectrum are available commercially, and a few are adapted for use underwater. Exemplary data using such an instrument are shown in Fig. 2.2 and demonstrate the selective attenuation of irradiance in a mesotrophic lake.

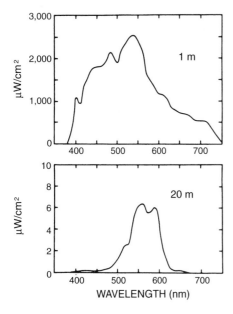

Figure 2.2. Comparison of the spectral distributions of energy using a scanning spectroradiometer, Gull Lake, Kalamazoo-Barry counties, Michigan, November 16, 1975, at 1 m (*upper*) and at 20 m (*lower*). 1 μW/cm^2 $\cong$ 0.046 μmol quanta/m^{-2}-sec.

Analyses of spectral energy distribution are especially useful in fresh waters. Large variations in concentrations of organic compounds occur among different lakes and streams and seasonally within the same water. Absorption of light energy increases drastically, particularly of UV and violet wavelengths, with increased concentrations of dissolved organic compounds, particularly humic and fulvic acids (Morris et al., 1995; Wetzel, 1999). The spectral energy distribution also is altered markedly. Distilled water has a very high absorption in the red and infrared wavelengths but absorbs very little of the energy of shorter wavelengths (UV, blue). Even very low concentrations of dissolved organic compounds greatly increase absorption of the shorter wavelengths of the spectrum. The distribution of spectral energy at depth can be analyzed by comparison of vertical profiles of total irradiance to those in which narrow band filters are used to exclude specific portions of the spectrum. Profiles and mean extinction coefficients of each of these spectral distributions can be compared.

Turbidity

Turbidity in water is caused by suspended inorganic and organic matter, such as clay, silt, carbonate particles, fine organic particulate matter, and plankton and other small organisms. This suspended matter causes light to be scattered and absorbed rather than transmitted. However, measurements of turbidity, although simple to perform, do not always yield significant correlations with concentrations of suspended materials. The size and refractive characteristics of particulate matter, while important optically, bear little direct relationship to the specific gravity and concentrations of the suspensoids. Nonetheless, turbidity measurements are very useful for evaluating the microstratification of microorganisms in density layers of lakes [e.g., Whitney (1937), Stewart et al. (1965), Baker and Brook (1971), Baker et al. (1985)]. Turbidity is a major concern of water treatment operations where suspended particulate matter must be removed prior to use for domestic or industrial purposes. Many aquatic organisms, especially filter feeders, cannot tolerate appreciable concentrations of inorganic particulate matter.

The measurement of turbidity has a long history. Most methods compare the intensity of light scattered by the sample under defined conditions to that scattered by a standard reference suspension under the same conditions. Early visual methods employed silica, clay (e.g., Kaolin), and other materials as the standard reference suspension; all have variable particulate characteristics. Formazin polymers have greater reproducibility and are now commonly employed as a standard reference suspension. When compared in a simple spectrophotometer, or turbidimeter with a standard light source, the measurement is referred to as the nephelometric method, from the root meaning "cloudiness."

Turbidimeters are available commercially that permit direct measurements of in situ turbidity. These instruments incorporate an underwater light source, a lens system (to collimate the light beam as it passes through a known and adjustable distance of water), and a photocell. The photocell is located behind a diaphragm to exclude nearly all scattered sunlight.

Transparency Estimates with a Secchi Disc

Although the measurement of extinction coefficients and spectral characteristics of underwater irradiance is done commonly with a photometer, an approximate evaluation of transparency of water can be made with a Secchi disc. Devised by an Italian oceanographer in the nineteenth century, this method still is used widely because of its simplicity. The Secchi disc is a weighted white disc, 20 cm in diameter. Transparency, estimated in this way, is the mean of the depths at which the Secchi disc disappears when viewed from the shaded side of the boat and at which it reappears upon raising after it has been lowered beyond visibility.

Secchi disc transparency is basically a function of reflection of light from the surface of the disc and therefore is affected by the absorption characteristics of the water and of dissolved and particulate matter contained in the water. Although high concentrations of dissolved organic matter decrease transparency in a nonlinear way, as measured with a Secchi disc (Wetzel, 1983, 1999), reduction in light transmission as evaluated by the disc is influenced strongly by increased scattering of light by suspended particulate matter. Depth of transparency is largely independent of surface light intensity but becomes erratic near dawn or dusk. Determinations are best made near midday.

Observed Secchi disc transparencies range from a few centimeters in very turbid lakes to over 40 m in a few clear lakes; most are in the range of 2 to 10 m. Marked seasonal fluctuations occur in response to variations in concentrations of plankton or inorganic particles [e.g., Wetzel (1983, 1999)].

Secchi disc transparencies correlate well with percentage transparency. In comparison to measurements with underwater photometers, the depths of Secchi disc transparency can vary from 1 to 15% transmission, 10 to 15% being most common (Štěpánek, 1959; Beeton, 1958; Tyler, 1968). Differences are related to size of suspensoids and to variations in the sensitivities of underwater photometers.

Color and Color Scales

The apparent color of lake, reservoir, or river water results from light scattered upward after it has passed through the water to various depths and undergone selective attenuation en route. In clear water, blue dominates because of molecular scattering of light by the water. Dissolved organic matter, however, selectively absorbs the shorter wavelengths and shifts the apparent color to yellow and red. Particulate matter also scatters light in variable ways depending on its composition and size, e.g., becomes less selective with increasing particle size. When suspended particulate matter becomes dense, the color of these particles often is imparted to the water in spite of the relatively nonselective scattering properties of these particles. Blue-green, yellow-brown, and red colorations commonly are observed when dense populations of algae or certain bacteria develop. Heavy suspensions of inorganic materials, such as calcium carbonate or clays, can yield blue-green or yellowish brown-red colorations, respectively.

Color of water is, at best, a general criterion of limited value in limnology. Color (shade or tint) consists of color intensity (brightness) and light intensity (lightness). The visual memory of humans is poor and the ability to discriminate colors is highly subjective under different environmental conditions. Empirical color scales have been devised against which the color of water can be compared, after filtration to remove suspensoids. Use of these scales was more common in earlier times before accurate analytical techniques were available for determining concentrations of dissolved organic matter. A positive correlation exists between brown organic color and the concentration of dissolved organic carbon.

   In Europe, the Forel-Ule color scale is used to compare the color of water to alkaline solutions of cupric sulfate, potassium chromate, and cobaltus sulfate. In the United States the most commonly used scale is a serial dilution of an acidic solution of platinum and cobaltic compounds (see p. 30). Subjectivity is reduced even further in color evaluations made by optical analyses and comparisons to standardized chromaticity coordinates [e.g., Smith et al. (1973)].

TEMPERATURE

   The greatest source of heat in water is solar irradiance by direct absorption. Some transfer of heat from the air and the sediments does occur, but this input is usually small compared to direct absorption of solar radiation by water, dissolved organic compounds, and suspended particulate matter. Although terrestrial heat derived from surface drainage and groundwater often is a small portion of the total in lakes, it is highly significant to the heat content of streams [e.g., Johnson et al. (1985)]. As will be demonstrated subsequently (Exercise 4), temperature is a measurement of the intensity, not the quantity, of heat.

Measurement of Temperature

Thermometers. Mercurial thermometers are useful for measuring surface temperatures by direct immersion but are of limited use for subsurface measurements. Such thermometers can be enclosed in tall, clear bottles. The corked bottle, harnessed and weighted to keep it vertical, can then be lowered to depth and the cork removed by a separate line to allow the bottle to fill with water from that depth. After a few minutes, the bottle is retrieved rapidly and the temperature is read through the bottle before appreciable temperature changes occur. This method obviously is tedious and of limited accuracy. Good grade maximum–minimum thermometers also can be used on a weighted, calibrated line to determine subsurface temperatures. Inexpensive laboratory and maximum–minimum thermometers often can be in error by 1°C or more. Careful calibration against standardized thermometers (registered against National Bureau of Standards reference thermometers) is mandatory for accurate work.

Reversing Thermometers. Among the most accurate devices for measuring subsurface temperatures is the reversing thermometer. After this instrument is lowered to depth and has equilibrated to temperature, a release mechanism is tripped by a messenger (metal weight) sent down the line from the surface. Upon release, the thermometer pivots vertically 180°. The thermometer is constructed in such a way that the stem is coiled and constricted. When the position of the thermometer is reversed (inverted), the mercury column separates at the point of constriction. As a result, the position of the detached mercury column will not change if subjected to different temperatures. With corrections for slight thermal expansions of mercury in glass, most calibrated reversing thermometers are accurate to at least 0.01°C. These thermometers are expensive and fragile, requiring extreme care during their use. A full description of their characteristics and use is given in Welch (1948).

Thermistor Electrical Thermometers. The electrical resistance of many substances changes as temperature changes. Early electrical thermometers utilized this relationship to measure temperature. In combination with a simple Wheatstone-Bridge, the change in resistance (or current) of a conductor, usually ceramic, was measured and calibrated in units of temperature. More recent instruments are all of the thermistor type in which, by similar principle, resistance decreases with increasing temperature. Thermistor response is rapid, requiring only a few seconds to come into equilibrium, and is quite accurate within the range of temperatures encountered in fresh waters. All such instruments must be calibrated against reference thermometers periodically.

Bathythermograph. The bathythermograph is an instrument for measuring and recording continuous vertical profiles of temperature with depth. The torpedo-shaped instrument consists of a heavy nose-piece to which a cable shackle and swivel are attached, a body tube housing pressure and thermal elements, and a protecting tail cage housing the capillary tubing of the thermal element.

The thermal element of traditional, older bathythermographs is a temperature-compensated Bourdon tube, actuated by a xylene-filled capillary about 15-m long, that is exposed to uniform flow of water over its surface. A pen and stylus are attached to the Bourdon tube. The pressure element consists of an evacuated bellows, movable at one end and fixed at the other, which compresses with increasing water pressure. An accurately wound calibration spring resists this compressing action and is attached to a slide holder at the end of the bellows.

In operation, a glass slide (smoked or coated with an extremely thin layer of a gold compound) is placed into the slide holder on the free end of the bellows assembly, and the stylus is positioned to make contact with the slide. As the bathythermograph is lowered in the water, the slide moves vertically as increasing water pressure compresses the bellows assembly. The stylus is fixed in the vertical direction. Horizontal movement of the stylus with respect to the slide, which is fixed in position in this direction, is produced by movement of the Bourdon tube caused by temperature change. The resulting trace is a composite of depth on the vertical ordinate and temperature on the horizontal ordinate. The accuracy of the temperature and depth values is about $\pm 1°C$ and $\pm 1\%$ of instrument depth range. New oceanographic bathythermographs are "disposable" in that data are transmitted electronically to the ship as the instrument descends through the water column to the sediments.

EXERCISES

OPTION 1. FIELD ANALYSIS

In a series of lakes of differing productivity (e.g., oligotrophic, eutrophic) and a lake with highly stained water from dissolved organic matter, make the following analyses.

1. Determine the water temperatures with a thermistor thermometer at meter intervals from the surface to the bottom in the deepest area of the lake. Near the surface, especially under ice cover, and in areas of large thermal discontinuities, measurements should be made at intervals of less than 1 m of depth. Plot these data (abscissa) versus depth (ordinate).

2. Using the data provided in Table 2.3 and procedures discussed below, calculate and plot isopleths of uniform temperature in a depth-time seasonal diagram.

3. Compare the temperatures in the upper strata of the open water with (a) water strata in the littoral zone in a wave-swept area and among dense stands of large aquatic plants, and (b) in stream inflow areas to the lake.

4. With an underwater photometer, measure the solar irradiance at the surface and at depths of 10 cm, 50 cm, 1 m, and meter intervals thereafter. Measurements should be made from the unshaded side of the boat. A separate but identical photometer, a "deck cell," should be used to measure surface irradiance concurrently with underwater measurements when light conditions change rapidly, such as under rapidly moving, broken cloud cover.

   Caution: Be certain to switch the meter to a high resistance scale before raising the underwater sensor. Exposure of the sensor to high light intensities when set to maximum sensitivity can damage the meter.

5. Place a blue filter over the sensor of the photometer and repeat the vertical depth profile of light attenuation. Repeat with green, red, and other available filters.

6. When a double hemispheric 4π photometer is available, measure the underwater irradiance as follows: Use only the upward-directed hemisphere to measure incident light as the sensor is lowered into the lake. At maximum depth, switch the instrument to the lower hemisphere and measure the reflected and scattered light at each depth while raising the sensor to the surface.

7. Determine the Secchi disc transparency by lowering, on the shaded side of the boat, the disc to the point where it is no longer visible. Lower the disc below this point and slowly raise it until it just reappears. Do not wear sunglasses while making measurements. Record the average of these two measurements and record the variations.

8. Using a Van Dorn (see p. 86) or Kemmerer water sampler, collect water samples from
   (a) the surface and near the sediments at the open water, central station of each lake,
   (b) among dense stands of large aquatic plants of a quiescent littoral zone, and
   (c) inlet water.

   In the laboratory use a spectrophotometer with narrow bandwidth capabilities (preferably 1 nm or less) and a 10-cm quartz cell to determine the absorption of unfiltered samples of lake water at 5-nm intervals from 285 to 750 nm. When a ratio-recording scanning spectrophotometer is available, the continuous records will facilitate analyses greatly. Filter the water samples using glass or membrane filters of 0.5-μm porosity. Repeat the spectrophotometric analyses with the filtered water. Compare these data graphically by plotting absorption versus wavelength.

9. Determine the turbidity of the samples with a spectrophotometer by comparison with results from a series of standards (see Apparatus and Supplies, p. 30). Prepare a reference curve from the standards.

10. Determine the relative color of the water samples by comparison to platinum-cobalt standard solutions (see "Apparatus and Supplies," p. 30).

11. Using your data, graph vertical profiles of temperature versus depth.

12. Using your field data, graph vertical profiles of spectral attenuation as a percentage of surface irradiance (0 m = 100%) versus depth using (a) a linear scale and (b) a logarithmic scale.

13. With your data, calculate using the formulas given on p. 18 the extinction coefficients of light at each meter interval and determine the mean extinction coefficient (all depth intervals below 1 m) for total light and observed light absorption with blue, green, and red filters.

OPTION 2. LABORATORY ANALYSES

1. Using the water samples provided by your laboratory instructor, perform the spectrophotometric analyses on unfiltered and filtered water as outlined on Option 1, section 8. Compare graphically the data of absorption versus wavelength.

Table 2.2. Percentage transmission of light underwater, Gull Lake, Kalamazoo-Barry counties, Michigan, June 28, 1976.

Depth (m)	Neutral filter	Blue filter (460–500 mnm)	Green filter (500–580 nm)	Yellow filter (580–590 nm)	Red filter (640–720 nm)
0	100%	100%	100%	100%	100%
0.1	94.05	87.32	94.74	88.42	87.50
1	73.81	63.38	70.53	50.00	60.00
2	50.00	53.52	55.79	35.26	35.50
3	33.04	32.25	44.21	19.74	22.50
4	25.00	27.46	35.79	16.58	15.00
5	17.86	22.68	28.42	12.37	8.75
6	13.10	18.73	19.37	7.89	6.15
7	9.82	13.24	14.21	6.58	4.50
8	5.33	7.75	10.16	5.39	3.55
9	4.02	5.92	6.84	3.95	2.50
10	2.83	3.66	4.84	2.32	1.60
11	1.96	1.72	3.47	1.56	0.70
12	1.19	1.11	1.84	1.00	0.44
13	0.52	0.63	1.16	0.66	0.26
14	0.33	0.44	0.74	0.42	0.17
15	0.22	0.24	0.47	0.21	0.12
16	0.15	0.17	0.33	0.14	0.08
17	0.10	0.13	0.21	0.10	0.06
18	0.07	0.08	0.15	0.07	0.04
19	0.05	0.02	0.12	0.05	0.03
20	0.04	—	0.09	0.04	0.01
21	0.03	—	0.02	0.02	—
22	0.02	—	0.01	0.01	—
23	0.01	—	—	—	—

2. Using the data provided (Table 2.2), calculate using the formulas given on p. 18 the extinction coefficients and determine the mean extinction coefficient (all depth intervals below 1 m) for total light and that for light absorption with blue, green, and red filters.
3. Graph the vertical profiles (Table 2.2) of spectral attenuation as a percentage of surface irradiance (0 m = 100%) versus depth using (a) a linear scale and (b) a logarithmic scale.
4. Graph the vertical profiles of temperature versus depth (Table 2.3).
5. Using the data provided (Table 2.3) and procedures discussed below, plot isopleths of uniform parameter values in depth-time diagrams of:
   (a) Temperature (isotherms) seasonally.
   (b) Underwater incident light (isophots).
   (c) Underwater reflected-scattered light.

Construction of Depth-Time Diagrams

Presentation of data as a vertical depth profile is satisfactory for any given time. However, when the same parameter is measured vertically versus depth repeatedly over a time period of, say, a year, a large number of vertical profiles would be needed. Not only is this consumptive of space, but often the seasonal trends are obscured by the many lines in graphs extending for pages. Graphs should be used to clarify relationships between complex variables.

An effective way to present such data is to determine the depth of some uniform value of the parameter. The uniform value of temperature might be 2°C; of underwater light it might be 20% of attenuation. The depth of these uniform data, say 14°C, can be estimated from

Table 2.3. Temperature and underwater irradiance, Lawrence Lake, Barry County, Michigan.

Depth (m)	April 9, 1974			April 23, 1974			May 7, 1974			May 21, 1974			June 4, 1974		
	°C	Inc. light (%)	Refl. light (%)	°C	Inc. light (%)	Refl. light (%)	°C	Inc. light (%)	Refl. light (%)	°C	Inc. light (%)	Refl. light (%)	°C	Inc. light (%)	Refl. light (%)
0	4.9	100	100	10.9	100	100	12.2	100	100	16.9	100	100	20.8	100	100
0.1	—	43.2	87.5	—	65.6	70.0	—	52.7	16.6	—	52.3	11.3	—	54.1	18.8
1	4.9	28.2	75.0	10.9	32.2	65.6	12.1	35.2	12.1	16.9	31.4	9.4	20.5	35.3	16.3
2	4.9	16.8	42.0	10.8	25.6	56.1	12.0	20.9	9.3	15.2	22.1	7.1	19.6	23.5	11.7
3	4.9	10.5	31.0	10.8	18.9	43.9	11.9	15.4	5.9	13.1	18.4	4.8	17.4	15.3	8.3
4	4.9	6.6	20.5	10.8	13.3	32.2	11.8	13.6	3.8	12.7	13.5	3.5	14.9	13.8	5.4
5	4.9	3.5	13.5	10.7	8.9	22.2	11.1	10.1	2.4	12.2	9.8	2.3	13.1	8.9	3.0
6	4.9	2.6	8.8	9.4	6.1	13.3	9.8	7.3	1.8	12.0	6.5	1.8	12.1	6.0	2.4
7	4.9	1.8	5.0	7.8	4.4	7.8	8.7	4.5	1.4	10.5	5.1	1.2	10.8	4.6	1.8
8	4.9	1.2	3.0	6.9	3.2	5.0	7.4	3.0	1.0	8.3	3.4	0.9	8.7	2.9	1.2
9	4.9	0.8	2.0	6.1	2.3	3.3	6.3	2.1	0.7	7.1	2.4	0.7	7.0	2.1	0.8
10	4.9	0.5	1.3	5.2	1.6	2.8	5.7	1.4	0.6	6.2	1.7	0.6	6.2	1.5	0.7
11	4.9	0.3	0.8	4.9	1.1	2.2	5.1	1.2	0.5	5.8	1.3	0.6	5.9	1.2	0.7
12	4.8	0.2	0.3	4.8	0.7	1.7	5.0	1.0	0.5	5.5	1.0	0.6	5.5	0.9	0.7
12.5	4.8	<0.1	—	4.8	<0.3	—	5.0	<0.5	—	5.5	<0.5	—	5.5	<0.4	—

Inc. Light = incident-scattered light of upward-directed hemisphere; Refl. light = reflected-scattered light of downward-directed hemisphere.

observed data by interpolation, assuming a linear increase or decrease, using a simple proportion.

Lex x = depth of measurement at upper depth, y = depth of measurement at lower depth,
m = value of measurement at upper depth, n = value of measurement at lower depth, o =
interpolated measurement of uniform value between the upper and lower measurements, and
p = unknown desired depth of o from upper depth. Then

$$\frac{(y-x)}{(n-m)} = \frac{p}{(o-m)}$$

and

$$\text{depth of } o = x + p.$$

For example, given the following data, what is the depth of o for 12.0 and for 14.0 mg oxygen
per liter?

Depth (m)	mg O_2/l
2	11.3
4	14.3
6	10.8

and

$$\frac{(4-2)}{(14.3-11.3)} = \frac{p}{(12.0-11.3)}$$
$$3p = (2)(0.7)$$
$$3p = 1.4$$
$$p = 0.47\text{m}$$
$$\text{depth of } o_{12} = 2+0.47 = 2.47\text{m}$$
$$\frac{(4-2)}{(14.3-11.3)} = \frac{p}{(14.0-11.3)}$$
$$3p = (2)(2.7)$$
$$3p = 5.4$$
$$p = 1.80\text{m}$$
$$\text{depth of } o_{14} = 2+1.80 = 3.80\text{m}$$

And, progressing from 4 to 6 m,

$$\frac{(6-4)}{(10.8-14.3)} = \frac{p}{(14.0-14.3)} \quad \text{and} \quad \frac{(6-4)}{(10.8-14.3)} = \frac{p}{(12.0-14.3)}$$
$$-3.5p = -0.6 \qquad\qquad\qquad -3.5p = -4.6$$
$$p = 0.17\text{m} \qquad\qquad\qquad p = 1.31\text{m}$$
$$\text{depth of } o_{14} = 4+0.17 = 4.17\text{m} \quad \text{depth of } o_{12} = 4+1.31 = 5.31\text{m}$$

Such a simple relationship can be easily programmed on a hand calculator or computer.

The depths of uniform values then are plotted on graph paper against time for each sampling date (Fig. 2.3). Special graph paper with a linear vertical scale and a horizontal scale
divided into days for a year greatly facilitates construction of such illustrations. The uniform
values then are connected by smooth lines. Such isopleth lines can never cross one another
but commonly exit and begin at the boundaries of the system (surface or bottom sediments)
during periods of rapid change. Often concentrations increase and decrease internally within
the water column; graphically, these changes are represented by cells of increasing or decreasing value.

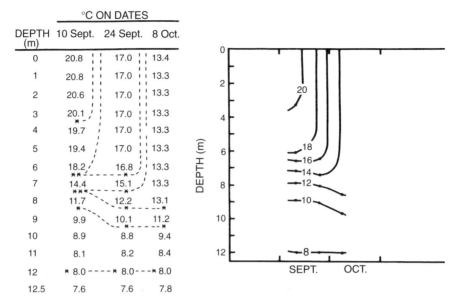

DEPTH (m)	°C ON DATES		
	10 Sept.	24 Sept.	8 Oct.
0	20.8	17.0	13.4
1	20.8	17.0	13.3
2	20.6	17.0	13.3
3	20.1	17.0	13.3
4	19.7	17.0	13.3
5	19.4	17.0	13.3
6	18.2	16.8	13.3
7	14.4	15.1	13.3
8	11.7	12.2	13.1
9	9.9	10.1	11.2
10	8.9	8.8	9.4
11	8.1	8.2	8.4
12	8.0	8.0	8.0
12.5	7.6	7.6	7.8

Figure 2.3. Method of plotting isopleths on depth-time diagrams.

Any parameter may be expressed by isopleths in depth-time diagrams: intensity factors such as temperature, concentrations of nutrients or organisms, or percentage relationships. Use judgment about which isopleth values to express between maximum and minimum values. When parameters change gradually, such as dissolved oxygen concentrations, isopleths for every 2 mg/l may demonstrate adequately the trends without severe crowding. Sodium concentrations, which conservative and change little, may require 20-μeq/l interval isopleths to show any seasonal trends. Functions that change exponentially, such as light attenuation, may require isopleths of 80, 60, 40, 20, 10, 5, 1, 0.1, and 0% of surface irradiance. Algal populations rapidly may attain very high levels, which persist for only a short time, and then remain at low concentrations for the remainder of the year. During the peak, isopleth lines may be 100, 500, 1000, 5000 cells/l, while during the rest of the year they might be 2, 5, and 10 cells/l.

Minimum and maximum values shuld be indicated. Low values rarely are zero, but rather should be reported as below some detectable limit of sensitivity of the assay or apparatus employed. For example, while water may be clearly anaerobic with oxygen concentrations at zero mg/l, manganese concentrations at the minimum isopleth should reflect the lower limit of measurement (e.g., <0.02 μeq/l). Computer software is available for computing and plotting isopleth distributions.

Questions and Problems

1. Define the euphotic zone in the lakes you studied or from data provided.
2. When the compensation point is defined as that point where net photosynthesis (gross photosynthesis minus respiration) is zero, and for a particular algal population the compensation depth was determined to be 1% of surface light, how much volume of water would be available for its growth in the different lakes?
3. How would this situation change for a particular shade-adapted blue-green alga that can survive at <0.1% of surface irradiance?
4. If the crustacean *Mysis* were to require cold temperatures (<10°C) and were adapted to a strictly blue photoenvironment, would this organism likely survive in the lake conditions you observed?

5. Compare the Secchi disc transparencies using several colored discs to vertical extinction coefficients within several spectral ranges as determined with an underwater photometer with narrow-band spectral filters. Compare your results with those of Štěpánek (1959) and Elster and Štěpánek (1967). The spectral analysis of paint often is available from paint manufacturers.
6. Why do clean lakes appear blue on a clear day?
7. Which wavelengths are the most important in heating the surface waters of a lake? What percentage of the total spectrum is found at these wavelengths?
8. What wavelengths penetrate most deeply in a dystrophic bog lake heavily stained with large concentrations of dissolved organic compounds?
9. Why do aquatic plants appear green?
10. What is the "greenhouse effect"?

Apparatus and Supplies

1. Thermometers and maximum-minimum thermometers, calibrated against a standard reference thermometer; calibrated, weighted line.
2. Reversing thermometer, if available.
3. Underwater electrical thermometer with calibrated cable of sufficient length to reach maximum depths of lakes and streams under study.
4. Underwater photometer with various sensor units (e.g., Li-Cor, Inc., Lincoln, NE, or Kahl Scientific Instrument Corp., EI Cajon, CA); narrow-band spectral filters.
5. Four-π underwater photometer, if available (e.g., Li-Cor Inc., Lincoln, NE).
6. Secchi disc, exactly 20 cm in diameter, weighted, with calibrated, nonstretchable line. White on upper surface or divided into quadrants, painted so that two quadrants directly opposing each other are white and the intervening quadrants are black. (Little difference has been found between either type of Secchi disc; most workers use all-white discs.) Lower surface should be painted black.
7. Van Dorn or Kemmerer water sampler and clean polyethylene bottles.
8. Spectrophotometer, ratio-recording and scanning if available, with quartz, long pathlength cells (5 to 10 cm).
9. Turbidity solutions:
   a. Stock 1. 1.000 g hydrazine sulfate [$(NH_2)_2 \cdot H_2SO_4$] dissolved in distilled water in a volumetric flask to make 100 ml.
   b. Stock 2. 10.00 g hexamethylenetetramine [$(CH_2)_6N_4$] dissolved in distilled water in a volumetric flask to make 100 ml.
   c. Standard: To a 100-ml volumetric flask, mix 5.0 ml of Stock Solution 1 with 5.0 ml of Stock 2. Allow to stand 24 h at room temperature and then dilute to 100 ml and mix. The turbidity of this suspension is 400 units. Store no longer than a week. A dilution of 10.00 ml of this standard to 100 ml with distilled water yields the standard turbidity suspension of 40 units. A series of more dilute standards are then made to establish the calibration curve.
10. Turbidimeter, if available (e.g., Hach Company, Loveland, CO).
11. Color solutions: Serial dilution from 5 to 1000 platinum color units, sealed in permanent vials or glass ampules. 1000 Pt units = the color resulting from 2.492 g potassium hexachloroplatinate (K_2PtCl_6), 2.000 g cobaltic chloride hexahydrate ($CoCl_2 \cdot 6H_2O$), 200 ml concentrated hydrochloric acid (HCl), and 800 ml distilled water.
12. Graph paper: linear, semi-logarithmic (4-cycle), and seasonal (1 year by 180 divisions, e.g., K & E 47-2730).

References

Åberg, B. and W. Rodhe. 1942. Über die Milieufaktoren in einigen südschwedischen Seen. Symbol. Bot. Upsalien. *5*(3). 256 pp.

Baker, A.L. and A.J. Brook. 1971. Optical density profiles as an aid to the study of micro-stratified phytoplankton populations in lakes. Arch. Hydrobiol. *69*:214–233.

Baker, A.L., K.K. Baker, and P.A. Tyler. 1985. Fine-layer depth relationships of lakewater chemistry, planktonic algae and photosynthetic bacteria in meromictic Lake Fidler, Tasmania. Freshwat. Biol. *15*:735–747.

Beeton, A.M. 1958. Relationship between Secchi disc readings and light penetration in Lake Huron. Trans. Amer. Fish. Soc. *87*:73–79.

Birge, E.A. 1915. The heat budgets of American and European lakes. Trans. Wis. Acad. Sci. Arts Lett. *18*(Part 1):166–213.

Birge, E.A. 1916. The work of the wind in warming a lake. Trans. Wis. Acad. Sci. Arts Lett. *18*(Part 2):341–391.

Elster, H.J. and M Štěpánek. 1967. Eine neue Modifikation der Secchischeibe. Arch. Hydrobiol. (Suppl.) *33*:101–106.

Hutchinson, G.E. 1957. A treatise on Limnology. I. Geography, Physics, and Chemistry. Wiley, New York. 1015 pp.

Johnson, N.M., G.E. Likens, and J.S. Eaton. 1985. Stability, circulation and energy flux in Mirror Lake. pp. 108–127. *In*: G.E. Likens, Editor. An Ecosystem Approach to Aquatic Ecology. Mirror Lake and Its Environment. Springer-Verlag, New York.

Karentz, D. and 13 others. 1994. Impact of UV-B radiation on pelagic freshwater ecosystems: Report of working group on bacteria and phytoplankton. Arch. Hydrobiol. Beih. Ergebn. Limnol. *43*:31–69.

Kerr, J.B. and C.T. McElroy. 1993. Evidence for large upward trends in ultraviolet-B radiation linked to ozone depletion. Science *262*:1032–1034.

Kirk, J.T.O. 1994a. Optics of UV-B radiation in natural waters. Arch. Hydrobiol. Beih. Ergebn. Limnol. *43*:1–16.

Kirk, J.T.O. 1994b. Light and photosynthesis in aquatic ecosystems. 2nd Ed. Cambridge Univ. Press, Cambridge, England.

Likens, G.E. and F.H. Bormann. 1972. Nutrient cycling in ecosystems. pp. 25–67. *In*: J. Wiens, Editor. Ecosystem Structure and Function. Oregon State Univ. Corvallis.

Morris, D.P., H. Zagarese, C.E. Williamson, E.G. Balseiro, B.R. Hargreaves, B. Modenutti, R. Moeller, and C. Queimalinos. 1995. The attenuation of solar UV radiation in lakes and the role of dissolved organic carbon. Limnol. Oceanogr. *40*:1381–1391.

Rich, P.H. and R.G. Wetzel. 1969. A simple, sensitive underwater photometer. Limnol. Oceanogr. *14*:611–613.

Sauberer, F. 1962. Empfehlungen für die Durchfuhrung von Strahlungsmessungen an und in Gewässern. Mitt. Int. Ver. Limnol. *11*. 77 pp.

Smith, R.C. and W.H. Wilson, Jr. 1972. Photon scalar irradiance. Appl. Optics *11*:934–938.

Smith, R.C., J.E. Tyler, and C.R. Goldman. 1973. Optical properties and color of Lake Tahoe and Crater Lake. Limnol. Oceanogr. *18*:189–199.

Štěpánek, M. 1959. Limnological study of the reservoir Sedlice near Želiv. IX. Transmission and transparency of water. Sci. Pap. Inst. Chem. Technol., Prague, Fac. Technol. Fuel Water *3*(Part 2):363–430.

Stewart, K.M., K.W. Maleug, and P.E. Sager. 1965. Comparative winter studies on dimictic and meromictic lakes. Verh. Int. Ver. Limnol. *16*:47–57.

Strickland, J.D.H. 1958. Solar radiation penetrating the ocean. A review of requirements, data and methods of measurement, with particular reference to photosynthetic productivity. J. Fish. Res. Bd. Canada *15*:453–493.

Talling, J.F. 1957. Photosynthetic characteristics of some freshwater plankton diatoms in relation to underwater radiation. New Phytol. *56*:29–50.

Tyler, J.E. 1968. The Secchi disc. Limnol. Oceanogr. *13*:1–6.

Welch, P.S. 1948. Limnological Methods. Blankiston, Philadelphia. 381 pp.

Westlake, D.F. 1965. Some problems in the mesurement of radiation under water: A review. Photochem. Photobiol. *4*:849–868.

Wetzel, R.G. 1983. Limnology 2nd Ed. Saunders Coll., Philadelphia. 858 pp.

Wetzel, R.G. 1999. Limnology: Lake and River Ecosystems. 3rd Ed. Academic Press, San Diego (in press).

Wetzel, R.G., P.G. Hatcher, and T.S. Bianchi. 1995. Natural photolysis of ultraviolet irradiance of recalcitrant dissolved organic matter to simple substrates for rapid bacterial metabolism. Limnol. Oceanogr. *40*:1369–1380.

Whitney, L.V. 1937. Microstratification of the waters of inland lakes in summer. Science *85*:224–225.

Physical Characteristics: Lake Models

The hydrodynamics of water movement is an integral component of a functional lake system. The importance of water movements and associated turbulence either has been underestimated or neglected in a large majority of limnological investigations. Consideration must be given to the effects of water movements on stratification and on the distribution of temperature, dissolved gases and nutrients, and biota.

The various laboratory exercises outlined below are based on a simple and rather naive model of a lake. Thus you should bear in mind constantly that you are studying only a model and not an actual lake. Furthermore, many of the formulas are simplified accordingly. Some virtues of models are:

1. Size is reduced to manageable proportions.
2. Complexity is reduced to such an extent that mechanisms and processes may be appreciated readily and intuitively understood.
3. Experimentation is facilitated, whereas it may be difficult and/or expensive with a natural system.

On the other hand, unjustifiable extrapolations to natural conditions can be made from the model when the observer is not careful to distinguish between the two. In the experiments outlined here, for example, when the sides and bottom of the aquarium are not insulated, there is more rapid heat exchange between the model "lake" and its surroundings than ever would be the case in nature. Likewise, the amount of energy supplied by the heat lamp in relation to the depth of water is much higher than in natural situations. The reason for the modification is, of course, to accelerate the process. These and similar points should be stressed in your report(s).

These exercises are based on and modified from material prepared by Mortimer (1951) and Vallentyne (1967).

Caution: Take care not to splash water on the heat lamp—otherwise, EXPLOSION!

LAKE MODEL 1: TEMPERATE LAKE

Purpose

Thermally induced density stratification commences in the spring and persists through the summer for moderately deep lakes in the Temperate Zone. These conditions will be illustrated in this lake model. It is instructive in this exercise to separate the actions of sun and wind, and to that extent the model is unnatural. The exercise also illustrates the phenomena of internal seiches and convective sinking during night or autumnal cooling. Calculations will be made of the "annual" heat budget, the stability of the "lake," the thermal resistance to mixing for different layers, and internal seiche periods.

Procedures

1. Prepare a chart with columns for time (minutes elapsed) and temperatures at depths of 1, 3, 5, 9, 13, 17, and 21 cm. Leave a space at the extreme right for inserting temperatures at other (variable) depths.
2. Fill the aquarium with cold tap water and mix to 4°C with ice. Allow the system to come to rest (3 to 5 min) with a small amount of ice floating on the surface. Then introduce a moderate wind (first wind) on the top, blowing from one side and then the other. After a minute or two you will notice that the temperature at 1 cm has dropped to 2 to 3°C. Stop the wind and record the temperatures at the surface and at each fixed depth. Next, remove the excess ice as gently as possible and turn on the heat lamp. Record this time as the beginning of the experiment. With the light on and on wind, record temperatures after approximately 5 min and every 5 min thereafter for approximately 50 min, taking special care to get additional temperature readings in regions where temperatures change rapidly with depth (e.g., upper 5 cm). Plot temperature changes as a function of depth on graph paper as they are recorded. You will observe an exponential decrease in temperature with depth. Why?
3. With the lamp still on, drop some crystals of catechol violet more or less randomly over the surface. Allow them sufficient time to sink so that most of the descending trails are in the upper third of the aquarium with a few extending into the middle third, and still fewer to the bottom. Observe the positions of the trails and the complexities of the water motion, particularly near the surface. What can you infer about the water currents in the "lake"?
4. Now introduce a light to moderate wind (second wind) blowing first from one side and then the other so as to mix the surface waters into one homothermal mass. Avoid too strong a wind, which could destroy the thermocline that you are now creating. The epilimnion (with homogeneous distribution of color) should now be about 5- to 7-cm thick. Turn off the light and allow the currents to slack; record the depth of the homogeneous red layer and temperatures as a function of depth.
5. Leave the light off. Using the same wind speed (third wind), impart a slope of 10 to 15° to the thermocline by blowing from one side only. This slope can be achieved best in stages, allowing successive winds with intervening periods of quiescence to build up the internal seiche. When the desired slope to the thermocline has been obtained, stop the wind and measure the periods of the internal seiche (average approximately five measurements). When an electrical thermometer is suspended at the end of the lake at the depth of the upper portion of the metalimnion, the periodic temperature changes of the oscillating internal seiche can be recorded directly. Allow the system to come to rest, record the temperatures at each depth, and measure the depth of the homogeneous red layer.
6. Next, carefully add a layer of ice cubes to the surface with minimal disturbance of the water. It will be necessary to cover about three-fourths of the surface if small cubes or crushed ice are used. Observe the descending convection cells sinking through the thermocline region, which in turn cause the thermocline to descend. When the thermocline has descended to about mid-depth in the tank,

carefully remove the remaining ice with as little turbulence as possible. With a wind (fourth wind) from one side, build up a 10 to 15° slope on the thermocline and redetermine the periods of the seiches. Record temperatures and the depth of the homogeneous red layer.

7. Allow the system to rest and then with a strong wind (DO NOT SPLASH WATER ONTO THE HEAT LAMP!) turn the lake over. When mixing is complete, record the temperatures.

Apparatus and Supplies

1. An aquarium, corresponding to a lake basin, internally about $40 \times 20 \times 24$ cm high, and insulated on the sides and bottom with 5 cm of polyurethane foam. The aquarium should be filled will cold water and ice to cool the water to 4°C. Some ice should be floating at the surface.
2. A heat source (e.g., 250-W lamp) corresponding to the sun. This lamp should be located directly over the middle of the aquarium, 25 cm above the water surface.
3. Seven mercury thermometers (−20 to 110°C) attached by masking tape or other means to two plastic rulers. Match the thermometers or intercalibrate the thermometers relative to each other. The rulers should be shortened to 23 cm and the thermometers positioned horizontally at depths of 1, 3, 5, 9, 13, 17, and 21 cm. The string of thermometers is suspended by masking tape on the inside of the front wall of the aquarium. In this manner temperatures can be read directly and quickly at the stated depths. An additional thermometer is required to read temperatures at intermediate depths. This thermometer should be inserted into the water with as little turbulence as possible.
4. Catechol violet or methylene blue crystals. Before the dye is added, crystals should be placed in the fold of a small piece of paper and *gently* crushed by pressing the two sides together. In this way the crystals will be small enough so that most will dissolve completely in the upper quarter of the water column. Use care in handling the catechol violet. Avoid excessive contact with skin.
5. A fan or blower to generate wind-driven surface currents.
6. A stopwatch.
7. A tea strainer to remove ice.
8. (Optional) An electrical thermometer connected to a data recorder. Relatively inexpensive recording thermistor thermometers with an accuracy of ±0.5°C are available from Onset Computer Corp. (Pocasset, MA). The data from these microloggers are conveniently downloaded or reprogrammed through a standard serial port on a personal computer with Windows-based software from Onset.

Calculations

1. *Density versus temperature.* Using data on water density as a function of temperature [e.g., a handbook of physics and chemistry, or pp. 204–205 in Hutchinson (1957)], make a plot from which you can read density to the sixth decimal place for various temperatures from 0 to 35°C. Retain this graph for future use in other experiments. On a computer or on a single sheet of graph paper, plot thermally inferred density versus depth, and temperature depth after the second wind, at the time of maximum thermal stratification. Note the difference.

2. *Heat budget.* There are various ways of expressing heat budgets (see Exercise 4). The one used here (Birgean heat budget) corresponds to the heat gained by a

lake between the times of its lowest and highest heat contents (usually late winter to midsummer). The simplest method of calculating the heat budget is to plot depth (z) on the vertical axis versus the product $A_z (t_{sz} - t_{wz})$ on the horizontal axis, where A_z is the area of the stratum at depth z and t_{sz} and t_{wz} are the maximum (summer) and minimum (winter) temperatures, respectively, at depth z. The integral is measured either planimetrically (see Exercise 2) or by counting squares, and the sum thus arrived at is divided by A_0 (surface area of the lake). Since the area at all depths is the same in the simplified case of the aquarium, the heat budget will be given directly as the integral from $z = 0$ to $z = $ max of $(t_{sz} - t_{wz})\, dz$. When the measurements of depth are recorded in centimeters and the heat capacity of water at all temperatures is taken as $1\,g$ cal cm^3, the integral will, with small error, be given in gram-calories per cm^2, or langleys, the usual units for heat budget values.* (Be certain that the origins for both axes of your plot are set at zero. Why is this a problem in limnology?) What then is the "annual" Birgean heat budget for the aquarium lake?

3. *Thermal resistance to mixing.* This concept, developed by Birge in 1910 and 1916, is a function of the density difference between the top and bottom of a defined thickness of water. Birge defined "thermal resistance" as the amount of work required to completely mix such a column of water:

$$\text{Work in ergs} = \frac{AC^2}{12}(\rho_2 - \rho_1)$$

where $A = $ area, $C = $ height of column, $\rho_1 = $ density at upper surface, and $\rho_2 = $ density at lower surface. It was convenient to assume that A and C were constants, i.e., $A = 1\,cm^2$ and $C = 100\,cm$, and thus make comparisons rather than absolute measurements. Comparisons were made against the difference in density of water at 4°C and that at 5°C, or 0.000008 g cm^{-3}. So, for any column of water with a uniform temperature gradient and dimensions specified, *relative* units of thermal resistance (work) could be determined from

$$\frac{(\rho_2 - \rho_1)(10^6)}{8}$$

In our case, since the area of layers at all depths is the same, we can compare simply the density difference of successive layers of 1-cm thickness with the difference in density of water at 4°C and 5°C (0.000008 g cm^{-3}).

Graph the temperature as a continuous curve versus depth using your data (1) for the aquarium "lake" just prior to the second wind (i.e., after 45 min when only the "sun" was shining), and (2) for the "lake" after the second wind then a good thermocline had developed. Prepare a table as follows:

Depth interval (cm)	t_2	ρ_2	t_1	ρ_1	$(\rho_2 - \rho_1)$	$\dfrac{(\rho_2 - \rho_1) \times 10^6}{8}$
0–1	—	—	—	—	—	—
1–2	—	—	—	—	—	—
2–3	—	—	—	—	—	—
etc.	—	—	—	—	—	—

---

*1 gcal (mean) $\times$ 4.1862 = 1 J; 1 J $\times$ 0.2389 = 1 gcal (mean).

In the preceding table, t_2 and t_1 are the temperatures at the bottom and top, respectively, of the 0- to 1-cm layer, etc., and ρ_2 and ρ_1 are the corresponding densities for pure water at temperatures t_2 and t_1. We are, of course, not dealing with distilled water here, but so long as the water is homogeneous in total dissolved solids, the difference in density $(\rho_2 - \rho_1)$ will be determined solely by temperature. Finally, plot the results in graphic form. Note the pronounced difference between the two situations of no wind and wind. What do you infer about turbulence and vertical currents in the thermocline region? Recall the dye trails you saw near the surface when catechol violet crystals were first added (before the wind was introduced). How do you now explain them?

4. *Periods of internal seiches.* The period of a seiche is the time required for an oscillation to complete one cycle of up-and-down motion. You have determined such periods for two separate situations, one with a high (near-surface) thermocline and the other with a lower thermocline. Were the periods identical? Why?

The period (T) of a uninodal, internal seiche in a rectangular basin of uniform depth, when the two layers have thicknesses of z_e and z_h (e = epilimnion, h = hypolimnion) and mean densities of ρ_e and ρ_h, is given by

$$T = \frac{2L}{\sqrt{\dfrac{g(\rho_h - \rho_e)}{\rho_h / z_h + \rho_e / z_e}}}$$

where L = length of the basin in the direction of the wind, and g = the acceleration due to gravity (980 cm sec^{-2}). Using your temperature and density data, calculate the theoretical periods for the seiches you observed. Calculate the theoretical periods using (1) the thermocline as a plane, and (2) the thermocline in the Birgean sense (thermocline = metalimnion). Do the values agree? How do these periods compare with the observed periods? Why do you think they differ? On the basis of the results above, which do you consider to be the more meaningful interpretation of "thermocline"—i.e., is the thermocline planar or zonal?

5. *The stability of a lake.* The "stability" of a lake as used by physical limnologists is the inertial resistance to complete mixing caused by vertical density differences [see Schmidt (1928), Reed (1970), Johnson et al. (1985)]. Measuring the stability (S) in g-cm cm^2,

$$S = \frac{1}{A_0} \int_{z_0}^{z_m} A_z (z - z_g)(\rho_z - \rho_m) dz$$

where A_0 = surface area of the lake, z = depth under consideration, z_0 = surface of lake, z_m = maximum depth, A_z = area at depth z, z_g = depth of the center of gravity of the unstratified lake, ρ_z = density at depth z, ρ_m = density at complete mixing (unstratified lake),

$$z_g = \frac{1}{V} \int_{z_0}^{z_m} z A_z dz$$

$$V (\text{lake volume}) = \int_{z_0}^{z_m} A_z dz$$

$$z_{gs} (\text{stratified lake}) = \frac{1}{\rho_m V} \int_{z_0}^{z_m} z \rho_z A_z dz$$

$$\rho_m = \frac{1}{V} \int_{z_0}^{z_m} \rho_z A_z dz$$

since

$$\text{the weight of the lake } = \rho_m V = \int_{z_0}^{z_m} \rho_z A_z dz$$

Cgs units should be used throughout. In our case, where $A_z = A_0$ and z_g = mean depth of a lake basin, or $z_m/2$ for the aquarium,

$$\rho_m = \frac{1}{z_m} \int_{z_0}^{z_m} \rho_z dz$$

(for our purposes it is sufficient simply to sum densities for every 1 cm of depth and divide the total by z_m).

The value of S is determined most easily by plotting z on the vertical axis (set the origin at zero) and the values for $(z = -z_g) \times (\rho_z - \rho_m)$ on the horizontal axis. The area enclosed by the curve is equal to S and may be determined planimetrically or by counting squares. Calculate stability for your model at the time of maximum thermal stratification (p. 34, #4). How do you regard the efficacy of this calculation as a measure of "stability"?

Note: It is suggested that the class divide into thirds and each group use Lake Models 2, 3, and 4. The results for each model then can be presented orally to the entire class.

LAKE MODEL 2: TROPICAL LAKE

Purpose

The physical properties of a tropical lake will be illustrated. It will be demonstrated how small temperature differences at high temperatures can generate the same stability in a lake as much greater temperature differences did at lower temperatures. The physical importance of temperature profiles thus is not temperature itself, but rather the thermally produced density differences.

Procedures

1. Allow the water (uniformly mixed at a temperature of about 20°C) to stand until no major currents are visible. Distribute over the surface of the water *a few* particles of catechol violet and allow the streaks to form. Introduce a moderate wind (lamp is off at this time) and observe that the waters will mix completely.
2. Record temperatures as a function of depth. Turn on the lamp, noting this time as the beginning of the experiment. Measure temperatures after 5 min, 15 min, and 30 min. Plot temperatures as a function of depth at the time they are taken, using the spare thermometer to obtain carefully the intermediate readings in the near-surface region.
3. Spread particles of catechol violet liberally over the surface as in Lake Model 1 and allow the streaks to descend.
4. Turn off the light. Introduce a moderate wind so as to bring the zone of homogeneous red color down to a depth of about 5 to 7 cm. Record temperatures and the depth of the red layer. Measure the periods of the internal seiche as in Lake Model 1.

5. With a slightly more forceful wind, bring the homogeneous red layer down to include the upper two-thirds of the "lake." Record temperatures and the depth of the homogeneous red layer. Now, with forced oscillation from the wind, cause another internal seiche to form and measure the period. Record temperatures and the depth of the homogeneous red layer again.
6. With a strong wind, bring about complete mixing. Record the final temperatures.

Apparatus and Supplies

1. Aquarium filled with tap water adjusted uniformly to about 20°C from top to bottom.
2. Heat lamp, thermometers, etc., positioned as in Lake Model 1. For more rapid heating, the lamp may be positioned nearer the water surface, but use caution.

Calculations

As in Lake Model 1, calculate (a) the stability of the lake, (b) the thermal resistance to mixing, (c) the heat budget, and (d) the theoretical periods of the internal seiches. Also make plots of temperature versus depth, and the thermally produced density versus depth for the values determined after the two winds. Compare with the results from Lake Model 1.

LAKE MODEL 3: MEROMICTIC LAKE

Purpose

The phenomenon of meromixis will be illustrated. The exercise will also show salt-produced density currents sinking and moving along the upper part of the chemocline and will simulate the phenomenon of "la bataillere." ("La bataillere" is the name given to a subsurface density current in the lake of Geneva where cold, silt-laden, glacial meltwaters move as a discrete stream along the thermocline region far out into the lake.) Being cold and containing suspended clay, the waters should be dense and descend as density currents along the bottom of the lake. However, the content of dissolved solids is low, producing an intermediate density, and the water flows into the lake as a subsurface stream along the bottom of the epilimnion. This phenomenon has no particular connection with meromixis—it merely is convenient to illustrate it at this same time when we are using a highly stratified lake.

Procedures

1. Fill the aquarium two-thirds full with tap water. Add ice liberally and uniformly cool the water to 4°C. Allow the water mass to become quiet.
2. Prepare a siphon arrangement from a 0.3% salt (NaCl) solution (at room temperature) containing methylene blue. An L-shaped glass tube with a short and *unpolished* end on the horizontal part is attached to the lower end of the siphon. The vertical part of the glass tube must be long enough so that it can rest diagonally on one side of the aquarium with the bottom of the tube resting in a corner at the bottom of the other side. The hose connection is provided with a screw clamp attached near the glass tube for regulation of the flow. Start the

siphon, allowing the blue liquid to come close to the end of the glass tube. Close the clamp and then lower the tube in such a way that the dense, salt-laden liquid does not flow into the clear aquarium water. Position the tube so that the opening is pointed in the direction of the long axis of the aquarium. Slowly open the screw clamp, allowing the liquid to flow out in a nonturbulent stream. Approximately 20 min will be required to fill the aquarium one-third full with the salt layer. (The flow may be increased somewhat as the level rises in the tank.) When the surface water in the tank has reached the desired level (just above the base of the metal rim), close the clamp and carefully remove the glass tube, keeping the terminal end uppermost so as to avoid mixing of the salt solution with the clear water.

3. Carefully remove excess ice from the surface, avoiding turbulent motions that might extend into the region of the chemocline. Record water temperatures and make additional measurements with the spare thermometer or electrical thermometer in the region of the chemocline.

4. Turn on the lamp and record this time as the beginning of the experiment. Leave the lamp on without any wind for 40 min.

5. Distribute catechol violet liberally over the surface so as to cause the upper quarter of the aquarium to become well colored with descending streaks of red dye.

6. With a moderate wind, cause the dye to be distributed in a homogeneously colored layer at the top (the epilimnion). The red layer should now occupy the upper third of the aquarium, the blue layer the lower third, and the middle layer will be essentially colorless. Record temperatures with extra measurements as necessary.

7. With forced oscillations from the wind source, create an internal seiche in the thermocline region. Note that another seiche develops in the chemocline region. Measure the period of both seiches simultaneously (different observers). Take care not to use too strong a wind that would destroy the three-layered system. Describe the motions.

8. When the seiches have died down, on one side of the aquarium introduce slowly from a beaker, whose liquid is level with the "lake" surface, about 50 ml of the thick bentonite-Congo red suspension. Try to obtain an even flow from the beaker to the "lake" water. Observe the vertically descending density current and watch it move along the upper part of the chemocline as a discrete layer. Why?

9. Dilute the suspension of bentonite-Congo red with cold distilled water to a point where you think it might flow along the thermocline. With a medicine dropper or pipet, add a portion to the end of the aquarium opposite that at which the thick bentonite-Congo red solution was added. When the solution sinks below the thermocline, it should be diluted further. Now *slowly* add about 25 ml of the *dilute* bentonite-Congo red suspension to the appropriate end of the aquarium. The solution will flow as a discrete layer along the thermocline. Watch what happens with time. The larger clay particles will begin to settle out in discrete, fingerlike projections sometimes separating into separate "blobs." How do you explain this? What does it mean with respect to studies made on sinking rates of small particles?

10. Enclose an Alka-Seltzer® tablet in wire mesh and allow it to drop to the bottom of the aquarium. Note the great turbulence generated by the bubbles and the mixing of the water masses.

Apparatus and Supplies

1. Aquarium filled two-thirds full with tap water and excess ice.
2. Heat lamp, thermometers, catechol violet, etc., as in Lake Model 1.
3. Approximately 10 liters per aquarium of a 0.3% salt solution containing methylene blue or other non-red dye.
4. A thick suspension of bentonite clay in distilled water with addition of Congo red dye to give a strong red color.
5. Alka-Seltzer® tablets enclosed in wire screening.
6. Hydrometers covering the range 1.0000 to 1.0050 sp. gr.

Calculations

With the hydrometers provided, measure the specific gravity of the 0.3% salt solution and the tap water used to fill the aquarium. After adding the salt-produced density to the thermally produced density following the first wind (remember to correct the densities of the salt and tap water solutions at the time of measurement for temperature), proceed as in Lake Model 1 to calculate (a) the stability of the lake, (b) the thermal resistance to mixing , and (c) the theoretical periods of the internal seiches. Also, make separate plots of density versus depth (a) as a result of temperature variations and (b) as a result of variations in salt concentration. Compare the results with those from Lake Model 1. See Hutchinson (1957, pp. 341–346) for additional information on the three-layered system.

LAKE MODEL 4: TEMPERATE LAKE WITH DENSITY DIFFERENCES BY SALINITY

Purpose

The thermally dependent density stratification of moderately deep lakes of the Temperate Zone can be simulated effectively with three strata separated by small salinity differences. With this model the effects of wind on stratification patterns and water movements can be observed without changing temperatures.

Procedures

1. Place the clear epilimnetic water into the model basin. Next, carefully siphon first the red metalimnetic water ($\rho \cong 1.015$) into place below the epilimnetic water; observe all of the precautions noted in Lake Model 3 (p. 39). Then carefully siphon the blue hypolimnetic water ($\rho \cong 1.050$) into place below the metalimnetic water. Without disturbance the stratified model can be held for at least 24 h without appreciable diffusion at the interfaces. Why?
2. Distribute a few crystals of methylene blue over the surface and observe the descending streaks of dye through the strata.
3. Apply a moderate wind to the surface and note the general size of surface waves (amplitude and wave length). Simultaneously apply a few crystals of methylene blue to the surface and follow the movement of water within the epilimnion at the surface and at the epilimnetic-metalimnetic interface.
4. With strong, forced oscillations from the wind source (see p. 34), create internal seiches in the metalimnion. Stop the wind and follow the amplitude and periodicity of the internal seiches. With care, this exercise can be repeated without disrupting the stratification appreciable. Why?

5. Again apply strong wind and impart a 10 to 15° slope to the epilimnion-metalimnion. Stop the wind and measure the (a) periods of the internal seiche, and (b) the magnitude of internal waves on the metalimnion. Continue to intensify the movements until the internal progressive waves form breakers. Compare the wavelengths and amplitudes of the internal waves to those of the surface waves.

6. Allow the system to return to rest. With trails from dye crystals, follow the water movements in the strata above and below each of the two density interfaces. How do these to-and-fro movements horizontally change in magnitude with increasing density and at the sediment-water interface? What can you infer about the relative turbulence generated and entrainment of water (and contained nutrients, gases, organisms) from one stratum to another?

7. With continued strong winds, bring about complete mixing. While applying dye crystals, observe the patterns of destratification and, in general, how much energy is required for this disruption. Determine the density of the fully mixed "lake" with the hydrometer.

Apparatus and Supplies

1. The experiments of this exercise should be done with a lake model constructed longer and narrower than typical aquaria. An excellent model has been made of 1-cm Plexiglas® acrylic sheeting with dimensions of $180 \times 60 \times 20\,cm$. Smaller models, made with redwood framing and glass sides, also have been used effectively. The ends of the "lake" are built up with inert materials (e.g., styrofoam) to produce an ellipsoidal morphology common in natural basins. The volumes of the final epilimnion, metalimnion, and hypolimnion can be calculated readily from the resulting trapezoidal shapes.

2. In the example given, the epilimnion consists of approximately 60 l of water without sugar additions ($\rho \cong 1.000$); ca. 30 l forms the metalimnion, to which red food coloring and sugar are added carefully to yield a density of 1.015 to $1.018\,g/cm^3$; and the hypolimnion contains ca. 63 l of water with blue food coloring and sufficient sugar to yield a density of 1.035 to $1.040\,g/cm^3$.

3. A strong blower, such as a vacuum cleaner, to generate wind-driven surface currents.

4. A stopwatch.

5. Catechol violet or methylene blue crystals (as in Lake Model 1).

6. Hydrometer in the range of 1.000 to 1.070.

Calculations

Proceed as in Lake Model 1 to calculate (a) the stability of the lake, (b) the resistance to mixing, and (c) the theoretical perods of the internal seiches. Make plots of the density versus depth and compare all of these results with those of the other lake models.

ADDITIONAL MODELS

A supply of polyurethane pieces will be available for those wishing to construct basins of various morphologies. This material can be cut easily with a hot knife and the rough parts smoothed with the flat blade of the hot knife. Care should be taken to have the dimensions of any form you cut about 2 cm wider and longer than those of the aquarium so that it will hold itself at any desired level without additional support. As example studies, the following are suggested as both instructive and feasible:

1. Simultaneous action of wind and "sun" on the thermal properties of lakes.
2. Production of multiple thermoclines.
3. Effect of sills and other bottom configurations (e.g., multiple depressions, submersed lakemounts) on water movements.
4. Determination of eddy conductivity coefficients from temperature measurements with continuous wind in the absence of the heat lamp.
5. Determination of eddy diffusivity coefficients from dye concentrations with wind in the absence and presence of the heat lamp.
6. Effects of humic substances on the thermal properties of lakes (dense tea solutions can be used).
7. Effect of varying the angle of the heat lamp on the heat budget (e.g., sun's angle).
8. Calcium carbonate precipitation with high surface temperatures from an initially saturated solution of cold water.
9. Effect of identical wind and "sun" conditions on lakes of varying depth.
10. Add colored "rain" to see the effects of precipitation on the mixing processes.
11. Effects of movements of organisms may be observed grossly by adding *Daphnia*, or some small fish, such as a guppy.

References

Birge, E.A. 1910. An unregarded factor in lake temperatures. Trans. Wis. Acad. Sci. Arts Lett. *16*:989–1004.

Birge, E.A. 1916. The work of the wind in warming a lake. Trans. Wis. Acad. Sci. Arts Lett. *18* (Part 2):341–391.

Hutchinson, G.E. 1957. A Treatise on Limnology. Vol. 1. Geography, Physics, and Chemistry. Wiley, New York. 1015 pp.

Johnson, N.M., G.E. Likens, and J.S. Eaton. 1985. Stability, circulation and energy flux in Mirror Lake. pp. 108–127. *In*: G.E. Likens, Editor. An Ecosystem Approach to Aquatic Ecology. Mirror Lake and its Environment. Springer-Verlag, New York.

Mortimer, C.H. 1951. The use of models in the study of water movement in lakes. Verh. Int. Ver. Limnol. *11*:254–260.

Reed, E.B. 1970. Annual heat budgets and thermal stability in small mountain lakes, Colorado, U.S.A. Schweiz. Z. Hydrol. *32*:397–404.

Schmidt, W. 1928 Über Temperatur und Stabilitätsverhältnisse von Seen. Geographiska Annaler *10*:145–177.

Vallentyne, J.R. 1967. A simplified model of a lake for instructional use. J. Fish. Res. Bd. Canada *24*:2473–2479.

The Heat Budget* of Lakes

One of the most important and interesting characteristics of a lake is its thermal structure. The heat content of a body of water is of vital importance in limnology. The metabolism, physiology, and behavior of aquatic organisms are related directly to the temperature of the aquatic environment. Extreme temperatures restrict the growth and distribution of plants, animals, and microbes. Because of the high specific heat of water, large volumes of water change temperature relatively slowly. Therefore, large lakes tend to moderate local climates, provide longer growing seasons for aquatic life, and serve as integrated recorders of recent climatic phenomena. For these and other reasons, the thermal structure and heat content of a body of water must be known with some degree of accuracy in limnological studies.

Temperature is related to the heat content of a body of water but does not measure it. Temperature is a measure of the intensity of heat stored in a volume of water, not a measure of the amount of heat stored. Heat (H) is measured in calories and is a function of the mass (M) of the substance in grams, temperature (t) in °C, and the specific heat (s) in cal/g-°C[†]

$$H = M \times t \times s$$

A bathtub filled with water at 0.5°C has a much greater heat content than does a glass filled with water at a temperature of 25°C.

HEAT CONTENT

The general equation which applies to the heat balance for a body of water in a given time interval is

$$\theta_R + \theta_E + \theta_L + \theta_V + \theta_S + \theta_B = 0$$

where θ_R = net radiation, θ_E = latent heat exchange, θ_L = sensible heat exchange with the atmosphere, θ_V = net advective exchange, θ_S = change in heat storage, and θ_B = conductive heat exchange through the bottom sediments.

Energy added is considered positive, while energy lost is negative. Over the course of an annual cycle the total heat gain balances the total heat loss. That is, in general and assuming no change in climate, natural lakes are not becoming warmer or colder

---

*The use of the word "budget," as normally defined, is incorrect but the term is firmly entrenched in the literature of limnology. Reference is to the heat storage capacity of a lake.
[†] 1 g cal (mean) × 4.1862 = 1 J.

from year to year. Values are usually measured in cal/cm², or langleys (1 langley = 1 cal/cm³).

Often it is difficult, requires sophisticated equipment, or both to measure all of the terms in the above equation (Fig. 4.1). Moreover, the addition of an ice and snow cover presents special problems (Fig. 4.2). However, the largest terms, θ_R and θ_S, can be measured readily, and several of the remaining terms can be estimated with some reliability (see Johnson et al., 1985).

Net Radiation

Net radiation can be measured directly by a net radiometer positioned over the surface of the lake. These data can be recorded continuously. Alternatively, net radiation may be estimated from

$$\theta_R = (\theta_s - \theta_r) + \theta_{\ln}$$

where θ_s = total incoming short wave radiation (~0.3 to 2.2 μm), θ_r = radiation reflected from the lake surface (albedo), and $\theta_{\ln}$ = net long-wave radiation (~6.8 to 100 μm).

For our purposes, θ_s can be obtained from a nearby climatological station, and θ_r will vary between about 3 and 90% of θ_s, depending on the nature of the lake surface (water, snow, or ice), and will be estimated using the following as a guide to the albedo.

Nature of the surface	% reflected
Water	3–10
Clear ice	20–50
Ice with bubbles	50–70
Old melting snow	40–60
Old dry snow	60–80
Fresh dry snow	70–90

$$\theta_{\ln} \cong 11(t_s + t_a)$$

where t_s = surface temperature in °C, and t_a = temperature of air in °C [after Johnsson (1946)].

Latent Heat Exchange

Freezing, melting, evaporation, condensation, and sublimation may occur at the surface of a lake. Freezing and melting processes especially are important during autumn and spring. Accurate measurement of a heat budget component such as evaporation for an entire lake's surface is difficult to do (Brutsaert, 1982; Winter, 1985).

Sensible Heat Transfer

The *sensible heat transfer* term refers to conduction and convection of heat from the surface of the lake to the air, or vice versa. The net transfer can be estimated by measuring the temperature gradient in the water or ice and computing the heat flow. Because this value is large it may be estimated by difference using the heat budget equation.

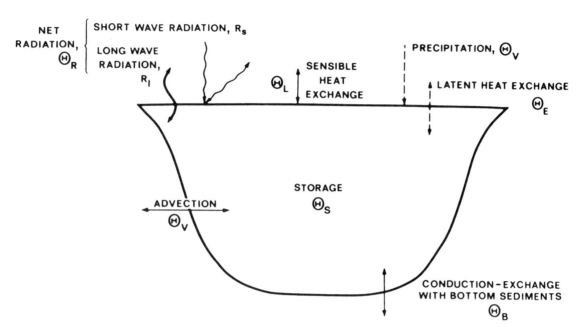

NET
RADIATION, Θ_R

SHORT WAVE RADIATION, R_s

LONG WAVE
RADIATION, R_l

SENSIBLE
HEAT
EXCHANGE Θ_L

PRECIPITATION, Θ_V

LATENT HEAT EXCHANGE Θ_E

STORAGE Θ_S

ADVECTION Θ_V

CONDUCTION-EXCHANGE
WITH BOTTOM SEDIMENTS Θ_B

Figure 4.1. Heat budget components for a lake during summer.

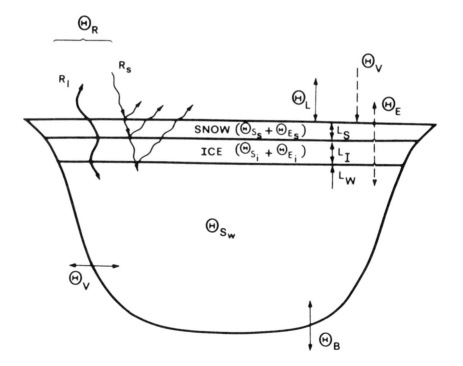

Θ_R

R_l R_s Θ_V

Θ_L Θ_E

SNOW $(\Theta_{S_s} + \Theta_{E_s})$ L_S

ICE $(\Theta_{S_i} + \Theta_{E_i})$ L_I L_W

Θ_{S_w}

Θ_V

Θ_B

$\Theta_{S_s} + \Theta_{E_s}$ = STORED HEAT PLUS LATENT HEAT OF FUSION FOR SNOW

$\Theta_{S_i} + \Theta_{E_i}$ = STORED HEAT PLUS LATENT HEAT OF FUSION FOR ICE

Figure 4.2. Heat budget components for a lake during winter in the temperate zone.

Advection

Subsurface flow, stream flow, rain, and snow are the primary advective sources. Advective loss occurs at the outlet and through deep seepage. For our purposes we may assume that $1 cm^3$ of fresh water has a mass of $1 g$. Thus, volume in cm^3 times temperature in °C is approximately equal to calories, since the specific heat of water is $1 cal/g$-°C. Heat gain or loss in calories then can be computed simply be multiplying the volume of water in cm^3 entering or leaving the basin by the temperature of the water in °C.

Storage

The total heat content of a lake consists of the caloric content of the water minus the amount of heat necessary to warm and then melt any ice or snow ("negative heat"). Any change in water temperature or in ice thickness constitutes a change in the heat storage term. The total heat content (storage) of the water on any sampling date may be determined from:

$$\theta_w = \sum_{z_0}^{z_m} t_z A_z h_z$$

where θ_w = heat content of the lake water in calories, z_0 = surface of the lake, z_m = maximum depth of the lake, t_z = average temperature in °C of a unit layer of water of thickness h_z (in cm), with the midpoint at depth z, and A_z = the area at depth z in cm^2.

The heat content usually is expressed on a unit area basis, θ_w/A_0, in cal/cm^2. A_0 = surface area in cm^2.

Since the specific heat of water is $1 cal/g$-°C, and since $1 cm^3$ of water has a mass of about $1 g$, it is convenient to use a constant area of $1 cm^2$ in calculations of heat content for water. Thickness or depth in cm [depth (cm) × area (cm^2) = cm^3] then can be multiplied directly by temperature in °C to obtain caloric content. However, because most natural lakes do not have basins with perpendicular walls, it is necessary to correct for heat content, which varies with depth. This correction can be done by using a ratio of the area at depth z to the area of the surface of the lake. The ratio usually is obtained from a hypsographic or hypsometric curve (see Exercise 1).

This method is based on the following rationale (Scott, 1964): "The amount of heat depends on temperature, heat capacity, and total mass of water. Since the mass of water in a lake varies with depth, the hypsometric curve of the lake can be used to obtain the amount of heat stored in a unit layer at any depth relative to the amount in the surface layer."

Bottom Conduction

Bottom conduction can be evaluated by measuring the temperature gradient in the bottom sediment and then computing the heat flow. Often it is difficult to measure the temperature gradient in deep lakes or lakes with sandy or rocky bottoms. In shallow lakes with soft sediments the thermal gradient can be measured with a thermal probe (Likens and Johnson, 1969). Depending on thermal conditions in the lake it may be necessary to penetrate the bottom sediments to a depth of several meters to obtain the data required. Based on a model and assumptions described by Likens and Johnson (1969), the daily flux of heat (q_b) can be estimated by:

$$q_b(\text{cal/cm}^2\text{ - day}) = (8.66 \times 10^{-3})\theta_B \sin\frac{2\pi}{365}(D+318)$$

where $\theta_B(\text{cal/cm}^2\text{-year}) = (4.5 \times 10^3)A\alpha^{1/2}$, A = amplitude of seasonal temperature change in water overlaying the bottom in °C, α = thermal diffusivity (this may be approximated at 0.002 cm^2/sec for most gyttja-type lake sediments), and D = number of days since the maximum sediment-water interface temperature was observed.

Sample data sheet for the heat storage calculation

$$\left(\begin{matrix}\text{Thickness}\\\text{in cm}\end{matrix}\right)\left(\begin{matrix}\text{Average}\\\text{area of}\\\text{layer in cm}^2\end{matrix}\right)\left(\begin{matrix}\text{Average}\\\text{temperature of}\\\text{stratum}\end{matrix}\right) = °C\text{ - cm}^3/\text{layer} = \text{cal/layer}$$

(1) Depth in lake (m)	(2) [h_z] Layer thickness (cm)	(3) [A_z] Average area for layer (cm^2)	(4) Layer volume (cm^3) [columns 2×3]	(5) [t_z] Average temperature for stratum (°C)	(6) Weighted calorie content per layer (°C-cm^3) [columns 4×5]
0–0.5					
0.5–1.5					
1.5–2.5					
etc.					

Notes
1. Partition the lake into layers of appropriate thickness according to the thermal profile, i.e., use thinner layers where the temperature change is rapid, use thicker layers where the temperature is constant or nearly so. Use common sense and remember that you are integrating the area under a curve.
2. Sum the values in column 6 and divide by the surface area of the lake in cm^2 to obtain the caloric content of the entire lake in cal/cm^2.

HEAT BUDGET

The difference in heat content during some time interval, as calculated above, is referred to as a *heat budget*. An alternate method of approximating the change in heat content with time is to construct a plot with depth z on the Y axis and the products $A_z (T_{z0}-T_{z1})$ on the X axis [T_{z0} and T_{z1} are the temperatures at the beginning and end of the period, respectively, at depth z; note that $(T_{zmax}-T_{zmin})$ is the formulation for the Birgean heat budget]. The integral is measured either with a digitizer, planimetrically, or by counting squares, and the sum thus produced is divided by A_o.

SPECIAL CONSIDERATIONS FOR SNOW AND ICE

Since specific heat and latent heat are related to mass rather than volume (for water the unit values for mass and volume are conveniently equal!), it is important to determine the density of snow or ice. For example,

$$\text{Latent heat of ice (calories)} = \text{thickness (cm)} \times \text{area (cm}^2\text{)}$$
$$\times \text{density (g/cm}^3\text{)} \times \text{latent heat of fusion (cal/g)}$$

When a hole is made in the ice cover of a lake (with no snow), it can be observed that the water level in the hole does not rise to the surface of the ice. This distance between the surface of the ice and the water level is referred to as the *water-level depression* and is related to the weight of the ice and the buoyant force of water. (Archimedes' Principle).

These relationships may be used to estimate the density of ice in the field

$$\rho_i = \frac{i_z - h_z}{i_z}$$

where ρ_i = density of ice in g/cm³, i_z = depth of ice in cm (assuming 1-cm² area), and h_z = water level depression in cm (assuming 1-cm² area).

When the density is known the thickness of the ice or mass snow (or other objects) on the ice may be calculated from these relationships. In practice, the mean density of an ice sheet may be appreciably different from the maximum density of ice (0.917 g/cm³) because of the presence of snow, slush, lenses of meltwater, and different types of ice. These factors must be considered in calculating the latent heat exchange in the heat budget equation for an ice-covered lake [cf., Adams (1976) and Adams and Lasenby (1978)].

If there were a layer of snow on the ice, then it would be necessary to estimate the densities of the snow and ice separately in order to calculate their respective latent heat contents. By obtaining a core of the snow layer with a transparent tube of known diameter (e.g., 5.08 cm) with removable stoppers at both ends, and by measuring the height of the snow in the tube and the weight (or volume) of the melted snow water, it would be possible to calculate the density of the snow. Alternatively, the density of the snow can be estimated crudely from Table 4.1.

Once the density of the snow has been determined, the mass of snow and ice and the latent heat of both snow and ice can be determined separately (assuming 1-cm² area):

Table 4.1. Typical values of density for new snow, old snow, firn, and ice.[a]

Material	Density (g/cm³)
New snow	0.01–0.300
Old snow	0.20–0.600
Firn	0.40–0.840
Glacier ice	0.84–0.917
Lake ice	0.80–0.917

[a] Modified from Meier (1964).

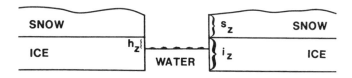

$$\text{Mass of snow } (g/cm^2) = \rho_s \times \bar{s}_z$$

$$\text{Calculated mass of ice } (g/cm^2) = (i_z - h_1) - (\rho_s \times \bar{s}_z)$$

$$\rho_i = \frac{\text{calculated mass of ice}}{i_z}$$

Then

$$\text{Latent heat of snow } (cal) = \rho_s \times \bar{s}_z \times (-80\ cal/g)$$

$$\text{Latent heat of ice } (cal) = \rho_i \times i_z \times (-80\ cal/g)$$

where $\bar{s}_z$ = average depth of snow in cm (assuming 1-cm² area) and ρ_s = density of snow in g/cm³.

Detailed considerations of the theories for the heat balance equation, as well as a thorough analysis of the terms, may be found in Juday (1940), Sverdrup et al. (1942), Neumann (1953), Saur and Anderson (1956), Hutchinson (1957), Hanson et al. (1961), Ragotzkie and Likens (1964), Wetzel (1983, 1999), and Johnson et al. (1985).

EXERCISES

This exercise is divided into three options.

OPTION 1. FIELD TRIP

Visit a lake two or more times, separated by at least two weeks, and obtain the following data.*

---

* A field trip to an ice-covered lake is a cold, but worthwhile, experience. Unfortunately, too little work has been done in limnology during the winter (see McKnight et al., 1996).

1. Total incoming, short wave radiation (nearby climatological station).
2. Air temperature at 1 m above the lake's surface.
3. Water temperatures at 1-cm intervals from the surface to the bottom of the lake. Where the temperature changes rapidly, measure at 0.5-m intervals. In winter, avoid mixing water below ice cover (temperature change in the water immediately below the ice is very large).
4. When ice and snow are present:
   a. measure ice thickness at several locations.
   b. measure snow depth at several locations and collect cores of the snow layer for measurement of snow density.
   c. measure the water level depression (avoid groups of people near the hole!).
5. Measure temperature of bottom sediments when possible.
6. Measure water temperature and flow in inlets and outlets and calculate advective gain or loss (see Exercise 5).
7. Prepare a hydrometric map of the lake with 1-m depth contours (see Exercise 1).
8. Planimeter areas of depth contours within the lake and calculate volumes. Construct a hypsometric graph from these data (see Exercise 1).
9. Calculate the heat content for the lake.

OPTION 2. LABORATORY EXERCISE

Using the data provided on Mirror Lake (Tables 4.2 to 4.4; Fig. 4.3), calculate heat budgets and answer the questions.

Table 4.2. Temperature profiles and other physical characteristics in Mirror Lake, New Hampshire.

Depth (m)	Jan. 25, 1970 Temp. (°C)	Feb. 21, 1970 Temp. (°C)	May 17, 1970 Temp. (°C)	June 24, 1970 Temp. (°C)
0	−7.7 (ice surface)	−4.7 (ice surface)	15.45	21.80
0.51	0	—	—	—
0.62	—	0	—	—
1	3.82	3.35	15.45	21.35
2	3.91	4.15	13.58	20.74
3	3.98	4.15	12.15	20.41
4	4.00	4.18	9.73	16.53
5	4.00	4.20	8.46	12.38
6	4.05	4.25	7.06	10.27
7	4.13	4.38	6.20	8.69
8	4.49	4.54	6.06	7.38
9	4.63	4.70	5.45	6.52
10	5.00	5.00	5.62	6.48
10.3 (mud surface)	5.05	5.25	5.75	6.50
Ice thickness (cm)	51	62	—	—
Snow depth (cm)	0	8		
Water level Depression below ice Surface (cm)	4.2	5.0	—	—
Air temperature (°C)	−7.7	−4.7	12.1	22.2

Table 4.3. Hydrographic data for Mirror Lake, New Hampshire.

43° 56.5′ N, 71° 41.5′ W

Elevation	213 m			Average depth	5.75 m
Maximum effective length	610 m			Length of shore line	2,247 m
Maximum effective width	370 m			Shore development	1.64
Area	15.0 ha			Volume development	1.57
Maximum depth	11.0 m			Relative depth	2.5%

				Volume	
Depth (m)	Area (m² × 10⁴)	% of total	Stratum (m)	(m³ × 10³)	(% of total)
0	15.0	100.0	0–1	142.9	16.6
1	13.6	90.5	1–2	130.0	15.1
2	12.4	82.9	2–3	119.5	13.9
3	11.5	76.5	3–4	110.0	12.8
4	10.5	70.1	4–5	101.8	11.8
5	9.86	65.7	5–6	94.1	10.9
6	8.96	59.7	6–7	78.5	9.1
7	6.79	45.2	7–8	48.9	5.7
8	3.21	21.4	8–9	23.6	2.7
9	1.61	10.7	9–10	10.7	1.2
10	0.609	4.06	10–11	2.0	0.2
11	0	0		——	——
			Total 862.0		100.0

OPTION 3. LABORATORY EXERCISE

Combine this exercise with Exercise 3 as an elaboration of the section on heat contents of the model systems. Calculate heat budgets based on data from the model systems and compare them with calculations based on data from Mirror Lake (Tables 4.2 and 4.3; Fig. 4.3).

Questions

1. Compare graphically the temperature profiles for Mirror Lake during 1970 (Table 4.2). Plot isolines of temperature using coordinates of depth and time and also plot tempera-

Table 4.4. Thermal properties of ice and snow.[a]

Latent heat	79.8 cal/g (or 334 × 10⁴ J/g)
Thermal conductivity (ice)	2.3 W/m-°C
Thermal conductivity (snow)	0.31 W/m-°C
Attenuation coefficient (ice)	1.5/m visible band; 20/m infrared band
Attenuation coefficient (snow)	6/m visible band; 20/m infrared band
Specific heat (ice)	0.506 cal/°C-g[b]
Linear coefficient of expansion (ice)	52.7 × 10⁻⁶/°C[c]
Volume coefficient of expansion (ice)	153 × 10⁻⁶/°C[c]

[a] Modified from Meier (1964), Patterson and Hamblin (1988).
[b] At 0°C.
[c] At melting temperature.

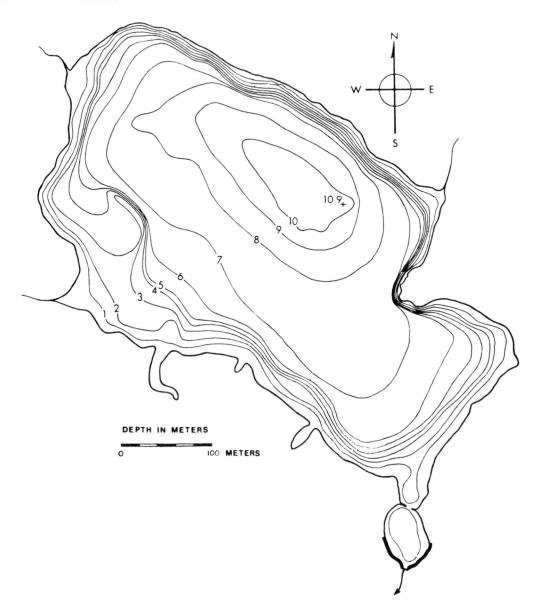

Figure 4.3. Bathymetric map (1975) of Mirror Lake, West Thornton, Grafton county, New Hampshire. Contour intervals in meters.

tures using coordinates of depth and temperature (see Exercise 1 for details on how to make these plots).

2. Construct a hypsograph for Mirror Lake. Use the data in Table 4.3. If you want, you can planimeter the basin contours for Mirror Lake (Fig. 4.3) and compare your values with those in Table 4.3 (see Chapter 1 for an explanation of hypsographic curve).

3. Calculate the change in heat stored in Mirror Lake from January 25 to February 21. Contrast this with the value for the period May 17 to June 24, 1970. Compute the values in terms of cal/cm²-day.*

4. Discuss the effects of an ice and snow cover on the heat budget of a lake.

---

* Assume that $\rho_i = 0.91 \, g/cm^3$ and ρ_s (density of snow) = 0.07 g/cm³ on February 21, 1970.

Table 4.5. Equations for linear and volumetric expansion of ice.[a]

Linear expansion:

$$l_t = l_0(1 + \alpha t)$$

where l_t = length at $t°C$, l_0 = length at $0°C$, α = coefficient of linear expansion, and t = temperature in $°C$.
  Volume expansion:

$$V_t = V_0(1 + \beta t)$$

where V = volume at $t°C$ and $0°C$, and β = coefficient of volumetric expansion.

[a] See Hutchinson (1957, pp. 532–533) for reference to ice push.

5. Assume that one-half of the net solar energy flux affects the water and that one-half affects the ice during winter. What would have been the thickness of the ice 10 days after January 25, 1970, in Mirror Lake? (Assume an average θ_R value during this period of –0.01 ly/min.)
6. What was the density of the ice on Mirror Lake on January 25, 1970?
7. What could be the reason for the increasing water temperatures near the bottom in ice-covered Mirror Lake during 1970 (Table 4.2)?
8. How would you explain the increase in water temperature near the bottom in Mirror Lake during the summer (Table 4.2)?
9. In Antarctic lakes the water level is often depressed as much as 35 cm in a hole drilled through the ice. Assuming a density of $0.917 \, g/cm^3$, what would be the thickness of the ice cover?
10. If the temperature were to rise from –40 to $0°C$ overnight at Mirror Lake, theoretically how far might the ice push onto the shore (see Table 4.5)?
11. How would you predict the ice-out date for a lake? List the evidence you would use as a basis for this "guestimate."

Apparatus and Supplies

1. Underwater thermometer.
2. Ice-thickness measure (meter stick with a hook at one end).
3. Ice auger or ice spud.
4. Underwater light meter.
5. Water bottle (Kemmerer or nonmetallic), line, and messengers.
6. Ice chest, Sterno, and matches.
7. Steel measuring tape (50 m or longer).
8. Stopwatch.
9. Floats (oranges) for stream velocity measurements (see Exercise 5).

References

Adams, W.P. 1976. Field determination of the densities of lake ice sheets. Limnol. Oceanogr. *21*:602–608.

Adams, W.P. and D.C. Lasenby. 1978. The role of ice and snow in lake heat budgets. Limnol. Oceanogr. *23*:1025–1028.

Brutsaert, W. 1982. Evaporation into the atmosphere. D. Reidel Publ Co., Dordrecht, Netherlands. 299 pp.

Hanson, E.B., A.S. Bradshaw, and D.C. Chandler. 1961. The physical limnology of Cayuga Lake, New York. Cornell Univ. Agric. Expt. Sta., Mem. No. 378:1–63.

Hutchinson, G.E. 1957. A Treatise on Limnology. Vol. 1. Wiley, New York. 1015 pp.

Johnson, N.M., G.E. Likens, and J.S. Eaton. 1985. Stability, circulation and energy flux in Mirror Lake. pp. 108–127. *In*: G.E. Likens, Editor. An Ecosystem Approach to Aquatic Ecology. Mirror Lake and Its Environment. Springer-Verlag, New York.

Johnsson, H. 1946. Termisk-hydrologiska studier i Sjon Klämmingen. Geogr. Ann. Stockholm *28*:1–154.

Juday, C. 1940. The annual energy budget of an inland lake. Ecology *21*:439–450.

Likens, G.E. and N.M. Johnson. 1969. Measurement and analysis of the annual heat budget for the sediments in two Wisconsin lakes. Limnol. Oceanogr. *14*(1):115–135.

McKnight, D.M., E.R. Blood, and C.R. O'Melia. 1996. Fundamental research questions in inland aquatic ecosystem science. pp. 257–278. *In*: Freshwater Ecosystems: Revitalizing Educational Programs in Limnology. National Academy Press, Washington, D.C.

Meier, M.F. 1964. Ice and glaciers. pp. 16-1 to 16-32. *In*: C.T. Chow, Editor. Handbook of Applied Hydrology. McGraw-Hill, New York.

Neumann, J. 1953. Energy balance and evaporation from sweetwater lakes of the Jordan Rift. Bull. Res. Council Israel 2:337–357.

Patterson, J.C. and P.F. Hamblin. 1988. Thermal simulation of a lake with winter ice cover. Limnol. Oceanogr. *33*:323–358.

Ragotzkie, R.A. and G.E. Likens. 1964. The heat balance to two Antarctic lakes. Limnol. Oceanogr.*9*(3):412–425.

Saur, J.F.T. and E.R. Anderson. 1956. The heat budget of a body of water of varying volume. Limnol. Oceanogr. *1*(4):247–251.

Scott, J.T. 1964. A comparison of the heat balance of lakes in winter. Tech. Rep. *13*. Dept. of Meteorol., Univ. of Wisconsin, Madison. 138 pp.

Sverdrup, H.U., M.W. Johnson, and R.H. Fleming. 1942. The Oceans: Their Physics, Chemistry and General Biology. Prentice-Hall, Englewood Cliffs, N.J. 1087 pp.

Wetzel, R.G. 1983. Limnology. 2nd Ed. Saunders Coll., Philadelphia. 858 pp.

Wetzel, R.G. 1999. Limnology: Lake and River Ecosystems. 3rd Ed. Academic Press, San Diego (in press).

Winter, T.C. 1985. Approaches to the study of lake hydrology. pp. 128–135. *In*: G.E. Likens, Editor. An Ecosystem Approach to Aquatic Ecology. Mirror Lake and Its Environment. Springer-Verlag, New York.

Morphology and Flow in Streams

Streams are characterized by a continual downstream movement of water, dissolved substances, and suspended particles. These components are derived primarily from the drainage basin or watershed*, which is the total land area draining into a given stream channel. Thus the hydrological, chemical, and biological characteristics of a stream reflect the climate, geology, and vegetational cover of the drainage basin [cf., Beaumont (1975), Likens and Bormann (1995), Hynes (1970), Oglesby et al. (1972) and Whitton (1975)]. Water from rain or snow, falling on hilly or mountainous terrain, actually follows diverse routes in moving downhill (Fig. 5.1). Precipitation first may be intercepted by vegetation, then by litter on the surface of the ground. When water is added to the surface of a soil more rapidly than it can soak in (i.e., the infiltration capacity is exceeded) it will run off overland. Normally, most of the water from precipitation infiltrates into the soil. Soils have variable capacity to store water depending on depth, structure, composition, and other factors. Before stream flow can occur, this storage capacity must be exceeded. Storage capacity continually is made available by evaporation and transpiration (evapotranspiration). Until recently, limnologists have ignored, for the most part, the importance of hydrologic flow paths in regulating the metabolism and biogeochemistry of streams and lakes, as well as their role in the historical generation and accumulation of lake sediments [see Likens (1984)].

Flow through the soil may be channeled by macropores, often produced by cracks, worm or other animal burrows, and old root channels. Impermeable layers can impede the vertical movement of water and cause lateral flow at intermediate depths in the soil (Fig. 5.1). The chemistry of precipitation may be altered considerably as the water passes through the terrestrial ecosystem(s) comprising the drainage basin (Likens and Bormann, 1995; Likens 1984). The surface of the saturated zone of permeable soil is called the *water table*; water in the soil above the water table is termed *vadose water* and that below the water table *ground water*. Ground water provides the relatively stable base flow component in streams. Overland flow, in addition to water that infiltrates the soil and then flows laterally to the stream channel (i.e., subsurface storm flow), are the main components of peak flows or floods [cf., Dunne (1978), Chorley (1978), Winter (1985b)].

---

*In American usage, watershed is equivalent to drainage basin or the European term, catchment, all of which refer to the region or area drained by a river system.

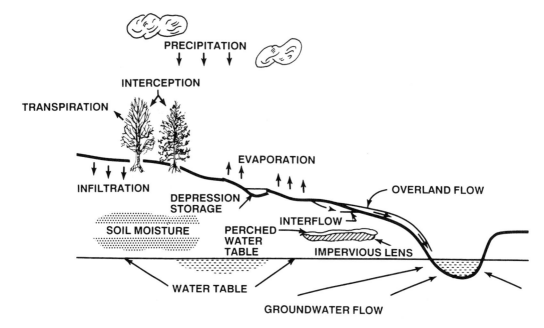

Figure 5.1. Major pathways of precipitation and the runoff phase of hydrologic cycle.

DRAINAGE AREA

A variety of methods have been proposed for ordering the tributary streams in a drainage network [cf., Gregory and Walling (1973)]. The Horton-Strahler method (Horton, 1945; Strahler, 1952) is probably the most widely used in the United States. In this method each headwater or "finger-tip" tributary is designated as the first order. Two first-order tributaries combine to produce a second-order stream, two second-order tributaries produce a third order, and so on (Fig. 5.2). The order of the trunk stream is not changed by the addition of a stream of lower order. Only when two tributaries of equal order are joined is the order increased. Practical limitations for this method include: (1) hydrological and ecological conditions may not be represented adequately since numerous tributaries can enter the main stream without changing order, and (2) ephemeral headwater streams or topographic maps of different scales can change significantly the order within the drainage network.

A variety of patterns may be observed in drainage networks (Fig. 5.3). Boundaries of drainage basins usually are determined from surface features (topographic divides) obtained from aerial photographs, field surveys, or topographic maps. However, subsurface flow may have different boundaries (phreatic divides), particularly in areas underlain with relatively soluble or permeable rocks. The drainage density for a watershed is defined as:

$$\frac{\text{total stream length}}{\text{area of drainage basin}}$$

When the log of the stream length (successive or cumulative distance), from headwaters to mouth, is plotted against the log of total drainage area for each point along the stream, a linear relationship is observed:

$$L = jA^m$$

Figure 5.2. Stream orders according to the Horton-Strahler method.

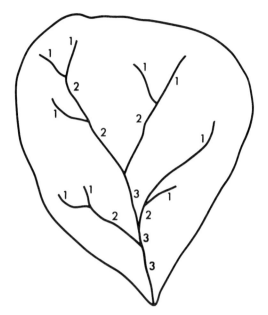

where L is length, A is area, and j and m are constants derived from a regression of the available data. In the northeastern United States, the coefficients j and m average about 1.4 and 0.6, respectively; that is, a drainage basin of $1\,mi^2$ ($2.59\,km^2$) will have a stream channel 1.4 mi (1.61 km) long (Leopold et al., 1964). For several regions in the United States, the exponent m varies between 0.6 and 0.7, indicating that, as drainage basins increase in size, they tend to elongate (Leopold et al., 1964).

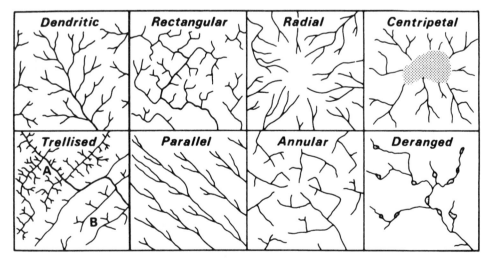

Figure 5.3. Various types of drainage networks. [From K.J. Gregory and D.E. Walling, 1973. Drainage Basin Form and Process. By permission of Edward Arnold (Publisher) Ltd., London.]

CHANNEL MORPHOLOGY

The basin or trough containing the stream water is the *stream channel*. At any moment in time, the channel is characteristic of the stream. The channel may be described by length, width, depth, cross-sectional area, slope, aspect, and so on. Length, width, slope, and aspect may be determined from simple measurement or, except for width, from a topographic map. Average depth is $\bar{z}$ = cross-sectional area/width, or

$$\frac{z_1 + z_2 \cdots z_n}{N+1}$$

where $z_1, z_2, \ldots, z_n$ are measurements of water depth or bankfull depth at equal distances across the stream channel and N is the number of measurements. The channel usually is bordered on one or both sides by a flat area called the *flood plain*. The channel rarely has an even gradient and is never straight over an appreciable distance. However, with time there is a gradual tendency to produce a more even gradient or more even dissipation of potential energy per unit length of stream. Meanders or bends play an integral role in this process, since meandering streams tend to approach the theoretical equilibrium of an even energy gradient much more closely than do straight channels. In addition, meanders ultimately increase the total capacity of the channel to transport water over a given distance.

The channel of a stream can contain a moderate, but not an excessive, discharge without overflowing its banks. A bankfull discharge is one that entirely fills the cross-sectional stream channel but does not overflow onto the flood plain. Long-term studies have shown that the bankfull stage (Bfs) will be equaled or exceeded on the average of once every 1.5 y (Leopold et al., 1964). A stream flowing at bankfull will not be at overflow stage at every point along the channel, due to local variations in bank height, channel depth, or both. Nevertheless, for a given stretch of stream, the bankfull channel and adjacent flood plain can be identified. Although bankfull discharge may occur as an event every year or two, most of the time discharges are relatively much smaller.

With time, stream channels move laterally within the flood plain because of erosion and redeposition of sediment. Old, abandoned channels usually can be observed on the flood plain. As new channels are formed, oxbow lakes may be cut off from the main channel; these serve as rich nursery areas for fish and invertebrates.

Small stream and large river ecosystems commonly have soils adjacent to the main channel that are water saturated some distance from the channel. In stream/river channels with gentle slopes this area, termed the *riparian zone*, can involve a region many times larger (>30 times) than the area of the open stream water (Fig. 5.4). Stream ecosystems are "bounded" by the subsurface, ground-water–stream–water interface rather than the main channel per se (Fig. 5.5).

The *hyporheic zone* is a portion of the groundwater interface in streams where a mixture of surface water and ground water occur both within the active channel and within the riparian zone (e.g., Findlay, 1995; Dahm and Valett, 1996; Boulton et al., 1998). The lateral boundaries are dynamic both spatially and over time. Flow into and out of the porous hyporheic sediments occur both from the ground water as well as from the channel zone, depending upon the morphological characteristics of the sediments within the channel (Fig. 5.4).

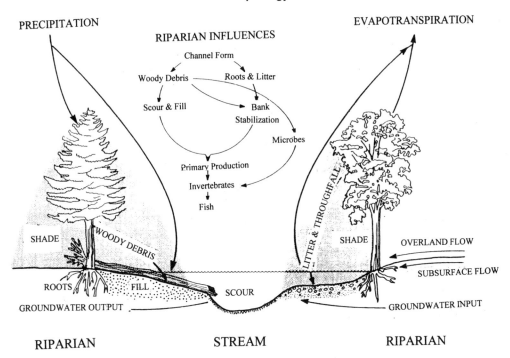

PRECIPITATION

EVAPOTRANSPIRATION

RIPARIAN INFLUENCES

Channel Form

Woody Debris ——— Roots & Litter

Scour & Fill → Bank
Stabilization

Microbes

Primary Production

Invertebrates

Fish

SHADE

WOODY DEBRIS

LITTER & THROUGHFALL

SHADE

OVERLAND FLOW

SUBSURFACE FLOW

ROOTS   FILL   SCOUR

GROUNDWATER OUTPUT

GROUNDWATER INPUT

RIPARIAN   STREAM   RIPARIAN

Figure 5.4. Diagram of the riparian zone of streams and rivers. (From Likens, 1992; modified extensively from Cummins, 1986.)

The directions and intensities of hydrological exchange into and from the porous sediments are important. As substances move downward along the stream gradient (advection) and as they enter the hyporheic zone, they can influence the biota directly by altering redox potential, organic matter content, and nutrient exchanges into and from the interstitial waters. As a nutrient enters and is utilized by the biota,

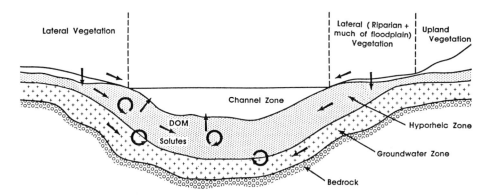

Lateral Vegetation

Lateral (Riparian + much of floodplain) Vegetation

Upland Vegetation

Channel Zone

DOM

Solutes

Hyporheic Zone

Groundwater Zone

Bedrock

Figure 5.5. Lateral and vertical boundaries of lotic ecosystems (from Wetzel and Ward, 1992; modified extensively from Triska et al., 1989). The stream ecosystem boundary is defined as the hyporheic/groundwater interface and thereby includes a substantial volume beneath and lateral to the main channel. Herbaceous vegetation rooted in the hyporheic zone is therefore part of stream ecosystem primary production. Arrows indicate flow pathways of dissolved organic matter and inorganic solutes derived from plant detritus within the stream ecosystem.

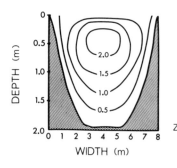
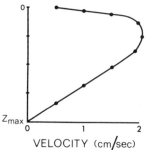

Figure 5.6. Idealized current velocity in cm/sec in a channel cross section (*left*) and in profile at the midpoint of the cross section (*right*).

is stored or transformed, and then is returned to the water, the cycle is completed some distance downstream. This so-called spiraling movement involves both cycling and downstream transport (cf., detailed discussion of Newbold, 1992; Webster and Ehrman, 1996). The rate of cycling and distance moved depends on the retentiveness of the biota and chemical reactions. The spiraling length consists of two components: (1) the distance traveled while in abiotic dissolved form in the water before being immobilized, called the uptake length, and (2) the distance traveled before being remobilized and returned to the water, called the turnover length, which is the inverse of the uptake rate. All of these gradients and exchanges affect the metabolism, growth, and distributions of bacteria, algae, and higher organisms (benthic invertebrates, aquatic plants, etc.) in and above the sediments.

CURRENT

Downstream water movement in the stream channel is referred to as *current*. The current erodes the channels, determines the degree and type of particle deposition, and thus the nature of the sediments. Current is the single most important factor affecting organisms in the channel.

Current tends to be faster in rapids and riffle areas and slower in pools. As a result, riffles are areas of erosion and scouring, whereas pools are areas of deposition along the length of a stream.

Current velocity may be measured using various types of meters and devices, e.g., Gurley Pigmy Current Meter or a Pitot tube. A float (oranges are good!) can be used to estimate the velocity of the surface currents. Apparatus and procedures are described in detail in Welch (1948) and Boyer (1964). Maximum velocity occurs at some point below the surface, and at substrate surfaces the velocity is zero [Fig. 5.6; see also, for example, Morisawa (1968)]. Mean velocity in this example occurs at a depth 0.6 of the distance from the water surface to the bed. Thus, mean velocity must be measured with some type of underwater current meter.

Larger particles may be transported by higher current velocities, and the sediment composition reflects the effects of current in erosion and transport (Table 5.1) [cf., Cummins (1962, 1964)].

FLOW OR DISCHARGE

Stream flow or *discharge* is the volume of water passing through the cross-sectional area of the stream channel per unit time and is calculated as $Q = A\bar{v}$, where

Table 5.1. Relationship of current velocity to sediment composition.[a]

Velocity range (cm/sec)	General bottom composition	Approximate diameter (mm)
3–20	silt, mud, (organic debris)	<0.02
20–40	fine sand	0.1–0.3
40–60	coarse sand to fine gravel	0.5–8
60–120	small, medium to large gravel	8–64
120–200	large cobbles to boulders	>128

[a] Modified from Einsele (1960).

Q = discharge in m³/sec, A is cross-sectional area in m², and $\bar{v}$ is mean velocity in m/sec.

A simple relationship between drainage area and (stream flow) discharge has been used to predict discharge, particularly for flood events [see Strahler (1964)]:

$$Q = jA^m$$

where Q is a measure of discharge, for example in liters/sec, A is area, and j and m are constants as above.

Discharge may be approximated from simple measurements using a float (Davis, 1938): $Q = wdla/t$, where w is width in m, d is mean depth in m, l is distance in m over which a float travels in time, t, in sec, and a is a coefficient that varies with the nature of the sediment [0.8 if rough and 0.9 if smooth (mud, sand, bedrock)]. However, discharge is most accurately measured by any one of a variety of calibrated flumes or weirs located at a gauging site on a stream [see, e.g., Boyer (1964) and Wisler and Brater (1949)].

Following a rainstorm or snow-melt period, stream discharge increases to a peak and then decreases. These changes in discharge can be graphed in the form of a stream hydrograph (Fig. 5.7). Hydrographs provide useful information on response

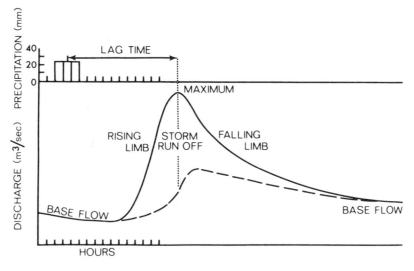

Figure 5.7. Stream hydrograph resulting from a single storm event. [Redrawn from data of Beaumont (1975).]

of stream discharge to storm events and for comparisons among streams. The hydrograph depends on various features of the storm and of the watershed. The rising limb (Fig. 5.7) of the curve is usually concave, reflecting the infiltration and storage capacity of the watershed. The peak of the curve represents maximum runoff. The falling limb indicates flow from storage within the soil after precipitation has stopped.

Discharges that exceed the bankfull capacity of the channel are called floods. Floods essentially are random events (Leopold et al., 1964). The mean annual flood is used widely as a hydrological measurement. It corresponds to the average value for the largest annual discharge during a number of years.

The mean annual flood is an important event in producing the flood plain. The return period, or recurrence interval, of flooding is calculated by

$$Tr = \frac{N+1}{M}$$

where Tr is the return period in yr, N is the number of years of record, and M is the rank of an individual event in the set. The mean annual flood has a return period of 2.33 yr (Leopold et al., 1964). Thus, once every 2.33 yr, on the average, the highest discharge of the year will equal or exceed the value of the mean annual flood (Fig. 5.8).

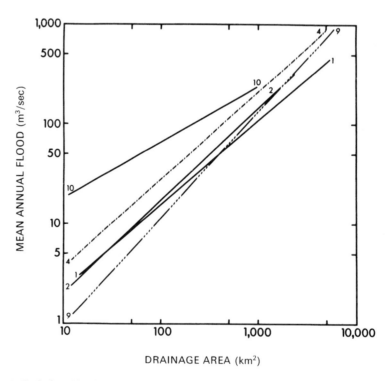

Figure 5.8. Relationships between mean annual flood and size of drainage area for some of the hydrologic regions given within the area identified as 1-B in Fig. 5.7. [Redrawn from data of Tice (1968).]

EXERCISES

OPTION 1

In the Laboratory

1. Locate and draw the drainage basin boundaries (B) on a U.S. Geological Survey map for the stream you plan to study.
   a. Draw drainage basin boundaries (divide lines) perpendicular to contour lines (elevations); overland flow and shallow subsurface water flows perpendicularly to contour lines.
   b. Draw divide lines through the center of "saddles" and closed contour lines.

   c. Water flows to the opening of the V's of contour lines. Draw divide lines accordingly.

2. Determine basin and channel characteristics as described in Table 5.2 (see also Exercise 1).
3. Plot the average stream length, number of streams, and drainage area versus stream order on semilogarithmic graph paper (or with appropriate software and a personal computer) for the drainage network.
4. Plot the percentage of total channel length against order.
5. Using the data in Table 5.3, make a regression of length and average discharge versus drainage area each on a log-log plot.
6. Calculate the regression constants for the above relationships.

In the Field

1. Locate a section of a stream for study.
2. Make a sketch map of the principal berms (flat levels of the bank) along the chosen section of stream. This operation can be done with an alidade and plane table (see Exercise 1) or roughly approximated by careful mesurement of the stream banks with a measuring tape and meter stick.
3. Measure three cross-sectional areas in this section of the stream. Compare riffle and pool areas. Depth measurements should be made at meter intervals across the channel. Width measurements should include the two principal berms, when present. Calculate the cross-sectional area of the channel based on the current water level, lowest berm, and the highest berm.
4. Compute the average depth of water and channel for upper and lower berms for each cross-section.
5. Determine at least three current velocity profiles across the stream channel for each of the three cross-sections. Compare riffle and pool areas.
6. Use a float to estimate surface velocity by measuring the time for its passage over a measured distance (50 to 100 m). Be certain that the float does not become lodged in a back

Table 5.2. Basin and channel characteristics.[a]

	Definitions of topographic characteristics
Basin length (L_B)	Straight-line distance from outlet to the point on the basin divide used to determine the main channel length, L_C.
Basin width (W_B)	Average width of the basin determined by dividing the area, A, by the basin length, L_B:

$$W_B = A/L_B$$

Basin perimeter (P_B)	The length of the line that defines the surface divide of the basin.
Basin land slope (S_B)	Average land slope calculated at points uniformly distributed throughout the basin. Slopes normal to topographic contours at each of 50 and preferably 100-grid intersections are averaged to obtain S_B. The difference in altitude for the two topographic contours nearest a grid intersection is determined, and the normal distance between these contours is measured.
Basin diameter (B_D)	The diameter of the smallest circle that will encompass the entire basin.
Basin shape (SH_B)	A measure of the shape of the basin computed as the ratio of the length of the basin to its average width:

$$SH_B = \frac{(L_B)^2}{A}$$

Compactness ratio (CR_B)	The ratio of the perimeter of the basin to the circumference of a circle of equal area. Computed from A and P_B as follows:

$$CR_B = \frac{P_B}{2(\pi A)^{1/2}}$$

Main channel length (L_C)	The length of the main channel from the mouth to the basin divide.
Main channel slope (S_C)	An index of the slope of the main channel computed from the difference in stream-bed altitude at point 10% (E_{10}) and 85% (E_{85}) of the distance along the main channel from the outlet to the basin divide. It can be computed by the equation

$$S_C = \frac{(E_{85} - E_{10})}{0.75 L_C}$$

Sinuosity ratio (P)	The ratio of the main channel length to the basin length:

$$P = \frac{L_C}{L_B}$$

[a] Modified from Winter (1985).

Table 5.3. Drainage network characteristics for a hypothetical drainage basin.

Order	Number of streams	Average length (km)	Average drainage area (km²)	Average discharge (1/sec)
1	200,000	0.02	0.00018	0.005
2	65,000	0.03	0.00091	0.025
3	20,000	0.06	0.00414	0.12
4	5,500	0.16	0.0129	0.36
5	1,500	0.40	0.0906	2.5
6	400	1.0	0.388	11
7	150	2.4	2.20	62
8	40	5.6	9.06	250

eddy or behind an obstruction. Repeat this estimate five times and determine the standard error of the mean (see Appendix 2 for this estimate).

7. Calculate the discharge for all three cross sections based on present water level and the lower and upper berm levels. These data will provide estimates of the discharge at bankful stage (Bfs).

8. Look up appropriate flood frequency data, including bankfull stage and mean annual flood (MAF) discharges for your area (Fig. 5.9) in the U.S. Geological Survey publications on *Magnitude and Frequency of Floods in the United States*. Remember that the MAF recurrence interval is 2.3 yr and the Bfs recurrence interval is 1.5 yr.

9. Designate the appropriate berm for Bfs by comparing the theoretical Q_{Bfs} with your values for Q_{Bfs}. Compute the appropriate recurrence intervals for each of the estimated discharge values measured in 7 above. Calculate the ratio:

$$\frac{\text{estimated discharge } (Q)}{\text{mean annual flood (MAF) discharge}}$$

Then determine the recurrence interval from a graph for the area, such as that in Figure 5.8. The discharge that corresponds to the flood plain generally will have a recurrence interval of 1.5 ± 0.5 yr.

For example, Figs. 5.8 and 5.9 are applicable to streams in the North Atlantic Slope (area 1-B of Fig. 5.9). The Chenango River at Sherburne, New York, is in section A of hydrologic region 4 of this area (Tice, 1968). The drainage area of the Chenango River at this point is 684 km². Thus, the MAF discharge can be approximated from Fig. 5.8 at about 200 m³/sec. The recurrence interval for a discharge of 300 m³/sec would be determined from the ratio

$$\frac{Q}{\text{MAF}} = \frac{300 \, \text{m}^3/\text{sec}}{200 \, \text{m}^3/\text{sec}} = 1.5$$

Then, using Fig. 5.10, the recurrence interval for a discharge of 300 m³/sec would be about 10 yr.

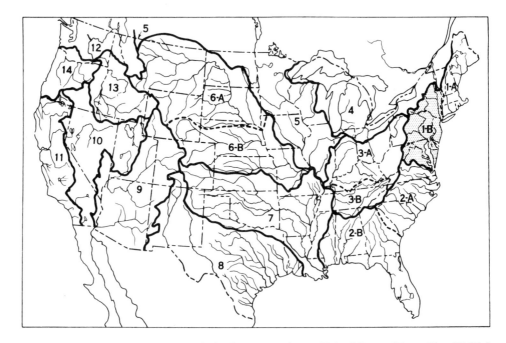

Figure 5.9. Hydrologic regions within the conterminous United States. [From Tice (1968).]

Figure 5.10. A composite flood-frequency curve for section A of hydrologic region 4 (Fig. 5.6) of area 1-B (Fig. 5.7) where once every 2.3 yr, on the average, the highest flow of the year will equal or exceed the mean annual flood. [Redrawn from data of Tice (1968).]

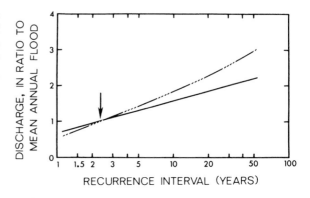

10. Characterize the substrate composition on the streambed as a function of different current velocities.
    a. Collect a random sediment sample in three different locations, such as a pool, a riffle area, near the bank, or behind a boulder or log. If possible, collect three replicate samples at each location. Collect approximately $10 \times 10 \times 10$ cm of sediment with a coring device that can be pushed into the sediments. Determine accurately the volume of sample actually collected. A section of stovepipe is a convenient corer for relatively soft sediments. Be careful to avoid loss of fine particles while the corer is being pushed into the sediment and when the sample is removed from the streambed. Dig away the sediments from outside of the coring device so that a flat plate can be slid along the bottom of the core. With this plate in position the core can be removed from the streambed.
    b. Separate the silt and clay from the samples by elutriation (cf., p. 186)." Allow this sample of silt and clay to settle overnight. Obtain a subsample of the supernatant and centrifuge it to collect the clay fraction. Carefully decant or siphon the water from the silt sample. Determine dry weights (105°C) for these two fractions. Dry the remainder of the sediment sample and sieve (a mechanical shaker is convenient). Use a standard set of sieves (U.S. sieve no. 5 through 230).

U.S. sieve no.	Mesh opening (mm)
230	0.0625
120	0.125
60	0.25
35	0.5
18	1
10	2
5	4

Record dry weights of sediment fractions.
    c. Determine organic matter by difference in weight after combustion of a sample at 550°C for 2 h.
    d. Establish a grid over the stream bed and measure the diameter of stones >4 mm in diameter at 100 locations on the grid.
11. Construct a histogram of dry weight according to particle size fraction.
12. Plot particle size against percentage of total sample weight in that size fraction.
13. Compare the estimated velocity range based on sediment composition (Table 5.1) with measured values for current velocity.
14. Answer the following questions.

OPTION 2

1. Using the data provided in Table 5.3 and Fig. 5.11, complete the laboratory portion of Option 1 for the Bear Brook drainage area. Calculate the cross-section area and discharge based on data in Fig. 5.6. Complete part 9 of Option 1 using Figs. 5.8 and 5.10.
2. Answer the following questions.

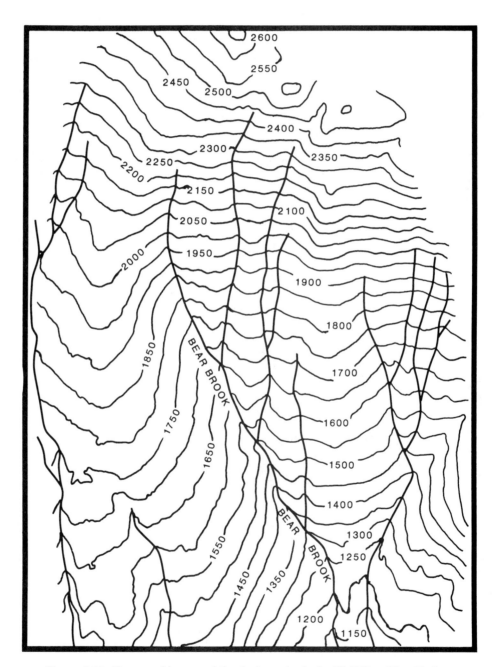

Figure 5.11. Topographic map of the drainage basin for Wolf Run, New York.

Questions

1. What difficulties exist in constructing a drainage network from a topographic map? What are the advantages?
2. Why are oranges good floats to use for estimating surface current velocity?
3. What are the advantages of using a Pitot tube to measure current velocity?
4. How does sediment composition reflect current velocity when velocity at the sediment boundary layer is zero?
5. What is the ecological significance of the velocity profile in streams?
6. Why are flood plains less common in mountainous regions?
7. What is the difference between suspended load and bed load?
8. How are bars formed in a stream channel?
9. How might basin characteristics be responsible for streamflow variations (see Winter, 1985b)?
10. How could the water draining into a lake from nonchannelized runoff (Fig. 5.1) be measured?
11. How would you expect a storm hydrograph to appear following a long drought? Following a rainy period?
12. What are the pitfalls of using sediment composition to estimate current velocity?
13. Who do humans live and build structures on a flood plain even though flood recurrence intervals are so predictable?

Apparatus and Supplies

1. Two steel measuring tapes, each 30 m long.
2. Meter sticks.
3. Plane table(s) and alidade(s).
4. Compass.
5. U.S.G.S. topographic map of area.
6. Planimeter or electronic digitizer.
7. Stopwatch.
8. Current meter (e.g., Gurley Pigmy meter, Weathermeasure Corp., P.O. Box 41257, Sacramento, CA 95841; Marsh McVirney Model 201 Portable Water Current Meter, 2281 Lewis Ave., Rockville, MD 20851).
9. Floats (oranges).
10. Standard soil sieves (U.S. sieve nos. 5, 10, 18, 35, 60, 120, 230).
11. Sediment coring device.
12. Drying oven (105°C).
13. Muffle furnace.
14. Jars or heavy plastic bags for sediment samples.
15. Semi-log and log-log graph paper or appropriate software and a personal computer.

References

Beaumont, P. 1975. Hydrology. pp. 1–38. *In*: B.A. Whitten, Editor. River Ecology. Studies in Ecology. Vol. 2. Univ. of California Press. Berkeley.

Boulton, A.J., S. Findlay, P. Marmonier, E.H. Stanley, and H.M. Valett. 1998. The functional significance of the hyporheic zone in streams and rivers. Annu. Rev. Ecol. Syst. *29*:59–81.

Boyer, M.C. 1964. Section 15. Streamflow measurement. pp. 15–1 to 15–41. *In*: V.T. Chow, Editor. Handbook of Applied Hydrology. McGraw-Hill, New York.

Chorley, R.J. 1978. The hillslope hydrological cycle. pp. 1–42. *In*: M.J. Kirkby, Editor. Hillslope Hydrology. Wiley, Chichester.

Cummins, K.W. 1962. An evaluation of some techniques for the collection and analysis of benthic samples with special emphasis on lotic waters. Amer. Midland Nat. *67*:477–504.

Cummins, K.W. 1964. Factors limiting the microdistribution of larvae of the caddisflies *Pycnopsyche lepida* (Hogen) and *Pycnopsyche guttifer* (Walker) in a Michigan stream (Trichoptera: Limnephilidae). Ecol. Monogr. *34*(3):271–295.

Cummins, K.W. 1986. Riparian influence on stream ecosystems. pp. 45–55. *In*: I.C. Campbell, Editor. Stream Protection: The Management of Rivers for Instream Uses. Water Studies Centre, Chisholm Institute of Technology, East Caulfield, Australia.

Dahm, C.N. and H.M. Valett. 1996. Hyporheic zones. pp. 107–119. *In*: F.R. Hauer and G.A. Lamberti, Editors. Methods in Stream Ecology. Academic Press, San Diego.

Davis, H.S. 1938. Instructions for conducting stream and lake surveys. Fishery Circular Bur. Fish. *26*. 55 pp.

Dunne, T. 1978. Field studies of hillslope flow processes. pp. 227–293. *In*: M.J. Kirkby, Editor. Hillslope Hydrology. Wiley, Chichester.

Einsele, W. 1960. Die Strömungsgeschwindigkeit als beherrschender Faktor bei der limnologischen Gestaltung der Gewasser. Österreichs Fischerei *2*:1–40.

Findlay, S. 1995. Importance of surface-subsurface exchange in stream ecosystems: The hyporheic zone. Limnol. Oceanogr. *40*:159–164.

Gregory, K.J. and D.E. Walling. 1973. Drainage Basin Form and Process. A Geomorphological Approach. Wiley, New York. 456 pp.

Horton, R.E. 1945. Erosional development of streams and their drainage basins: hydrophysical approach to quantitative morphology. Bull. Geol. Soc. Amer. *56*:275–370.

Hynes, H.B.N. 1970. The Ecology of Running Waters. Univ. of Toronto Press. 555 pp.

Leopold, L.B., M.G. Wolman, and J.P. Miller. 1964. Fluvial Process in Geomorphology. Freeman, San Francisco. 522 pp.

Likens, G.E. 1984. Beyond the shore line: A watershed-ecosystem approach. Verh. Int. Ver. Limnol. *22*:1–22.

Likens, G.E. 1992. The Ecosystem Approach: Its Use and Abuse. Excellence in Ecology, Vol. 3. Ecology Institute, Oldendorf-Luhe, Germany. 167 pp.

Likens, G.E. and F.H. Bormann. 1995. The Biogeochemistry of a Forested Ecosystem. 2nd Edition. Springer-Verlag, New York. 159 pp.

Morisawa, M. 1968. Streams: Their Dynamics and Morphology. McGraw-Hill, New York. 175 pp.

Newbold, J.D. 1992. Cycles and spirals of nutrients. pp. 379–408. *In*: P. Calow and G.E. Petts, Editors, The River Handbook. I. Hydrological and Ecological Principles. Blackwell Science, Oxford, England.

Oglesby, R.T., C.A. Carlson, and J.A. McCann (eds.). 1972. River Ecology and Man. Academic Press, New York. 465 pp.

Strahler, A.N. 1952. Hypsometric (area–altitude) analysis of erosional topography. Bull. Geol. Soc. Amer. *63*:1117–1142.

Strahler, A.N. 1964. Section 4-II Geology, Part II. Quantitative geomorphology of drainage basins and channel networks. pp. 4–39 to 4–76. *In*: V.T. Chow, Editor. Handbook of Applied Hydrology. McGraw-Hill, New York.

Tice, R.H. 1968. Magnitude and frequency of floods in the United States. Part 1-B. North Atlantic Slope Basins, New York to York River. U.S. Geol. Survey Water-Supply Paper 1672, Washington, D.C. 585 pp.

Webster, J.R. and T.P. Ehrman. 1996. Solute dynamics. pp. 145–160. *In*: F.R. Hauer and G.A. Lamberti, Editors. Methods in Stream Ecology. Academic Press, San Diego.

Wetzel, R.G. and A.K. Ward. 1992. Primary production. pp. 354–369. *In*: P. Calow and G.E. Petts, Editors. The Rivers Handbook. I. Hydrological and Ecological Priniples. Blackwell Science, Oxford, England.

Welch, P.S. 1948. Limnological Methods. Blakiston, Philadelphia. 379 pp.

Whitton, B.A. (ed.). 1975. River Ecology. Studies in Ecology. Vol. 2. Univ. of California Press, Berkeley. 725 pp.

Winter, T.C. 1985a. Physiographic setting and geologic origin of Mirror Lake. pp. 40–53. *In*: G.E. Likens, Editor. An Ecosystem Approach to Aquatic Ecology: Mirror Lake and Its Environment. Springer-Verlag, New York.

Winter, T.C. 1985b. Approaches to the study of lake hydrology. pp. 128–135. *In*: G.E. Likens, Editor. An Ecosystem Approach to Aquatic Ecology. Mirror Lake and Its Environment. Springer-Verlag, New York.

Wisler, C.O. and E.F. Brater. 1949. Hydrology. 2nd Ed. Wiley, New York. 408 pp.

Dissolved Oxygen

The measurement of dissolved oxygen is one of the most frequently used and the most important of all chemical methods available for the investigation of the aquatic environment. Dissolved oxygen provides valuable information about the biological and biochemical reactions going on in waters; it is a measure of one of the important environmental factors affecting aquatic life and of the capacity of water to receive organic matter without causing nuisance.

Oxygen gas dissolves freely in fresh waters. Oxygen may be added to the water from the atmosphere or as a by-product of photosynthesis from aquatic plants and is utilized by many respiratory biochemical, as well as by inorganic chemical, reactions. The concentration of dissolved oxygen in water depends also on temperature, pressure, and concentrations of various ions [cf., Hutchinson (1957), Wetzel (1999)].

To be successful, a method for measuring dissolved oxygen must meet two requirements. First, owing to the small amount of substance to be determined (a few mg/l), it must be exact; second, it must be done with apparatus suited for field operation.

The method least subject to chemical errors, and probably the first to be proposed, is that of Bunsen, in which the gases are boiled out under either atmospheric pressure or diminished pressure. The amount of gas collected then is measured by absorption methods. However, the Bunsen method is too cumbersome for field work and requires considerable skill for accurate manipulation.

A few colorimetric methods have been proposed, but most have been found to be quite inaccurate, particularly at low concentrations. Although a number of chemical methods have been employed for dissolved oxygen measurement, the Winkler method, or some modification of it, is the most frequently used in limnology.

In recent times, major advances have occurred in the development and application of oxygen-sensitive electrodes for the rapid and sensitive measurement of dissolved oxygen. The Clark-type polarographic oxygen sensors often consist of platinum anode and a gold-plated cathode, encased in an electrolyte-filled housing and separated from the water by an oxygen-permeable membrane. Oxygen must diffuse through the membrane and electrolytic solution to the electrodes. The quantity of oxygen reduced per unit time is directly proportional to the oxygen concentration in the water, and the resulting electrical current is measured with a meter [cf., Gnaiger and Forstner (1983)].

Oxygen electrodes have the advantages of speed of measurement and the potential for continuous measurement in remote places. Commerically available macroelectrodes (ca. 1–5-mm diameter) require a rapid flow of water across the membrane; without such exchange, measurements are inaccurate and unreliable. Simple up and down movements of the electrodes in the water are insufficient to provide the conditions necessary for accurate measurements. Nearly all macroelectrodes are unreliable at dissolved oxygen concentrations between 0 and 1 mg/l. This low range is critical for many major chemical transformations and dissociation reactions, as well as crucial for microbial metabolism. Macroelectrodes are not satisfactory for studies of oxygen distribution near interfaces, particularly in zones of steep oxygen gradients, as at the sediment-water interface. Problems of slow diffusion rates and inaccuracy at low oxygen concentrations are circumvented to a significant extent with oxygen microelectrodes [less than 100-μm diameter; cf., Revsbech and Jorgensen (1986)]. The advantages of small size, where sensing surfaces are only a few micrometers in diameter, and rapid response times are counterbalanced by the difficulty of construction and fragility. Nonetheless, microelectrodes are providing unprecedented understanding about the distribution and dynamics of oxygen microgradients and about the ecology of the organisms generating these gradients [e.g., Carlton and Wetzel (1987, 1988)]. Recently fiber-optic oxygen microsensors based on chemical quenching of luminescence allow measurements with great stability without chemical consumption of oxygen and no need for flow about the sensing tip (Klimant et al., 1995). These optical electrodes circumvent many of the problems associated with the electrochemical oxygen sensors. Replicated calibration of oxygen sensors by chemical methods of analysis is required with solutions containing known quantities of dissolved oxygen; calibration in air is not satisfactory. Thus, although oxygen sensors are being improved constantly and will dominate measurements of dissolved oxygen in the future, the need still exists for chemical methods of measuring dissolved oxygen.

THE WINKLER METHOD

The Winkler method for measuring dissolved oxygen was introduced in 1888 by L.W. Winkler of Budapest and is a cleverly conceived as well as a very accurate procedure. The method depends on the oxidation of manganous hydroxide (bivalent manganese) by the oxygen dissolved in the water, resulting in the formation of a tetravalent compound. When the water containing the tetravalent compound is acidified, free iodine is liberated from the oxidation of potassium iodide. The free iodine is equivalent chemically to the amount of dissolved oxygen present in the samples and is determined by titration with a standard solution of sodium thiosulfate (e.g., 0.005 M). The reactions involved by the addition of reagents (KI, KOH, $MnSO_4$, and H_2SO_4) to the water are as follows.

Manganous sulfate reacts with the potassium hydroxide-potassium iodide mixture to produce a white flocculent precipitate of manganous hydroxide:

$$MnSO_4 + 2KOH \rightarrow Mn(OH)_2 + K_2SO_4$$

When the white precipitate is obtained, there is no dissolved oxygen in the sample. A brown precipitate indicates that oxygen was present and reacted with the manganous hydroxide, forming manganic basic oxide:

$$2Mn(OH)_2 + O_2 \rightarrow 2MnO(OH)_2$$

Upon the addition of sulfuric acid, this precipitate is dissolved, forming manganic sulfate:

$$2MnO(OH)_2 + 4H_2SO_4 \rightarrow 2Mn(SO_4)_2 + 6H_2O$$

There is an immediate reaction between $Mn(SO_4)_2$ and the potassium iodide added previously, liberating iodine and resulting in the typical iodine coloration (brown) of the water:

$$2\,MnO(SO_4)_2 + 4KI \rightarrow 2MnSO_4 + 2K_2SO_4 + 2I_2$$

The number of moles of iodine liberated by this reaction is equivalent to the number of moles of oxygen present in the sample. The quantity of iodine is determined by titrating a portion of the solution with a standard solution of sodium thiosulfate:

$$4Na_2S_2O_3 + 2I_2 \rightarrow 2Na_2S_4O_6 + 4NaI$$

Four moles of thiosulfate are titrated for each mole of molecular oxygen (O_2). Thus 1 ml of 0.025 M sodium thiosulfate is equivalent to 0.025 meq of oxygen. This value commonly is multiplied by 8 mg/meq to convert to mg O_2. When 200 ml of the original sample is titrated, then 1 ml 0.025 M $Na_2S_2O_3$ = 1 mg dissolved oxygen/l.

The iodine should be distributed uniformly throughout the bottle before decanting the amount needed for titrating. The volume decanted should coorespond to the volume of the original sample. Experienced analysts can detect the visual endpoint of the titration with a precision of $\pm 50\,\mu g/l$ (American Public Health Association, 1998). A correction for the dilution of sample with the reagents can be made when high accuracy is required. Thus, when a total of 2 ml, 1 ml each of the manganous sulfate and alkaline-iodide reagent, is added to a 300-ml bottle, the volume taken for titration should be

$$100 \times \frac{300}{300-2} = 100.7\,ml \quad or \quad 200 \times \frac{300}{300-2} = 201.3\,ml$$

If the sodium thiosulfate were other than 0.025 M, if the bottles were of another size, or if a different volume were to be titrated, the appropriate correction factor must be computed and applied to the calculation.

If the results were desired in cubic centimenters of oxygen gas, the value in mg/l should be multiplied by 0.698.

The amount of dissolved oxygen in distilled water (100 ml) can be calculated with a precision expressed as a standard deviation of about 0.043 ml of 0.025 M sodium thiosulfate; in sewage and secondary effluents the precision is about 0.058 ml. In the presence of appreciable interferences, even with the proper Winkler modification, the standard deviation may be as high as 0.1 ml. Even greater errors may occur in waters containing suspended organic solids or in heavily polluted waters. Large errors may be introduced in the measurement of dissolved oxygen by neglecting proper precautions in the presence of interfering substances such as nitrites, iron salts, and organic matter, which are common in natural waters, or by an improper application of the various modifications of the Winkler method that are designed to overcome these interferences.

Errors from nitrites are introduced at the time the solution is made acidic with sulfuric acid. In an acid medium, nitrites react with the potassium iodide, liberating iodine:

$$2KI + H_2SO_4 \rightarrow 2HI + K_2SO_4$$
$$2HNO_2 + 2HI \rightarrow 2H_2O + N_2O_2 + I_2$$

If the reaction were complete at this point, the error due to the presence of nitrites in most cases would not be significant. However, if the sample were allowed to stand exposed to the air, the dissolved oxygen would react with N_2O_2, again producing the nitrite:

$$2N_2O_2 + 2H_2O + O_2 \rightarrow 4HNO_2$$

This reaction will again liberate more iodine. Should this cycle be repeated a significant number of times, the error introduced would soon become very large. The continuous reaction can be minimized by an immediate and rapid titration of the sample after exposure to the air. The effect of nitrites is eliminated when sodium azide is added to the sample along with the alkaline potassium iodide. The reactions are as follows:

$$2NaN_3 + H_2SO_4 \rightarrow 2HN_3 + Na_2SO_4$$
$$HNO_2 + HN_3 \rightarrow N_2O + N_2 + H_2O$$

Iron affects the measurement of dissolved oxygen in two ways [e.g., Buswell and Gallaher (1923)]. First, when present in large amounts, iron requires so much of an even stronger oxidant to oxidize it from the ferrous condition that the dissolved oxygen in the reagent itself is appreciable; second, when oxidized to the ferric state it serves as an oxidizing agent itself and will oxidize the iodide ion to free iodine, which in turn will give high values of dissolved oxygen:

$$Fe^{3+} + I \rightleftarrows Fe^{2+} + \tfrac{1}{2}I_2$$

In the Rideal-Stewart modification, the ferrous iron and some of the nitrites and organic matter are first oxidized by potassium permanganate, and then excess permanganate is removed with potassium oxalate. Care must be taken not to add too great an excess of the oxalate or error will be introduced into the final result. The addition of potassium fluoride as the first reagent, or after the Rideal-Stewart treatment for ferrous ion, will eliminate this interference as the more active fluoride ions will be liberated rather than iodine.

The extent of interference by organic matter, which causes low results, varies with the technique of the procedure. If the precipitate of manganous and manganic hydroxide settles and compacts on the bottom of the bottle, and if the acid is added without prompt and complete mixing, local depletion of iodide occurs, thus permitting more extensive oxidation of organic matter. Interference is minimized by adding the acid promptly and mixing rapidly.

In addition to the Winkler method, which should be used in measuring the dissolved oxygen of relatively pure water, there are several modifications which are described in detail in American Public Health Association et al. (1998):

1. *Winkler method, unmodified*: It should be noted that 1 ml of each reagent is recommended in 250- to 300-ml bottles (Welch, 1948).
2. *Alsterberg (Azide) modification*: This modification is used for water containing more than 0.05 ml/l ferrous iron and can be used in waters containing up to 1 mg/l ferrous iron. Other reducing or oxidizing agents should be absent. If 1 ml of potassium fluoride solution were added to the sample before acidifying and there were no delay in titration, the method also would be applicable in the presence of 100 to 200 mg/l ferric iron. Interference by nitrite nitrogen up to 0.05 mg/l NO_2- N is eliminated by use of sodium azide.
3. *Rideal-Stewart (permanganate) modification*: This modification should be used only on samples containing ferrous iron. It is ineffective for the oxidation of sulfates, thiosulfates, and polythionates, or of the organic matter in sewage.
4. *Alum flocculation modification*: Samples high in suspended solids may absorb appreciable quantities of iodine in acid solution. This interference may be removed by alum flocculation.

5. *Copper sulfate-sulfamic acid flocculation modification*: This modification is used for biological flocs such as activated sludge mixtures, which have high rates of oxygen consumption.
6. *Pomeroy-Kirschman-Alsterberg modification*: This modification is designed for samples containing more than 15 mg/l dissolved oxygen or which have a high content of organic matter, such as domestic sewage. It uses an alkaline-iodide solution that is 6N in sodium iodide and 10N in sodium hydroxide. This solution is saturated and provides sufficient iodide for oxygen-enriched samples. KI cannot be used because of limited solubility.

There are several arguments in favor of increasing the iodide concentration as is done in the Pomeroy-Kirschman-Alsterberg modification above. It is known that an excess of oxalate in the permanganate method causes low results, since at the time of acidification the manganic hydroxide acts on any reducing agents that may be present. Iodide, oxalate, and other reducing agents compete with one another. When oxalate is present in substantial amounts and the iodide concentration is low, a considerable amount of manganic hydroxide will react with oxalate, causing low results. When a high concentration of iodide is present, more iodide and less oxalate will be used. After adding the acid, there may be a slow reaction of iodine with organic matter. This problem is diminished by a high concentration of iodide.

Loss of iodine vapors may cause appreciable error; this error is diminished by increasing the iodide concentration, since iodide holds the iodine in solution by the equilibrium equation: $I^- + I_2 \rightleftarrows I_3$.

An additional advantage is that the high concentration of iodide results in a sharper end point in titration.

7. In saline waters with high carbonate concentrations, the Winkler method cannot be used because of CO_2 effervescence upon acidification. Variations of the Miller method of oxygen determination should be used [cf., Walker et al. (1970) and Ellis and Kanamori (1973)].

A simple, accurate, and precise spectrophotometric method is available for measuring dissolved oxygen in fresh waters (Roland et al. 1999). Because of the speed and simplicity of this spectrophotometric method, it is especially useful for replicate measurements and for laboratory incubation experiments. Interferences include turbidity, color, and chlorophyll in the samples, but these can be corrected (Roland et al. 1999).

Method: 1. Treat samples with manganous sulfate, alkaline iodide-azide solution, and acid according the Winkler procedure (Alsterberg modifications, above). Be as precise as possible in adding the correct volume of reagents.
2. Make spectrophotometric measurements at 430 nm using 1-cm pathlength cuvettes and a spectrophotometer (e.g., Shimadzu UV-160).
3. Calibrate spectrophotometer readings using standard values determined by titration and/or oxygen saturated samples. In the absence of significant interference from organic color, chlorophyll, or turbidity: DO (mg/liter) $\cong 0.0093 \times Abs_{430} - 1.6$.
4. The working range for this method is 4 to 12 O_2/liter.

PROCEDURES

The primary difficulty in determining the concentration of dissolved oxygen in fresh waters is related to the collection of the water sample. Of necessity many samples must be taken at a particular time, place, and depth, under a specified set of conditions, which may or may not represent the average conditions over a period of time or the average cross section of the lake, stream, or other body of water.

Moreover, the samples should represent the actual conditions at the point of sampling and not be aerated or de-aerated artificially. Samples should not be allowed to stand for long periods or at high temperature to permit algae, bacteria, and other organisms to change the dissolved oxygen content by their metabolism, nor should the dissolved gases be allowed to escape.

1. Using a Van Dorn (p. 86) or similar in situ water sampler, collect unmodified water samples from depth.
2. Remove the sample from the water bottle immediately.
3. Transfer the sample from the water bottle by placing the delivery tube to the bottom of a 250- or 300-ml BOD (= biological oxygen demand) bottle. Allow the water to flow continuously through the bottle, excluding all entrapped bubbles, until at least three times the capacity of the sample bottle has overflowed (count the seconds needed to fill the BOD bottle initially and then repeat twice). Withdraw the delivery tube gently at the end of sample flushing, without stopping flow, so that the bottle is filled completely. Do not stopper the bottle when the following step can be done *immediately*.
4. *Immediately* and gently add, just below the surface, 1 ml of $MnSO_4$ reagent and 1 ml of NaOH + KI. Be careful not to mix the automatic pipets (*do not mouth pipet these reagents*). Carefully stopper the bottle without introducing any air bubbles and mix vigorously by inversion.
5. Allow the precipitate to settle. Shake vigorously again and allow the precipitate to settle to at least the bottom third of the bottle volume.
6. Add 1 ml of concentrated H_2SO_4 with an automatic pipet by inserting the tip just below the surface of the sample. Carefully stopper without introducing air bubbles and shake the bottle until all of the precipitate has dissolved.
7. Measure 100 ml of the sample with a volumetric pipet (a graduated cylinder is much too inaccurate and volumetric flasks are not designed to deliver accurately) and then transfer to a 250-ml Erlenmeyer flask or porcelain casserole dish. Touch the tip of the pipet to the side of the flask during delivery.
8. With a 50-ml burette filled with 0.005 M standardized sodium thiosulfate solution, titrate with mixing until a very pale yellow ("straw") color is observed and then proceed with the next step.
9. Add 2 drops of stabilized starch mixture, mix to get a uniform blue color, and titrate carefully but rapidly to a colorless end point. The blue coloration should return upon standing in 15 to 20 sec and can be ignored. When the blue coloration does not return, the end point has been overshot. Record the volume of titrant used in ml to two decimal places.

10.
$$mgO_2/l = \frac{(\text{ml titrant})(\text{molarity of thiosulfate})(8000)}{\left(\begin{array}{c}\text{ml sample}\\\text{titrated}\end{array}\right)\left(\dfrac{\text{ml of bottle} - 2}{\text{ml of bottle}}\right)}$$

Report values to two decimal places.

When it is not possible to titrate the sample shortly after dissolving the precipitate, care must be taken to avoid exposure to light. Samples should not be stored for more than 8h before titration.

PERCENTAGE SATURATION OF DISSOLVED OXYGEN

Since the solubility of oxygen in water is influenced by temperature, pressure, and ionic concentrations, it is often important to calculate percentage saturation:

$$\%\text{saturation} = \frac{[O_2]\text{ measured} \times 100}{O_2\text{ solubility at saturation}}$$

The concentration of dissolved oxygen at saturation for specified temperatures is given in Table 6.1. These values are applicable to a sea level barometric pressure of 760mm Hg. For high altitude and resulting barometric pressure other than that of sea level, correction for barometric pressure at the time of sampling must be made when precise results are desired. The correction is made as follows:

$$S_p = S\left(\frac{P}{760}\right)$$

where S_p = solubility at pressure P, S = solubility at 760mm Hg, and P = barometric pressure in mm Hg. Alternatively, with a slight reduction in accuracy, a nomogram can be used (Fig. 6.1). Older nomograms of Ricker, of Rawson, and of Mortimer (1956) are inaccurate and should not be used.

EXERCISES

OPTION 1. FIELD TRIPS

1. From the central depression of a lake, collect water samples for oxygen analysis and fix the dissolved oxygen chemically by the techniques outlined earlier (pp. 78–79). Collect replicates at each depth. Place samples in an ice chest for transport to the shore or laboratory.
2. Determine the temperature of the water at each depth throughout the vertical profile with an underwater thermometer.
3. Working in pairs, determine the dissolved oxygen content of each sample. Analyze duplicate samples from the same sample bottle and determine the standard error and variance (cf., Appendix 2). When prestandardized sodium thiosulfate solution is not available, standardize the $Na_2S_2O_3$ solution according to the directions given below (p. 82) and calculate the molarity of your solution.
4. Calculate the percentage saturation of dissolved oxygen from the temperature data and barometric pressure or altitude. Use the nomogram provided in Fig. 6.1. From the collective class data, plot the dissolved oxygen concentrations, percentage saturation, and temperature data versus depth.
5. Collect oxygen in the littoral zone among dense strands of aquatic macrophytes and from inlet sources. Compare these values with those from comparable depths in open water.

Table 6.1. Solubility of oxygen in pure water in relation to temperature in equilibrium with air at standard pressure saturated with water vapor.[a]

Equilibrium temperature (t) (°C)	Oxygen (C*) (mg l^{-1})	Oxygen ($C^{\dagger}$) (μgat kg^{-1})
0	14.621	913.9
1	14.216	888.5
2	13.829	864.4
3	13.460	841.3
4	13.107	819.2
5	12.770	798.2
6	12.447	778.0
7	12.139	758.7
8	11.843	740.2
9	11.559	722.5
10	11.288	705.5
11	11.027	689.2
12	10.777	673.6
13	10.537	658.6
14	10.306	644.1
15	10.084	630.3
16	9.870	616.9
17	9.665	604.1
18	9.467	591.7
19	9.276	579.8
20	9.092	568.3
21	8.915	557.2
22	8.743	546.5
23	8.578	536.1
24	8.418	526.1
25	8.263	516.4
26	8.113	507.1
27	7.968	498.0
28	7.827	489.2
29	7.691	480.7
30	7.558	472.4
31	7.430	464.4
32	7.305	456.6
33	7.183	449.0
34	7.065	441.6
35	6.949	434.4
36	6.837	427.3
37	6.727	420.5
38	6.620	413.7
39	6.515	407.2
40	6.412	400.8

[a] From Mortimer (1981) after data of Benson and Krause (1980). Unit standard concentrations (C* by volume in mg l^{-1} or mg dm^{-3} and $C^{\dagger}$ by weight in μgat kg^{-1}) of oxygen (1 mole = 31.9988 g) in pure water in equilibrium with air at standard pressure (1 atm = 760 mm Hg = 101.325 Pa) saturated with water vapor and with the dry air mole fraction for oxygen taken as 0.20946. To convert mg to cm^3 (ideal gas S.T.P.) or to μgat, multiply mg by 0.70046 or 62.502, respectively. To convert μgat to μmol, divide by 2. To convert C* to $C^{\dagger}$, at temperature (t)° and with the accuracy of this table, multiply C* by the density of pure water at t°.

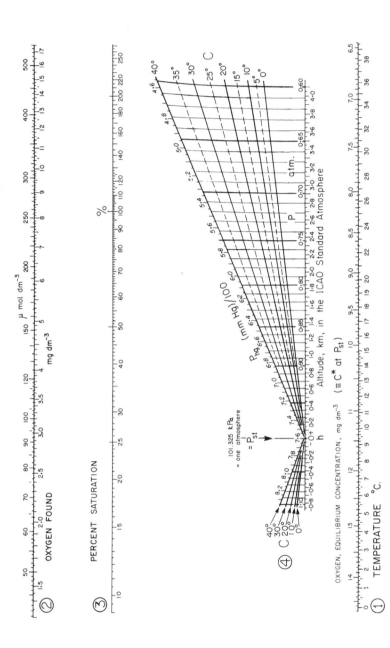

Figure 6.1. Oxygen saturation nomogram. The nomogram provides estimates of: (a) solubility of oxygen in fresh water in equilibrium with the atmosphere (at 100% humidity), and (b) percentage saturation of oxygen in samples at observed temperatures and oxygen content. *Scale 1*: Bottom scale = water temperature (°C); upper scale = corresponding saturated concentration (C^*) oxygen in equilibrium with the atmosphere (at 100% humidity and standard pressure of 760 mm Hg or 101.325 kPs). *Scale 2*: Oxygen concentration (s) in mgl^{-1} (lower) and μmolm^{-1} (upper)/ *Scale 3*: Percentage saturation of dissolved oxygen. *Scale 4*: Scales to estimate equilibrium concentrations at pressures (P) other than standard (P_{st}). Place one end (point) of a scale (or calipers) at zero h and extend the other end (point) into the fan of scales corresponding to the altitude (in km) or pressure (lower or upper scale, in mm Hg) to intersect with the water temperature among the temperature lines extending from the fan ends. The distance from zero h to the pressure (altitude)-temperature point within the fan Scale 4 is then brought down onto Scale 1. This distance is extended from the temperature of the water from which oxygen concentration was determined to the right (extended to the left of the water temperature on Scale 1 when the lake surface is below sea level). Where a straight line between this point and the oxygen concentration (Scale 1) intersects Scale 3 gives the percentage saturation of dissolved oxygen. When the oxygen concentration (s) is below the range of Scale 2, multiply s by a convenient factor, e.g., 10. Determine the percentage saturation as described above, but divide that estimate by the factor to give the correct result. [From Mortimer (1981)]. The oxygen content of air-saturated fresh waters over ranges of temperature and atmospheric pressure of limnological interest. Mitt. Int. Ver. Limnol. 22, 23 pp.]

OPTION 2. LABORATORY EXERCISE

1. From the water sources provided by your instructor, fill the sample bottles carefully with the rubber tubing, taking care to avoid the inclusion of air bubbles. Measure the temperature of the water source. Fix the dissolved oxygen of the water samples chemically by the techniques outlined earlier (p. 78).
2. Working in pairs, determine the dissolved oxygen content of each sample. Analyze duplicate samples from the same sample bottle and determine the standard error and variance (cf., Appendix 2). When prestandardized sodium thiosulfate is not available, standardize the $Na_2S_2O_3$ solution according to the directions given below and calculate the molarity of your solution.
3. Calculate the percentage saturation of dissolved oxygen from the temperature data and the barometric pressure or altitude. Use the nomogram provided in Fig. 6.1.
4. From the data provided in Table 6.2, plot the dissolved oxygen concentration, percentage saturation, and temperature data versus depth.

Questions

1. What do the data for concentrations of dissolved oxygen tell you about solubility of this gas in relation to temperature?
2. How does the percentage saturation of oxygen clarify the relationship of solubility and temperature of the water?
3. How would you expect the oxygen concentrations of epilmnetic water to fluctuate throughout the course of a day? Why? In an unproductive versus a productive lake?
4. How would you expect the oxygen concentrations of epilimnetic water to vary in horizontal distribution from the shore line to open water?
5. How would you expect the dissolved oxygen content to vary above and below the sediment-water interface? [cf., Carlton and Wetzel (1988).]

Apparatus and Supplies

1. Manganous sulfate reagent (400 g $MnSO_4 \cdot 2H_2O$ or 364 g $MnSO_4 \cdot H_2O$ dissolved in 1 liter of distilled water).
2. Alkaline-iodide-azide reagent (500 g NaOH or 700 g KOH and 135 g NaI or 150-g KI dissolved in 1 liter of distilled water; 10 g NaN_3 (*Caution*: extremely poisonous) in 40 ml distilled water, added to 960 ml of alkaline-iodide reagent).
3. Standardized sodium thiosulfate solution (either (a) or (b)):
    a. 0.005 M, prepared from 0.1 M stock solution (kept refrigerated). Remove 0.1 M $Na_2S_2O_3$ and allow to come to room temperature; dilute 50 ml to 1 liter with distilled water.
    b. 0.025 M (6.205 g $Na_2S_2O_3 \cdot 5H_2O$ dissolved in 1 liter of freshly boiled and cooled distilled water). 1.00 ml of 0.025 N $Na_2S_2O_3$ solution = 0.200 mg dissolved oxygen. Standardization:
        i.   Dissolve 2 g KI in 100 to 150-ml distilled water.
        ii. Add 10 ml of 1:9 H_2SO_4.
        iii. Add exactly 20.00 ml 0.0021 M $KH(IO_3)_2$. Dissolve 812.4 mg KH $(IO_3)_2$ in 1 liter of distilled water.
        iv. Dilute to 200 ml with distilled water.
        v.   Titrate with the thiosulfate solution using procedures identical to those for dissolved oxygen.
        vi. Calculate molarity of the thiosulfate solution.

$$(Vol_1)(M_1) = (Vol_2)(M_2) [cf., Appendix 1]$$

4. Stabilized starch solution [an excellent stabilized starch solution that will not deteriorate for years is given by Van Landingham (1960)].

Table 6.2. Oxygen concentrations and temperatures in the central depression of Lawrence Lake, Barry Co., Michigan. Altitude = 274 m.

Depth (m)	April 17, 1972 °C	April 17, 1972 O₂ (mg/l)	May 1, 1972 °C	May 1, 1972 O₂ (mg/l)	July 11, 1972 °C	July 11, 1972 O₂ (mg/l)	August 22, 1972 °C	August 22, 1972 O₂ (mg/l)	October 16, 1972 °C	October 16, 1972 O₂ (mg/l)	October 30, 1972 °C	October 30, 1972 O₂ (mg/l)	November 13, 1972 °C	November 13, 1972 O₂ (mg/l)
0	7.8	11.35	12.9	10.55	22.2	9.22	25.2	8.28	12.9	8.80	9.3	8.89	7.8	9.57
1	7.8	—	12.8	—	22.2	—	25.2	—	12.9	—	9.4	—	7.8	—
2	7.6	—	11.1	—	21.6	—	24.6	—	12.8	—	9.3	—	7.8	—
3	7.5	11.20	10.0	11.61	20.8	8.91	22.5	10.80	12.8	8.84	9.3	8.86	7.8	9.37
4	7.5	—	9.0	—	19.3	—	21.2	—	12.8	—	9.3	—	7.8	—
5	7.5	11.68	8.3	11.52	16.9	10.32	19.7	9.93	12.8	8.72	9.3	10.01	7.8	9.47
6	6.6	—	8.0	—	13.9	—	16.7	—	12.8	—	9.3	—	7.8	—
7	5.3	11.89	7.8	11.14	11.7	11.76	13.6	10.69	12.8	8.72	9.3	8.91	7.8	9.45
8	5.0	—	7.3	—	9.7	—	11.1	—	12.7	—	9.3	—	7.8	—
9	4.8	—	5.3	—	8.6	—	9.6	—	10.6	—	9.3	—	7.8	—
10	4.7	11.40	4.9	10.66	7.8	7.89	8.6	2.81	9.3	0.83	9.3	8.61	7.8	9.78
11	4.6	—	4.9	—	7.2	—	8.0	—	8.7	—	9.2	—	7.8	—
12	4.4	10.42	4.9	7.29	6.9	2.49	7.7	0.31	8.4	0.06	8.7	1.28	7.8	9.44
12.5	4.4	—	5.0	—	6.7	—	7.3	—	8.1	—	8.1	—	7.9	—

5. Concentrated sulfuric acid.
6. Automatic dispensing pipets, 2 ml, adjustable.
7. Burettes (50 ml) and volumetric pipettes (100 ml, 20 ml).
8. Erlenmeyer flasks or porcelain casserole dishes (250 to 300 ml).
9. BOD oxygen bottles with tapered stoppers.
10. Underwater thermometer.
11. Mercurial barometer (optional).
12. Planimeter or computerized digitizer.

References

American Public Health Association et al. 1998. Standard Methods for the Examination of Water and Wastewater. 20th Ed. Water Environment Federation, Alexandria, VA. 1183 pp.

Benson, B.B. and D. Krause, Jr. 1980. The concentration and isotopic fractionation of gases dissolved in fresh water in equilibrium with the atmosphere. 1. Oxygen. Limnol. Oceanogr. 25:662–671.

Buswell, A.M. and W.U. Gallaher. 1923. Determination of dissolved oxygen in the presence of iron salts. Ind. Eng. Chem. 15:1186–1188.

Carlton, R.G. and R.G. Wetzel. 1987. Distribution and fates of oxygen in periphyton communities. Can. J. Bot. 65:1031–1037.

Carlton, R.G. and R.G. Wetzel. 1988. Phosphorus flux from lake sediments: Effect of epipelic algal oxygen production. Limnol. Oceanogr. 33:562–570.

Ellis, J. and S. Kanamori. 1973. An evaluation of the Miller method for dissolved oxygen analysis. Limnol. Oceanogr. 18:1002–1005.

Gnaiger, E. and H. Forstner (ed). 1983. Polarographic Oxygen Sensors: Aquatic and Physiological Applications. Springer-Verlag, New York. 370 pp.

Hutchinson, G.E. 1957. A Treatise on Limnology. I. Geography, Physics, and Chemistry. Wiley, New York. 1015 pp.

Klimant, I., V. Meyer, and M. Kuhl. 1995. Fiber-optic oxygen microsensors, a new tool in aquatic biology. Limnol. Oceanogr. 40:1159–1165.

Mortimer, C.H. 1981. The oxygen content of air-saturated fresh waters over ranges of temperature and atmospheric pressure of limnological interest. Mitt. Int. Ver. Limnol. 22, 23 pp.

Revsbech, N.P. and B.B. Jorgensen. 1986. Microelectrodes: Their use in microbial ecology. Microbial Ecol. 9:293–352.

Roland, F., N.F. Caraco, J.J. Cole, and P. del Giorgio. 1999. Rapid and precise determination of dissolved oxygen by spectrophotometry: Evaluation of interference from color and turbidity. Limnol. Oceanogr. 44:1148–1154.

Van Landingham, J.W. 1960. A note on a stabilized starch indicator for use in iodometric and iodimetric determinations. Limnol. Oceanogr. 5:343–344.

Walker, K.F., W.D. Williams, and U.T. Hammer. 1970. The Miller method for oxygen determination applied to saline waters. Limnol. Oceanogr. 15:814–815.

Welch, P.S. 1948. Limnological Methods. Blakiston, Philadelphia. 381 pp.

Wetzel, R.G. 1983. Limnology. 2nd Ed. Saunders Coll. Philadelphia. 860 pp.

Wetzel, R.G. 1999. Limnology: Lake and River Ecosystems. 3rd Ed. Academic Press, San Diego (in press).

Inorganic Nutrients: Nitrogen, Phosphorus, and Other Nutrients

Compounds of nitrogen, and especially those of phosphorus, are major cellular components of organisms. Since the availability of these elements may be less than the biological demand, environmental sources can regulate or limit the productivity of organisms in freshwater ecosystems. Other elements such as iron and sulfur are essential cellular constituents but are required in relatively low concentrations in relation to availability in fresh waters. The major base cations, calcium, magnesium, sodium, and potassium, usually are required in very low quantities, but their concentrations in fresh water can influence the osmoregulation of organisms.

It should be remembered that chemical concentration in water is a static variable, that is, the chemical mass (weight) of an element or compound per unit volume at a particular place or time within the lake. Certain nutrients, such as magnesium and sodium, are relatively conservative in concentration in that their solubility is high, they usually are abundant in relation to metabolic demands, and their concentrations are relatively unaffected by metabolically altered reduction-oxidation conditions of the water. Concentrations of nitrogen and phosphorus compounds, on the other hand, are highly dynamic because they may be utilized, stored, transformed, and excreted rapidly and repeatedly by the various aquatic organisms.

Measurements are further complicated by the chemical form in which the element occurs.

Ionic concentrations are often exceedingly low, requiring great care in collection and analysis of water samples to avoid contamination. In general, all equipment and glassware should be cleaned thoroughly with acid, such as 6N HCl, and only high-quality deionized or glass-distilled water and analytical grade reagents should be used.

Storage of water samples before analysis should be avoided. If storage is necessary, the time must be minimized if reliable results are to be obtained. Prompt freezing of water samples or the addition of certain preservatives may be satisfactory for certain analyses but not for others. Storage methods will be discussed specifically in the following procedures.

It is often meaningful to the biologist, from the standpoint of availability, to separate analyses of total nutrient concentrations into those fractions that are dissolved in inorganic and organic states from those that occur in particulate form. This separation is done generally by filtration through glass fiber filters, taking great care to make certain that contamination is not introduced from the filters and filtration apparatus. Separate chemical analyses for the inorganic and the organic components then are performed on each of these fractions.

The following discussion of chemical analyses assumes that samples were obtained with suitable sampling apparatus. By far the most commonly used device is a nonmetallic Van Dorn sampler (Fig. 7.1). This device (Van Dorn,

1956) has the important advantages of being completely nonmetallic at points of contact with the sample and of being simple and relatively foolproof to operate. Available commercially in several sizes, Van Dorn water samplers are lowered slowly to the depth to be sampled. The weighted messenger is dropped down the line to trigger closure of the end cups. Horizontally positioned Van Dorn samplers are also available for sampling in sharply stratified layers, but these tend to disturb the microstrata by their physical movement. Battery-driven nonmetallic or peristaltic pumps also are available to sample at specific depths.

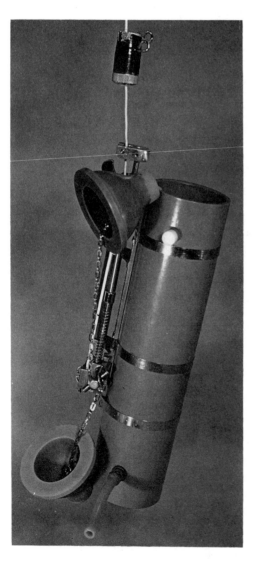

Figure 7.1. The Van Dorn nonmetallic underwater sampler in the open position.

NITROGEN COMPOUNDS

Ammonium Nitrogen

Except under very alkaline conditions (pH > 9.0), most of the ammonia (NH_3) in fresh water exists in the ionic form (NH_4^+). Ammonium is an important source of nitrogen for bacteria, algae, and larger plants in lakes and streams, and, because concentrations commonly are low, the content of water samples can change quickly and markedly. Thus, samples of lake and stream water should be analyzed as soon as possible after collection. If analyses are to be delayed for more than a few hours, samples should be preserved (see below).

A phenol-hypochlorite method using nitroprusside as a catalyst is recommended for the measurement of ammonium in water (Solorzano, 1969; Harwood and Kühn, 1970). Ammonium reacts with phenol and hypochlorite under alkaline conditions to form indophenol blue. The color development is proportional to the concentration of ammonium within a given range (0 to $1000\,\mu g$ NH_4-N/liter). Sensitivity is good, with a standard deviation under carefully controlled conditions of ± 2 to $5\,\mu g$ NH_4-N/liter.

Sample Storage. Samples should be analyzed within a few hours of collection. If stored, maintain at 5°C in the dark. When hydrogen sulfide is present, acidify to pH 3 and bubble with an inert gas such as helium until free of the H_2S odor.

Reagents. Distilled water that has been passed through an acidic deionization cation exchange resin should be used throughout. (Scrupulously avoid atmospheres containing dust, smoke, volatile cleaning agents, etc. that can contain NH_3, as serious contamination of samples could occur.)

Caution: Most of the reagents are caustic and very *toxic*; never pipet by mouth and wear appropriate safety gear.

1. *Buffer*: Na_3PO_4 (or K_3PO_4) in a 5% (w/v) solution.
2. *Phenol stock*: 500 g phenol dissolved in methanol and diluted to 800 ml with methanol. Store under refrigeration in an amber bottle.
3. *27% NaOH*: 270 g NaOH pellets dissolved in water, cooled, and diluted to 1 liter.
4. *Phenate reagent A*: 15 ml of phenol stock and 0.02 g sodium nitroprusside are diluted to 100 ml with water.
5. *Reagent B*: Mix equal volumes (15 ml) of sodium hypochlorite (*fresh* commercial bleach of 5% chlorine will suffice) and 27% NaOH solution and dilute to 50 ml with distilled water.
6. Store both reagents A and B in amber bottles in a refrigerator; allow to reach room temperature before using.
7. *Ammonium standard*: Dissolve 3.819 g of dry NH_4Cl in distilled-deionized water and bring to 1 liter. 1.00 ml = $1000\,\mu g$ NH_4-N. Dilute serially for standards.

Procedures. Using 50-ml graduated cylinders,

1. Dispense 50.0 ml of distilled-deionized water (blank), ammonium standards, and samples into 50-ml graduated cylinders.
2. Add 2 ml phosphate buffer; mix (a vortex mixer works well). Ignore any precipitate that may form.
3. Add 5 ml of reagent A; mix well.
4. Add 2.5 ml of reagent B; mix well.
5. Allow 1 h for color development (should be stable at room temperature for 24 h).

6. Read the optical density at 630 nm, subtracting the O.D. of the blank, using cells of appropriate path lengths for the color intensity observed (1-cm path length for high ammonium levels; 5-cm path length for the range 0 to 200 μg NH$_4$-N/l; 10-cm for <20 μg NH$_4$-N/l).

7. Prepare a standard curve of O.D. versus concentration of your corrected standards. Sample concentrations then can be read directly from the graph. When the standard curve is linear, as it should be in this case, sample concentrations can be calculated directly, where

OD$_b$ = absorbance of distilled water plus reagents (blank)

OD$_s$ = absorbance of standards and samples plus reagents

Concentration of standard

$$F = \frac{(\mu g\ NH_4\text{-}\ N/l)}{OD_s\ of\ standard - OD_b} = extinction\ factor$$

Then

$$\mu g\ NH_4\text{-}\ N/l = F(OD_s\ of\ sample - OD_b)$$

Note: Such standard curves can be generated easily on a computer or a hand calculator, as well as pertinent statistics such as a correlation coefficient (r^2). Standard curve relationships should be either linear or quadratic. If 3rd order, the standard curve should be redone.

Nitrate and Nitrite Nitrogen

By far the best method for analysis of NO$_3$-N in water is to reduce the nitrate in alkaline-buffered solution to nitrite by passing the sample through a column of copperized cadmium metal filings (Wood et al., 1967). The nitrite then is measured by a sensitive diazotization method that results in a stable pink azo dye whose absorbance obeys Beer's Law up to about 500 μg NO$_3$-N or NO$_2$-N/l. Nitrite concentrations of the original sample can be determined very accurately (S.D. ± ca. 0.5 μg NO$_2$-N/l) prior to reduction of NO$_3$ to NO$_2$ by the same diazotization technique.

The cadmium-reduction method requires special glass columns for efficient analysis. Once the equipment is assembled and operational, however, this method permits analysis of about eight samples per hour per column. Four columns easily can be operated simultaneously.

Sample Storage. Samples should be filtered through combusted (500°C) glass fiber filters (Whatman GF/F, 0.6–0.7-μm pore size; other sizes, such as GF/C, are *not* acceptable). The filtrate is assumed to contain only dissolved materials. The filtrate can be stored at 4 to 5°C but should be analyzed within 24 h.

Apparatus and Reagents

1. *Reduction column.* Columns constructed of borosilicate glass of the approximate dimensions illustrated in Fig. 7.2 are recommended. The bore of the portion retaining the copperized cadmium filings should not be less than 8 mm. Very fine "wool" turnings of copper are recommended (Strickland and Parsons, 1968) in place of glass wool at the bottom of the column; if copper turnings are used, however, they must be very fine.

Figure 7.2. Cadmium-reduction columns used in the analysis of nitrate.

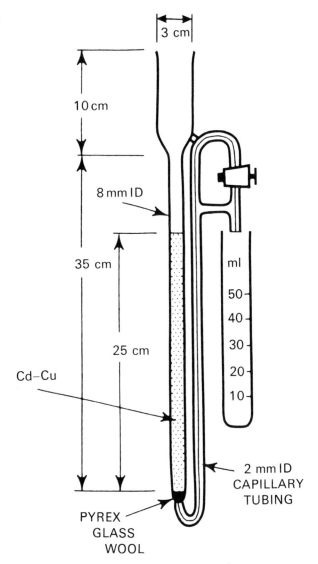

2. *Cadmium-copper filings*: Filings of cadmium (>99.9% purity) should be sieved to retain particles of a size between 0.5 and 2 mm. Avoid ingesting or inhaling the toxic cadmium particles! Perform the sieving of filings and the filling of columns in a hood. Approximately 40 g per column of the Cd filings are washed with 2N HCl (85 ml of concentrated HCl diluted to 500 ml with distilled water) in a separatory funnel and rinsed with distilled water. The Cd then is treated with ca. 100 ml of 0.08 M cupric sulfate solution (20 g $CuSO_4 \cdot 5H_2O$ dissolved in 1 liter of distilled water) and swirled until the blue color disappears as the copper is deposited on the cadmium. The Cd-Cu filings then are rinsed with distilled water carefully; avoid exposure to air. With the columns filled with water, add the Cd-Cu filings as indicated in Fig. 7.2. No entrapped air bubbles should remain in the columns.

3. Flush each column with a solution of 50 ml of distilled water plus 5 ml of buffer solution of the following composition:

100 g ammonium chloride, 20 g sodium tetraborate, and 1 g of disodium dihydrate EDTA dissolved in distilled water to make 1 liter.

Flow should be adjusted to a delivery rate of 6 ml/min.

4. *Sulfanilamide solution*: Dissolve 5 g of sulfanilamide in a solution of 50 ml of concentrated HCl (sp. gr. 1.18) and 300 ml of distilled water. Dilute with water to 500 ml.

5. *N-(1-naphthyl) ethylene diamine dihydrochloride solution*: Dissolve 0.5 g of dihydrochloride in 500 ml distilled water. Store in amber or low actinic glass bottle; renew once a month or earlier if dark brown coloration develops.

6. *Nitrate standard*: Dissolve 0.722 g of KNO_3 (dried for 1 h at 110°C) in distilled water and make to 1 liter:

$$1.00 \, ml = 100 \, \mu g \, NO_3\text{-}N.$$

7. *Nitrite standard*: Dissolve 0.4929 g of $NaNO_2$ (dried for 1 h at 110°C) in distilled water and make to 1 liter:

$$1.00 \, ml = 100 \, \mu g \, NO_2\text{-}N$$

Procedures for Nitrate Nitrogen

1. Prepare duplicate aliquots of nitrate standard to yield 50 ml from each Cd-Cu reduction column. The concentrations used should approximate those of the samples to be analyzed and include concentrations both lower and higher than those of the samples.
2. Place 50 ml of distilled water (blank), nitrate standards, and samples into graduated cylinders. Add 5 ml of buffer solution and mix thoroughly (with stoppered cylinders, invert; if baseless cylinders are used, a vortex mixer is very effective and rapid).
3. Add 10 ml of the buffered sample to the column, allow it to run through, and discard the effluent.
4. Add the remainder of the buffered solution to the column, collect 2 ml of effluent in the same cylinder, rinse the walls of the cylinder, discard the effluent, and shake the cylinder as dry as possible. Collect exactly 25 ml of the column effluent. A distilled water blank and a standard solution must be carried through each column used.
5. As soon as possible, add 0.5 ml of the sulfanilamide solution to the 25-ml sample of effluent from the column and mix well.
6. After 5 min, but not exceeding 8 min, add 0.5 ml of the naphthyl ethylenediamine solution and mix immediately.
7. Between 10 min and 2 h later, measure the extinction coefficient at a wavelength of 543 nm of the solution in a 1-cm (or longer) cell. Use distilled water as a reference. When the extinction coefficient is greater than about 1.2, dilute by exactly one-half with distilled water and remeasure. Remember to double the value obtained from the one-half-strength solution. Path lengths of 1 cm, 5 cm, and 10 cm are appropriate for nitrate concentrations of about 60 to 300, 30 to 60, and 0 to 30 μg NO_3-N/l, respectively.

If the samples of water have a visible natural coloration, a sample blank without the addition of naphthyl ethylenediamine reagent should be processed. The following extinction values then are obtained:

OD_b = absorbance of distilled water plus reagents; the "blank"
OD_o = absorbance of samples without naphthyl ethylenediamine reagent (if brown coloration exists)
OD_s = absorbance of standards or samples plus reagents

8. Intersperse water samples, standards, and blanks. At least three standard solutions should be used and the range of the samples should be bracketed.

9. Prepare a standard curve of OD versus concentration for the corrected standards ($OD_{corr} = OD_s - OD_b$). Sample concentrations then can be read directly from the graph, calculator, or computer. When the standard curve is linear, as it should be, sample concentrations can be calculated by a unit extinction factor (F):

$$F = \frac{\text{standard concentration } (\mu g\ NO_{3^-}\ N/l)}{OD_s \text{ of standard} - OD_b}$$

Then

$$\mu g\ NO_{3^-}\ N/l = F[OD_s \text{ of sample} - (OD_b + OD_o)]$$

10. Reduction efficiency of the Cd-Cu columns can be calculated by determining the extinction coefficient of a 100-μg NO_3-N/l standard and a 100-μg NO_2-N/l standard. When the efficiency is less than 95%, the columns should be repacked.

11. The combined nitrogen measured by this Cd-Cu reduction technique is a combination of NO_3-N and NO_2-N in the sample. The NO_2-N concentrations of the samples must be determined separately without the reduction procedure, as outlined in the next section, and subtracted from the NO_3-N + NO_2-N values determined here, accounting for the reduction column efficiency:

Given $p = NO_3$-N + NO_2-N concentration
   $q = NO_2$-N concentration
   r = efficiency of reduction column, %

Then $\mu g NO_{3^-}\ N/l = p - \left(\dfrac{100}{r}\right)q$

12. Storage of the columns: Pass 100 ml of distilled water through the columns, followed by 50 ml of the ammonium chloride buffer solution. Leave the columns covered with the latter solution and cover the reservoir tightly (e.g., Parafilm®).

Procedures for Nitrite Nitrogen

1. Prepare duplicate aliquots of several nitrite standards with concentrations below, within, and above those of the samples to be analyzed.

2. Place 50-ml samples of distilled water blanks, standards, and samples into graduated cylinders. Add 5.0 ml of the ammonium chloride buffer solution and mix thoroughly.

3. After 5 min, but not exceeding 8 min, add 1.0 ml of naphthyl ethylenediamine solution and mix immediately.

4. Between 10 min and 2 h later, compare the extinction of the solution against a distilled water reference at 543 nm. Path lengths of 1 cm, 5 cm, and 10 cm are appropriate for nitrite concentrations of about 60 to 300, 30 to 60, and 0 to 30 μg NO$_2$-N/liter, respectively.

5. Calculations are identical with those for nitrate nitrogen (section 9 above), but now for nitrite alone:

$$F = \frac{\text{standard concentration } (\mu g \ NO_3^- \ N/l)}{\text{OD}_s \text{ of standard} - \text{OD}_b}$$

and

$$\mu g \ NO_2^- \ N/l = F[\text{OD}_s \text{ of sample} - (\text{OD}_b + \text{OD}_o)]$$

Organic Nitrogen

A large fraction of the total nitrogen in fresh waters may occur as organic nitrogen in particulate, and especially in dissolved, form. While these organic nitrogen compounds occur largely in particulate and dissolved organic detritus and generally are not available to photosynthetic organisms, they represent a major reservoir of nitrogen in aquatic ecosystems. The large pool of dissolved organic nitrogen compounds was formerly believed to be relatively recalcitrant to rapid microbially mediated remineralization. Recent evidence indicates that these compounds can be remineralized relatively rapidly, particularly if exposed to photochemical alteration by UV of natural sunlight.

A number of analytical techniques are available for the determination of particulate and dissolved organic nitrogen. However, many of the modern methods employ elaborate instrumentation (see the "Automated Analyses" section below).

Particulate and dissolved organic nitrogen are usually separated by filtration in all-glass apparatus with glass fiber filters (Whatman GF/F, 0.6–0.7-μm pore size; other sizes, such as GF/C, are *not* acceptable). The fractions then are analyzed by different techniques.

The nitrogen in the particulate matter is converted to ammonia by the classical Kjeldahl method, after treatment with sulfuric acid. The resulting ammonia then is concentrated in a closed system by distillation and titrated or analyzed absorptiometrically by complex formation with ninhydrin without prior distillation (Strickland and Parsons, 1968). Alternatively, the particulate matter is combusted at high temperature in a C-N-S or C-H-N analyzer; in this case the organic nitrogen is converted to free N$_2$ and then analyzed by gas chromatography.

Dissolved organic nitrogen concentrations are more difficult to determine in water (Scudlark et al. 1998). The micro-Kjeldahl technique has been employed [e.g., Stadelmann (1971)], but sensitivity limits often exceed the low concentrations found in many natural waters. More recently organic nitrogen compounds and ammonia have been analyzed by photolytic decomposition and oxidation with high intensity, ultraviolet (UV) radiation (Armstrong and Tibbitts, 1968; Manny et al., 1971; Stainton et al., 1977). Duration of combustion, pH, and adequate oxygen affect the photolytic oxidation and must be controlled carefully. The products of photooxidation then are measured as ammonium, nitrate, and nitrite. When the concentrations of the inorganic nitrogen compounds in the samples prior to UV irradiation are known, an estimate of dissolved organic nitrogen can be made. This method is especially useful when the samples contain low levels of organic nitrogen. (See also the

"Automated Analyses" section below for additional procedures for nitrogen species.)

NITROGEN FIXATION

Investigations during the past two decades have demonstrated that fixation of molecular nitrogen (N_2) by microflora can serve as a major input of nitrogen to aquatic ecosystems under certain conditions [cf., review in Wetzel (1983, 1999)]. Nitrogen fixation is primarily light-dependent in that it requires hydrogen acceptors and adenosine triphosphate (ATP), both of which are generated in photosynthesis. Nitrogen fixation in aquatic systems is restricted almost exclusively to heterocystous cyanabacteria [but see Gallon et al. (1975)] and photosynthetic bacteria.

Several methods have been employed to assay rates of N_2 fixation [reviewed in Hardy et al. (1973)]. While the use of the stable isotope, $^{15}N_2$, in enrichment experiments is preferred, $^{15}N_2$ is expensive, and detection by mass or optical emission spectrometry requires elaborate instrumentation. Another approach is to provide an alternate substrate, such as acetylene, for the active enzyme, nitrogenase. The assay is based on nitrogenase-catalyzed reduction of acetylene (C_2H_2) to ethylene (C_2H_4), gas chromatographic isolation of the two hydrocarbons, and their quantitative measurement by a hydrogen-flame analyzer [e.g., Hardy et al. (1968) and Stewart et al. (1967)]. The C_2H_2-C_2H_4 assay is 10^3 to 10^4 times more sensitive than $^{15}N_2$ methods. Ethylene transformation by methane-oxidizing bacteria can occur in some natural systems and sometimes confounds the method. This metabolism can be evaluated by controls in which the rate of utilization of added ethylene is determined. Since the needed instrumentation for the gas chromatography is now relatively simple, portable, and inexpensive, in situ assays of N_2 fixation are readily possible.

PHOSPHORUS

Intense ecological interest in phosphorus stems from its major role in metabolism in the biosphere. In comparison to the relatively rich supply of other major nutritional and structural components of the biota (C, N, O, S), phosphorus is least abundant and commonly limits biological productivity in aquatic ecosystems.

The cycling of phosphorus is complex. Most of the phosphorus of fresh waters is in the particulate phase of living biota, primarily algae and higher aquatic plants. Labile compounds of low molecular weight are secreted by this particulate phase as a transitory, high-molecular weight colloidal fraction. Part of this colloidal fraction, as well as a portion of the phosphorus of the particulate fraction, is lost from the productive zone by sedimentation, and part is hydrolyzed to soluble orthophosphate. The latter may be assimilated rapidly by the biota; therefore the concentration of orthophosphate at any given time is usually very low in the tropogenic zone of fresh water [cf., Lean (1973)]. For this reason, concentrations of orthophosphate are not very diagnostic for evaluating phosphorus dynamics in aquatic ecosystems. Losses of colloidal and particulate phosphorus are replaced by regeneration of solubilized phosphorus from decomposition; by release of phosphorus from sediments, macrophytes, algae, animals, and bacteria; and by phosphorus contained in precipitation (rain and snow) and in surface and deep drainage to lakes and streams.

Phosphorus occurs in a number of inorganic and organic compounds in both particulate and dissolved forms [see Strickland and Parsons (1968) for a detailed discussion]. Differentiation of forms is based on their reactivity with molybdate, ease of hydrolysis, and particle size. Only two fractions will be discussed here to illustrate certain analytical techniques commonly used to determine phosphorus in aquatic ecosystems. The biological significance of many of the phosphorus compounds is still unclear and under investigation, and current theories are subject to modification.

Soluble Reactive Phosphate-Phosphorus (PO_4-P)

In this procedure, the filtered water sample is allowed to react with a composite reagent of molybdate, ascorbic acid, and trivalent antimony. The molybdic acids formed then are converted by reducing agents to a blue-colored complex [Murphy and Riley (1962)]. While the conditions used are specific for group Va elements (P and As), the method cannot discriminate between phosphate (PO_4^{-3}) and arsenate (AsO_4^{-3}) compounds. Although arsenate occurs in concentrations much below phosphate in most natural waters, the former is a common contaminant from pesticide treatment of terrestrial vegetation. While this method is otherwise specific for PO_4-P, it can hydrolyze labile organic compounds containing phosphorus and hence overestimate biologically available phosphorus.

The method is applicable in a range of about 1 to $500 \mu g$ PO_4-P/l, with a precision level of $\pm 1 \mu g$ PO_4-P/l. If natural concentrations are low, an extraction method is necessary to increase sensitivity [see below and Strickland and Parsons (1968)].

Sample Storage. The water samples should be filtered immediately through glass fiber filters (Whatman GF/F, 0.6–0.7-μm pore size; other sizes, such as GF/C, are *not* acceptable) and stored at 5°C. Analyses must be performed within 2 h, preferably within a half-hour of collection. When the analyses must be delayed for more than a few hours, they should be quick frozen (40% glycol bath at –20°C) in screw-capped polyethylene bottles.

Reagents

1. *Ammonium molybdate solution*: Dissolve 15 g of ammonium paramolybdate $(NH_4)_6Mo_7O_{24}\cdot4H_2O$ in distilled water and dilute to 500 ml. Store in an amber polyethylene bottle.
2. *Sulfuric acid*: Add 140 ml of concentrated sulfuric acid to 900 ml of distilled water. Store in a glass-stoppered bottle.
3. *Ascorbic acid solution*: Dissolve 27 g of L-ascorbic acid in 500 ml of distilled water. Make fresh daily or store solution frozen in a plastic bottle; thaw for use and refreeze immediately. The solution is stable for a few days if refrigerated.
4. *Potassium antimonyl-tartrate solution*: Dissolve 0.34 g of potassium antimonyl-tartrate $[K(SbO)C_4H_4O_6\cdot\frac{1}{2}H_2O]$ in 250 ml of distilled water, warming if necessary. Store in a glass bottle.
5. *Composite reagent*: Mix together 100 ml ammonium molybdate, 250 ml sulfuric acid, 100 ml ascorbic acid, and 50 ml potassium antimonyl-tartrate solutions. Prepare daily for use and discard any excess.
6. *Phosphate standard solution*: Dissolve 0.2197 g of oven-dried (105°C, 24 h) potassium dihydrogen phosphate (KH_2PO_4) in distilled water and dilute to 1 liter. Store in dark bottle with 1 ml of chloroform:

$$1.00\,\text{ml} = 50.0\,\mu\text{g PO}_4\text{-P}$$

Dilute with distilled water as necessary to make appropriate standards.

Procedures for Concentrations Greater than $10\,\mu$g PO$_4$-P/l

1. Obtain water samples of greater than 100 ml and warm, if necessary, to between 15 and 30°C.
2. Because certain waters are colored or turbid, measure the absorbance (OD_{turb}) of a subsample of the water to obtain a turbidity correction.
3. To 100 ml of the sample add 10 ± 0.5 ml of the composite reagent from a 25-ml graduated cylinder; mix thoroughly at once.
4. After 10 min, but within 2 h, measure the extinction coefficient of the solution in a 1- to 10-cm cell at a wavelength of 885 nm. Adjust the photometer to zero reading using distilled water before measuring the extinction coefficient of the sample.
5. Measure the absorbance of a reagent blank (OD_b) (steps 3 and 4 above using distilled water in place of the sample).
6. Subtract the extinction value of the reagent blank and the turbidity-color correction from the value for the sample extinction to obtain a corrected sample extinction:

$$OD_{corr} = OD_s - (OD_b + OD_{turb})$$

7. Prepare a standard curve by determining the absorbance (OD_{std}) of four standard solutions (e.g., 10, 50, 100, and $500\,\mu$g/l) diluted from the stock solution. The standards are treated in the same fashion as the samples except that the turbidity-color correction is unnecessary.

Alternatively, a unit extinction factor (F) for PO$_4$-P can be calculated:

$$F = \frac{\text{standard concentration } (\mu\text{g PO}_4\text{- P}/\text{l})}{OD_{std} - OD_b}$$

and

$$\mu\text{g PO}_4\text{- P}/\text{l} = F[OD_s - (OD_b + OD_{turb})]$$

Procedures for Concentrations Less than $10\,\mu$g PO$_4$-P/l: Alternative A

1. For concentrations of phosphate-phosphorus of less than about $10\,\mu$g P/l, the assay sensitivity may be increased by doubling the sample size and extracting the blue complex with an organic solvent.
2. Samples (ca. 250 ml) should be obtained and stored as discussed above. All glassware used in the following methods should be cleaned thoroughly with concentrated sulfuric acid and rinsed with double-distilled water. When not in use, glassware should be kept covered with 0.1% sulfuric acid-distilled water solution.
3. Pipet 200 ml of the water sample (temperature between 15 and 30°C) into a 250-ml separatory funnel. Add 20 ± 1 ml of the composite reagent from a 25-ml graduated cylinder and mix immediately.
4. After 10 min, but within 2 h, add 28 ml of isobutanol. Stopper tightly and shake vigorously for 60 sec.

5. Allow the funnel to stand for 5 min and separate off the lower aqueous phase.
6. Drain the isobutanol fraction into a clean, dry 10-ml graduated cylinder, allowing 10 sec for it to drain.
7. Bring the volume of the isobutanol to 10 ± 0.1 ml with absolute ethanol and mix.
8. Measure the extinction of the alcoholic extract in a 10-cm absorption cell against isobutanol as a reference blank at a wavelength of 690 nm.
9. Repeat the extractions and measurements using appropriately diluted standards (1 to $10 \mu g$ PO_4-P/l) and reagent blanks with distilled water.
10. Proceed with the computation of a standard curve and calculations of concentrations as in the previous method.

Procedures for Concentrations Less than $10 \mu g$ PO_4-P/l: Alternative B

1. This alternative procedure is identical to Alternative A except that butyl acetate is utilized as an organic solvent to extract the blue complex instead of isobutanol. Butyl acetate is much less soluble than isobutanol, final additions of absolute ethanol are unnecessary, and reproducibility among replicate samples can be improved.
2. Pipet 200 ml of water sample (temperature between 15 and 30°C) into a 250-ml separatory funnel. Add 20.0 ml of the composite reagent and mix thoroughly. Wait about 20 min for color development.
3. Add 10.0 ml of butyl acetate. Stopper tightly and shake vigorously for 5 min.
4. Allow the aqueous and organic phases to separate.
5. Drain and discard the lower aqueous phase. Remove about 5 ml of the butyl acetate phase from the separatory funnel with a pipet, being extremely careful not to include any water droplets.
6. Measure the extinction of the butyl acetate solution with a 1-cm (or 10-cm) cell at a wavelength of 885 nm. Use pure butyl acetate for a blank.
7. Standard solutions are prepared and measured as above, to prepare a standard curve.
8. Since very small concentrations are being measured, reagent and double distilled water contamination must be substracted. Separate analyses using double-distilled water and the reagents yield a blank (OD_b). Then,

$$OD_{corr} = OD_{sample} - OD_b$$

9. Proceed with the computation of a standard curve and calculations of concentrations as in the previous methods.

Dissolved Organic Phosphorus (DOP)

Soluble reactive phosphate is determined before and after irradiation with light of a wavelength less than 250 nm (Strickland and Parsons, 1972; Ridal and Moore, 1990). Samples (3 ml) are placed in fused quartz cuvettes and irradiated for two hours after the addition of $7.5 \mu l$ of 30% H_2O_2 in a Rayonet photochemical reactor equipped with 16 ultraviolet lamps (~21 watts) (Southern N.E. Ultraviolet Co.). Orthophosphate is determined by the ascorbic acid method on a technicon auto-analyzer (Collos and Mornet, 1993; Camarero, 1994; American Public Health Association et al., 1998). Dissolved organic phosphorus is the difference between the irradiated and nonirradiated sample.

Total Phosphorus

In this method the total phosphorus in a water sample is oxidized by persulfate, which liberates organic phosphorus as inorganic phosphate (Menzel and Corwin, 1965). The total phosphate then is determined by the methods described above for inorganic phosphate.

Total phosphorus concentration of an unfiltered water sample minus that of the dissolved PO_4-P fraction, as determined above, approximately equals the concentration of organic phosphorus. The organic component may be separated further into dissolved and particulate fractions by filtration glass fiber filters (Whatman GF/F, 0.6–0.7-μm pore size; other sizes, such as GF/C, are *not* acceptable). Here we shall deal only with the total phosphorus content of the sample.

Any filtration should be done within 1 h of sample collection. When analysis cannot be performed immediately, the filtrate should be frozen at once to –20°C in polyethylene bottles. All analyses should be done immediately once the samples are quick-thawed; do not thaw and refreeze the samples.

Reagents

1. All reagents as used for inorganic phosphate analyses discussed above.
2. *Persulfate solution*: Mix 5% w/v $K_2S_2O_8$ in distilled water; prepare daily.

Procedures

1. Add 16 ml of the 5% persulfate solution to 100-ml samples in 250-ml borosilicate flasks.
2. Place flasks in a boiling water bath for 1 h (or autoclave for $\frac{1}{2}$ h at 15 lbs/in^2 or 1055 g/cm^2).
3. Cool and adjust the volume to 120 ml with accurate 150-ml graduated cylinders.
4. The liberated PO_4-P then is analyzed according to the methods used for inorganic PO_4-P above (use procedures for concentrations > 10 μg PO_4-P/l). When you suspect that you have concentrations < 10 μg P/l, the extraction procedures discussed above can be used to measure the liberated PO_4-P by increasing the sample and reagent volumes appropriately.
5. PO_4-P standards and reagent blanks should be made and submitted to identical boiling and volume adjustment procedures.
6. Proceed with the computation of a standard curve and calculations of sample concentrations.

   Note: See also the "Automated Analyses" section below.

DISSOLVED SILICA

Dissolved silica (SiO_2) usually occurs in moderate abundance in fresh water. Although essentially unionized and relatively unreactive chemically, dissolved silica is assimilated in large quantities by diatoms and a few other algal groups in the synthesis of their cell walls or frustules. Since diatoms are major algal components in many lakes, diatom utilization can modify greatly the concentrations and flux rates of dissolved silica in lakes and streams. The availability of dissolved silica can have a marked influence on the productivity and succession of algal populations [cf., Wetzel (1983, 1999)].

The solubility of particulate amorphous silica is related inversely to pH between approximately 3 and 7 (Marshall, 1964) and increases somewhat at alkaline values to a pH of 9. Carbonic acid in water reacts with silicate materials to form dissolved silica and carbonates. Utilization of dissolved silica in the trophogenic zone of lakes by diatoms often markedly reduces epilimnetic concentrations and contributes to a progressive seasonal succession of diatom and other algal species. Below a concentration of about 0.5 mg SiO_2/l, most diatoms cannot compete effectively with non-siliceous algae.

Silicon in solution as silicic acid (H_4SiO_4) or silicate (SiO_3^{2-}) reacts with acidic ammonium molybdate to form a yellow silicomolybdate complex. This complex then is reduced by sodium sulfite to form the molybdate blue color. The following method is accurate to at least ±0.02 mg/l in the ranges generally encountered in fresh waters (Rainwater and Thatcher, 1960). The method is not applicable where the dissolved silica content is greater than 100 mg/l.

Sample Storage. Samples in clean polyethylene bottles should be filtered to remove particulate matter (nonglass fiber filters, 0.5-μm pore size) immediately after collection. The filtrate is relatively stable, but analyses should be performed without appreciable delay. When water samples are frozen, the dissolved silica may be converted to an unreactive form. However, if thawed samples are allowed to stand for a period of time, the silicon will again become reactive. These reactions are dependent on salinity of the sample (Burton et al., 1970), and evaluation of these effects should be made in each case when samples are stored in the frozen state.

Reagents

1. *Hydrochloric acid, 0.25 N*: Mix 22 ml of concentrated HCl (sp. gr. 1.18) with water and dilute to 1 liter.
2. *Ammonium molybdate, 5%*: Dissolve 52 g $(NH_4)_6Mo_7O_{24} \cdot 4\,H_2O$ in water and dilute to 1 liter.
3. *Disodium EDTA, 1%*: Dissolve 10 g disodium EDTA in water and dilute to 1 liter.
4. *Sodium sulfite, 17%*: Dissolve 170 g Na_2SO_3 in water and dilute to 1 liter.
5. *Silica standard*: Stabilize sodium silicate pentahydrate by placing approximately 3 g $Na_2SiO_3 \cdot 5H_2O$ in a desiccator for 3 h. Dissolve 1.765 g in water and dilute to 1 liter. Dilute 100.0 ml (volumetric pipet) of this intermediate solution with deionized-distilled water to 1 liter; 1.00 ml = 0.050 mg SiO_2.

Procedures

1. Pipet 10.0 ml of each sample into 25- or 50-ml Erlenmeyer flasks.
2. Pipet 10.0 ml of deionized-distilled water into a flask to serve as a blank.
3. Prepare a series of standards as follows using the 0.05-mg SiO_2/ml stock solution; make up to 10.0 ml with deionized-distilled water:

$$0.5\,ml = 0.025\,mg/10\,ml = 2.5\,mg/l$$
$$1.0\,ml = 0.05\,mg/10\,ml = 5.0\,mg/l$$
$$2.0\,ml = 0.10\,mg/10\,ml = 10.0\,mg/l$$
$$3.0\,ml = 0.15\,mg/10\,ml = 15.0\,mg/l$$
$$4.0\,ml = 0.20\,mg/10\,ml = 20.0\,mg/l$$

4. Add 5 ml 0.25 N HCl to each flask; swirl.
5. Add 5 ml 5% ammonium molybdate; swirl.
6. Add 5 ml 1% disodium EDTA; swirl.
7. After 5 min have elapsed following the addition of the molybdate, add 10 ml of 17% sodium sulfite.
8. Mix and allow to stand approximately 30 mins. The color is stable for several hours after this time. (Do not use rubber stoppers to close the flasks.)
9. Using a wavelength of 700 nm and a 1-cm cell, read the absorbance of the samples and standards against the blank.
10. Determine the mg dissolved silica in the samples from a plot of absorbancies of the standards versus their known concentrations.

Note: See also the "Automated Analyses" section below.

SALINITY: MAJOR IONS

The salinity of inland waters is usually dominated completely by four major cations (calcium, Ca^{2+}; magnesium, Mg^{2+}; sodium, Na^+; and potassium, K^+) and the major anions (carbonates, CO_3^{2-} and HCO_3^-; sulfate, SO_4^{2-}; and chloride, Cl^-). The concentrations of these ions generally constitute over 99% of the total salinity. The salinity of the water is variable and is governed by contributions from rock sources of the drainage basin, atmospheric wet and dry deposition, and balances between evaporation and precipitation. Concentrations of Mg, Na, and Cl ions are relatively conservative and exhibit minor spatial and temporal fluctuations within lakes and streams from biotic utilization or biotically mediated changes in the environment. On the other hand, concentrations and forms of Ca, K, inorganic carbon, and SO_4 are highly dynamic and are influenced markedly by metabolic activities.

A number of satisfactory methods exist for the analyses of concentrations of the major ions [cf., Mackereth (1963), Golterman and Clymo (1969), and the American Public Health Association et al. (1998)]. Application of the most appropriate methods depends on the concentrations found in the waters under study and the sensitivity of methods. See the "Automated Analyses" section below.

Carbonates-bicarbonates

The methods for analysis of inorganic carbon are discussed separately in Exercise 8.

Sulfate

Several gravimetric, turbidimetric, complexometric, and potentiometric methods are available for the analysis of sulfate ions in water [cf., Golterman and Clymo (1969) and the American Public Health Association et al. (1998)]. The most reliable methods are thought to be turbidimetric, in which the sulfate ion is precipitated in acidic solution with barium chloride [e.g., Tabatăbai (1974)]. The barium sulfate crystals then are stabilized in suspension. The absorbance of the suspension then is measured spectrophotometrically, and the SO_4 concentration is determined by comparison with known concentrations. At concentrations less than about 2 mg SO_4/l, the turbidimetric method is highly unreliable, and other procedures such as the Thorin method should be used [e.g., Rainwater and Thatcher (1960, p. 279)].

Chloride

Concentrations of chloride ions in natural waters are commonly determined by the titration of chloride with mercuric nitrate in acidic solution to an end point (violet color) where $HgCl_2$ forms a complex with diphenylcarbazone-bromphenol blue mixed indicator [cf., Golterman and Clymo (1969) and the American Public Health Association et al. (1998)].

Note: See the "Automated Analyses" section below.

SPECIFIC CONDUCTANCE

In some chemistry or physics class you undoubtedly observed that a light bulb would not light when an electrical current was passed through a circuit which had two separate electrodes immersed in pure water; no current passed through the water. When salt was added to the water, the electrical current passed and the bulb glowed. The resistance of the water to electrical current was reduced by the addition of ions of salt and the electron flow was increased. Specific conductance (conductivity) of fresh water is based on the same principle. The purer the water (i.e., the fewer the dissolved electrolytes in the water), the greater will be the resistance to electrical current.

By definition, specific conductance of an electrolyte is the reciprocal of the specific resistance of a solution and is expressed in micro-Siemens* per distance ($\mu S/cm$). The amount of current conducted is proportional to the concentration of ions in solution and, hence, to both the concentration and extent of dissociation of the dissolved salts. The standard conductance cell consists of two black-platinum electrodes $1 cm^2$ in area and $1 cm$ apart.

The temperature of the electrolyte affects ionic velocities and must be controlled carefully. Conductance increases about 2 to $3\%/°C$. The international standard of $25°C$ is taken as the standard temperature of conductance measurements. A custom-made water bath is very convenient for routine analyses of conductance (Fig. 7.3). When samples are not maintained at $25°C$ during measurements, an appropriate factor for temperature compensation must be applied (Table 7.1).

It has been found that the specific conductance of water in bicarbonate-dominated lakes and streams is closely proportional to the concentrations of major cations (Ca^{2+}, Mg^{2+}, Na^+, K^+) (Rodhe, 1949; Likens and Johnson, 1968). Once the concentrations of these cations are known for an individual body of water, changes in the specific conductance can be used to estimate the proportional concentrations of the major cations with relatively small error. This relationship does not hold, however, for minor cations and anions and especially is not true for PO_4^{-3} concentrations (Rodhe, 1951).

---

* Also expressed as $\mu mhos/cm$ (reciprocal of ohms).

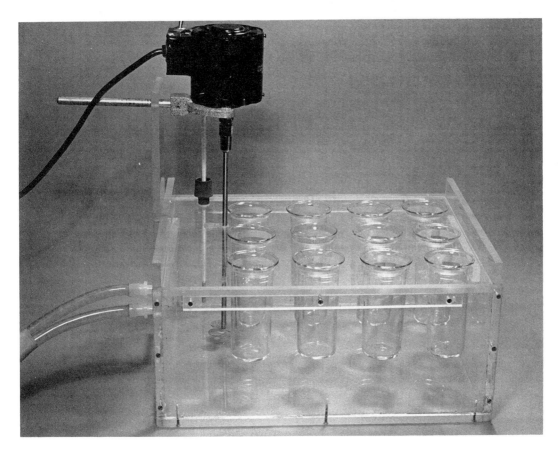

Figure 7.3. A custom-made water bath with stirrer and electrolytic beakers (180 ml) for the measurement of specific conductance.

Table 7.1. Factors for converting specific conductance of water to values at 25°C (based on 0.01 M KCl and 0.01 M NaNO$_3$ solutions).

°C	Factor	°C	Factor	°C	Factor
32	0.89	22	1.06	12	1.30
31	0.90	21	1.08	11	1.33
30	0.92	20	1.10	10	1.36
29	0.93	19	1.12	9	1.39
28	0.95	18	1.14	8	1.42
27	0.97	17	1.16	7	1.46
26	0.98	16	1.19	6	1.50
25	1.00	15	1.21	5	1.54
24	1.02	14	1.24	4	1.58
23	1.04	13	1.27	3	1.62

Table 7.2. Specific conductances of KCl solutions at 25°C.

Concentration (N)	Specific conductance (μS/cm)
0.0001	14.94
0.0005	73.90
0.001	147.0
0.005	717.8
0.01	1,413.0
0.02	2,767
0.05	6,668
0.1	12,900
0.2	24,820

Procedures

1. Place water samples into clean, high-form beakers (or electrolytic high-form beakers). Bring samples to 25 ± 0.2°C in an appropriate water bath or record temperature of each sample at the time of measurement of specific conductance.
2. Immerse conductance cell into the solution, taking precautions to exclude all air bubbles. Measure the resistance in ohms per the manufacturer's instructions for the particular instrument. Rinse the cell in a beaker of distilled water between samples. For samples of very dilute water, rinse the cell with a portion of the water to be measured and discard this rinse water.
3. Repeat the measurement in a 0.00702 N KCl solution; this solution has a specific conductance of 1000 μS/cm at 25°C. [Dissolve 0.5232 g of oven-dried KCl in distilled water and dilute to 1000 ml; equals 0.00702 N.] The conductance of other normalities of KCl solutions at 25°C are given in Table 7.2.
4. Read the conductance directly, if the instrument is so calibrated. If the readings are in resistance (R, in ohms):

$$\text{Specific conductance } (\mu\text{S/cm}) = \frac{(1/R \text{ of } 0.00702\text{N KCl})(1000)}{(1/R \text{ of sample})}$$

Correct for temperature (Table 7.1) to 25°C, if necessary.

TOTAL "DISSOLVED" RESIDUES

Large quantities of particulate or dissolved compounds can render fresh water unsuitable for domestic or industrial use. A simple quantitative method for evaluating the total inorganic and organic solids (residue) content of water involves evaporation to dryness. Although such results are of limited value since they indicate nothing of the qualitative composition of the material, the measurement of total "dissolved" residues (formerly termed total dissolved solids) is still widely used. "Dissolved" is a misnomer since the values include both dissolved and particulate (inorganic and organic) matter.

Total residue includes nonfilterable residue (suspended) and filterable residue (dissolved). It is important to monitor and regulate the temperature at which the sample is dried because variable occlusion of water in different minerals and the volatilization of certain organic compounds can occur at relatively low temperatures

(American Public Health Association et al. 1998). Generally 105°C or 180°C is used to dry the samples.

Procedures

1. Combust porcelain evaporating dishes of 150- to 200-ml capacity in a furnace at 550°C for 1 h. Cool and obtain an accurate tare weight.
2. Filter samples of water through glass fiber filters (Whatman GF/F, 0.6- to 0.7-μm pore size).
3. Measure a volume of the sample filtrate, sufficient to yield between 100 and 250 mg of residues, into the tared evaporating dishes.
4. Dry the samples to a constant weight at either 105°C or 180°C. While still warm, place the dishes into a desiccator to cool.
5. Weigh the samples as soon as they have cooled on an analytical balance accurate to at least 0.5 mg.
6. Report data as total "dissolved" residue on drying at the °C used in mg/l to the nearest whole mg:

$$\text{Total ``dissolved'' residue (mg/l)} = \frac{(\text{mg filterable residue})(1000)}{\text{ml sample}}$$

7. Repeat the analyses on unfiltered samples. Compare these values for *total residues* with the results of total dissolved residues from the filtered samples.

AUTOMATED ANALYSES

Many of the standard procedures discussed above have been adapted for automated analysis. In some cases new methods have been developed to facilitate automation. Here some of the more common approaches to automated analyses will be described briefly. The instruments required for these analyses are very expensive but, if available, would allow samples to be analyzed more rapidly and would reduce random errors associated with contamination, measurement of volumes, and reaction times.

Atomic Absorption and Flame Emission Spectroscopy

Flame Emission Spectroscopy. When a metal atom is subjected to a source of energy such as a flame, electrons in orbit about the nucleus are raised to a higher energy level. As the electron returns to its former conditions, energy is given off in the form of light at a specific wavelength.

To analyze for a particular metal in solution using this method, the solution is aspirated into a high temperature flame. The energy of the flame raises some of the metal atoms to the excited condition; on their return to the ground (neutral) state, the light emitted is resolved by a monochromator and isolated at a specific wavelength. The increase in intensity at this specific wavelength is proportional to the original concentration of metal ion in solution.

Atomic Absorption Spectroscopy. Atomic absorption spectroscopy is an extremely sensitive method for the analysis of metals in solution, based on the capability of atoms to absorb radiation of well-defined (resonant) wavelengths. When a liquid sample is aspirated into a flame, only a small fraction (about one in 10^{10}) of the atoms present ever attain sufficient energy to emit light. The overwhelming majority are in a ground or neutral state. As such, they are capable of absorbing resonant wavelengths of light.

In the atomic absorption spectrophotometer, light from a hollow cathode lamp, the filament of which is made from the element of interest, is passed through a flame and into a spectrometer where a resonant wavelength is isolated. When a sample is atomized into the flame, the atoms in ground state absorb this wavelength and reduce its intensity. The degree of absorption is proportional to the concentration of absorbing atoms in the flame. Specificity is assured since an atom will absorb only at its resonant wavelength.

Interference is a common problem. For example, PO_4 ions will bind with calcium to form $Ca_3(PO_4)_2$, which is a very stable molecule. To overcome the bonding between calcium and the phosphate, either a high temperature flame is required or some type of preferential binding agent must be added. In the case of calcium, lanthanum ($LaCl_3$) commonly is added to the solution. In this case the PO_4 ions will be freed to bind preferentially with the lanthanum and the chloride with calcium. $CaCl_2$ is readily dissociated by a low energy flame, and the preponderance of ground state atoms permits the analysis of calcium in minute concentrations. A high energy flame is not satisfactory in this case because ionization of Ca will occur and reduce the sensitivity.

Atomic absorption spectroscopy is useful for the quantitative analysis of calcium, magnesium, potassium, sodium, manganese, aluminum, and a number of trace metals. Sodium is particularly well suited to analysis by flame emission. Several analytical instruments have the capability to analyze by both atomic absorption and flame emission spectroscopy. Both methods of analysis require a working (standard) curve in which absorbance or emission is plotted against concentration for known values. Standards should be prepared in the same solvent as the unknown [see Beaty (1978)].

Inductively Coupled Plasma Emission Spectroscopy (ICP). ICP is based on the measurement of specific wavelengths of light from excited atoms by optical spectroscopy. "Samples are nebulized to produce an aerosol. The aerosol is transported by an argon carrier stream to an inductively coupled argon plasma, which is produced by a radio frequency generator. In the plasma (which is at a temperature of 6,000 to 10,000°K), the analytes in the aerosol are atomized, ionized, and excited. The excited ions and atoms emit light at their characteristic wavelengths. The spectra from all analytes are dispersed by a grating spectrometer, and the intensities of the lines are monitored by photomultiplier tubes" [from Shugar et al. (1981)]. As a result, numerous analytes can be measured in a single sample.

The ICP is used commonly for measuring major and trace metal concentrations (e.g., Ca, Mg, Fe, and Mn) as well as SiO_2 [e.g., Willard et al. (1988) and Shugar et al. (1981)].

Ion Chromatography

Ion chromatography (IC) is an ionic separation technique that depends on the affinity of an ion for an exchange site. Ions in a liquid passing through an exchange column will have different rates of migration through these columns because of their different affinities for exchange sites. Therefore each ion can be identified in the eluent by conductimetric detection of its retention time.

Both anions and cations can be measured with IC [e.g., Gjerde and Fritz (1987) and Small et al. (1975)], but IC is particularly valuable for the analysis of ionic sulfur and nitrogen compounds.

High-Performance Liquid Chromatography (HPLC)

In high-performance liquid chromatography (HPLC) liquid samples are forced under high pressure through a column packed with an appropriate stationary phase. Components separated from the liquid by interaction with the stationary phase emerge from the column in the order of their reaction, i.e., the least retarded component elutes first. Several different detectors can be used to quantify elutants, including fluorescence and electrochemical.

HPLC is used commonly to measure NH_4, NO_2, NO_3, SO_4, plant pigments, and high molecular weight organic compounds [e.g., Lodge (1989) and Snyder and Kirkland (1986)].

Continuous Flow Spectrophotometric Procedures

Continuous Flow Analysis (CFA) and Flow Injection Analysis (FIA) bring sample and reagents together under carefully controlled, automated conditions. With CFA, such as an Auto-Analyzer (AA), the chemical reactions occur in continuously flowing, air-segmented streams of liquid (sample plus reagents). In FIA, the sample and reagents are mixed by laminar flow in narrow-bore tubing without air segmentation (Valcárcel and Luque de Castro 1987; Ruzicka and Hansen 1988; Karlberg and Pacey 1989). Color formation resulting from the chemical reactions is measured with a spectrophotometer.

These methods are used commonly for the measurement of NO_2^-, NO_3^-, NH_4^+, PO_4^{3-}, dissolved Si, and total dissolved N [e.g., Shugar et al. (1981); Willard et al. 1988; McKelvie et al. 1995; Robards et al. 1994)].

ION SELECTIVE ELECTRODES

Much attention has been given in recent years to the use of ion selective electrodes for measuring the concentrations or activities (or apparent concentrations) of ions in solutions. The principle of ion selective electrodes is identical to that of pH electrodes: In a glass pH electrode the thin glass bulb acts as a membrane specific for hydrogen ions. Other ion specific electrodes function similarly as electrochemical cells, where two solutions containing the same ion in differing concentrations are separated by a membrane sensitive to this ion. A voltmeter or a potentiometer is used to measure an electrical potential generated across the active surface of the membrane. The difference in potential between the internal electrode in the sample solution and a reference electrode is related exponentially to the activity of the selected ion in the sample solution. Detailed discussions of construction and membrane properties of ion selective electrodes are given in Durst (1969), Carson and Keeney (1971), and especially Whitfield (1971).

Electrode response is a function of ion activity, not of ion concentration. The membrane potential is logarithmically related to the measure of ionic activity by the Nernst equation. The Nernst equation modified for membrane electrodes is

$$E = E' + S \log \frac{A_s}{A_i}$$

where E is the net membrane potential, S is an empirical slope obtained from calibration, A_s and A_i are the ion activities in the sample solution and internal solution of the membrane electrode, respectively, and E' is the potential under known conditions of A_s and A_i. When the electrode gives 100% Nernstian response, then the slope is

$$S = 2.303 \frac{RT}{nF}$$

where R is a gas constant, T is the absolute temperature, n is the valence of the ion, and F is the Faraday constant.

The ionic strength of an aqueous solution is a quantity that describes the intensity of electrostatic interactions and can be used to find the thermodynamic activity from measured concentrations. Ionic strength (I_s) is defined as

$$I_s = \tfrac{1}{2} \sum_i C_i \cdot Z_i^2$$

where C = concentration of ion i and Z = charge of ion i. The ionic strength increases as the solutions become more concentrated and the activity of the given ion decreases. At very high concentrations ($\gtrsim 1\,M$), the activity increases with an increase in ionic strength [cf., Butler (1964)].

The relationship between ion activity and ion concentration is, using the hydrogen ion as an example:

$$(H^+) = \gamma [H^+]$$

where (H^+) = hydrogen ion activity, γ = the activity coefficient of H^+, and $[H^+]$ = hydrogen ion concentration. The activity coefficient γ of an ion in dilute solutions [<0.005 M; for solutions of higher ionic strength, an extended form of the Debye-Hückel equation should be used; cf., Butler (1964)] is given by the Debye-Hückel equation as:

$$\log \gamma = -AZ^2 (I_s)^{1/2}$$

where $A = 0.509$ at 25°C (a constant for all aqueous solutions), Z = charge of ion i, and I_s = ionic strength.

While ionic activity is related to ion concentration, ionic strength and complex ion formation affect activity coefficients. Linearity cannot be assumed, even if total ionic strength is controlled. Ion complexing, in addition to the effects of ionic strength, tends to reduce activity and must be treated separately in calculating relationships between activity and concentration [cf., Garrels and Christ (1965)]. Nevertheless, ion selective electrodes will measure the free ion activity in spite of problems of ionic strength and formation of ion complexes. For certain applications, measures of ion activity provide valuable information that cannot be obtained by other analytical methods. For example, measurements of calcium ion concentrations in hard waters can be confounded by particulate and colloidal $CaCO_3$ suspensoids, both of which are solubilized upon acidification for routine analysis by atomic absorption spectroscopy. Ionic activity permitted differentiation of those portions of total calcium in ionic and combined forms (White and Wetzel, 1975). Cation exchange equilibria in sediment interstitial water can be determined by ionic activity and are important to plant assimilation analyses.

Membranes differ in construction for different ions. Various electrodes employ liquid ion exchangers held by membrane discs, solid ion exchange polymer membranes, crystalline solid-state membranes, and precipitates impregnated in a matrix such as silicone rubber. Practically none of the electrodes is completely specific, and all are subject to interference by other ions.

EXERCISES

OPTION 1. FIELD TRIPS

1. From the central depression of a lake or at several stations in, above, and below tributary inlets to a stream, collect water samples for chemical analyses with a scrubbed Van Dorn or similar water sampler. Dispense with minimal agitation and aeration into clean, amber polyethylene bottles. Fill the bottles to the brim and cap tightly. Store samples in an ice chest for transport to the laboratory. Collect replicates at each depth.
2. Determine the temperature of the water at each depth throughout the vertical profile with an underwater thermometer (see Exercise 2).
3. Collect water samples for analyses of concentrations of dissolved oxygen throughout the water column using the procedures in Exercise 6.
4. Collect water samples for chemical analyses from (a) inlet sources and (b) the outlet of a lake, and in the littoral zone among dense stands of aquatic macrophytes. Compare values obtained from these sources with those from comparable depth in the center of the lake (open water).
5. Return to the laboratory as promptly as possible and perform the chemical analyses as discussed above for:
   a. Ammonium nitrogen.
   b. Nitrate and nitrite nitrogen.
   c. Soluble reactive phosphorus.
   d. Total phosphorus (particulate and dissolved fractions).
   e. Dissolved silica.
   f. Major ions, if equipment is available.
   g. Specific conductance.
   h. Total "dissolved" residues.

Depending on the amount of time available for such analyses, the content of these exercises may require two or more class periods. Once you are familiar with the methods, the work load can be divided easily among teams to perform all of the analyses simultaneously.

OPTION 2. LABORATORY EXERCISES

1. From the water samples provided by your instructor, perform the chemical analyses as discussed earlier in this exercise for:
   a. Ammonium nitrogen.
   b. Nitrate and nitrite nitrogen.
   c. Soluble reactive phosphate.
   d. Total phosphorus (particulate and dissolved fractions).
   e. Dissolved silica.
   f. Major ions, if equipment is available.

g. Specific conductance.

h. Total "dissolved" residues.

Work in teams to perform the analyses simultaneously, then rotate to learn the techniques of all analyses.

Questions

1. In the vertical profile of a stratified lake, you may observe an inverse relationship between concentrations of nitrate and ammonium nitrogen. What are the probable biotic factors responsible for such a relationship?

2. Nitrate and ammonium concentrations are often very low in the epilimnion. Why? What alternatives do phytoplanktonic algae have for a source of inorganic nitrogen under these conditions?

3. Nitrite concentrations are usually very low in natural waters. Why?

4. What are potential pathways for loss and transformation of nitrogen from and within lakes? List all of the sources for nitrogen you can think of for a lake ecosystem. How is nitrogen lost from the ecosystem?

5. What does the concentration of soluble reactive phosphate mean in terms of utilization by photosynthetic organisms?

6. If concentrations of soluble reactive phosphate were very low, would this mean that algae are likely to be phosphate-limited? Why?

7. What alternative phosphate sources do rooted aquatic plants have?

8. What conclusions can you draw from your data on the differences between concentrations of total phosphorus?

9. In stratified lakes, phosphorus concentrations commonly increase in the hypolimnion. What mechanisms are likely to cause this pattern? Why?

10. Were there any relationships between your observed concentrations of phosphorus and those of dissolved oxygen?

11. In stratified lakes, concentrations of dissolved silica commonly are lower in the epilimnion and higher in the hypolimnion. Discuss probable causes for such relationships.

12. The calcium concentrations of stratified, hard water lakes often are reduced markedly in the epilimnion but increase in the hypolimnion. Why do you think this stratification occurs?

13. Magnesium and sodium concentrations are relatively "conservative" in lakes. What is meant by this statement and why?

14. Discuss your observed measurements of specific conductance. If an increase were observed with increasing depth, what would this indicate?

15. Of what potential value are measurements of total "dissolved" residues if the water were to be used for domestic purposes?

Apparatus and Supplies

1. Ammonium nitrogen:
   a. 50-ml graduated cylinders.
   b. Vortex mixer.
   c. Reagents: Na_3PO_4 (or K_3PO_4), phenol, NaOH, nitroprusside, sodium hypochlorite (fresh commercial bleach), and NH_4Cl.
   d. Pipets and propipets.
   e. Spectrophotometer(s), with 1-, 4-, or 5-, and 10-cm path length cells.
2. Nitrate-nitrite nitrogen:
   a. Cadmium reduction columns (see Fig. 7.2).
   b. Cadmium filings, cupric sulfate, 2N HCl, and glass wool.
   c. Reagents: NH_4Cl, $Na_2B_4O_7$, disodium dihydrate EDTA, sulfanilamide, N-(1-naphthyl) ethylenediamine dihydrochloride. KNO_3, and $NaNO_2$.

d. 50-ml graduated cylinders, pipets, and propipets for distribution.

e. Spectrophotometer(s), with 1-, 4-, or 5-, and 10-cm path length cells.

3. Phosphorus:
   a. Clean (nonphosphate containing detergents!) 130-ml screw-capped polyethylene bottles.
   b. Sulfuric acid (sp. gr. 1.82), ammonium paramolybdate, L-ascorbic acid, potassium anti-monyl-tartrate, potassium dihydrogen phosphate, isobutanol, absolute methanol, and potassium persulfate.
   c. Volumetric pipets, graduated cylinders (10 and 25 ml), separatory funnels (250 ml), and borosilicate glass flasks.
   d. Spectrophotometer (1-, 4- or 5-, and 10-cm cells).
   e. Water bath capable of boiling, or autoclave.

4. Dissolved silica:
   a. Nonglass microfilters, $0.5\text{-}\mu m$ pore size.
   b. 25- or 50-ml Erlenmeyer flasks; pipets.
   c. Reagents: hydrochloric acid (0.25 N), ammonium paramolybdate, disodium EDTA, sodium sulfite, and sodium silicate pentahydrate.
   d. Spectrophotometer, 1-cm cells.

5. Specific conductance:
   a. Conductance bridge and meter (e.g., Yellow Springs Instrument Co., Model 31, Yellow Springs, Ohio).
   b. Water bath for 25°C (e.g., Fig. 7.3) for temperature measurement apparatus.
   c. Potassium chloride for standard solutions.

6. Total "dissolved" residues:
   a. Evaporating dishes, porcelain, 150- to 200-ml capacity.
   b. Filtration apparatus, glass fiber filters.
   c. Drying oven.
   d. Analytical balance.

References

American Public Health Association et al. 1998. Standard Methods for the Examination of Water and Wastewater. 20th Ed. American Public Health Association, Washington, D.C. 1183 pp.

Armstrong, F.A.J. and S. Tibbitts. 1968. Photochemical combustion of organic matter in seawater for nitrogen, phosphorus and carbon determination. J. Mar. Biol. Assoc. U.K. *48*:143–152.

Beaty, R.D. 1978. Concepts, Instrumentation and Techniques in Atomic Absorption Spectrophotometry. Perkin-Elmer Corp., Norwalk, CT.

Burton, J.D., T.M. Leatherland, and P.S. Liss. 1970. The reactivity of dissolved silicon in some natural waters. Limnol. Oceanogr. *15*:463–476.

Butler, J.N. 1964. Ionic Equilibrium. A Mathematical Approach. Addison-Wesley, Reading, Mass. 547 pp.

Camarero, L. 1994. Assay of soluble reactive phosphorus at nanomolar levels in nonsaline waters. Limnol. Oceanogr. *39*:707–711.

Carlson, R.M. and D.R. Keeney. 1971. Specific ion electrodes: Techniques and uses in soil, plant, and water analysis. pp. 39–65. *In*: L.M. Walsh, Editor. Instrumental Methods for Analysis of Soils and Plant Tissue. Soil Sci. Soc. America, Madison, Wis.

Collos, Y. and F. Mornet. 1993. Automated procedure for determination of dissolved organic nitrogen and phosphorus in aquatic environments. Mar. Biol. *116*:685–688.

Durst, R.A. (ed). 1969. Ion selective electrodes. Spec. Publ. National Bureau Standards, Washington, D.C. *314*. 452 pp.

Gallon, J.R., W.G.W. Kurz, and T.A. LaRue. 1975. The physiology of nitrogen fixation by a *Gloeocapsa* sp. pp. 159–173. *In*: W.D.P. Stewart, Editor. Nitrogen Fixation by Free-living Micro-organisms. Academic, New York.

Galloway, J.N., J. Cosby, and G.E. Likens. 1979. Acid precipitation: The measurement of pH and acidity in acid precipitation. Limnol. Oceanogr. *24*:1161–1165.

Garrels, R.M. and C.L. Christ. 1965. Solution, Minerals, and Equilibria. Harper and Row, New York. 450 pp.

Gjerde, D.T. and J.S. Fritz. 1987. Ion Chromatography. 2nd Ed. Dr. Alfred Huthig Verlag, Berlin.

Golterman, H.L. and R.S. Clymo (ed). 1969. Methods for Chemical Analyses of Fresh Waters. IBP Handbook No. 8. Blackwell, Oxford. 172 pp.

Hardy, R.W.F., R.C. Burns, and R.D. Holsten. 1973. Applications of the acetylene-ethylene assay for measurement of nitrogen fixation. Soil Biol. Biochem. 5:47–81.

Hardy, R.W.F., R.D. Holsten, E.K. Jackson, and R.C. Burns. 1968. The acetylene-ethylene assay for N_2 fixation: Laboratory and field evaluation. Plant Physiol. *43*:1185–1207.

Harwood, J.E. and A.L. Kühn. 1970. A colorimetric method for ammonia in natural waters. Water Res. *4*:805–811.

Karlberg, B. and G.E. Pacey. 1989. Flow Injection Analysis—A Practical Guide. Elsevier, Amsterdam. 372 pp.

Lean, D.R.S. 1973: Phosphorus dynamics in lake water. Science *179*:678–680.

Likens, G.E. and P.L. Johnson. 1968. A limnological reconnaissance in interior Alaska. U.S. Army Cold Regions Research and Engineering Laboratory, Research Rept. *239*. Hanover, N.H. 41 pp.

Lodge, J.P., Jr. (ed). 1989. Methods of Air Sampling and Analysis. 3rd Ed. Lewis Publ. Chelsea, Mich.

Mackereth, F.J.H. 1963. Some Methods of Water Analysis for Limnologists. Sci. Publ. Freshw. Biol. Assoc. U.K. *21*. 71 pp.

Manny, B.A., M.C. Miller, and R.G. Wetzel. 1971. Ultraviolet combustion of dissolved organic nitrogen compounds in lake waters. Limnol. Oceanogr. *16*:71–85.

Marshall, C.E. 1964. The Physical Chemistry and Minerology of Soils. Vol. I. Soil Materials. Wiley, New York. 388 pp.

McKelvie, I.E., D.M.W. Peat, and P.J. Worsfold. 1995. Techniques for the quantification and speciation of phosphorus in natural waters. Analytical Proceedings *32*:437–445.

Menzel, D.W. and N. Corwin. 1965. The measurement of total phosphorus in seawater based upon the liberation of organically bound fractions by persulfate oxidation. Limnol. Oceanogr. *10*:280–282.

Murphy, J. and J.P. Riley. 1962. A modified single solution method for the determination of phosphate in natural waters. Anal. Chim. Acta *27*:31–36.

Rainwater, F.H. and L.L. Thatcher. 1960. Methods for Collection and Analysis of Water Samples. U.S. Geol. Surv. Water-Supply Pap. *1454*. 301 pp.

Ridal, J.J. and R.M. Moore. 1990. A re-examination of the measurement of dissolved organic phosphorus in seawater. Mar. Chem. *29*:19–31.

Robards, K., I.D. McKelvie, R.L. Benson, P.J. Worsfold, N. Blundell, and H. Casey. 1994. Determination of carbon, phosphorus, nitrogen, and silicon species in waters. Anal. Chim. Acta *287*:143–190.

Rodhe, W. 1949. The ionic composition of lake waters. Verh. Int. Ver. Limnol. *10*:377–386.

Rodhe, W. 1951. Minor constituents in lake waters. Verh. Int. Ver. Limnol. *11*:317–323.

Ruzicka, J. and E.H. Hansen. 1988. Flow Injection Analysis. John Wiley and Sons, New York. 498 pp.

Scudlark, J.R., K.M. Russell, J.N. Galloway, T.M. Church, and W.C. Keene. 1998. Organic nitrogen in precipitation at the mid-Atlantic U.S. coast—Methods evaluation and preliminary measurements. Atmos. Environ. *32*:1719–1728.

Shugar, G.J., R.A. Shugar, L. Bauman, and R.S. Bauman. 1981. The Chemical Technician's Ready Reference Handbook. McGraw-Hill, New York.

Small, H., T.S. Stevens, and W.C. Bauman. 1975. Novel ion exchange chromatographic method using conductimetric detection. Anal. Chem. *47*:1801–1804.

Snyder, L.R. and J.J. Kirkland. 1986. An Introduction to Modern Liquid Chromatography. 3rd Ed. Wiley, New York.

Solorzano, L. 1969. Determination of ammonia in natural waters by the phenolhypochlorite method. Limnol. Oceanogr. *14*:799–801.

Stadelmann, P. 1971. Stickstoffkreislauf and Primärproduktion im mesotrophen Vierwald-stättersee (Horwer Bucht) und im eutrophen Rotsee, mit besonderer Berüchsichtigung des Nitrats als limitierenden Faktors. Schweiz. Z. Hydrol. *33*:1–65.

Stainton, M.P., M.J. Capel, and F.A.J. Armstrong. 1977. The Chemical Analysis of Fresh Waters. 2nd Ed. Misc. Spec. Publ. Dept. Environment Canada *25*. 166 pp.

Stewart, W.D.P., G.P. Fitzgerald, and R.H. Burris. 1967. *In situ* studies on N_2 fixation using the acetylene reduction technique. Proc. Nat. Acad. Sci. U.S.A. *58*:2071–2078.

Strickland, J.D.H. and T.R. Parsons. 1968. A Practical Handbook of Seawater Analysis. Bull. Fish. Res. Bd. Canada *167*. 311 pp.

Strickland, J.D.H. and T.R. Parsons. 1972. A Practical Handbook of Seawater Analysis. 2nd Edition. Bull. Fish. Res. Board Can. 167. 311 pp.

Tabatǎbai, M.A. 1974. Determination of sulfate in water samples. Sulfur Inst. J. *10*:11–13.

Valcárcel, M. and M.D. Luque de Castro. 1987. Flow Injection Analysis—Principles and Applications. Ellis Horwood Ltd., Chichester. 400 pp.

Van Dorn, W.G. 1956. Large-volume water samplers. Trans. Amer. Geophys. Union *37*:682–684.

Wetzel, R.G. 1983. Limnology. 2nd Ed. Saunders Coll. Philadelphia. 860 pp.

Wetzel, R.G. 1999. Limnology: Lake and River Ecosystems. 3rd Ed. Academic Press, San Diego (in press).

White, W.S. and R.G. Wetzel. 1975. Nitrogen, phosphorus, particulate and colloidal carbon content of sedimenting seston of a hard-water lake. Verh. Int. Ver. Limnol. *19*:330–339.

Whitfield, M. 1971. Ion Selective Electrodes for the Analysis of Natural Waters. Handbook 2, Australian Marine Sciences Association, Sydney. 130 pp.

Willard, H.H., L.L. Merritt, Jr., J.A. Dean, and F.A. Settle, Jr. 1988. Instrumental Methods of Analysis. Wadsworth, Belmont, Calif.

Wood, E.D., F.A.J. Armstrong, and F.A. Richards. 1967. Determination of nitrate in sea water by cadmium-copper reduction to nitrite. J. Mar. Biol. Assoc. U.K. *47*:23–31.

The Inorganic Carbon Complex: Alkalinity, Acidity, CO_2, pH, Total Inorganic Carbon, Hardness, Aluminum

Precipitation falling on the surface of the Earth as rain or snow contains a variety of gases. Aerosols or dust particles also may be dissolved or picked up from the air. Thus, by the time the water reaches the Earth's surface, it is no longer pure. As it flows over or penetrates into plants or the ground, water dissolves more gases, notably carbon dioxide, and various mineral substances with which it comes into contact. When CO_2 dissolves in water, carbonic acid is formed:

$$H_2O + CO_2 \rightleftarrows H_2CO_3 \rightleftarrows H^+ + HCO_3^-$$

and pH is lowered.

Most minerals are only slightly soluble in water, and appreciable amounts usually are not dissolved. For example, limestone (calcium carbonate, $CaCO_3$) is but slightly soluble in pure water. Limestone is much more soluble, however, in water containing carbonic acid, whereby the $CaCO_3$ is changed to calcium bicarbonate:

$$CaCO_3 + H_2CO_3 \rightleftarrows Ca^{2+} + 2HCO_3^-$$
(insoluble)

The dissolved CO_2 has a marked effect on the properties of the water. It forms a weak, carbonic acid solution that can change the pH of the water and dissolve minerals, which in turn can increase alkalinity and impart hardness to the water. The amount of dissolved salts in water is of major importance to the maintenance of life and in the treatment of water for domestic and industrial use.

In addition to bicarbonates, carbonates, and hydroxides, other minerals that often dissolve in water in moderate amounts are silica and the chlorides, sulfates, and nitrates of calcium, magnesium, sodium, and potassium. With increased emission of sulfur dioxide (SO_2) into the atmosphere from industrial activities, SO_4^{2-} is becoming the dominant anion in precipitation of large geographical areas. In those areas the geochemical relationships as described above may be altered appreciably.

Inorganic carbon as dissolved CO_2 and HCO_3^- is the primary source of carbon for photosynthesis by algae and larger aquatic plants in natural waters. This utilization is balanced by respiratory production of CO_2 by most organisms and by influxes of CO_2 and HCO_3^- from incoming water and from the atmosphere. The amounts of inorganic carbon available for use in photosynthesis are adequate in most natural fresh waters; only under special conditions of soft waters and intensely productive situations does inorganic carbon become a limiting factor to photosynthesis of planktonic algae.

Natural waters exhibit wide variations in this relative acidity and alkalinity, not only in actual pH values, but also in the total amount of dissolved material producing the acidity or alkalinity. The concentrations of these compounds and the ratios of one to another determine the observed pH and the capacity for buffering of a given body of water. The lethal effects of most acids appear when pH < 5.5 and of most alkalis near pH 9.5, although the tolerances of many organisms are considerably more restricted within these pH extremes. Thus, the buffering capacity of natural waters to resist changes in pH can be of great importance to the maintenance of life.

pH

Ions in solution are capable of conducting an electrical current. Even pure water will conduct a current to a slight degree because a small number of water molecules dissociate into ions:

$$H_2O \rightleftarrows H^+ + OH^-$$

In this case an equal number of hydrogen and hydroxylions is formed. Thus the water is neither acidic nor alkaline, but is neutral.

By careful measurements it has been found that, when pure water at 25°C ionizes, 0.0000001 g of H^+ is liberated per liter. Since 1 mole of protons weighs 1 g, the concentration of protons in pure water is also 1×10^{-7} mol/l, which is conveniently described on a negative logarithmic scale simply as 7. The pH scale is a series of numbers ranging from 0 to 14 that denote various degrees of acidity or alkalinity. Values below 7 and approaching 0 indicate increasing acidity, while values from 7 to 14 indicate increasing alkalinity. Mathematically,

$$pH = \log \frac{1}{H^+ \text{ concentration}} \text{ and } \log \frac{1}{10^{-7}} = 7$$

the pH of pure water with a H^+ concentration of 10^{-7} at 25°C. By Le Chatelier's Principle:

$$\frac{(H^+ \text{ concentration}) \times (OH^- \text{ concentration})}{H_2O \text{ concentration}} = \text{a constant}$$

Because the concentration of H_2O is very large relative to the equilibrium concentration of H^+ and OH^-, it may be taken as a constant, and the equation becomes $[H^+][OH^-] = $ a constant with a value of 10^{-14}. According to this equation, if the H^+ concentration were to increase, the OH^- concentration would decrease proportionally, and vice versa. With an equal concentration of each ion, the solution is neutral and its pH = 7. If OH^- ions (a base) were added to such a solution, there would be a decrease in the H^+ ion concentration, the solution would become more alkaline, and the pH value would increase. Similarly, if H^+ ions (an acid) were added to a neutral solution, the OH^- concentration would decrease, the solution would become more acidic, and the pH value would decrease.

The farther the pH value is from the neutral point of 7, the greater is the concentration of either H^+ or OH^- ions. Since these are logarithmic values, each integer represents a H^+ concentration ten times that of the next higher number. A pH of 2 represents 0.01 mol of H^+ ions per liter, ten times the H^+ concentration at pH 3 (0.001 mol H^+/1). The "p" of pH refers to the potential (puissance) of the hydrogen ion activity. Therefore, by definition, the average of pH values cannot be determined by simple arithmetic, but must be calculated from the antilogarithms (concentration units).

For very dilute solutions (<0.001 M), activity coefficients approximate unity, and thus activity approximates concentration (see Exercise 7, pp. 105–106). Rigorously, however, pH should be defined as:

$$pH = -\log[H^+]\gamma$$

where γ = the activity coefficient for hydrogen ion (H^+).

Measurement of pH

Ostensibly, pH is one of the easiest measurements to make; in fact, it is extremely difficult to obtain accurate measurements, particularly under field conditions. Measurement of pH in limnology can be made either by color reactions of organic compounds (pH indicators) or by electrometric methods (pH meters). Since electrometric pH meters are available so commonly and permit much more reliable measurements than do color indicators, electrometric measurements should be used whenever possible.

Method 1: Colorimetric (Indicators). Certain organic compounds change color as the pH of the surrounding medium varies. If a given indicator were added to a series of buffer solutions of known pH values, a series of color standards would be formed for each pH. The pH of an unknown sample would be estimated by adding the indicator and comparing the color with the standards. Each of the commonly used indicators has a limited pH range; therefore, several sets of standards must be prepared, each with a different indicator. Details of common methods employed are given in Golterman and Clymo (1969). Such methods are much less accurate and convenient than electrometric measurements and, if more than a few measurements were to be made, would be more expensive when all costs are considered. Colorimetric measures are not suitable for waters of low buffer capacity nor for waters containing colored materials such as dissolved organic compounds.

Various sets of liquid color standards (sealed in glass) or stained glass are available for comparison to water samples treated with organic indicators. Various test papers, impregnated with a series of colored organic dyes. permit estimation of pH by reaction with a water sample and comparison to a standard color sequence.

Method 2: Electrometric (pH Meters). The pH of a sample should be measured by an electrometer connected to two electrodes (or one electrode in which both are combined) that are immersed in the solution. The electrode system consists of two components (Fig. 8.1). The glass electrode is sealed; the sensitive part is on the bottom. Extreme care must be exercised to avoid scratching this sensitive surface. Abrasions to this area cause sluggish, variable responses. The glass electrode is in contact with a layer of water adsorbed to its surface. The electrode potential is a function of the H^+ concentrations in the water and in the electrolyte within the electrode. This potential varies by 59.16 mv per pH unit (25°C). The electrical potential developed at this electrode depends on H^+ ion concentration. Therefore, sufficient time must be allowed for ions of this layer to come into equilibrium with those of the sample water. In dilute waters this equilibrium takes a relatively long time to

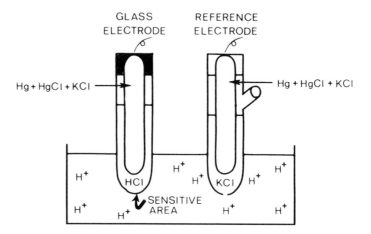

Figure 8.1. Electrode system employing the glass electrode for measurement of pH.

be reached, particularly if the electrode were stored in a strong buffer solution prior to use.

Often a calomel electrode is used as a reference electrode, which forms an electrical connection in the solution independently of the composition of the solution. The connection is by a liquid to liquid junction through a porous channel near the bottom of the electrode. The internal electrolyte (generally a saturated KCl solution) flows out of this opening. Therefore, the electrolyte must be replenished occasionally and the port through which the KCl solution is refilled be left open as a vent during operation.

The glass electrode discriminates H^+ ions and the calomel electrode completes the electrical cell by allowing a constant stream of saturated KCl to flow to the solution through a small pore in the base of the electrode. The pH of water samples measured by this method varies with temperature and pressure. Thus measurements made in the laboratory must be corrected for in situ temperature and pressure. Corrections for hydrostatic pressure are usually very small and may be considered negligible in most studies.

Combination electrodes have both a glass electrode and a reference electrode in a single probe. Although expensive, the combination electrode is very convenient in that it can be placed directly into small samples without appreciable disturbance.

Electrodes should be stored in distilled water at all times; a dry electrode must be soaked in water at least 24h before use. If the reference electrode becomes blocked or the glass electrode becomes fouled, the following separate techniques of cleaning, each of increasing severity, may be taken:

1. Immerse in hot distilled water.
2. Soak in 0.1 N HCl for 1h or more.
3. Immerse in 2% hydrofluoric acid for a few seconds and rinse immediately with water and then 0.1 N HCl to remove traces of fluoride.

Before use the pH meter should be standardized with a pH 7 buffer solution and the slope of the electrode response should be adjusted, typically with a pH 4 buffer. The electrodes should be rinsed well before being inserted into a sample. The meter should stabilize on the sample pH quickly, before the exchange of CO_2 with the atmosphere alters the pH.

Figure 8.2. Relation between pH and the relative proportions of inorganic carbon species of CO_2, HCO_3^-, and CO_3^{2-} in solution. [From Wetzel (1999).]

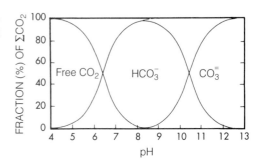

ALKALINITY

Historically, the term alkalinity referred to the buffering capacity of the carbonate system in water. Today, alkalinity is used interchangeably with acid neutralizing capacity (ANC), which refers to the capacity to neutralize strong acids such as HCl, H_2SO_4, and HNO_3. Alkalinity in water is due to any dissolved species (usually weak acid anions) that can accept and neutralize protons. Because CO_2 is relatively abundant in water in gaseous and dissolved form, and carbonates are common as primary minerals over wide areas of the Earth, most fresh waters contain bicarbonate alkalinity. Carbonate may be present when the pH is high (see Figs. 8.2 and 8.3). Hydroxides are negligible unless the pH is extremely high (see Fig. 8.3). Borate silicate, phosphate, and sulfide are usually present only in trace quantities and thus contribute negligible amounts of alkalinity. When waters contain large amounts of dissolved organic carbon, organic anions may add additional alkalinity.

Free CO_2 is often in equilibrium with dissolved CO_2 in surface waters. This equilibrium, together with the equilibria for the dissolved ions, may be represented by the following equations:

$$CO_2(air) \rightleftarrows CO_2(dissolved) + H_2O \rightleftarrows H_2CO_3 \rightleftarrows H^+ + HCO_3^- \rightleftarrows H^+ + CO_3^{2-} \quad (1)$$

The concentration of CO_2 in the air averages about 0.03% (about $10^{-3.5}$ atm) but currently is increasing and varies with location. Photosynthesis and respiration by

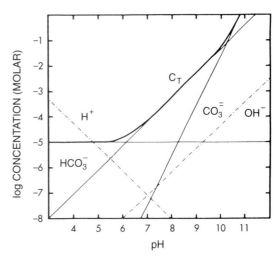

Figure 8.3. Model of distribution of carbon species of the carbonate system of natural waters. Pure water is equilibrated with atmospheric CO_2 at a constant partial pressure ($pCO_2 = 10^{-3.5}$ atm, 25°C). The pH can be varied by the addition of a strong acid or strong base, thereby keeping the solution in equilibrium with pCO_2. $C_T = [CO_2 (aq)] + [H_2CO_3] + [HCO_3^-] + [CO_3^{2-}]$ [Modified from Stumm and Morgan (1996).]

aquatic organisms can influence substantially the quantity of CO_2 in water at any given time and place.

For convenience, dissolved CO_2 and H_2CO_3 can be added together and called free CO_2 ($H_2CO_3^*$) such that $[H_2CO_3^*] = [CO_2] + [H_2CO_3]$. The equilibrium conditions in water are exemplified by the equations:

$$CO_2 \text{ gas} + H_2O \rightleftarrows H_2CO_3^* \tag{2}$$

$$H_2CO_3^* \rightleftarrows H^+ + HCO_3^- \tag{3}$$

$$HCO_3^- \rightleftarrows H^+ + CO_3^{2-} \tag{4}$$

These equilibrium equations represent a balance of ions in the reaction. For example, adding protons will cause an imbalance to the right in these equations, which is countered instantly by reactions that form ions on the left. Carbonate will react with the protons and form bicarbonate (Eq. 4), which will react with more protons to form carbonic acid (Eq. 3), which will convert back to CO_2 and water (Eq. 2). Likewise, adding hydroxide, which removes protons, will shift the reactions to the left and will be balanced by reactions to the right. In all cases, the concentration of protons will remain nearly constant, and thus pH is relatively stable until the available supply of bicarbonate and carbonate ions is exhausted.

Reaction 2 is not pH dependent, but when the partial pressure of CO_2 (pCO_2 in atmospheres) is known, then the concentration of $H_2CO_3^*$ can be calculated from

$$H_2CO_3^* = K_H pCO_2 \quad \text{where} \quad K_H = 10^{-1.5}$$

For air, $pCO_2 = 10^{-3.5}$; thus $H_2CO_3^*$ in equilibrium with the atmosphere is 10^{-5} mol/l.

Reaction 3 is pH dependent according to the equation

$$K_1 = [H^+][HCO_3^-]/[HCO_3^-] = 10^{-6.3}$$

Thus, at $pH = pK_1 = 6.3$; $[HCO_3^-] = [H_2CO_3^*]$, and buffering of this reaction is at its maximum.

Reaction 3 is also pH dependent according to

$$K_2 = [H^+][HCO_3^{2-}]/[HCO_3^-] = 10^{-10.3}$$

Thus, at $pH = pK_2 = 10.3$, $[CO_3^{2-}] = [HCO_3^-]$, and the buffering of this reaction is at its maximum.

When the total dissolved inorganic carbon (DIC) concentration and the pH are known, each carbonate species at equilibrium can be predicted:

$$H_2CO_3^* = DIC\alpha_0$$
$$HCO_3^- = DIC\alpha_1$$
$$CO_3^{2-} = DIC\alpha_2$$

where

$$\alpha_0 = \left[1 + \frac{K_1}{[H^+]} + \frac{K_1 K_2}{[H^+]^2}\right]^{-1}$$

$$\alpha_1 = \left[\frac{[H^+]}{K_1} + 1 + \frac{K_2}{[H^+]}\right]^{-1}$$

$$\alpha_2 = \left[\frac{[H^+]^2}{K_1 K_2} + \frac{[H^+]}{K_2} + 1\right]^{-1}$$

Since alkalinity includes HCO_3^- and CO_3^{2-}, alkalinity can be predicted from DIC and pH:

$$Alk = DIC[\alpha_1 + 2\alpha_2] + [OH^-] - [H^+]$$

When any compound dissolves in water it produces equal numbers of positive (cationic) charges and negative (anionic) charges. Therefore, electroneutrality is maintained, and this charge balance can be expressed mathematically as:

$$[H^+] + \{base\ cations\} = [HCO_3^-] + 2[CO_3^{2-}] + [OH^-] + \{strong\ acid\ anions\} \quad (5)$$

where

$$\{base\ cations\} = 2[Ca^{2+}] + [Na^+] + 2[Mg^{2+}] + [K^+] + [NH_3^+] + \cdots$$

and

$$\{acid\ anions\} = 2[SO_3^{2-}] + [Cl^-] + [NO_3^-] + \cdots$$

Alkalinity may be thought of as an "imbalance" between acid $[H^+]$ and the weak anions, mainly bicarbonate with some carbonate and hydroxide. Alkalinity represents the number of protons that can be neutralized by all of the weak anions after subtracting the protons already present and can be written as:

$$Alkalinity = [HCO_3^-] + 2[CO_3^{2-}] + [OH^-] - [H^+] \quad (6)$$

If electroneutrality were to be maintained, then the following also would be true:

$$Alkalinity = \{base\ cations\} - \{strong\ acid\ anions\} \quad (7)$$

These two equations are statements about equivalents. In practice, alkalinity can be measured by adding known amounts of strong acid until all of the weak anions have been used, at which point the pH will decline precipitously and all acid added subsequently will persist as protons. This point is often called "the equivalence point" and in fresh water usually occurs at a pH between about 4.2 and 5.1, depending on the amount of dissolved inorganic carbon present. It also is possible to calculate the alkalinity by measuring all of the base cations and strong acid anions and subtracting the moles of charges. Note that in acidic waters it is possible to have negative alkalinities.

The terms alkalinity, carbonate alkalinity, alkalinity reserve, titratable base, and acid neutralizing capacity (ANC) have been used commonly to express the total quantity of bases that can be titrated with a standard solution of a strong acid (e.g., $0.1\,N\,H_2SO_4$). Currently, several methods of measuring alkalinity are used, depending on conditions and on the degree of accuracy required. Since alkalinity is not affected by the exchange of CO_2 with the atmosphere, sampling procedures for most surface waters are straightforward, only requiring the use of clean bottles. If the water sample were anoxic and reduced substances were present, such as Fe^{2+} or H_2S, then care would need to be taken to insure that these substances do not become oxidized by contact with atmospheric oxygen. Because of the uptake and release of other ions during photosynthesis or decomposition, samples should be analyzed promptly, preferably within 3 h for high precision. Although exchange of CO_2 with the atmosphere does not affect alkalinity itself, it may affect the pH and the equivalence point if substantial changes in the total amount of dissolved inorganic carbon

Table 8.1. Example of titration data for 50-ml solution of sodium bicarbonate.[a]

Acid Added (ml)	pH	F1 (Eq.)
0	8.169	3.385 (E-10)
0.025	6.901	6.279 (E-9)
0.050	6.478	1.665 (E-8)
0.075	6.129	3.720 (E-8)
0.100	5.718	9.584 (E-8)
0.125	4.961	5.486 (E-7)
0.150	4.281	2.626 (E-6)
0.175	3.995	5.078 (E-6)
0.200	3.822	7.569 (E-6)
0.225	3.698	1.007 (E-5)
0.250	3.601	1.258 (E-5)
0.275	3.523	1.510 (E-5)
0.300	3.456	1.761 (E-5)

[a] The pH and Gran function are shown for increments of 0.025 ml of 0.1 N strong and additions.

were to occur during titration. Thus, it is best to titrate quickly and without undue surface agitation of the sample.

The most accurate procedure is the Gran titration (Gran, 1952), in which known increments of standard acid are used to titrate well beyond the equivalence point, to where proton accumulation is proportional directly to the titrant added, and then extrapolate back to the titrant volume of the equivalence point (Edmond, 1970).

An example of such a titration is given in Table 8.1 and shown graphically in Fig. 8.4. The accumulation of protons is calculated as the product of the hydrogen ion concentration and the sum of the sample volume plus the titrant volume. The accumulation of protons is referred to as the first Gran function, F1. Note that the

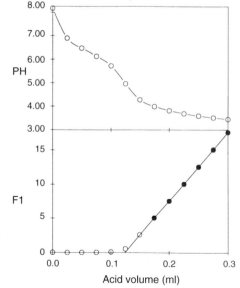

Figure 8.4. Alkalinity titration curves for data for Table 8.1. *Top panel*: pH versus volume of acid added. *Lower panel*: Gran function versus volume of acid added. Also shown is the best fit regression line for the last six data points. The line has a slope equal to the normality of the acid (0.0001 eq/ml = 0.1 N) and has an intercept of 0.125 ml.

accumulation of protons (F1) is very slow at pH values greater than 5, due to bicarbonate buffering; but the accumulation becomes rapid and linear below pH = 4, where bicarbonate buffering is exhausted. Nonlinearity in this region of the plot indicates either a malfunctioning electrode or the presence of noncarbonate buffers (e.g., organic acids).

Typically, the volume of acid required to reach the equivalence point is determined by extrapolation of the linear portion of the plot (between pH values of 4 and 3.5 or less). Alkalinity is computed as the microequivalents of acid added at the equivalence point divided by the sample volume in liters and reported as $\mu eq/l$:

$$\text{Alk}, \mu eq/l = (0.1N)(0.0001251)(10^6 \ \mu eq/l/N)/0.0501 = 250 \ \mu eq/l$$

Alternately, a plot of the pH values at each step in the titration versus the titrant volume can be used to estimate the equivalence point from the point where the change in the pH curve is most rapid. For noncritical analyses the pH of the equivalence point can be assumed, and then the sample is titrated to that point with the aid of a pH meter.

Historically, aquatic chemists used a somewhat different method of measuring and reporting alkalinity. With this method a pH indicator was used to indicate the equivalence and point pH. Typically, the "faintest pink" color of the methyl orange indicator was used [pH of approximately 4.25; Asbury et al. (1989)]. This method has now been replaced with a mixed bromcresol green methyl red indicator that has a sharper and less ambiguous end point. Terms such as total alkalinity have been used to describe alkalinity measured with color indicators, and the units are expressed frequently in terms of mg $CaCO_3/l$, because this was the common chemical referred to in the treatment of water and waste waters. The following equation may be used to convert between the two sets of alkalinity units:

$$\text{mg } CaCO_3/l \times 20 \ \mu eq/\text{mgCaCO}_3 = \mu eq/l \qquad (8)$$

Note that this calculation does not compensate for the overtitration bias caused by the lower end point of the methyl orange indicator as compared to the Gran procedure. For many soft water lakes, subtracting $55 \ \mu eq/l$ from the methyl orange alkalinity value will make them approximately comparable to Gran alkalinities (Asbury et al., 1989).

The term *phenolphthalein alkalinity* is used to express the portion of alkalinity contributed by the hydroxide and carbonate ions. The pH indicator phenolphthalein turns from pink to clear as the pH is lowered to 8.3, where essentially all of the carbonate is converted to bicarbonate (Figs. 8.2 and 8.3).

ACIDITY AND CARBON DIOXIDE

Acidity is, in a sense, the opposite of alkalinity. It is the total amount of acids that can be titrated with a strong base. Acidity commonly is not measured in limnological investigations, although acidity titrations sometimes are used to estimate the free CO_2 in the solution, as CO_2 is used more readily by phytoplankton than HCO_3, and it can become toxic to aquatic organisms if present in high concentrations. The amount of carbon dioxide (actually, CO_2 acidity) can be determined by adding phenolphthalein to a sample of water and titrating with a standard basic solution (free of CO_2) until the pink color just appears (pH approximately 8.3). There are numerous difficulties with this assay, and, in general, the results are not very accurate. The best methods employ sparging and analysis by gas chromatography (Schindler et al., 1972, 1973; Stainton 1973) or infrared CO_2 analysis (see Chapter 9). Older titrimetric methods are accurate when done carefully, but are tedious [e.g. Nygaard (1965)].

DISSOLVED INORGANIC CARBON

The dissolved inorganic carbon (DIC) of water is the sum of all carbon present as CO_2, H_2CO_3, HCO_3^-, and CO_3^{2-}. This quantity is of particular interest in evaluating the amount of inorganic carbon available for photosynthesis (see Chapter 14). In controlled light-dark experiments, estimates of photosynthesis and respiration rates can be done from changes in DIC.

Hypolimnetic accumulations of DIC also have been used to obtain whole-lake estimates of respiration (see Chapter 29). DIC is measured by acidifying a sample in an airtight syringe, equilibrating the CO_2 gas that evolves from the sample in a headspace of helium gas, and analyzing with either a gas chromatograph equipped with a suitable column (Porapak Q works well) and thermoconductivity detector (Schindler et al., 1972, 1973; Stainton 1973) or with an infrared gas analyzer (Menzel and Vaccaro, 1964). The results are compared to known standards of solutions of $NaHCO_3$ and expressed as $\mu M/l$ (see also Exercise 9).

Total inorganic carbon in fresh waters can be estimated from measurements of alkalinity and pH from:

$$DIC = [[Alk] - [OH^-] + [H^+]]/[\alpha_1 + 2\alpha_2] \tag{9}$$

Below pH = 8.3 the equation may be simplified to:

$$DIC = [[Alk] - [OH^-] + [H^+]]\left[\frac{H^+}{10^{-6.3}} + 1\right] \tag{10}$$

Any noncarbonate alkalinity (e.g., borates, organic alkalinity) should be subtracted from the alkalinity before using these equations. Further details can be found in Stumm and Morgan (1996). A diagram of the DIC, pH, and alkalinity relationship is shown in Fig. 8.5. These relationships may be inaccurate, particularly for waters with low DIC concentrations or when unexpected buffers are present.

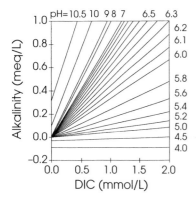

Figure 8.5. Alkalinity versus DIC at various pH levels. Data are from Eq. (9) (see text).

The total DIC of fresh water can be *estimated* from measurements of total alkalinity pH and temperature:

$$C \text{ available} = (\text{total alkalinity}) \times (\text{pH factor from Table 8.2})$$
$$\text{or}$$
$$= (\text{total alkalinity} - \text{phenolphthalein alkalinity, when}$$
$$\text{present}) \times (0.240)$$
$$= \text{mg C/l}$$

The pH factors, based on known dissociation constants of the carbonate complex at a given concentration of hydrogen ions and temperature, have been conveniently calculated into tabular form (Table 8.2). Interpolation of pH and temperature values may be necessary at certain points. The factors of Table 8.2 are not applicable to highly saline or acidic lakes.

HARDNESS

The hardness of water is caused by its concentration of polyvalent cations, principally calcium and magnesium, which tend to precipitate soap. Although this parameter is of relatively little interest to limnologists, it was measured frequently by early aquatic chemists and engineers. Hardness commonly is measured and adjusted by water treatment operators; it is expressed typically in terms of mg $CaCO_3$/l.

Hardness can be computed from known concentrations of calcium and magnesium (see Exercise 7). When other hardness-producing cations are present in significant amounts, their concentrations must be measured and included in the computations, however.

The concentration (mg/l) of each hardness-producing cation is multiplied by the appropriate factor (Table 8.3) to obtain equivalent calcium carbonate concentrations:

$$\text{Hardness } CaCO_3 \text{ equivalent}(\text{mg/l}) = [\text{cation, mg/l}] \times \text{factor}$$

These equivalents are then summed to obtain the total hardness. Conversion factors among the many different systems of expressing hardness are given in Wetzel (1999, Table 11.2).

ALUMINUM

Aluminum is the most abundant metallic element in the Earth's crust, but concentrations of dissolved Al^{n+} in natural waters with pH values between 6.0 and 8.0 are usually at trace levels. In more acidic (pH < 6.0), poorly buffered waters like those found in eastern Canada and the northeastern United States, Al^{n+} concentrations may be elevated by an order of magnitude to hundreds of μg/liter (e.g., Cronan and Schofield, 1979; Hongve, 1993, Driscoll and Postek, 1996). Aqueous aluminum chemistry is complex, as Al^{n+} can exist at several valences and can polymerize with both organic and inorganic ligands. In fresh waters, most aqueous aluminum is complexed with organic carbon, hydroxide, sulfate, fluoride, or phosphate.

Total reactive, monomeric aluminum (acid-soluble aluminum) can be separated into organic monomeric aluminum (Al_o) and inorganic monomeric aluminum (Al_i). Al_o represents organically complexed aluminum in solution, whereas Al_i includes

Table 8.2. pH factors for the conversion of total alkalinity to milligrams of carbon per liter.[a]

pH	Temperature (°C)					
	0	5	10	15	20	25
5.0	9.36	8.19	7.16	6.55	6.00	5.61
5.1	7.49	6.55	5.74	5.25	4.81	4.51
5.2	6.00	5.25	4.61	4.22	3.87	3.63
5.3	4.78	4.22	3.71	3.40	3.12	2.93
5.4	3.87	3.40	3.00	2.75	2.53	2.38
5.5	3.12	2.75	2.43	2.24	2.06	1.94
5.6	2.53	2.24	1.98	1.83	1.69	1.59
5.7	2.06	1.83	1.62	1.50	1.39	1.31
5.8	1.69	1.50	1.34	1.24	1.15	1.09
5.9	1.39	1.24	1.11	1.03	0.96	0.92
6.0	1.15	1.03	0.93	0.87	0.82	0.78
6.1	0.96	0.87	0.77	0.73	0.70	0.67
6.2	0.82	0.74	0.68	0.64	0.60	0.58
6.3	0.69	0.64	0.59	0.56	0.53	0.51
6.4	0.60	0.56	0.52	0.49	0.47	0.45
6.5	0.53	0.49	0.46	0.44	0.42	0.41
6.6	0.47	0.44	0.41	0.40	0.38	0.37
6.7	0.42	0.40	0.38	0.37	0.35	0.35
6.8	0.38	0.37	0.35	0.34	0.33	0.32
6.9	0.35	0.34	0.33	0.32	0.31	0.31
7.0	0.33	0.32	0.31	0.30	0.30	0.29
7.1	0.31	0.30	0.29	0.29	0.29	0.28
7.2	0.30	0.29	0.28	0.28	0.28	0.27
7.3	0.29	0.28	0.27	0.27	0.27	0.27
7.4	0.28	0.27	0.27	0.26	0.26	0.26
7.5	0.27	0.26	0.26	0.26	0.26	0.26
7.6	0.27	0.26	0.26	0.25	0.25	0.25
7.7	0.26	0.26	0.25	0.25	0.25	0.25
7.8	0.25	0.25	0.25	0.25	0.25	0.25
7.9	0.25	0.25	0.25	0.25	0.25	0.25
8.0	0.25	0.25	0.25	0.25	0.24	0.24
8.1	0.25	0.25	0.24	0.24	0.24	0.24
8.2	0.24	0.24	0.24	0.24	0.24	0.24
8.3	0.24	0.24	0.24	0.24	0.24	0.24
8.4	0.24	0.24	0.24	0.24	0.24	0.24
8.5	0.24	0.24	0.24	0.24	0.24	0.24
8.6	0.24	0.24	0.24	0.24	0.24	0.24
8.7	0.24	0.24	0.24	0.24	0.24	0.24
8.8	0.24	0.24	0.24	0.24	0.23	0.23
8.9	0.24	0.24	0.23	0.23	0.23	0.23
9.0	0.24	0.23	0.23	0.23	0.23	0.23
9.1	0.23	0.23	0.23	0.23	0.23	0.23
9.2	0.23	0.23	0.23	0.23	0.23	0.23
9.3	0.23	0.23	0.23	0.22	0.22	0.22
9.4	0.23	0.23	0.22	0.22	0.22	0.22

[a] From Saunders et al. (1962).

Table 8.3. Factors to estimate water hardness from concentrations of cations.

Cation	Factor	Cation	Factor
Ca	2.497	Al	5.564
Mg	4.116	Sr	1.142
Fe	1.792	Zn	1.531
Mn	1.822		

aquo (free) aluminum and inorganic complexes of aluminum (Driscoll 1984; McAvoy et al. 1992). Inorganic monomeric Al may be toxic to aquatic organisms, depending on pH, presence of other inorganic solutes, and life history stage (e.g., Driscoll et al. 1980; Baker and Schofield, 1982; Havas and Likens, 1985).

Traditionally, the ferron method had been used for measuring total aluminum (Al_{total}) (Rainwater and Thatcher 1960), but Al_{total} can also be measured directly with flameless atomic absorption spectrophotometry (graphite furnace) in aqueous samples. Fractionation by ion-exchange columns, reaction with complexing agents, dialysis, etc. have been used to discriminate among the various aluminum species in acidified fresh waters (e.g. Hodges 1987; Bloom and Erich 1996; McAvoy et al. 1992).

EXERCISES

These exercises are divided into two options.

OPTION 1. FIELD TRIP

Visit the central depression of a thermally stratified lake, collect water samples at meter intervals from the surface to the bottom, and make the following evaluations. Temperature measurements are needed from each collection depth.

1. Measurement of pH:
   a. Place the water samples into rigorously cleaned hard glass or polyethylene bottles. The bottles should be filled to the top with the sample water and then capped tightly. Ground-glass stoppered BOD bottles are recommended. The samples should not stand in sunlight; measurement should be made as soon as possible, never exceeding 2h from the time of collection. The samples should be at room temperature.
   b. To standardize the pH meter, place the electrodes in fresh buffer solution for 5 min and replace in fresh buffer. Standardize by manufacturer's instructions, allow 2 min, and restandardize. Repeat until no drift is apparent. Rinse copiously with distilled water, followed by a rinse or two with the sample, and then place in some sample water for 5 min.
   c. Rinse the electrodes with sample water several times, replace with fresh sample water, allow sufficient time for drift to stop, and make the reading. Remove the electrodes. Mixing of the sample during measurements should not be done, or done cautiously, as a moderate amount of stirring will reduce time for equilibration, but excessive mixing may generate spurious electrical potentials, or enough CO_2 may be exchanged with the air to significantly alter the pH.
2. Meaurement of alkalinity: Using a number of water samples collected from natural waters, e.g., a vertical profile in a stratified lake, determine the alkalinity of each sample. Replicate

the analyses from each sample and evaluate the variance (see Appendix 2). Compare your results with those of other members of your class. What are the sources of error?

a. *Method 1: Gran alkalinity*

    i. Pipet carefully 50 ml of the sample into a glass beaker. Add a small Teflon-coated stir bar.

    ii. Measure the pH of the sample as described previously.

    iii. Add 0.025 ml of 0.1 N HCl. (An automatic pipet is recommended.) Mix the solution at slow speed. Stop mixing and record the pH after it has stabilized.

    iv. Repeat step iii until the pH drops below 3.5. Depending on the volumes used, at least five data points should lie between pH 4 and 3.5.

    v. Calculate the first Gran function F1 from:

$$F_1 = (\text{volume of sample} + \text{volume of acid added}) * 10^{-pH}$$

    Example: the tenth pair of data in Table 8.1 were calculated after nine increments of 0.025 ml of acid were added. After converting to liters, the Gran function was calculated from:

$$F_1 = (0.0501 + 0.0002251) \times 10^{-3.698}$$
$$F_1 = 1.007 \times 10^{-5} \text{ equivalents}$$

    vi. Using only the data where pH is less than 4, regress F1 as the independent y variable versus the volume of acid added as the x variable. The correlation coefficient r should equal or exceed 0.999, and the slope should be within 10% of the normality of the acid used. The X intercept can be determined from the Y intercept and slope from:

$$X \text{ intercept} = -1 * \frac{Y \text{ intercept}}{\text{slope}}$$

    vii. Calculate alkalinity from:

$$\text{Alk} = (\text{normality of acid})(X \text{ intercept, liters})/(\text{sample volume, liters})$$
Multiply by 1×10^6 to convert units to $\mu eq/l$.

b. *Method 2: Titration with color indicator solutions.* Hydroxide and carbonate ions cause high pH in water and result in values above 8.3, while bicarbonate ions in solution cause pH values below 8.3. Hydroxide and bicarbonate ions do not exist together in the same solution since they will react to form carbonate ions. Carbonate and hydroxide ions can exist together, and bicarbonate and carbonate ions often are found together.

The organic indicator phenolphthalein changes color at about pH 8.3. When phenolphthalein is added to a solution that contains carbonate or hydroxide alkalinity, the solution turns pink. When acid then is added, the alkalinity will be reduced and the pink color will disappear as the pH falls below 8.3. At this point the hydroxide is neutralized.

When phenolphthalein is added to water and no pink reaction appears, only bicarbonate alkalinity is present. Thus, the phenolphthalein alkalinity measures all the hydroxide alkalinity and one-half of the carbonate alkalinity. To determine the other half of the carbonate alkalinity and the bicarbonate alkalinity, more acid is added until a pH change indicates that the bicarbonate has been dissociated and released as CO_2 to the atmosphere during titration.

Pipet carefully 50 ml of the sample into a Erlenmeyer flask over a white titration surface. Add four to five drops of phenolphthalein indicator. If pink, add N/50 (0.02 N) sulfuric acid slowly until the pink color disappears upon stirring. Note the volume of acid used; this volume corresponds to the equivalence point of pH 8.3 and is required to calculate phenolphthalein alkalinity.

Then add three to four drops of mixed bromcresol green-methyl red indicator (or methyl orange) to the same sample. Continue to slowly add N/50 sulfuric acid to the appropriate equivalence point.

It has been shown that methyl orange usually is unsuitable for the determination of low alkalinities. Errors arise largely from the fact that the first perceptible color change with methyl orange occurs at a pH of 4.6. Many individuals possess imperfect color perception, rendering it almost impossible to properly identify the faint orange color characteristic of methyl orange at a pH of 4.6. Often the sample is overtitrated to a deeper orange or faint pink, representing a pH as low as 4.25, with a consequent sacrifice in accuracy.

For most fresh waters the pH of the end point can be estimated from:

$$pH_{end} = \{-\log_{10}[(DIC)(10^{-6.3}) + 10^{-14}]\}/2 \tag{11}$$

[see Stumm and Morgan (1981)]. The following pH values are suggested as the equivalence points for the corresponding alkalinity concentrations as calcium carbonate (American Public Health Association et al., 1989):

Alkalinity (as mg/l $CaCO_3$)	pH
<30–30	5.1
150	4.8
500	4.5

Thus, methyl orange under the most favorable circumstances is justified only for solutions of alkalinity in excess of 150 ppm. A mixed indicator prepared from bromcresol green and methyl red is suitable for the high pH end points, while methyl orange can be used for those below 4.6. It is advisable to prepare buffer solutions of the applicable pH, add the proper volume of indicator, and use these solutions as standards for color comparison for the various indicator color transitions.

Sample volumes requiring less than 50 ml of titrant yield the sharpest color change at the end point and are, therefore, recommended.

The mixed indicator yields the following color responses:

pH	Color
>5.2	Greenish blue
5.0	Light blue with lavender gray
4.8	Light pink gray with bluish cast
4.6	Light pink

Note the total volume of acid used; this is required for the calculations of total alkalinity. Phenolphthalein alkalinity as μeq/l =

$$= \frac{\left(\begin{array}{c}\text{volume of standard acid}\\ \text{used to first end point (ml)}\end{array}\right) \times (\text{normality of acid} \times 10^6)}{\text{volume of sample (ml)}}$$

$$\text{Total alkalinity as } \mu eq/l = \frac{\left(\begin{array}{c}\text{total volume of standard}\\ \text{acid used (ml)}\end{array}\right) \times (\text{normality of acid} \times 10^6)}{\text{volume of sample (ml)}}$$

Note: When the total alkalinity is determined on the same solution used for phenolphthalein alkalinity, the volume of acid required for the phenolphthalein titration is included in the total ml of standard acid in the total alkalinity calculations (above).

Interpretation of Results. The results obtained from the phenolphthalein and total alkalinity determinations offer a means for the classification of the three principal forms of alkalinity present in many water types. The classification ascribes the entire alkalinity to bicarbonate, carbonate, and hydroxide and assumes the absence of other weak acids of organic and inorganic composition. It also presupposes the incompatibility of hydroxide and bicarbonate alkalinities in the same sample.

The alkalinity relationships are expressed by:

a. Carbonate alkalinity is present when the phenolphthalein alkalinity is not zero but is less than the total alkalinity.
b. Hydroxide alkalinity is present when the phenolphthalein alkalinity is more than one-half the total alkalinity.
c. Bicarbonate alkalinity is present when the phenolphthalein alkalinity is less than one-half the total alkalinity.

When P = the phenolphthalein alkalinity and T = the total alkalinity, we can differentiate between the hydroxide, carbonate, and bicarbonate alkalinity (all values in mg/l as $CaCO_3$). Five conditions are possible:

a. $P = 0$. All alkalinity is bicarbonate.
b. P is less than $\frac{1}{2}T$. Bicarbonate and carbonate alkalinity both exist.
　　Carbonate alkalinity = $2P$
　　Bicarbonate alkalinity = $T - 2P$
c. $P = \frac{1}{2}T$. All alkalinity is carbonate.
d. P is greater than $\frac{1}{2}T$. Hydroxide and carbonate alkalinity both exist.
　　Hydroxide = $2P - T$
　　Hydroxide = $2(T - P)$
e. $P = T$. All alkalinity is hydroxide.

The conversion of results, based on 0.02 N H_2SO_4 standard acid and a 50-ml sample, then are related as follows:

Alkalinites (expressed as mg/l as $CaCO_3$)[a]
based on acid titration

Result of titration (ml of acid added)	Hydroxide	Carbonate	Bicarbonate
$P = 0$	0	0	$T \times 20$
$P < \frac{1}{2}T$	0	$2P \times 20$	$(T - 2P) \times 20$
$P = \frac{1}{2}T$	0	$2P \times 20$	0
$P > \frac{1}{2}T$	$(2P - T) \times 20$	$2(T - P) \times 20$	0
$P = T$	$T \times 20$	0	0

[a]The factor of 20 in these computations results from conversion of ml of 0.02 N acid in a 50-ml sample to mg/l.

The most acceptable usage when alkalinity is reported is to express the results as *equivalents per units volume*, i.e., *in milliequivalents per liter*.

Acid that is 0.02 N contains 0.02 meq of acid in each ml, so that each ml of standard acid used in the titration is equal to 0.02 meq of bicarbonate in 100 ml of sample (0.2 meq/1 of sample) or 0.02 meq in 50 ml of sample (0.4 meq/1 of sample). Therefore, if 0.02 N acid and a 50-ml sample are used,

$$meq/1 = (0.4) \times (ml\ acid\ used)$$

Since the equivalent of $CaCO_3$ is 50, the meq/1 can be derived by dividing mg/l by 50. When alkalinity is expressed in milliequivalents per liter, it is an entirely unambiguous quantity that will not be altered by the addition of CO_2, which results in conversion of carbonate to bicarbonate.

3. Calculate the total inorganic carbon (DIC) theoretically available for photosynthetic utilization (See Fig. 8.5).
4. Collect water from the littoral zone among dense stands of submersed macrophytes and make similar measurements.
5. Collect water from inlet sources (inlet stream, springs) and from the outlet, and make similar measurements.
6. Answer the questions following Option 2 using your data.

OPTION 2. LABORATORY EXERCISES

1. Measure the pH of tap water and of samples of natural water provided by your instructor. Determine the pH of the water samples electrometrically and compare to estimates using pH indicator papers (see Option 1, Sect. 1).
2. Measure the phenolphthalein and total alkalinity (see methods of Option 1, Sec. 2) of the following samples:
   a. Tap water, full strength and serially diluted with distilled water to concentrations <15 mg/l.
   b. Lake or stream water provided.
   c. Analyze three samples using bromcresol green-methyl red mixed indicator and compute the standard deviation (see Appendix 2).
   d. Compare the above results to those of a sample analyzed using methyl orange indicator. Try this comparison on dilute samples.
3. Measure the alkalinity by Gran titration analysis (see Option 1, Section 2a). Interpret your results analogous to those illustrated in Fig. 8.3 and compare the results of this method with those obtained with color indicator solutions.
4. Using your data or those given in Table 8.3 and 8.4 answer the following questions.

Questions

1. Compare graphically the vertical profiles of the following parameters from your lake system (Option 1) or from the data given in Table 8.4 or Table 8.5 (Option 2):
   a. Total alkalinity.
   b. Temperature.
   c. pH values.
   d. CO_2 concentrations.
   e. Inorganic carbon (DIC) available for photosynthesis.
   f. Hardness equivalents.
2. From the collection of data from vertical profiles made in your lake at two or more intervals separated by at least a month (Option 1) or from data given in Table 8.4 or Table 8.5 (Option 2), calculate the changes in the various parameters evaluated. Discuss the potential causes of the losses from and gains to the strata. What are the potential effects of such seasonal changes on the distribution and availability of other ions and compounds on the biota?
3. Compare the horizontal variations of the concentrations of the measured parameters (Question 1) found in the littoral zone, inlet sources, and the outlet to those observed in the surface waters of the central station. Discuss the potential causes of the differences.
4. Construct depth-time diagrams of the isopleths of total alkalinity (meq/l), pH (interpolations must be done logarithmically—why?) and inorganic carbon (DIC) available for photosynthesis (see Exercise 2 for details on how to make these plots).
5. From the data given in Table 8.6 or Table 4.3 (p. 53), construct depth-volume curves for Lawrence or Mirror lakes in terms of depth versus cubic meters and depth versus percentage volume. Calculate the mass of inorganic carbon of the strata of water with increasing depth.

Table 8.4. Measurements of temperature, alkalinity, pH, and cations of Lawrence Lake, Barry County, Michigan.

Depth (m)	May 24, 1971					June 22, 1971				
	°C	Alkalinity (meq/l)	pH	Ca^{2+} (meq/l)	Mg^{2+} (meq/l)	°C	Alkalinity (meq/l)	pH	Ca^{2+} (meq/l)	Mg^{2+} (meq/l)
0	18.0	4.46	8.30	3.90	2.11	25.0	4.00	8.20	2.91	2.42
1	17.9	4.49	8.32	3.90	2.08	25.1	4.05	8.15	2.91	2.46
2	17.9	4.51	8.30	3.90	2.11	25.1	4.06	8.15	3.02	2.46
3	17.9	4.53	8.30	3.85	2.08	23.6	4.14	8.17	3.10	2.46
4	15.6	4.59	8.09	3.95	2.11	19.6	4.24	8.16	3.29	2.46
5	12.8	4.56	8.10	3.91	1.98	16.8	4.29	8.10	3.55	2.49
6	11.0	—	—	—	—	13.6	—	—	—	—
7	9.3	4.54	8.05	3.88	1.98	11.0	4.49	8.08	3.70	2.39
8	8.2	—	—	—	—	9.4	—	—	—	—
9	8.2	—	—	—	—	8.4	—	—	—	—
10	7.8	4.54	7.94	3.89	1.92	7.8	4.55	8.03	3.73	2.32
11	6.3	—	—	—	—	7.5	—	—	—	—
12	6.2	4.57	7.78	3.96	1.98	7.3	4.65	7.94	3.48	2.34
12.5	6.2	—	—	—	—	7.0				
Inlet 1	11.2	5.55	7.5	4.74	2.27	12.7	5.70	7.5	5.09	2.27
Inlet 2	9.8	5.80	7.5	4.44	2.20	11.8	6.20	7.5	4.89	2.20
Ground water spring	8.2	5.10	7.5	3.87	1.92	8.5	5.10	7.5	3.89	1.92

(Continued)

Table 8.4. (*Continued*)

Depth (m)	July 20, 1971					August 26, 1971				
	°C	Alkalinity (meq/l)	pH	Ca^{2+} (meq/l)	Mg^{2+} (meq/l)	°C	Alkalinity (meq/l)	pH	Ca^{2+} (meq/l)	Mg^{2+} (meq/l)
0	23.5	2.55	8.15	3.07	2.48	23.4	3.34	8.36	2.70	2.25
1	23.5	1.92	8.17	3.03	2.53	23.4	3.39	8.34	2.78	2.34
2	23.2	2.00	8.19	3.06	2.55	23.4	3.24	8.31	2.80	2.28
3	23.2	2.44	8.19	3.24	2.56	23.4	3.42	8.30	2.80	2.33
4	23.2	2.79	8.19	3.09	2.50	23.4	3.49	8.23	2.96	2.31
5	21.0	3.16	8.14	3.55	2.48	22.8	3.85	8.18	3.42	2.33
6	15.6	—	—	—	—	19.3	—	—	—	—
7	12.2	3.78	8.10	3.76	2.47	14.8	4.36	8.15	3.96	2.06
8	10.0	—	—	—	—	11.2	—	—	—	—
9	8.8	—	—	—	—	9.3	—	—	—	—
10	7.8	4.31	7.99	4.06	2.55	8.2	4.62	7.79	4.24	2.06
11	7.3	—	—	—	—	7.7	—	—	—	—
12	7.1	4.35	7.80	4.06	2.45	7.4	4.91	7.61	4.45	2.25
12.5	6.9	—	—	—	—	7.2	—	—	—	—
Inlet 1	13.1	5.60	7.5	4.89	2.27	11.8	5.70	7.5	5.09	2.22
Inlet 2	13.2	6.25	7.6	4.84	2.20	12.5	6.24	7.6	4.84	2.18
Groundwater spring	9.0	5.10	7.5	3.84	1.92	9.5	5.14	7.5	3.72	1.89

Table 8.5. Measurements of temperature, alkalinity, pH, and cations of Mirror Lake, Grafton County, New Hampshire.

Depth (m)	May 9, 1972					June 15, 1972				
	°C	Alkalinity (μeq/l)	pH	Ca^{2+} (μeq/l)	Mg^{2+} (μeq/l)	°C	Alkalinity (μeq/l)	pH	Ca^{2+} (μeq/l)	Mg^{2+} (μeq/l)
0	10.08	64.0	6.52	118	41.1	19.88	86.0	6.75	117	41.1
1	10.04	—	—	—	—	18.79	—	—	—	—
2	10.01	68.0	6.48	119	40.3	17.60	58.0	6.70	118	41.1
3	8.05	—	—	—	—	17.21	—	—	—	—
4	6.31	68.0	6.30	121	41.9	14.90	62.0	6.68	118	41.1
5	5.73	—	—	—	—	11.21	—	—	—	—
6	5.22	68.0	6.12	122	41.9	8.91	70.0	6.38	123	42.8
7	5.01	—	—	—	—	7.49	—	—	—	—
8	4.93	74.0	5.99	127	44.4	6.59	84.0	6.10	125	43.6
9	4.88	—	—	—	—	5.99	—	—	—	—
10	4.87	96.0	5.92	129	44.4	5.82	102	5.92	131	44.4
Inlet 1	5.2	—	5.00	120	38.0	9.2	152	5.72	127	41.1
Inlet 2	7.0	—	6.50	120	40.3	10.7	52.0	6.65	138	48.5
Inlet 3	7.0	—	6.30	122	41.1	9.9	96.0	6.65	143	48.5
Outlet	10.4	—	6.40	114	38.7	18.1	—	6.31	115	38.7

(*Continued*)

Table 8.5. (*Continued*)

Depth (m)	July 20, 1972					August 15, 1972				
	°C	Alkalinity (μeq/l)	pH	Ca²⁺ (μeq/l)	Mg²⁺ (μeq/l)	°C	Alkalinity (μeq/l)	pH	Ca²⁺ (μeq/l)	Mg²⁺ (μeq/l)
0	25.28	58.0	6.90	113	39.5	20.38	64.0	6.85	113	39.5
1	25.32	—	—	—	—	20.10	—	—	—	—
2	25.00	60.0	7.00	113	39.5	20.00	64.0	6.90	115	39.5
3	23.23	—	—	—	—	20.00	—	—	—	—
4	19.82	60.0	7.05	117	40.3	20.00	66.0	6.90	114	39.5
5	14.90	—	—	—	—	19.73	74.0	—	—	—
6	11.55	84.0	6.70	121	41.1	14.52	82.0	6.48	117	40.3
7	9.90	—	—	—	—	11.51	102	—	—	—
8	8.65	94.0	6.02	127	42.8	9.62	116	5.90	134	44.4
9	7.45	—	—	—	—	8.46	130	—	—	—
10	7.13	116	6.05	140	45.2	7.88	—	5.90	153	47.7
Inlet 1	14.6	78.0	5.73	139	48.5	12.75	132	6.20	144	52.6
Inlet 2	17.6	122	6.50	175	59.2	13.25	202	6.87	187	65.0
Inlet 3	15.8	162	6.70	179	57.6	13.00	256	7.02	208	69.9
Outlet	27.9	66.0	6.51	112	38.7	20.15	60.0	6.65	111	39.5

Table 8.6. Area and volume of depth strata of Lawrence Lake, Barry County, Michigan.

Depth (m)	Area (m^2)	Depth stratum (m^3)	Volume (m^3)
0	49,643	0–1	45,823
1	42,195	1–2	39,503
2	36,948	2–3	34,314
3	31,812	3–4	27,461
4	27,230	4–5	25,982
5	24,806	5–6	24,263
6	23,773	6–7	23,161
7	22,600	7–8	21,083
8	19,643	8–9	17,835
9	16,120	9–10	14,258
10	12,501	10–11	10,266
11	8,201	11–12	5,685
12	3,508	12–12.5	715
12.5	125	Total	292,350

Apparatus and Supplies

1. Underwater thermometer.
2. Water sampling bottle (nonmetallic), line, and messengers.
3. Small (120 ml or 4 oz) polyethylene bottles (amber) and ice chest to hold sample bottles.
4. Ground-glass stoppered BOD bottles.
5. 10-ml burettes, standardized 0.02 N H_2SO_4, 0.1 N HCl, phenolphthalein indicator, methyl red-bromcresol green indicator, methyl orange indicator, and glass beakers.
6. pH meters, buffers of pH 4, 6, 7, and 8.
7. Magnetic stirrer.
8. Automatic pipet capable of dispensing small volumes (e.g., 0.025 ml).

References

American Public Health Association et al. 1998. Standard methods for the Examination of Water, Sewage, and Wastewater. 20th Ed. American Public Health Association, Washington, D. C. 1185 pp.

Asbury, C.E., F.A. Vertucci, M.D. Mattson, and G.E. Likens. 1989. Acidification of Adirondack lakes. Environ. Sci. Technol. *23*:362–365.

Baker, J.P. and C.L. Schofield. 1982. Aluminum toxicity to fish in acidic waters. Water Air Soil Pollut. *18*:289–309.

Bloom, P.R. and M.S. Erich. 1996. The quantitation of aqueous aluminum. chapter 1, pp. 1–38. *In*: G. Sposito, Editor. The Environmental Chemistry of Aluminum. 2nd Ed. Lewis Publishers, Boca Raton, FL.

Cronan, C.S. and C.L. Schofield. 1979. Aluminum leaching response to acid precipitation: effects on high elevation watersheds in the Northeast. Science *204*:304–306.

Driscoll, C.T. 1984. A procedure for the fractionation of aqueous aluminum in dilute acidic waters. Internat. Four. Environ. Analy. Chem. *16*:267–283.

Driscoll, C.T., J.P. Baker, J.J. Bisogni, and C.L. Schofield. 1980. Effect of aluminum speciation on fish in dilute and acidified waters. Nature *284*:161–164.

Driscoll, C.T. and K.M. Postek. 1996. The chemistry of aluminum in surface waters. pp. 363–418. *In*: G. Sposito, Editor. The Environmental Chemistry of Aluminum, 2nd Ed. CRC Press, Boca Raton, FL.

Edmond, J.M. 1970. High precision determination of titration alkalinity and total carbon dioxide content of sea water by potentiometric titration. Deep-Sea Res. *17*:737–750.

Galster, H. 1991. pH Measurement: Fundamentals, Methods, Applications, Instrumentation. VCH Verlagsgesellschaft, Weinheim. 355 pp.

Golterman, H.L. and R.S. Clymo, (eds). 1969. Methods for Chemical Analysis of Fresh Waters. IBP Handbook No. 8. Blackwell, Oxford. 172 pp.

Gran, G. 1952. Determination of the equivalence point in potentiometric titrations. Part II. Analyst 77:661–671.

Hodges, S.C. 1987. Aluminum speciation: a comparison of five methods. Soil Sci. Soc. Am. J. 51:57.

Hongve, D. 1993. Total and reactive aluminum concentrations in non-turbid Norwegian surface waters. Verh. Internat. Verein. Limnol. 25:133–136.

McAvoy, D.C., R.C. Santore, J.D. Shosa, and C.T. Driscoll. 1992. A comparison between pyrocatechol violet and 8-hydroxyquinoline procedures for determining aluminum fractions. Soil Sci. Soc. Amer. J. 56:449–455.

Menzel, D.W. and R.F. Vaccaro. 1964. The measurement of dissolved organic and particulate carbon in sea water. Limnol. Oceanogr. 9:138–142.

Nygaard, G. 1965. Hydrographic studies, especially on the carbon dioxide system, in Grane Langsø. Biol. Skr. Dan. Vid. Selsk. 14(2). 110 pp.

Rainwater, F.H. and L.L. Thatcher. 1960. Methods for collection and analysis of water samples. USDI Geol. Surv. Water-Supply Paper 1454. 301 pp.

Saunders, G.W., F.B. Trama, and R.W. Bachmann. 1962. Evaluation of a modified ^{14}C technique for shipboard estimation of photosynthesis in large lakes. Publ. Great Lakes Res. Div., Univ. Mich. 8, 61 pp.

Schindler, D.W. and E.J. Fee. 1973. Diurnal variation of dissolved inorganic carbon and its use in estimating primary production and CO_2 invasion in Lake 227. J. Fish. Res. Bd. Canada 30:1501–1510.

Schindler, D.W., R.V. Schmidt, and R.A. Reid. 1972. Acidification and bubbling as an alternative to filtration in determining phytoplankton production by the ^{14}C method. J. Fish. Res. Bd. Canada 29:1627–1631.

Stainton, M.P. 1973. A syringe gas-stripping procedure for gas-chromatographic determination of dissolved inorganic carbon and organic carbon in fresh water and carbonates in sediments. J. Fish Res. Bd. Canada 30:1441–1445.

Stumm, W. and J.J. Morgan. 1981. Aquatic Chemistry. 2nd Ed. Wiley-Interscience, New York. 780 pp.

Stumm, W. and J.J. Morgan. 1996. Aquatic Chemistry. 3rd Ed. Wiley-Interscience, New York. 1022 pp.

Wetzel, R.G. 1999. Limnology: Lake and River Ecosystems. 3rd Ed. Academic Press, San Diego (in press).

Organic Matter

Organic matter in aquatic ecosystems ranges from dissolved organic compounds to large aggregates of particulate organic matter and from living to dead material. Most of the organic matter, whether dissolved or particulate, is detritus (i.e., organic matter from dead organisms). Metabolism of the organic matter and interactions of this material chemically and biologically are, to a significant extent, governed by the size and chemical composition of the organic matter. Little, if any, dissolved and colloidal organic matter is utilized by aquatic animals directly, whereas particulate organic matter of a given size range may be a major food source. Decomposition of dissolved organic matter results in gaseous end products; particulate organic matter must be converted enzymatically to soluble organic compounds prior to biochemical degradation to gaseous products.

Nearly all of the organic carbon in natural waters is in the form of dissolved organic carbon (DOC), colloidal organic carbon (COC), and dead particulate organic carbon (POC). This quantity of detrital (nonliving) organic carbon greatly exceeds that amount of carbon in all of the living biota combined. A ratio of ca. 10:1 for DOC/COC to POC is almost universal in the open water of both lake and stream ecosystems (Wetzel, 1984, 1995).

DOC/COC may be separated from POC by sedimentation, centrifugation, and filtration. Historically, DOC has been defined rather arbitrarily in most studies by the practical limit of filtration of natural water—that amount of organic matter retained on a filter of 0.5-μm pore size is the POC; the organic carbon of the filtrate is the DOC/COC fraction (cf., Gustafsson and Gschwend, 1997). The filter-passing materials are a mixture of truly dissolved organic compounds and of macromolecular colloids of a size of ca. 300–2000 Da to several thousand Kda (0.7–2 nm). Separation of dissolved/colloidal fractions from particulate components on the basis of size is not always consistent with thermodynamic chemical properties and reactivities. However, the physical separation described in this exercise, where the dissolved and colloidal components are not significantly affected by gravitational settling, is reasonably consistent with chemical delineations and is characteristic of transport in solution. Measured DOC concentrations, therefore, may contain a significant colloidal fraction in addition to truly dissolved organic carbon.

Particulate organic carbon of a size greater than 0.5 μm can originate from many sources, such as photosynthetic production of algae and aquatic plants, fragmentation of larger particles, animal maceration and egestion, microbial processes, flocculation of dissolved organic matter, and terrestrial inputs. Particulate organic matter (POM) is further differentiated on the basis of size into fine particulate organic matter (FPOM; >0.5 μm to <1000 μm or 1.0 mm) and coarse particulate organic matter (CPOM; any organic particle >1.0 mm). Each of these categories is further differentiated into smaller categories (cf. Lamberti and Gregory, 1996; Wallace and Grubaugh, 1996). Determinations of the average organic carbon content of the POM or fractions of POM allow an estimate of the amount of organic carbon potentially available for metabolism by organisms. Quantification of the amount of organic carbon in POM or DOM does not mean that these organic compounds are available, however, for immediate hydrolysis and assimilation, because of the large variability in organic composition.

In this exercise, the amounts and distribution of DOC/COC and POC will be evaluated in the pelagic zone of a lake. Variations in a stream ecosystem are analyzed in a subsequent exercise (Exercise 24). Organic nitrogen and phosphorus compounds have been discussed earlier (Exercise 7).

COLLECTION AND SEPARATION

Water samples should be obtained with a clean water sampler and immediately transferred to scrubbed, darkened bottles (e.g., 1-l bottles). The samples should be placed in cool storage, such as in an ice chest, and filtered as soon as possible.

Measured aliquots of the samples should be filtered using a series of clean all-glass filtration flasks. It is imperative that the filtration apparatus, and especially the sintered filter support, be cleaned thoroughly with acid, profusely washed and rinsed with low-carbon redistilled water, and allowed to air-dry without contamination by dust. Glass fiber filters (Whatman GF/F of 0.6- to 0.7-μm pore size) should be pre-combusted in a muffle furnace at 500°C (*Caution*: higher temperatures will cause fusion of the glass) inside folded packets of aluminum foil with the dull side of the foil against the filters. Use clean forceps in handling of the filters. The amount of water filtered varies with the concentration of the particulate organic matter. In lakes of moderate productivity, between 0.5 and 1.0 l of water can be filtered onto 47-mm diameter filters to give an adequate POC sample. Vacuum differential applied to the filtration apparatus should not exceed 0.3 atm.

The filters for POC analysis should be placed onto labeled, clean aluminum foil (dull side) and carefully stored under desiccation immediately. The filtrate should be processed immediately for DOC content.

DISSOLVED ORGANIC CARBON

Nearly all analyses of organic carbon of the dissolved/colloidal fraction are now performed by high-temperature combustion of a filtered water sample that has been acidified and sparged of dissolved inorganic carbon. The organic carbon is oxidized to CO_2 and quantified by nondispersive infrared analyses in comparison to known organic standards. Because of the excellent reproducibility and consistency of well-standardized instruments, this method is recommended over other methods. The dissolved organic carbon analyzers, however, are expensive and not available in all laboratories. An alternative earlier method of wet chemical oxidation is also presented, although this method can be subject to incomplete oxidation of more recalcitrant dissolved organic compounds. The methods generally agree, however, within 15% (Sharp et al., 1993).

Method 1: High-Temperature Combustion (HTC)

Water samples of the filtrate are placed into cleaned, combusted (500°C) vials and acidified to pH 3 (not lower) with ultrapurified 2N hydrochloric acid (e.g.,

EM Industries, Gibbstown, NJ). Samples can be stored if evaporation is totally prevented and organic contamination is not introduced, as for example, via the caps.

The high temperature combustion method is based on instruments (e.g., Shimadzu TOC-5000 analyzer) that utilize ultrapure oxygen gas as both a sparging and a carrier gas. Acidified water samples (10 ml) are placed into ultracleaned, combusted (500°C) vials (20 ml) and sparged for 6 minutes at a flow rate of 150 ml/min immediately before analysis. Sparged samples (100 μl) are injected into quartz combustion tubes containing catalysts (0.5% Pt on alumina overlain with oxidized combusted CuO wire mesh). The gas is further scrubbed through 25% H_3PO_4, an electronic dehumidifier, and a halogen scrubber [4,4'-methylenebis(N,N-dimethylaniline)] for saline samples of water prior to entering the infrared detector to measure the CO_2.

The carbon contents of environmental samples are compared to a series of standards prepared with ultrapure water. Often potassium hydrogen phthalate is used for the standard with dilutions to provide a range of dissolved organic carbon concentrations found in natural samples.

Method 2: Wet Chemical Oxidation

A sample of the filtrate is analyzed for DOC by first acidifying the water to pH ≈ 3 to convert all inorganic carbon to CO_2. The sample is then bubbled vigorously but briefly, to minimize evaporation, with an inert gas (O_2, N_2, or He) to sparge all inorganic CO_2 from the sample. The remaining dissolved organic carbon of the acidified sample is then treated with a strong oxidant such as potassium persulfate at elevated temperature and pressure. Organic carbon compounds are oxidized to CO_2 in sealed ampoules. The amount of CO_2 generated from the DOC is then measured by an infrared CO_2 analyzer and is a measure of the DOC of the sample.

Procedures

1. Obtain enough ampoules of 1-ml size for triplicate analyses of each water sample. Precombust for 8 h at 525°C in a muffle furnace. Store in a clean, closed container.
2. Place a known amount of potassium persulfate into each ampoule by means of a quantitative "microscoop" (which can be made from the tapered end of a plastic semimicro centrifuge tube). Add approximately 200 mg $K_2S_2O_8$ to 1-ml ampoules.
3. Prepare a series of organic carbon standards, using glucose, to extend over the range of DOC in samples to be analyzed (see "Apparatus and Supplies," p. 145).
4. Place a precise amount of sample (200 to 500 μl) into the ampoules using an accurate pipetting system (e.g., Eppendorf).
5. Add 3% phosphoric acid (see p. 145) to each sample at a ratio of 20 μl per 0.5 ml of sample.
6. Sparge the sample with purified oxygen for 10 mins, bubbling at a flow rate of about 200 ml/min. The gas should pass a U-trap of ascarite (a CO_2 absorbing compound) before entering a flow meter and stripping manifold consisting of Tygon tubing connected to 20-gauge, 15-cm stainless steel needles or cannulae.
7. With the oxygen gas still flowing into the ampoule, insert a second cannula, with flowing oxygen, affixed to a modified electrician's pliers (Fig. 9.1). Remove the

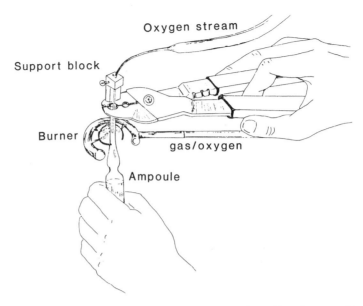

Figure 9.1. Sealing of ampoules under a continuous sparging of head space with a stream of oxygen. A support block of aluminum with a set screw to hold the cannula is affixed to the upper side of the pliers. The cannula is joined with Teflon or other heat-resistant tubing to the tubing that delivers the sparging gas. [From McDowell et al. (1987).]

initial sparging cannula. Holding the lower portion of the ampoule with a free hand, place the neck of the ampoule into a propane- or methane-oxygen microburner (Fig. 9.1) and immediately rotate while pulling downward to seal the ampoule. In this manner, the flow of oxygen continues uninterrupted throughout the sealing procedure (McDowell et al., 1987) and minimizes the entry of CO or CO_2 from the flame.

8. Autoclave the samples at 121°C for 45 min after the autoclave has come to full temperature and pressure. *Caution*: Reduction of pressure and temperature at the end of autoclaving must be done by slow exhaust. The samples may be stored indefinitely after autoclaving.

9. Place the ampoule firmly into the silicone tubing with the cannula sticking through the silicone tubing above it (see Fig. 9.2). Start the flow of N_2 gas (200 ml/min) to the infrared CO_2 spectrometer and maintain this flow until the recorder reads zero.

   Crush the end of the ampoule within the tubing with a pair of pliers. Quickly push the cannula down into the ampoule until the tip is well under the liquid. Continue passing N_2 until the recorder returns to zero. Remove the ampoule, shake out the broken glass into a waste beaker, and repeat the process with the next ampoule.

10. Measure the peak height of the curve recorded from the IR analyzer or integrate the area under the curve. The values of the glucose carbon standards and blanks (reagents only in organic-free water) should be determined similarly. The IR analyzer can be calibrated further by injecting a measured amount of CO_2 (e.g., 10 μl) with a gas-tight syringe into the flow system through the silicone tubing. With appropriate adjustments, a similar system could be used with a gas

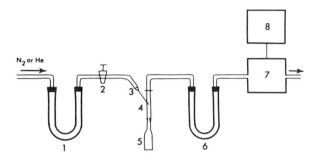

1. Ascarite   5. Ampoule
2. Precision Flowmeter   6. $CaCl_2$
3. Cannula   7. IR Analyzer
4. Silicone Tubing   8. Amplifier and Recorder

Figure 9.2. Closed system to sparge CO_2 of dissolved organic matter to an infrared spectrophotometer for analysis.

chromatograph [e.g., Stainton (1973)]. This method is potentially an underestimate (Sharp et al., 1993).

Calculations

Plot a calibration curve of CO_2 peak heights or curve areas for all standards minus blank corrections. The calibration curve should be linear for the lower concentrations and slightly curvilinear for the higher concentrations. Blank corrections should not exceed 0.5 mg C/l. Determine the concentrations of the water samples from this calibration curve.

When samples require dilution, their concentration is given by (Stainton et al., 1977):

$$C_1 = (C_2 \times D) - [C_3 - (D - 1)]$$

where C_1 = concentration of DOC in sample, C_2 = concentration of DOC in diluted sample, C_3 = concentration of DOC in dilution water, and D = dilution factor.

The smallest amount of DOC that can be detected with certainty is about 0.09 mg C/l within a range of 0.09 to 6.0 mg C/l (ca. 10 to 800 μmol C/l). Precision at the 1.0-mg C/l level lies in the range of: Mean of n determinations $\pm 0.06 n^{1/2}$ mg C/l (standard deviation of ca. $\pm 10 \mu$mol C/l).

PARTICULATE ORGANIC CARBON

A number of techniques have been developed to determine the organic carbon content of particulate organic matter [cf., Strickland and Parsons (1972), Golterman and Clymo (1969), and Stainton et al. (1977)]. Many methods involve dry oxidation under an oxygen atmosphere and measurement of the CO_2 liberated by infrared CO_2 spectrometry or gas chromatography. Sophisticated carbon-hydrogen-nitrogen-sulfur (CHNS) analyzers (e.g., Carlo Erba EA 1108, Fisons Instruments, Inc., Dearborn, MI) allow simultaneous analyses of organic C, H, N, and S in biological and detrital samples. Other techniques utilize wet oxidation where the amount of dichromate reacting with organic matter in a strongly acidic solution is determined by titration (Maciolek, 1962). The latter methods have been used for both POC and DOC, although they are much more tedious and subject to error for the analysis of DOC than is the method outlined above.

The wet oxidation procedure discussed here is simple and allows a large number of samples to be assayed without complex or specialized equipment. The method may give results slightly higher than carbon measured by dry oxidation and measurement of CO_2. Results are expressed in terms of glucose carbon (Strickland and Parsons, 1972). The true carbon content of particulate organic material would approach this value only if all the carbon were present as carbohydrate. The average composition of phytoplankton and particulate detritus is such that the true carbon content is within 10 to 20% of the oxidation value given by this procedure. Thus, the "oxidizable carbon" is a realistic measure of the energy stored in the particulate organic matter.

The particulate organic matter is filtered onto precombusted glass fiber filters, as discussed earlier. Carbon content then is determined by oxidation with a mixture of potassium dichromate and concentrated sulfuric acid. The decrease in extinction of the yellow dichromate solution resulting from reduction by the organic matter is measured spectrophotometrically.

Procedures

1. Place dry filters into the bottom of small (e.g., 50 or 125 ml) Erlenmeyer flasks with clean forceps. The filter should be flat on the bottom of the flask with the organic material upward.
2. Add suitable volumes (Table 9.1) of dichromate-sulfuric acid oxidant and distilled water from graduated pipets according to the anticipated amount of organic carbon. Immediately cover each flask by inverting a small, clean beaker (50 or 100 ml) over the neck of the flask.

    Highly eutrophic lakes and ponds seldom have less than 1500 µg POC/l, except in the winter (ca. 50 to 500 µg C/l). The POC content of moderately productive lakes usually falls into the range of 100 to 1000 µg C/l.
3. For each set of samples, prepare a duplicate blank using a clean, precombusted filter. The extinction coefficients of blanks measured against water should be between 1 and 1.1 and should be checked.
4. Calibrate the oxidant with a standard glucose solution as follows. Into five flasks place clean, precombusted filters. To two flasks add 10.0 ml of sulfuric acid-dichromate oxidant and 4.0 ml of distilled water, and to the remaining three flasks add 10.0 ml of oxidant and 4.0 ml of dilute glucose solution (see "Apparatus and Supplies," p. 145).
5. Place the flasks in a sand-bath heater or oven and heat the contents for 60 min at 100 to 110°C.
6. Cool the mixture and transfer the solution to a ground-glass stoppered graduated cylinder of a suitable volume (Table 9.1). Wash the sides of the flask with redistilled water and transfer to the cylinder; repeat the procedure again. Make the solution to final volume with distilled water and mix.
7. Allow the glass fibers to settle and cool the cylinders to room temperature by placing them in a water bath. Decant a suitable volume into labeled centrifuge

Table 9.1. Recommended oxidant and distilled water volumes and suitable spectrophotometer cell lengths.[a]

Anticipated carbon (μg/l)	Oxidant (ml)	Distilled water (ml)	Final volume (ml)	Cell length (cm)	Volume factor (F')
0–300	2.00	0.8	100	10	1
300–700	2.00	0.8	50	5	1
300–700	2.00	0.8	50	1	5
300–700	4.00	1.6	50	2.5	1
700–1000	5.00	2.0	50	1	2
1000–2000	10.00	4.0	50	1	1

[a] Modified from Strickland and Parsons (1972).

    tubes. Centrifuge at 2000 rpm for about 5 min to remove the remaining glass fibers and particles of undigested inorganic materials.

8. Pour the solution into a spectrophotometer cell and measure the extinction at a wavelength of 440 nm. The *blank solution* (with a higher extinction than the *sample solution*) should be placed in the spectrophotometer cell normally used for the samples and the *sample solution* in the cell normally used for the reference liquid. In this manner the *difference* between the sample and blank, i.e., a measure of the bleaching of the dichromate, is measured directly in the most sensitive range of the spectrophotometer. The result is a more precise value than would be obtained if two relatively large extinction coefficients were measured separately against water and then subtracted.

Calculations

1. Oxidant calibration with glucose standards:

$$F = 120/E_c$$

where F = calibration factor and E_c = mean corrected extinction of the three standards [i.e., mean absorption (OD) of distilled water blanks minus mean absorption of glucose standards]. F should be approximately 275 and need not be redetermined for the same batch of oxidant.

2. Samples POC:

$$\mathrm{mg\,C/m^3} = \frac{(1.1E)(F)(v)(F')}{V}$$

where E = mean extinction of samples at 440 nm (i.e., mean absorbance of blanks minus absorbance of samples); F = calibration factor of glucose standard solution; v = volume of oxidant used (ml); F' = factor for oxidant volume, volume of distilled water added, and cell length used, as given in Table 9.1; and V = volume of lake water filtered (1). This method can be used within a range of 10 to 4000 mg C/m^3 (Strickland and Parsons, 1972). Precision in the 800-μg C level lies in the range: mean of n determinations $\pm 120/n^{1/2}\,\mu$g carbon. Dividing by the volume of water sample used in liters yields the corresponding data in mg C/m^3, since 1 μg/l = 1 mg/m^3.

EXERCISES

OPTION 1. FIELD ANALYSES

1. Collect water samples with a Van Dorn or similar sampler from:
   a. A vertical depth series in the central depression of a lake or reservoir. When stratified, include samples within the epi-, meta-, and hypolimnion.
   b. Inlet stream water entering the lake and in the mouth of the outlet.
   c. If possible, several littoral areas, inside and lakeward from dense stands of aquatic macrophytes.
2. Transfer the water samples to clean, labeled, darkened bottles. Store in an ice chest.
3. As rapidly as possible, filter subsamples of known volume as outlined earlier in this chapter. *Gently* invert the bottles to mix the contents thoroughly before removing the samples. Repeat several times.
4. Determine the DOC and POC concentrations of the water sample as outlined earlier.
5. Plot the concentrations of DOC and POC comparatively and compare to similar analyses of aquatic ecosystems studied by other workers.
6. Answer the questions following Option 3.

OPTION 2. VARIANCE IN DOC AND POC CONCENTRATIONS

1. Collect water samples with a Van Dorn or similar sampler from the central depression of a lake or reservoir or from the same location in a stream as follows:
   a. From the same depth location, three or four replicate samples out of the same Van Dorn sampler.
   b. Obtain four replicate Van Dorn samples from exactly the same depth and remove a 1-liter sample from each for analyses.
   c. Repeat at several depths. If stratified, take replicate samples within the epi-, meta-, and hypolimnion.
2. Treat the collected samples as stated in Option 1.
3. Determine the concentrations of DOC and POC of the replicate water samples as outlined earlier.
4. Determine the variance of the values obtained from the same water sample and from replicate samplings of the same depth. Determine the standard deviation, standard error, and coefficient of variance among the replicates.
5. Answer the questions following Option 3.

OPTION 3. LABORATORY ANALYSES

1. With samples provided by your instructor, determine the concentrations of DOC and POC as outlined in Options 1 and 2.
2. Answer the following questions.

Questions

1. What is your opinion of using arbitrarily a 0.5-μm pore size filter to separate particulate and dissolved organic matter? What practical alternative methodology can you suggest?
2. The particulate organic fraction includes all organic matter greater than 0.5 μm up to and including pieces of detritus as large as logs. From the standpoint of feeding studies, such a composite category is not very meaningful. How might one subdivide the particulate detritus further? What is meant by fine and coarse particulate organic matter?

3. If one obtains the dry weight of particulate organic matter, how would the organic carbon content be estimated by direct analyses?

4. What do concentrations of DOC or POC mean energetically within an aquatic ecosystem? [See Rich and Wetzel (1978) and Wetzel (1995).]

5. A common observation in aquatic systems is that the vertical concentrations of DOC change relatively little over an annual period. What do these observations imply about the quality (i.e., the chemical composition) of the organic compounds?

6. How might the rates of utilization (turnover) of dissolved organic matter be determined? Of particulate organic matter?

7. In the water samples you obtained, what do you believe is the primary origin of the DOC measured? Of the POC?

8. What are the regions of the lake where decomposition of the DOC is likely to occur? Of the POC? Why? How might these sites change seasonally?

9. What is the difference between decomposition of organic matter by microflora and digestion-assimilation in the gut of an animal? Which process is most important in decomposition of organic matter of the whole system? Why? [See Cole (1985).]

10. What is coprophagy? How might it assist in the "processing" of particulate organic matter?

11. Some workers believe that most of the decomposition of organic matter occurs in the sediments rather than in the water column. What factors would enhance benthic respiration? Enhance decomposition in the water column?

12. Some DOC and POC leave a lake in outflow. In which form would the greatest export occur?

Apparatus and Supplies

1. Scrubbed water sampler, e.g., Van Dorn type.
2. All-glass filtration apparatus, adjustable vacuum source.
3. Glass fiber filters, e.g., Whatman GF/F (0.6- to 0.7-μm pore size; GF/C size are *not* acceptable), precombusted at 500°C for at least 1 h at temperature; forceps; micropipets (e.g., Eppendorf).
4. Dissolved organic carbon:
   a. 1-ml ampoules; muffle furnace.
   b. Potassium persulfate (analytical reagent quality).
   c. 3% phosphoric acid. Dilute 30 ml of analytical reagent grade 85% (syrupy) H_3PO_4 to 1000 ml with organic-free water. Add 10 g $K_2S_2O_8$ and heat the bottle (stopper loosened) for 4 h in a boiling water bath. Cool and store in the same bottle; stable indefinitely.
   d. Glucose standards: Dissolve 1.25 g of dry glucose in 250 ml of distilled water. Dilute 5.0 ml of this solution to 100 ml with organic-free water. Finally, dilute 5.0 ml of the latter solution to 100 ml with organic-free water:

   $$1\,ml = 1\,mg\,C/l \text{ when diluted to } 5.0\,ml$$

   Prepare duplicate standards within the range of DOC anticipated, using 0.5, 1.0, 2.0, 3.0, 4.0, and 5.0 ml of the standard solution. Determinations above about 6 mg C/l may deviate from linearity. Samples with greater concentrations of DOC can be diluted with organic-free water prior to analysis.
   e. N_2 and O_2 gas of high quality; ascarite CO_2 absorbant; silicone grease (e.g., Dow Corning); flow meters; long, 20-gauge cannulae; torch for sealing ampoules; autoclave.
   f. Nondispersive infrared CO_2 spectrometer with recorder [a gas chromatograph can also be used; see Stainton et al., (1977)]; sample gas train (as indicated in Fig. 9.2); anhydrous $CaCl_2$.
5. Particulate organic carbon:
   a. Clean 125-ml Erlenmeyer flasks, graduated 10-ml pipets, 50-ml beakers, 50- or 100-ml stoppered graduated cylinders, centrifuge tubes.
   b. Sand-bath heater or oven, 100 to 110°C.
   c. Clinical centrifuge.

d. Sulfuric acid-dichromate oxidant: Dissolve 4.84 g of potassium dichromate ($K_2Cr_2O_7$) in 20 ml of distilled water. Add this solution *slowly* to about 500 ml of concentrated (sp. gr. 1.82) AR sulfuric acid in a 1000-ml volumetric flask. Cool to room temperature and make to volume with more acid. Store in a glass-stoppered bottle and protect from dust. Stable indefinitely.

e. Glucose standard solution:
   i. Dissolve 7.50 g of pure glucose (dextrose) in water in a 100-ml volumetric flask and bring up to volume with distilled water.
   ii. Dilute 10.0 ml of the concentrated solution to 1 liter with distilled water in a volumetric flask. Use within one day.

$$1.00\,ml = 300\,\mu g \text{ of carbon.}$$

f. Spectrophotometer.

References

Cole, J.J. 1985. Decomposition. pp. 302–310. *In*: G.E. Likens, Editor. An Ecosystem Approach to Aquatic Ecology: Mirror Lake and its Environment. Springer-Verlag, New York.

Golterman, H.L. and R.S. Clymo (eds). 1969. Methods for Chemical Analysis of Fresh Waters. IBP Handbook No. 8, Blackwell, Oxford. 172 pp.

Gustafsson, Ö. and P.M. Gschwend. 1997. Aquatic colloids: Concepts, definitions, and current challenges. Limnol. Oceanogr. *42*:519–528.

Jordan, M.J., G.E. Likens, and B.J. Peterson. 1985. Organic carbon budget. pp. 292–301. *In*: G.E. Likens, Editor. An Ecosystem Approach to Aquatic Ecology: Mirror Lake and its Environment. Springer-Verlag, New York.

Lamberti, G.A. and S.V. Gregory. 1996. Transport and retention of CPOM. pp. 217–229. *In*: F.R. Hauer and G.A. Lamberti (eds). Methods in Stream Ecology. Academic Press, San Diego.

Maciolek, J.A. 1962. Limnological organic analyses by quantitative dichromate oxidation. Res. Rept. U.S. Fish. Wildl. Service *60*. 61 pp.

McDowell, W.H., J.J. Cole, and C.T. Driscoll. 1987. Simplified version of the ampoule-persulfate method for determination of dissolved organic carbon. Can. J. Fish. Aquat. Sci. *44*:214–218.

Menzel, D.W. and R.F. Vaccaro. 1964. The measurement of dissolved organic and particulate carbon in sea water. Limnol. Oceanogr. *9*:138–142.

Rich, P.H. and R.G. Wetzel. 1978. Detritus in the lake ecosystem. Amer. Nat. *112*:57–71.

Sharp, J.H. 1973. Size classes of organic carbon in seawater. Limnol. Oceanogr. *18*:441–447.

Sharp, J.H., R. Benner, L. Bennett, C.A. Carlson, R. Dow, and S.E. Fitzwater. 1993. Re-evaluation of high temperature combustion and chemical oxidation measurements of dissolved organic carbon in seawater. Limnol. Oceanogr. *38*:1774–1782.

Stainton, M.P. 1973. A syringe gas-stripping procedure for gas-chromatographic determination of dissolved inorganic and organic carbon in fresh water and carbonates in sediments. J. Fish. Res. Board Can. *30*:1441–1445.

Stainton, M.P., M.J. Capel, and F.A.J. Armstrong. 1977. The Chemical Analysis of Fresh Water. 2nd Ed. Misc. Special Publ. 25, Fish. Environ. Canada. 180 pp.

Strickland, J.D.H. and T.R. Parsons. 1972. A Practical Handbook of Seawater Analysis. 2nd Edn. Bull. Fish. Res. Bd. Canada *167*. 310 pp.

Wallace, J.B. and J.W. Grubaugh. 1996. Transport and storage of FPOM. pp. 191–215. *In*: F.R. Hauer and G.A. Lamberti (eds). Methods in Stream Ecology. Academic Press, San Diego.

Wetzel, R.G. 1984. Detrital dissolved and particulate organic carbon functions in aquatic ecosystems. Bull. Mar. Sci. *35*:503–509.

Wetzel, R.G. 1995. Death, detritus, and energy flow in aquatic ecosystems. Freshwat. Biol. *33*:83–89.

Composition and Biomass of Phytoplankton

The structure of photosynthetic populations in aquatic ecosystems is dynamic and constantly changing in species composition and biomass distribution. An understanding of community structure is dependent on an ability to differentiate between true population changes and variations in spatial and temporal distribution. Changes in species composition and biomass may affect photosynthetic rates, assimilation efficiencies, rates of nutrient utilization, grazing rates, and so on.

SAMPLING

Phytoplankton in the open water of a lake or stream often is sampled by means of water bottles such as the Van Dorn model (Fig. 7.1). These samplers are lowered open to specific depths and then are closed by means of a weight messenger that is dropped along the cable to trip the closing mechanism. Since a number of analyses may be done on the same water sample, it is important that the surfaces of the sampling device in contact with the water sample be chemically inert.

A Van Dorn or similar sampler may be too large to use in very shallow habitats or in specialized situations where acute microstratification of phytoplankton populations is of interest. Horizontally held Van Dorn samplers frequently cause major water disturbance by creating turbulence. Small pumps driven by battery-powered motors or simple hand-pump vacuum siphoning systems can be effective with minimal disturbance at shallow depths. Sampling of phytoplankton populations near the sediments may require coring devices (see Exercise 12), from which samples can be removed carefully from the overlying water.

The manner in which sampling is done should conform to the objectives of the studies. Analyses of species composition and biomass of algal populations are time-consuming. Ecologically useful evaluations of species composition and biomass may be made more effectively by less precise analyses of replicated samples from the same place rather than by very precise analyses of a single sample.

Tubular or integrating water samplers [e.g., Schröder, (1969) and Straškraba and Javornický (1973)] permit the collection of water from the surface to some preselected depth for a composite plankton sample. Such integrated samples may give more meaningful data than single depth samples and are a practical necessity when investigating horizontal variations in populations, as in large reservoirs where marked physical and chemical gradients exist.

Plankton nets are not recommended for either qualitative or quantitative analyses of phytoplankton. A large percentage of important algal species is much smaller than the mesh opening dimensions of nets even of the finest mesh size. Furthermore, the volume of water passing through most nets is very difficult to measure.

Preservation

Whenever possible, phytoplankton species should be examined while alive, particularly delicate species of flagellated algae. Algae may be maintained for several hours without appreciable deterioration when kept cold during transportation to the laboratory. Usually, however, preservation is necessary for longer storage periods.

Samples can be preserved in a 0.5 to 2% buffered formalin solution and should be used for preservation of picophytoplankton (see p. 156), although formaldehyde tends to cause rupture, deformation, and shrinkage of certain algae. A better preservative is Lugol's solution, added to samples to yield a 1% final concentration (1:100). The absorption of iodine from Lugol's solution by the cells also promotes settling when the sedimentation-inverted microscopy technique, discussed below, is used. A small amount of glycerol also is added commonly to samples preserved in Lugol's solution to prevent drying during long periods of storage. A 3% solution of glutaraldehyde in final concentration, neutralized to pH 7 with NaOH, is also a good preservative that results in minimal shrinkage and distortion of cells. See "Apparatus and Supplies" (p. 171) for further details.

QUANTITATIVE EVALUATION OF NUMBERS AND BIOMASS OF SPECIES

Detailed analyses of phytoplanktonic populations require estimation of numbers and volume of each species. Phytoplankton consist of individual cells, filaments, and colonies. It is best to count cells, although this procedure is impractical in the case of many multicellular colonies. When colonies of species are counted, it is important to determine the average number of cells per colony. The number of cells per colony can vary spatially within a lake and seasonally with changes in the population vigor of a species. In filamentous algae, the average length of the filaments should be determined.

Cell numbers often do not represent true biomass because of considerable variation in sizes of cells among algal species. This disparity can be evaluated by multiplying the number of cells of a given species by its average cell volume and then summing these volumes over all species. Cell volume is estimated from knowledge of mean cell dimensions and correspondence of cellular shape to geometric solids or combinations of simple solids [spheres, cones, truncated cones, cylinders, etc.; cf., Sicko-Goad, et al., (1977)]. The volume of physiologically inert wall material ranges from nil, in some flagellates, to over 20% of the total cell volume in certain diatoms (Sicko-Goad, et al., 1977). Although a number of tables have been published [e.g., Table 8.1 in Wetzel (1983, 150)] listing the cell volumes of various algal species, these values must be viewed as being only approximate. Cell dimensions of a species can vary greatly from season to season or from lake to lake. The cell volume of each important algae should be determined for each sample.

Biovolume

Cell volumes are calculated for each species from formulae for solid geometric shapes that most closely match the cell shape based on cell dimensions. The extent of change in cell dimensions caused by preservatives should also be evaluated for important species by comparison with living specimens [e.g., Borsheim and Bratbak (1987)]. A microscope fitted with an internally reflecting drawing prism, allowing for simultaneous binocular viewing and projection of an enlarged image of the specimens on a table top, enhances the accuracy of measurements [cf., Tyler (1971)]. Computers with appropriate software and other electronic devices are available to digitize and measure two-dimensional distributions of data such as line graphs and irregular objects (e.g., Sigma Scan measurement systems, Jandel Scientific, Sausalito, CA).

When counting and evaluating the size of algae, some nondestructive method of concentrating the organisms in a water sample is usually necessary. Methods commonly used include sedimentation of algae onto a surface, filtration onto a surface that can be rendered transparent, and centrifugation. By far the best method is the sedimentation technique, because the algae settle by gravity onto a glass surface in a random distribution and are not subjected to potentially disrupting vacuum or pressure forces.

SEDIMENTATION AND ENUMERATION BY INVERTED MICROSCOPY

Phytoplankton samples to be observed by the sedimentation technique should be preserved with Lugol's solution and stored in clear glass bottles, and in darkness, until analyzed. The absorption of iodine from Lugol's fixative stains the algae but also adds weight, which serves to accelerate their sedimentation in the chambers. When too little Lugol's solution is added, gas occlusions of cyanobacteria may not be discharged, and sedimentation of these forms may be incomplete as a result.

Sedimentation chambers, originally developed by Utermöhl in the 1930s [reviewed in Utermöhl (1958)], are available commercially from a number of manufacturers (e.g., Zeiss, Germany; Wild Instrument Co., Switzerland; Prior and Co., England; Unitron Inc., U.S.A.; and others). Chambers also can be constructed from cast acrylic tubing (not extruded tubing, which can have ripple imperfections) and large coverslips. In the latter case, the volume of each chamber must be determined carefully.

Sedimentation chambers are normally manufactured with volumes of 1, 5, 10, 25, 50, and 100 ml (Figs. 10.1 and 10.2). The chambers and accompanying coverslips must be cleaned thoroughly to avoid contamination with organisms from previous samples. The bottom coverslip is mounted carefully in the retention ring and screwed to the cylindrical chamber for a pressure fitting. (*Caution*: The coverslips are expensive and easily broken if overtightened.) Preserved samples in bottles must be mixed uniformly by very *gentle* inversion and then poured into the chambers. Samples from oligotrophic waters often require settling of 50 or 100 ml, while samples from more productive waters with increased plankton densities are

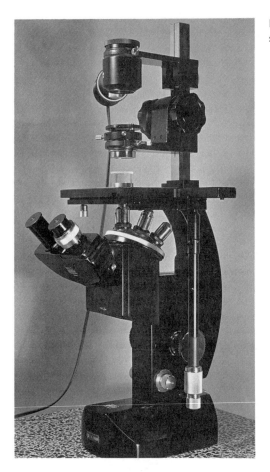

Figure 10.1. A Wild M40 inverted compound micro-scope and a sedimentation chamber.

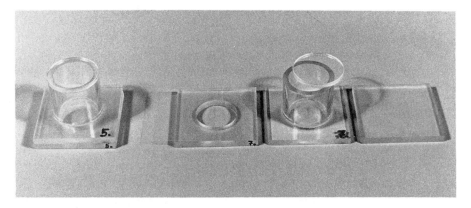

Figure 10.2. A compound algal sedimentation chamber, assembled (*left*) and disassembled (*right*).

sampled adequately with 5 or 10 ml. The suspected density of organisms should determine the size of the chamber selected.

Fill the sedimentation chamber to the top edge with sufficient excess to permit the water to "bead" upward above the edge. Slide the chamber cover cap across the top of the cylinder to remove any excess water and to enclose the sample of exact volume without entrapping any air bubbles. In order to ensure complete sedimentation of all organisms, sedimentation time in hours must be at least three times the height of the sedimentation chamber in centimeters. Place the chamber in a vibration-free area during the sedimentation process. Where wide fluctuations in room temperature occur, the chambers should be covered (e.g., with an inverted styrofoam box) during sedimentation in order to avoid convection currents.

Inverted Microscope. The basic difference between an erect and an inverted microscope is that, in the latter, the positions of the objective and condenser are reversed in relation to the stage (Fig. 10.1). The inverted design allows for observation through the bottom of containers with bright-field, dark-field, phase-contrast, and fluorescent procedures. Use of this microscope is limited by the thickness of the bottom cover glass and by the working distance of the condenser, which must be long enough to allow the chamber to rest between the stage and the condenser across the optical axis.

Because of these limitations in resolution, it is difficult to use the sedimentation chamber-inverted microscope at magnifications much greater than 500 to 600×. With special chambers, discussed below, and ultrathin coverslips, the resolution of the inverted microscope can be extended to include organisms smaller than 10 μm.

The sedimentation chambers are placed carefully on the stage of the microscope; great care is taken not to disturb the sedimented organisms. Identification, counting, and size measurements then are made according to procedures outlined below.

OTHER MODIFIED SEDIMENTATION CHAMBERS FOR USE WITH A COMPOUND MICROSCOPE

A number of sedimentation chambers have been designed that permit removal of the vertical cylinder after setting of the algae. The resulting base containing the sedimented algae in a very small, flat depression chamber is covered with a coverslip. The algae in this chamber can be viewed on either a conventional microscope or an inverted microscope.

The simplest compound chamber [after Lund (1951) and Utermöhl (1958)] has a base unit that contains a shallow cylindrical chamber of a volume of about 2cm^3 and a height of about 4 mm (Fig. 10.2). The top and bottom sides of the base unit are ground flat. The lower surface of the base plate is coated lightly with water-insoluble grease (e.g., Apiezon M stopcock grease) and accepts a large, thin coverslip. The upper tubular chamber is open below and is glued permanently to a square base of the same size as the base chamber. The lower surface of the base plate of the tubular chamber is ground flat.

During use, the upper chamber, lightly coated with grease at the base, is centered precisely over the hole of the lower chamber. After the algae have sedimented, the upper chamber is slid laterally onto a square of plastic of identical thickness as the base chamber, while the exposed upper surface of the base chamber simultaneously is covered with a coverglass. In this manner, the water of the upper chamber is removed and that of the lower chamber is intact and covered. The sedimented algae can be counted with an inverted microscope or a compound microscope at low to moderate magnification.

An excellent modification of this compound chamber has been devised to permit permanent mounting of the slide (Coulon and Alexander, 1972). These slides then can be viewed with a conventional microscope. The simple apparatus consists of a cylinder, permanently affixed by glass epoxy cement, to a standard microscope slide that has an off-center hole of diameter identical to that of the inside of the cylinder (Fig. 10.3). Another slide with a central hole of the same diameter and then an unmodified slide are placed under this chamber. In practice, the three slides are aligned so that the holes form a continuous chamber (Fig. 10.3). The plankton sample is added, covered, and allowed to sediment. Eight drops of 50% neutralized glutaraldehyde per 5 ml of sample can be added directly to the chamber before sedimentation to serve as a mounting medium. After settling, the top chamber slide is slid *gently* to the side and removed without disturbing the lower two slides. Simultaneously, a coverslip can be slid over the hole of the middle slide as the upper chamber is removed. Alternatively, the material in the hole of the middle slide can be evaporated to dryness and then the second slide can be separated from the third by gently inserting a razor blade between them at one corner. The algae then can be either counted directly or mounted permanently in ordinary fashion.

MICROSCOPE CALIBRATION PROCEDURES

It is necessary to calibrate each microscope against a known scale, since every instrument differs in magnification according to the combination of oculars and objectives. A marked glass disc, termed a Whipple ocular micrometer (or reticule, Whipple grid, or plankton counting square), is installed in the plane of the real image in the ocular (Fig. 10.4). As a result, the image of the specimen being examined appears to be immediately under the markings on the ocular micrometer. Using the micrometer, the dimensions of objects can be measured directly.

The Whipple ocular micrometer contains an accurately ruled grid that is subdivided into 100 squares (Fig. 10.4). One of the central squares is subdivided further into 25 smaller squares. The ocular micrometer is installed by carefully unscrewing the upper lens mounting of the ocular (usually the right one of a binocular microscope) and then placing the reticule on the circular diaphragm found about halfway down in the ocular. After the ocular is reassembled, observe the markings; when they are not in focus, remove the reticule and reverse it. Using a stage micrometer (Fig. 10.5), determine the dimensions of the ocular micrometer grid according to the procedure outlined in Fig. 10.6. These measurements apply only to the specific combination of oculars and objectives on the microscope for which they were determined.

Alternatively, a series of parallel and perpendicular cross hairs, made of molten glass drawn to very fine threads, can be affixed permanently to the internal flange within the ocular [see Lund et al. (1958) for details] or a Porton sizing reticle [see Brock (1983)]. Calibration procedures would be the same as discussed for the Whipple reticule.

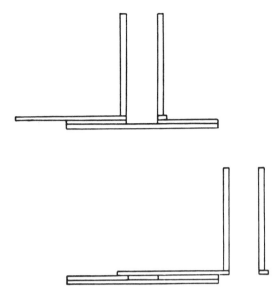

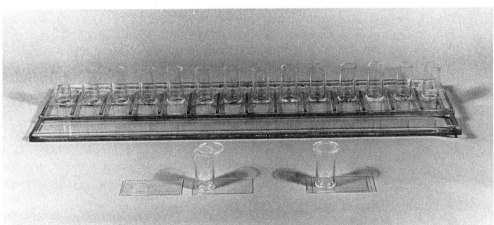

Figure 10.3. Diagram of a compound sedimentation chamber made from modified microscope slides. *Upper*: Assembled, ready for sedimentation of sample. *Lower*: With upper chamber moved laterally to remove excess water. The tubular chamber then can be removed and the lower pieces covered with a large coverslip, as indicated in the photograph. The large glass support base assists in removing chambers without disturbance of sedimented materials. [Modified from Coulon and Alexander (1972); see Crumpton and Wetzel (1981).]

COUNTING AND SIZE DETERMINATIONS

Counting procedures are similar whether sedimentation chambers with an inverted microscope or slides or counting cells such as those discussed below are used. Before counting, a portion of the edges and central areas of the chambers or slides should be examined to ensure that definition is good and distribution of the organisms is relatively uniform. Larger species tend to congregate near the edges of the chambers during sedimentation, whereas smaller species are slightly concentrated in the central part of the chamber.

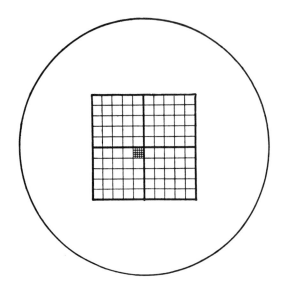

Figure 10.4. The rulings of a Whipple ocular micrometer reticule.

By moving the mechanical stage, the entire bottom of the sedimentation chamber, counting cell, or slide area can be examined in a systematic manner. Organisms lying between two parallel cross hairs are counted as they pass a vertical line. Proceed from left to right from edge to edge; move down to the next full field within the horizontal parallel lines, and then, in the next row, proceed from the right edge to the left. Only part of the optical field is covered by the first and the last traverses because of its circular nature. When organisms lie across a horizontal line, those along the upper one are counted as lying within the lined area, but those lying across the lower are not counted. When the next lower area is counted, those along the previous lower line will overlap with the upper one and be counted.

Often time constraints prohibit counting of all organisms of the entire area of the tube bottom, cell, or slide when the organisms are dense. In this case, portions of the total are counted, such as several diagonal rectangular strips near the center of

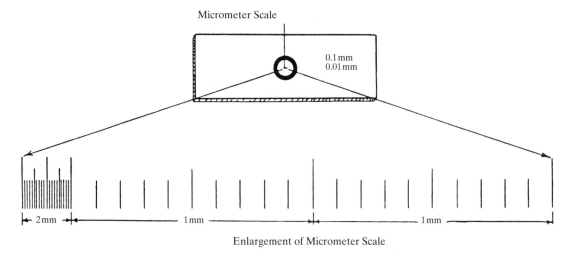

Enlargement of Micrometer Scale

Figure 10.5. A stage micrometer with the scale enlarged. [From Jackson and Williams (1962).]

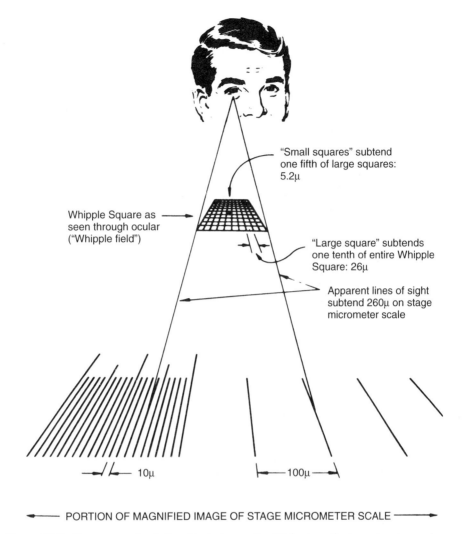

"Small squares" subtend
one fifth of large squares:
5.2μ

Whipple Square as
seen through ocular
("Whipple field")

"Large square" subtends
one tenth of entire Whipple
Square: 26μ

Apparent lines of sight
subtend 260μ on stage
micrometer scale

——— 10μ

|——100μ——|

—— PORTION OF MAGNIFIED IMAGE OF STAGE MICROMETER SCALE ——→

Figure 10.6. The apparent relationship between the Whipple reticule and a stage micrometer as seen through a microscope at a magnification of about 430 × (10 × ocular and 43 × objective). [From American Public Health Association (1989).]

the field from one edge to the other. Or, when the organisms are particularly dense, diagonals may be counted over specific distance, e.g., 10 cm, as determined by stage calibrations and mechanical stops on the stage of the microscope. Knowing the surface area of that portion counted in relation to that of the total, a factor can be determined to expand the average counts of several replicates to the total area of the counting surface. This total area represents the number of organisms per given volume of sample; this volume then can be expanded by an appropriate factor to yield the organisms per liter of water from the given depth of the lake.

Use of phase contrast optics is recommended. Phase contrast takes advantage of slight differences in refractive indices between the water and the cells and increases their contrast. Definition of refractile vacuoles, nuclei, eyespots, flagella, and other cellular components is enhanced, and thus their identification is made more certain.

Table 10.1 Accuracy obtained at 0.95 confidence limits at differing size of counts.[a]

Number of organisms	Approximate 0.95 confidence limits	
	As percentage of count	Range
4	±100%	0–8
16	±50%	8–24
100	±20%	80–120
400	±10%	360–440
1,600	±5%	1,520–1,680
10,000	±2%	9,800–10,200
40,000	±1%	39,600–40,400

[a] After Lund et al., 1958.

Sources of Error

The main sources of error in enumerating organisms include those associated with sampling and counting techniques. A particular problem arises as a result of counting colonies and then multiplying them by a mean number of cells per colony to derive an estimate of cell number (Lund et al., 1958). Many studies are concerned with generations of populations, or changes in abundance. In such studies, a method that estimates abundance to an accuracy of ±50% of the true value often is adequate, as the time required to obtain greater accuracy increases exponentially. Significance at the 95% level (one chance in 20 of being incorrect) is a reasonable working objective in most biological studies.

It generally is true in enumeration of plankton that the only important source of error is the random error associated with sampling, subsampling, counting, and conversion of colony counts to the estimates of cell population density (Lund et al., 1958; Javornický, 1958). Randomness generally is tested with the chi-square (χ^2) statistical test (see Appendix 2). When the distribution of organisms is random, it is possible to calculate the number of organisms that must be counted to obtain various accuracies at, say, the 95% confidence limits (see Table 10.1). Accuracy depends on the total size of the count, which may be comprised of a few large samples or many small samples. At least 100 individuals of each of the important species should be counted in each sample.

Picophytoplankton

Picophytoplankton are generally unicellular algae and cyanobacteria that range in size between 0.2 and $2\,\mu$m. The picophytoplankton can constitute a major portion of the phytoplankton and can be metabolically significant because of their rapid rates of reproduction. Because of their small size, they do not settle readily in sedimentation chambers; other techniques, such as the epifluorescence microscopy methods of MacIsaac and Stockner (1993; cf. also Olrik et al., 1998) that follow, are needed for enumeration.

Picophytoplankton samples must be preserved with formaldehyde to a yield a 0.2% final concentration. If picophytoplankton have been preserved with Lugol's fixative, the samples must be bleached with saturated sodium thiosulphate solution (ca. one drop per 5 ml of sample) until the yellow coloration disappears. The filtration procedure consists of prewetting the support filter with pure water and placing on the filter holder. A black-stained polycarbonate filter of 0.2-μm pore size then is placed on the support filter with the black side upward. After mounting the filtration tower, 1 to 25 ml, depending on the water quality, of thoroughly mixed (gentle inversion of sample bottle at least 20 times) sample is filtered with a vacuum differential of less than 0.3 atm (<5 kPa). Immediately after filtration, the filter is removed with forceps and held in the air to dry for about one minute. The filter, with the black side onto which the plankton sample has been filtered facing upward, is placed directly *onto* one drop of immersion oil on a slide. Another drop of immersion oil is added on top of the filter and overlain with a cover slip. The sample is now ready for examination but could be frozen for long-term storage (up to 6 months).

With an epifluorescence microscope at a magnification of 1000–1250× with a 100× objective with immersion oil, analyze picophytoplankton cells under green-yellow waveband (520–560 nm) for excitation pigments of cyanobacteria, where those species with dominance of phycocyanins appear red and those with type II phycoerythrin appear orange. Then examine cells under violet-blue excitation (400–500 nm) for excitation of chlorophyll of eukaryotic algae, yielding deep red, and for type I phycoerythrin of cyanopicoplankton (bright yellow-orange). Approximately 100 to 200 individuals of each of the most common types are counted within about 20 random microscope fields. Dimensions of counted cells should be measured with the calibrated ocular micrometer. Then:

$$\text{Cells/ml} = \frac{(C)(A_e)}{(n)(v)(A_s)}$$

where C = number of organisms counted; A_e = effective filter area (mm²), determined by measuring the diameter of the filtration tower; n = number of fields enumerated; v = sample volume corrected for preservative addition; and A_s = area of counting field (mm²). If colony-forming cyanopicoplankton are encountered, they should be counted separately from unicellular forms.

Palmer-Maloney (P-M) Nannoplankton and Other Counting Cells

The P-M cell consists of a circular chamber (17.9 mm in diameter, 0.4 mm deep) and two narrow channels through the wall (Fig. 10.7). When covered with a coverslip (20 to 22 mm diameter or square), the disc-shaped area holds 0.1 ml of sample. The cell is filled with the coverslip in place by introducing the sample into one of the side channels.

The primary advantage of the P-M and similar cells for counting organisms is that their small thickness permits use of 43 to 45× microscope objectives. The major disadvantage is the small volume, usually 0.1 ml. Since, generally, it is unconcentrated plankton samples that are counted, a rather high plankton density (range of ca. 10 plankters per field) is required to yield statistically reliable data. Small phytoplankton (less than 15 μm) often constitute the major component of algal populations, and their densities and biovolumes must be evaluated. As alluded to earlier, the rate of production by small algae having short generation times (high turnover rates) easily can exceed that of more numerous slower growing algae of a larger size. In the absence of available sedimentation chambers in which small as well as large phytoplankton algae can be counted, the P-M cell (or similar cells) should be used.

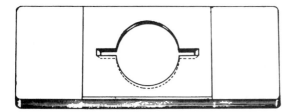

Figure 10.7. The Palmer-Maloney plankton cell for counting phytoplankton. [From Palmer and Maloney (1954).]

Either strips or fields are counted; often two perpendicular strips that cross at the center or a number (e.g., 30) of randomly selected fields are counted. Then,

$$\text{No.}/\text{ml} = \frac{(C)(1000\,\text{mm}^3)}{(A)(D)(F)}$$

where C = number of organisms counted; A = area of a field in mm^2 (Whipple grid image); D = depth of a field in mm (P-M cell depth, usually 0.4 mm); and F = number of fields counted. Multiply or divide the number of cells/ml by a dilution or a concentration factor of the sample, when necessary.

Medical hemocytometers, which have a ruled grid on the counting base and are fitted with a ground coverslip, also can be used as counting cells for small algae. The chambers are usually 1 mm in depth and the grid is divided into 1-mm squares. The Petroff-Hausser counting chamber, designed for bacterial counts, also can be used for phytoplankton enumeration at high magnification. Detailed instructions for use and conversion of counts to number per milliliter are provided for each device by the manufacturer.

FILTRATION ONTO MEMBRANE FILTERS

A number of methods of similar principle have been devised to concentrate phytoplankton onto membrane filters. The filter material then is cleared by various means and the filter can be mounted permanently on a microscope slide. The advantages of this method are that high magnification can be used and the samples are permanent. Severe disadvantages exist, however. By applying pressure or a vacuum, even at low pressure differentials of 0.3 atm, which is necessary to pass water through membranes of a pore size of 0.5 μm, severe rupture and deformation of small, delicate algae often occur. Flagella are almost always lost as microflagellates strike the filter surface. Certain algae with heavy cell wall structures, such as diatoms, are preserved relatively well. When using this method, examine separate samples by other techniques as well.

Procedures

Well-mixed, preserved water samples of known volume should be passed through gridded, white membrane filters of a pore size of ca. 0.5 μm (e.g., Millipore HA, gridded, 0.45-μm pore size). Vacuum of less than 0.5 atm should be applied to the filtration system.

Labeled microscope slides are prepared, and two or three drops of microscope immersion oil are placed onto each slide. The gridded filters are removed from the filtration apparatus while they are still wet and placed directly *onto* the small pool of immersion oil. A small ring weight of sufficient diameter to hold down the edges of the filter but not to come into contact with the algae is then placed carefully on the filter. Add two drops of oil on the top of the filter. Under normal conditions of humidity, the oil will displace the water and render the filter transparent within 12 to 24h at room temperature. A mounting medium of good refractive index (e.g., Fisher Permount) then can be applied, and the cleared filter is covered with a coverslip. A number of modifications, particularly in relation to staining techniques, are discussed in McNabb (1960), Holmes (1962), Moore (1963), Clark and Sigler (1963), and deNoyelles (1968).

Enumeration of membrane filters is confounded by the distribution of the algae. The distribution of organisms on the filter is nonrandom, since there is a tendency for algae to be concentrated near the edges of the filters (Holmes, 1962). Therefore it is recommended that the entire filter be counted or at least on strips covering the entire diameter of the filter. Knowing the size of the area examined, the total area of the circle onto which the algae were filtered ($A = \pi r^2$), and the volume of water filtered, it is simple to convert to number of organisms per milliliter.

FLOW CYTOMETRY

A flow cytometer senses individual particles of a suspension flowing through a sensing laser beam. The design restricts the particles to individual cells that pass through the beam. The resultant scattering and absorption of light allows rapid counts and measures of individual cells (Olson et al., 1993). Flow cytometry allows physical separation of cells based on combinations of optical characteristics, such as specific fluorescence of different groups of phytoplankton. These techniques are evolving rapidly and are being combined with specific molecular probes and dyes. For example, bacterial abundance in plankton can be examined with flow cytometry and specific fluorescent nucleic acid stains (del Giorgio et al., 1996).

OTHER CONCENTRATION METHODS

Centrifugation

Concentration by centrifugation is not recommended as a quantitative technique to collect algae from water. Removal is not complete at normal centrifugal speeds, damage to many algae is great, and it is difficult to transfer algae quantitatively from the centrifugation apparatus to counting chambers.

Water Displacement

A plastic tube (e.g., Plexiglas) to which a membrane filter is bonded with ethylene dichloride, may be immersed into a flask containing a preserved sample of known volume (Fig. 10.8). Water flows slowly upward through the filter and can be removed with a pipet or by gentle aspiration (Dodson and Thomas, 1964). The weight of the tube itself usually is adequate to ensure a steady but nondamaging flow through the filter. The pore size of the filter should be as large as possible while still not permitting the algae to pass through. Entrapment of algae on the filter should be checked, although usually it is minimal.

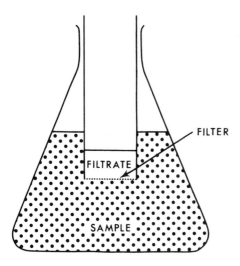

Figure 10.8. Tubular device for concentrating plankton without the use of vacuum or pressure. [Modified from Dodson and Thomas (1964).]

The volume of the resulting concentrate must be determined. Subsamples from the concentrate then may be enumerated by methods previously discussed, e.g., the Palmer-Maloney cell.

BIOMASS

Fresh and Dry Weight

Population numbers of algal species, multiplied by their respective volumes, weights the contributions of individual species on the basis of size. These values can be converted to fresh weight biomass by multiplying population number × cellular volume × density.

The measurement of dry weight of natural populations of algae is not practical since no reasonable method exists to separate algae from particulate detritus in the water. The weight of particulate detritus in fresh water usually greatly exceeds that of the algae and often constitutes well over 80% of the total seston (cf., Lundgren, 1978; Wetzel, 1999). Moreover, drying by heat results in an appreciable loss of volatile organic constituents.

Cell Volumes

As noted earlier, cell volume (biovolume) provides a much more accurate evaluation of cellular biomass because of the great differences in cell dimensions among species and sometimes seasonally among the same species under different growth conditions. Cell volumes are calculated for each species by applying cellular dimensions to formulae for solid geometric shapes most closely matching the shape of the cells.

Examples of geometric shapes are given in Fig. 10.9. A much more comprehensive set of geometric shapes and mathematical equations for calculating biovolumes of >850 pelagic and benthic fresh water and marine microalgal genera are presented in the detailed article by Hillebrand et al., 1999. This article should be consulted in comprehensive comparative analyses of microalgal biovolumes.

Shape	Diagram	Formula	Representative species
Sphere	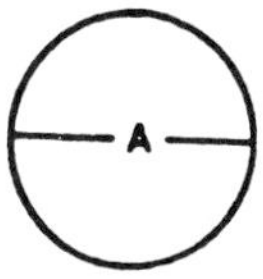	$\pi A^3/6$	*Sphaerocystis schroeteri*
Ellipsoid		$\pi AB^2/6$	*Scenedesmus bijuga* *Crypomonas* *Euglena*
Rod		$\pi AB^2/4$	*Melosira granulata* *Cyclotella* *Asterionella*
Two cones		$\pi AB^2/12$	*Ankistrodesmus falcatus*
One cone	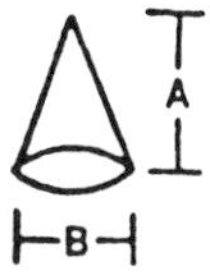	$\pi AB^2/12$	(Horn of *Ceratium*)
Rectangular box	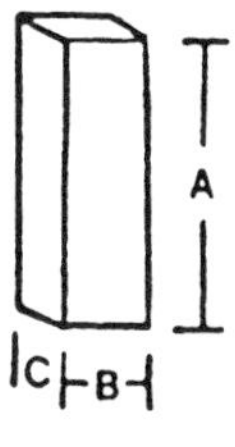	ABC	

Figure 10.9. Representative formulae to estimate biovolume from dimensions of cells [Modified from Kellar et al. (1980)].

Other approaches for estimating biomass are to determine the concentration of a particular chemical constituent of the phytoplankton per unit volume or per unit surface area of water. The composite total may be reported without a breakdown by species, such as total weight, or it may be reported as a concentration of an algal-specific chemical constituent, such as photosynthetic pigments, carbon, or nitrogen.

Shape	Diagram	Formula	Representative species

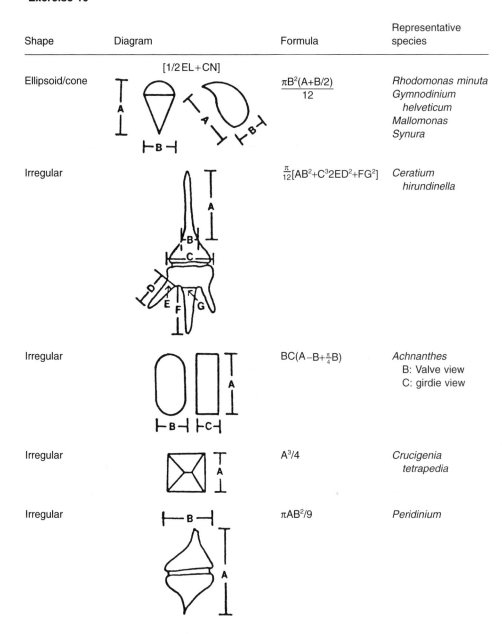

Ellipsoid/cone	[1/2 EL+CN]	$\dfrac{\pi B^2(A+B/2)}{12}$	*Rhodomonas minuta* *Gymnodinium helveticum* *Mallomonas* *Synura*
Irregular		$\frac{\pi}{12}[AB^2+C^3 2ED^2+FG^2]$	*Ceratium hirundinella*
Irregular		$BC(A-B+\frac{\pi}{4}B)$	*Achnanthes* B: Valve view C: girdie view
Irregular		$A^3/4$	*Crucigenia tetrapedia*
Irregular		$\pi AB^2/9$	*Peridinium*

Figure 10.9. (*Continued*)

Organic Carbon of the Algae

The determination of total particulate organic carbon is discussed in Exercise 9. Here again, no practical method exists to separate particulate detritus from the phytoplankton.

An estimate of organic carbon content of algae can be determined by the general ratio of cellular carbon to cell volume. A ratio of cellular organic carbon (in μg) to cell volume (in μm^3/l) of 0.10 has been found to be relatively constant for numerous algae (Mullin et al., 1966; Srathmann, 1967). This value, however, can be viewed as no more than an approximate one [cf., Sicko-Goad et al. (1977)]. An approximate cell volume to cell carbon conversion is $0.2\,\mathrm{pg\,C}/\mu m^3$ for preserved nondiatom algal species ($0.1\,\mathrm{pg\,C}/\mu m^3$ for living flagellates) (Strathmann, 1967; Redalje and Laws, 1981; Borsheim and Bratbak, 1987). Cellular carbon often correlates better with surface area than cell volume for diatoms (Strathmann, 1967; Bellinger, 1974).

Carbon biomass of algae is often of interest in developing carbon budgets in relation to food-web analyses and organic matter fluxes. In spite of great uncertainty, carbon biomass of algae can be estimated from volume-based biomass, with the assumption that the density of the organisms equals that of water ($1\,mm^3/l = 1\,mg/l$ or $1\,\mu m^3/l = 1\,\mu g/l$). Based on estimates by a number of workers (Sicko-Goad et al., 1977; Lundgren, 1978; Ahlgren, 1983; Olrik et al., 1998), approximate values for carbon biomass can be made from volume biomass:

C = phytoplankton carbon (μg C/l)

B = phytoplankton biomass ($\mu g/l$ wet mass, from biovolume)

Then:

Cyanobacteria	$C = B \times 0.22$
Thecate dinoflagellates	$C = B \times 0.13$
Diatoms	$C = B \times 0.11$
Chlorophytes	$C = B \times 0.16$
All other phytoplankton species	$C = B \times 0.11$

Pigment Concentrations

Measurements of the concentration of photosynthetic pigments of algae and cyanobacteria can be used to estimate the composite biomass of phytoplanktonic populations. Although pigment concentrations generally are correlated to the biomass of phytoplankton, pigment concentrations of algae can vary widely depending on metabolism, light, temperature, nutrient availability, and many other factors. In addition, pigments of certain bacteria, especially those of photosynthetic and non-photosynthetic sulfur bacteria, can interfere with the analysis of chlorophyll in algae and cyanobacteria. Chlorophyllous pigments also degrade to relatively stable phaeophytin products, which interfere with the spectrophotometric or fluorometric determinations of chlorophyll. Phaeophytic concentrations, however, can be estimated separately on the same samples for which chlorophyll is determined. Thus, pigment analyses can yield a sensitive approximation of algal and cyanobacterial biomass but, because of physiological variability and differences in the efficiencies of analyses, interpretation of these data must be done with care.

Plant pigments of algae and cyanobacteria consist of the chlorophylls and carotenoids (carotenes and xanthophylls). The three major chlorophylls, a, b, and c, absorb light maximally at specific wavelengths when dissolved in organic solvents. From these absorption characteristics an estimate can be made of the concentrations of the pigments. Chlorophyll a is by far the most dominant chlorophyllous pigment and occurs in greatest abundance. Thus often chlorophyll a alone is used to estimate algal biomass. The spectrophotometric estimate of concentrations of the chlorophyll c moieties as have been estimated in the past (e.g., Jeffery and Humphrey, 1975) by a trichromatic method of analysis is at best only approximate and has been replaced by high performance liquid chromatography (discussed below). The monochromatic spectrophotometric method for estimation of chlorophyll *a* is included here as a very general method that yields approximations (Rott, 1980) but employs widely available instrumentation. The method is rapidly being superseded by the fluorometric method of greater sensitivity.

Degradation products of the chlorophylls, termed phaeopigments, are similar structurally to chlorophyll except that the magnesium is lost from the ring structure and side chains are altered. These phaeopigments include several chlorophyll *a* degradation products (several phaeophytins and phaeophorbides) that can be present in high concentrations. Accurate separation and quantification of these products can be obtained only by chromatographic methods. Phaeopigments absorb light at the same wavelengths as the chlorophylls but less strongly. Since both chlorophylls and phaeopigments occur in fresh waters in variable amounts, depending on environmental conditions, concentrations of both must be estimated for each sample. In both the spectrophotometric and fluorometric methods discussed below, the total amount of pigment is determined in alkaline acetone (chlorophyll plus phaeopigments). The same sample is then acidified, during which time the chlorophyll is degraded to phaeopigments, so that the chlorophyll concentration can be determined by difference.

Chromatographic Method

It is now widely accepted that the most accurate means of quantifying chloropigments and carotenoids is by high-performance (ion-pairing) liquid chromatography (HPLC). Pigments are extracted in an organic solvent, such as 100% acetone, and separated under high pressure with a reversed phase C_{18} column with a gradient of solvents (e.g., Riaux-Gobin and Klein, 1993; Schmid et al., 1998; Wiltshire et al., 1998). Separated products are compared against pigment standards, only a few of which (chlorophyll *a* and *b*, β-carotene) are available commercially; others must be extracted from cultures and identified by sophisticated chemical techniques. In addition the HPLC instrumentation is expensive, and the technique is time consuming (>30 min per analysis, once standardized).

Sample Preparation. Water samples should be filtered through either membrane or glass fiber filters. The pore size of the filters must be sufficiently small to retain all algae of the smallest dimensions. When membrane filters are used, a pore size of 0.45 μm (e.g., Millipore HA) is recommended; when glass fiber fiters are employed, a pore size of 0.5 to 0.7 μm should be used (e.g., Whatman GF/F). The pressure differential during filtration should not exceed 0.3 atm, to minimize damage to delicate organisms. The amount of sample required will vary according to the concentration of phytoplankton. While 0.2 l may be quite adequate for water of very productive ecosystems, it may be necessary to filter one liter from oligotrophic waters. Filtration should be done as rapidly as possible, avoiding exposure to bright light or high temperature. Addition of a small amount of magnesium carbonate slurry during the filtration has been suggested to assist in filtration and to prevent acidity development during extraction, but no filtration advantages have been observed in practice (Lium and Shoaf, 1978). Moreover, alkaline acetone is recommended for the extraction (see below).

After the filtration is completed, the filter should be removed while moist and immediately used in the extraction procedure. When extraction cannot be done right away, the filters should be folded carefully in half with their inner surfaces coming against each other, placed into labeled, folded absorbent pads (e.g., Millipore), stored in a small darkened desiccator, and *immediately* frozen. Pigment samples may be stored for a few days when frozen at −20 to −60°C. Storage of samples in cold, darkened acetone is not recommended because significant degradation of pigments occurs.

Extraction. Although chlorophyllous and carotenoid pigments may be extracted readily in organic solvents, for many algae extraction is not complete unless the cells are ruptured. Therefore filters should be placed with forceps into the base of a glass/glass or preferably Teflon/glass tissue grinder; add approximately 3 to 4 ml of 90% *alkaline* acetone solution (see "Apparatus and Supplies," p. 163). Cyanobacteria are difficult to disrupt; the addition of a small amount of glass powder assists in the extraction. Millipore filters will dissolve immediately. Grind the sample vigorously but carefully *in a hood* for 45 sec; remove the pestle slowly while it is still rotating at the end of the period. Decant into a stoppered graduated centrifuge tube or cylinder. Add about 3 ml more of the acetone solution to the grinding tube and grind vigorously for another 15 sec. Add this rinse solution to the same centrifuge tube and record the total volume to at least the nearest tenth of a ml. Immediately stopper the centrifuge tube tightly; hold in darkness until an even number of filters has been extracted. Centrifuge at maximum speed in a clinical centrifuge (3000 to 4000 rpm; ca. 1000 g) for 5 min. Use a refrigerated centrifuge at 5 to 10°C, if available. When glass fiber filters are used, fibers may cling to the walls of the tube. Remove the tube in this case and gently splash the walls by flicking the tube, thus dislodging any fibers adhering to the sides. Centrifuge for another 2 min.

Carefully remove the clear supernatant so as not to disturb the sedimented residues. Use a rubber bulb and curve-tipped pipet. Be certain to expel the air from the bulb before placing the pipet into the centrifuge tube, otherwise air bubbles will cause turbulence. Place the liquid into clean (must be nonacidic) cuvettes of the spectrophotometer and measure the absorption-emission characteristics, as discussed below, without delay. Small-volume cuvettes holding 7 ml or less with light path lengths of 5 or 10 cm, are recommended strongly to avoid dilution of the pigments.

Spectrophotometric Estimation of Chlorophyll Pigments and Their Degradation Products

Procedures

1. Measure the extinction coefficient of your pigment extracts in 90% alkaline acetone solution at 750 (for turbidity blank), 665, 664, 663, 647, 630, and 480 nm. Absorption at 750 nm is subtracted from each of the other values to correct for the presence of turbidity and colored materials. The pigments and phaeopigments absorb very little light at 750 nm. The spectrophotometer should be brought to 100% transmission at each wavelength by referencing against a matched cell containing only 90% acetone. When the cells are exactly matched, fill both cells with 90% acetone and determine the "blank" correction at each wavelength. Correct all extinction coefficient values by this amount. Record extinction values to the nearest 0.001 in the range of 0 to 0.04 and to the nearest 0.005 for extinctions of coefficients exceeding 0.04.

2. Add 0.1 ml of 1N HCl per ml of extract directly to the spectrophotometric cell, reseal, and mix thoroughly. Allow to rest in the cuvette holder for 5 min.

3. Remeasure the extinction coefficient values at 750, 665, and 663 nm.
4. Rinse the cells copiously with alkaline acetone and shake dry prior to use with the next sample.

Calculations

Monochromatic Method (Lorenzen, 1967)

$$\text{Chl } a(\mu g/l \text{ or } mg/m^3) = \frac{(k)(F)(E_{665_0} - E_{665_a})(v)}{(V)(Z)}$$

where

E_{665_0} = turbidity-corrected absorption at 665 nm before acidification
$\quad = A_{665_0} - A_{750_0}$, where A = the absorption value
E_{665_a} = turbidity-corrected absorption at 665 nm after acidification = $A_{665_a} - A_{750_a}$
$\quad k$ = absorption coefficient of chlorophyll a, = 11.0
$\quad F$ = factor to equate the reduction in absorbancy to initial chlorophyll concentration, 1.7:0.7, or = 2.43
$\quad R$ = maximum ratio of $E_{665_0} : E_{665_a}$ in the absence of pheopigments, = 1.7
$\quad v$ = volume of extract in ml
$\quad V$ = volume of water filtered in liters
$\quad Z$ = length of light path through cuvette or cell in cm.

$$\text{Phaeopigments}(\mu g/l \text{ or } mg/m^3) = \frac{(k)(F)[(E_{665_0} - E_{665_0})](v)}{(V)(Z)}$$

Plant Carotenoids (Strickland and Parsons, 1972)

$$\text{Car } (\mu SPU)/l \text{ or } mSPU/m^3) = \frac{(10.0)(E_{480_0})(v)}{(V)(Z)}$$

where

$\qquad$ SPU = specified plant pigment units approximating the mg

$$E_{480_0} = A_{480} - [(3)(A_{750})]$$

Use the factor of 10.0 when the algae consist primarily of Chrysophyta, Pyrrophyta, or both. When the algae consist predominantly of members of Chlorophyta and/or Cyanophyta, then use the following equation:

$$\text{Car } (\mu SPU/l \text{ or } mSPU/m^3) = \frac{(4.0)(E_{480_0})(v)}{(V)(Z)}$$

where

$$E_{480_0} = A_{480_0} - [(3)(A_{750_0})]$$

$$\text{Phaeopigments } (\mu g/l \text{ or } mg/m^3) = \frac{(26.7)[1.7(E_{665a}) - E_{665_0}](v)}{(V)(Z)}$$

where

$$E_{665a} = A_{665a} - A_{750a}$$

etc., as above.

Spectrophotometric Determination of Chlorophyll Pigments and Their Degradation Products by Other Methods

A large number of comparative methodological studies have emerged during the past two decades for the spectrophotometric and fluorometric estimation of pigments. Other organic solvents, methanol and ethanol in particular, have been used for the extraction of pigments and compared to the removal efficiencies of acetone (Marker et al., 1980; Nusch, 1980; Riemann, 1980, 1982: many others). Acetone is the most straightforward to use as a solvent but is less efficient than methanol or ethanol for the extraction of chlorophyll. Acidification of methanol extracts for evaluation of phaeopigments is extremely sensitive; slight excesses of acidity can introduce large experimental errors (Marker et al., 1980; Marker and Jinks, 1982). Such acidification problems are less serious with ethanol [see Nusch (1980)] and least serious for acetone (Marker et al., 1980; Marker and Jinks, 1982). It is clear that, for precise research purposes, chromatography (HPLC) is the preferred method for pigment analyses [e.g., Jacobsen (1982) and Riaux-Gobin and Klein (1993)]. However, HPLC is expensive to perform and not conducive to rapid, multiple pigment analyses. Comparative analyses of spectrophotometric, fluorometric, and chromatograpic (HPLC) determinations of chlorophyll *a* found differences to be less than 10% in most cases [e.g., Schanz and Rai (1988)]. Because determinations of phaeopigment concentrations are of major importance, often exceeding 50% of the total pigment concentrations, pigment degradation products must be measured, and acetone is usually the preferred solvent for extraction of chlorophylls and phaeopigments. Because of the modest extraction efficiency of acetone, however, it is important that an effective grinding process in subdued light always be performed.

Fluorometric Determination of Chlorophyll *a* and Phaeopigments

The fluorometric assay is based on the fact that pigments fluoresce after they are extracted in an organic solvent and then excited by light of specific wavelengths. The fluoresced light emitted by the sample then is filtered selectively to obtain the peak, which is detected by a sensitive photomultiplier tube. Chlorophyll *a*, when excited by light between 430 to 450 nm, gives a maximum fluorescent emission between 650 and 675 nm. Extraction procedures are identical to those described earlier for the spectrophotometric methods, except that glass fiber filters (Whatman GF/F, 0.6 to 0.7-μm pore size) must be used. In the fluorometer, excitation is provided around 440 nm, and emission is detected at about 660 nm. The output of the fluorometer is in arbitrary units; therefore it must be calibrated using a chlorophyll *a* solution of known concentration. A brief description of procedures for doing this is given below. Details of the methodology are given in Yentsch and Menzel (1963), Holm-Hansen et al. (1965), Strickland and Parsons (1972), Stainton et al. (1977), Sterman (1988), and Riaux-Gobin and Klein (1993).

A number of additional fluorescence methods have been developed to estimate the pigment-based biomass of composite phytoplankton communities as well as of algal groups within the communities. Water containing phytoplankton can be pumped from depth and assayed directly in a flow-through cuvette of fluorometers calibrated against standards (Turner, 1985; Wiltshire et al., 1998). Submersible probes have also been used for in situ measurements (Beutler et al., 1998). By use of fluorescence excitation spectra with specific emission wavelengths from individual light-emitting diodes, it is possible to assign distinct spectral groups to different algal classes based on the absorption wavelengths of the different photosynthetic pigments (phycocyanin, phycoerythrin, fucoxanthin, peridinin, and chlorophylls).

Measurements of relative algal groups can be extended further by examining directly delayed fluorescence excitation where the photosystem of intact, living cells is illuminated with light and subsequently subjected to darkness. Photons of light are emitted for minutes after the original illumination. The delayed fluorescence occurs only in living cells and originates from recombination of positive charges with electrons from the electron transport chain at the reaction center P680 of photosystem II. The emitted photons, detected with a photomultiplier, are directly proportional to the number of intact photosystems in the sample (Friedrich et al., 1998; Gerhardt and Bodemer, 1998; Wiltshire et al., 1998). The delayed excitation fluorescence of phytoplankton show good correlations with extractive chlorophyll a and biovolumes.

A measure of physiological condition of phytoplankton can be further quantified by measurements of variable chlorophyll a fluorescence (Oliver et al., 1996; Oliver and Whittington, 1997). The technique examines the processes of light capture and electron transport and provides an estimate of the rate of photosynthesis. Active fluorometry uses a weak flash of light to evaluate the status of the photosystem and a bright flash to precondition the system. The energy of the fluorescence reaction is inversely proportional to capacity to photosynthesize and the effects of environmental conditions on the processes.

Standardization with Chlorophyll Solutions

Purified chlorophyll a and b can be purchased reliably from major chemical companies (e.g., Sigma) or from fluorescent instrument manufacturers (e.g., Turner Designs, Sunnyvale, CA). These standards should be employed wherever possible. An alternative method for deriving approximate standards from local plant materials is discussed below.

1. Using 90% alkaline acetone solution, extract from about $5\,cm^2$ of a fresh green lettuce leaf to obtain 25 ml of chlorophyll solution that is deep green in color. Filter and store briefly in darkness.
2. Dilute this stock, if necessary, so that E_{665_0} (1-cm path length) would be approximately 0.6 (= ca. 6 mg chl a/l).
3. Measure absorbances (1-cm cuvettes) at 750, 665, and 645 nm against a reference cell using 90% acetone.
4. Assuming that only chlorophylls a and b occur in lettuce, calculate the chlorophyll a concentration:

$$\text{Chl } a(\text{mg/l}) = 11.57E_{665_0} - 1.31E_{645_0}$$

where

$$E_{665_0} = A_{665_0} - A_{750_0}, \text{ etc., as above.}$$

5. Prepare accurate dilutions of the standard solution with 90% alkaline acetone to obtain at least three readings on each of the sensitivity settings of the fluorometer ($1\times, 3\times, 10\times,$ and $30\times$) in such a way that the readings approximate 25, 50, and 75 divisions on the 0 to 100 unit scale of the fluorometer.
6. Plot the concentrations of chlorophyll a (μg/l) against fluorometer readings for each sensitivity setting. The slope of these lines should provide linear calibration factors (F) for the fluorometer reading for each sensitivity setting:

$$F_{(1\times,3\times,10\times,30\times)} = \frac{\text{Chl } a(\mu g/l)}{\text{scale reading}}$$

Analyses

1. Zero the fluorometer using 90% acetone as a reference solution.
2. Using the alkaline acetone extracts of known volume as prepared for spectro-photometric analyses discussed earlier, measure the fluorescence to the nearest 0.2 unit with an appropriate sensitivity setting.

3. $$\text{Chl } a(\mu g/l \text{ or } mg/m^3) = \frac{(F)(\text{fluorometer reading})(v)}{(V)}$$

   where v = volume of extract in ml; and V = volume of water filtered in ml.
4. After the first reading in alkaline acetone (R_b) is taken, remove the tube and add two drops of 4N HCl. Mix the contents of the tube by inversion and allow to sit for 2 min.
5. Remeasure the fluorescence of the acidified sample (R_a) when a stable value has been reached.
6. Using serial dilutions of chlorophyll extracts from lettuce, as discussed earlier, determine the fluorescence in basic 90% acetone (R_{bstd}) and in the extracts after they have been acidified (R_{astd}). Then the ratio, τ, is obtained;

$$\tau = R_{bstd}/R_{astd}$$

These ratios τ are determined at different instrument sensitivities.
   Then,

$$\text{Chl } a(\mu g/l) \text{ or } mg/m^3 = (F)\left(\frac{\tau}{\tau-1}\right)(R_b - R_a)$$

$$\text{Phaeopigment}(\mu g/l \text{ or } mg/m^3) = (F)\left(\frac{\tau}{\tau-1}\right)(\tau R_a - R_b)$$

7. Calculate amount of pigment per cell:

$$\mu g \text{ chl } a/\text{cell} = \frac{(\mu g \text{ chl } a)(1 \text{ liter})(1 \text{ ml of initial sample})}{(\text{liter})(1000 \text{ ml})(\text{no.cells}/\text{ml})}$$

$$\mu \text{mol chl } a/\text{cell} = \frac{\mu g \text{ chl } a/\text{cell}}{(\text{molecular weight of chl } a, 894)}$$

EXERCISES

OPTION 1. FIELD TRIPS AND LABORATORY ANALYSES

1. Collect water samples at regular depth intervals from the central depression of a lake or reservoir with a scrubbed Van Dorn or similar water sampler. Dispense the water into clean, amber polyethylene bottles. Collect replicates at each depth.
2. If possible, collect other vertical series (a) within a littoral zone among dense stands of higher aquatic plants and (b) in the open water at the mouth of a major inlet stream.

3. If possible, make comparative collections and analyses of the phytoplankton from vertical profiles in (a) a relatively unproductive oligotrophic lake; (b) a productive, eutrophic lake or reservoir; (c) the open water of a bog; and (d) a stream at several points along its drainage (in the central stream and in backwaters from downstream and upstream stations, and directly below the outlet from a lake). In stratified waters, determine the temperature profile at meter intervals.

4. In the laboratory, work through fresh subsamples from several different sites to familiarize yourself with the kinds of algae present.

5. Preserve several samples with Lugol's solution, as discussed earlier, from each collection site and depth.

6. Make quantitative analyses of (a) the number of each species and (b) the size (dimensions) and volume of major algal species. Compare as many of the different methods discussed as possible, especially the sedimentation technique, the Palmer-Maloney counting cells, and the method of filtration onto membrane filters with subsequent clearing and mounting onto slides.

7. Estimate the amount of algal carbon biomass from your estimates of cellular volume.

8. Determine the concentrations of algal pigments (chlorophylls, carotenoids) and phaeopigments from water samples of a vertical profile from, or within, a lake or several lakes or along a stream gradient. From replicated samples, determine the coefficients or variance (see Appendix 2).

OPTION 2. LABORATORY EXERCISES

1. Using water samples provided by your instructor, work through fresh subsamples from several different sites to familiarize yourself with the general species of algae encountered.

2. Preserve a number of samples with Lugol's solution, as discussed earlier, from each collection site and depth.

3. Make comparative quantitative analyses of (a) the number of each species and (b) the size dimensions and volume of major algal species. Use as many of the different methods discussed earlier as possible, particularly to compare quantitative results obtained by the sedimentation technique, the Palmer-Maloney counting cell, and the method of filtration onto membrane filters with subsequent clearing and mounting onto slides.

4. Estimate the amount of algal carbon biomass from your estimates of cellular volume.

5. Using samples provided by your instructor, determine the amount of algal pigment biomass (chlorophylls, carotenoids) and phaeopigment concentrations from water samples of a profile from, or within, a lake or several lakes or along a stream gradient. From replicated samples, determine the coefficients of variance.

Questions

1. How does the vertical profile of algal species and biomass compare to the thermal stratification?

2. You may have observed a concentration of algae in the metalimnion. What factors may have caused this?

3. When significant numbers and biomass of algae are found in the hypolimnion, what does this indicate about their well-being and metabolic state? How might this question be addressed better?

4. How would you rate the various methods for enumeration of algae relative to their practicality, ease, precision, accuracy, and reliability?

5. If horizontal variations in distribution of biomass were observed between the open water, littoral areas, and inlet areas, how might you evaluate experimentally the causal factors regulating such a spatial pattern?

6. Is it reasonable to make a comparison of the algal species composition and biomass between two lake ecosystems sampled at this particular time of year? Support your answer.

7. How frequently would one have to sample a freshwater ecosystem to evaluate accurately the population dynamics of the phytoplankton over an annual period? Why? Which criteria are important in regulating change?

8. How do you explain the observed differences in the ratios of active algal pigments to phaeopigment degradation products?

9. How could the observed differences in the vertical comparison of algal numbers and pigment biomass (corrected for phaeopigments) be explained? Does the relationship between algal volume and corrected chlorophyll concentrations improve the relationship?

10. What is the relationship between the transmission and absorption of light?

11. How would you determine the sinking rates of algae in the counting chamber? Why should drafts be avoided when settling plankton?

12. What kind (species) of algae would you expect to find at depth in a brown-water, bog lake? Why?

13. Why are pigment extracts stored in the dark?

14. Why store algal samples in darkness? What is actually happening when an algal sample is "preserved"?

15. What are the major problems in attempting to determine the volume of an algal cell? Colony?

Apparatus and Supplies

1. Water samplers, e.g., Van Dorn water bottle.

2. 500-ml or 1-liter amber polyethylene bottles.

3. Preservatives:
   a. Lugol's fixative: Dissolve 10 g I_2 (pure iodine; *caution*: toxic) and 20 g KI in 200-ml distilled water and 20 ml concentrated glacial acetic acid. Store in ground glass-stoppered, darkened bottle.
   b. 50% glutaraldehyde filtered through glass fiber filters of small pore size (e.g., 0.5 μm) to remove common particulate contaminants. *Caution*: very toxic. Concentrated glutaraldehyde has a limited shelf-life (ca. six months).
   c. Buffered formalin solution: Buffer 40% formaldehyde with sufficient sodium acetate or hexamethylentetramine to yield a final pH of 7.5 to 8.0 in the preserved sample.

4. Sedimentation chambers of various designs (see text).

5. Inverted or conventional compound microscopes. For picophytoplankton analyses, epifluorescence microscope with a 100× aperture oil immersion objective, oculars 8 to 12× with micrometer and counting grid. Filter sets: 520 nm longwave pass filter coupled with one 560-nm shortwave pass filter; one 580 nm dichroic mirror, and one 590-nm longwave pass barrier filter (e.g., Nikon B2A DM580 set). One 400 to 500 nm filter, one 395-nm longwave pass filtered coupled with one 500-nm shortwave pass filter, one 510-nm dichroic mirror, and one 520-nm longwave pass barrier filter (e.g., Nikon G1B DM510 set).

6. Whipple ocular micrometer eyepieces (reticules) and stage micrometer.

7. Palmer-Maloney plankton counting cells.

8. Filtration apparatus, membrane filters (0.45-μm and 0.8-μm pore size), glass fiber filters (Whatman GF/F, 0.6 to 0.7-μm pore size), and black 0.2-μm pore size polycarbonate membrane filters (e.g., Nuclepore).

9. Acetone, concentrated NH_4OH, Teflon/glass grinding tubes, 15-ml graduated centrifuge tubes, polypropylene centrifuge tubes with caps, clinical centrifuge, 4N HCl [i.e., dilute 0.33 l of concentrated HCl (sp. gr. 1.18 or 37%) to 1.00 l].
   a. Alkaline acetone solution: Pipet 100 ml of redistilled water into a 1000-ml ground glass-stoppered volumetric flask. Bring up to mark with spectral grade redistilled acetone; mix thoroughly. Add two drops of reagent grade concentrated NH_4OH; mix thoroughly.

10. Spectrophotometer (preferably narrow-band wavelength, ±1.0 nm), Turner fluorometer with blue source lamp (Turner models 110–853, 111, or 112), primary filter for excitation

(Kodak Wratten 47B or Corning CS.5-60) and Corning CS.2-64 secondary filter for the emitted light, or equivalent.

References

Ahlgren, G. 1983. Comparison of methods for estimation of phytoplankton carbon. Arch. Hydrobiol. *98*:489–508.

American Public Health Association. 1998. Standard Methods for the Examination of Water and Wastewater. 20th Ed. Water Environment Federation, Arlington, VA. 1183 pp.

Bellinger, E.G. 1974. A note on the use of algal sizes in estimates of population standing crops. Brit. Phycol. J. *9*:157–161.

Beutler, M., K.H. Wiltshire, B. Meyer, and C. Moldaenke. 1998. Differenzierung spektraler Algengruppen durch computer-gestützte Analyse von Fluoreszenzanregungsspektren. Vom Wasser *91*:1–14.

Blomqvist, P. and E. Herlitz. 1998. Methods for quantitative assessment of phytoplankton in fresh waters. Part 2. Rapport 4861, Naturvårdsverket Förlag, Uppsala, Sweden. 68 pp.

Borsheim, K.Y. and G. Bratbak. 1987. Cell volume to cell carbon conversion factors for a bacterivorous *Monas* sp. enriched from seawater. Mar. Ecol. Progr. Ser. *36*:171–176.

Brock, T.D. 1983. Membrane Filtration: A User's Guide and Reference Manual. Science Tech. Inc. Madison. 381 pp.

Clark, W.J. and W.F. Sigler. 1963. Method of concentrating phytoplankton samples using membrane filters. Limnol. Oceanogr. *8*:127–129.

Coulon, C. and V. Alexander. 1972. A sliding-chamber phytoplankton settling technique for making permanent quantitative slides with applications in fluorescent microscopy and autoradiography. Limnol. Oceanogr. *17*:149–152.

Crumpton, W.G. and R.G. Wetzel. 1981. A method for preparing permanent mounts of phytoplankton for critical microscopy and cell counting. Limnol. Oceanogr. *26*:976–980.

deNoyelles, F., Jr. 1968. A stained-organism filter technique for concentrating phytoplankton. Limnol. Oceanogr. *13*:562–565.

Dodson, A.N. and W.H. Thomas. 1964. Concentration of plankton in a gentle fashion. Limnol. Oceanogr. *9*:455–456.

Friedrich, G., V. Gerhardt, U. Bodemer, and M. Pohlmann. 1998. Phytoplankton composition and chlorophyll concentration in freshwaters: Comparison of delayed fluorescence excitation spectroscopy, extractive spectrophotometric method, and Utermöhl-Method. Limnologica *28*:323–328.

Gerhardt, V. and U. Bodemer. 1998. Delayed fluorescence excitation spectroscopy: A method for automatic determination of phytoplankton composition of freshwaters and sediments (interstitial) and of algal composition of benthos. Limnologica *28*:313–323.

del Giorgio, P.A., D. Bird, Y.T. Prairie, and D. Planas. 1996. Flow cytometric determinations of bacterial abundance in lake plankton with the green nucleic acid stain SYTO 13. Limnol. Oceanogr. *41*:783–789.

Golterman, H.L. and R.S. Clymo (eds). 1969. Methods for Chemical Analysis of Fresh Waters. IBP Handbook No. 8 Blackwell, Oxford. 172 pp.

Hillebrand, H., C.-D. Dürselen, D. Kirschtel, U. Pollingher, and T. Zohary. 1999. Biovolume calculation for pelagic and benthic microalgae. J. Phycol. *35*:403–424.

Holmes, R.W. 1962. The preparation of marine phytoplankton for microscopic examination and enumeration on molecular filters. U.S. Fish. Wildl. Serv., Spec. Sci. Rep. Fish. *433*. 6 pp.

Holm-Hansen, O., C.J. Lorenzen, R.W. Holmes, and J.D.H. Strickland. 1965. Fluorometric determination of chlorophyll. J. Conseil Perm. Int. Explor. Mer *30*:3–15.

Jackson, H.W. and L.G. Williams. 1962. Calibration and use of certain plankton counting equipment. Trans. Amer. Microsc. Soc. *81*:96–103.

Jacobsen, T.R. 1982. Comparison of chlorophyll a measurements by fluorometric, spectrophotometric and high pressure liquid chromatographic methods in aquatic environments. Arch. Hydrobiol. Beih. Ergebn. Limnol. *16*:35–45.

Javornický, P. 1958. Revise některých metod pro zjišťování kvantity fytoplanktonu. (The revision of some quantitative methods for phytoplankton research.) (In Czech, with English summary.) Sci. Pap. Inst. Chem. Technol. Prague, Fac. Technol. Fuel and Water 2(Part 1):283–367.

Jeffrey, S.W. and G.F. Humphrey. 1975. New spectrophotometric equations for determining chlorophylls a, b, c_1 and c_2 in higher plants, algae and natural phytoplankton. Biochem. Physiol. Pflanzen *167*:191–194.

Kellar, P.E., S.A. Paulson, and L.J. Paulson. 1980. Methods for biological, chemical and physical analyses in reservoirs. Tech. Rep. *5*, Lake Mead Limnological Res. Center, Univ. Nevada, Las Vegas. 234 pp.

Lium, B.W. and W.T. Shoaf. 1978. The use of magnesium carbonate in chlorophyll determinations. Wat. Resources Bull. *14*:190–194.

Lorenzen, C.J. 1967. Determination of chlorophyll and pheo-pigments: Spectrophotometric equations. Limnol. Oceanogr. *12*:343–346.

Lund, J.W.G. 1951. A sedimentation technique for counting algae and other organisms. Hydrobiologia *3*:390–394.

Lund, J.W.G., C. Kipling, and E.D. LeCren. 1958. The inverted microscope method of estimating algal numbers and the statistical basis of estimations by counting. Hydrobiologia *11*:143–170.

Lundgren, A. 1978. Experimental lake fertilization in the Kuokkel area, northern Sweden: Changes in sestonic carbon and the role of phytoplankton. Verhand. Internat. Verein. Limnol. *20*:863–868.

MacIsaac, E.A. and J.G. Stockner. 1993. Enumeration of phototrophic picoplankton by autofluorescence microscopy. pp. 187–197. *In*: P.F. Kemp, B. Sherr, E. Sherr, and J.J. Cole, Editors. The Handbook of Methods in Aquatic Microbial Ecology. Lewis Publishers, Boca Raton.

Marker, A.F.H., C.A. Crowther, and R.J.M. Gunn. 1980. Methanol and acetone as solvents for estimating chlorophyll a and phaeopigments by spectrophotometry. Arch. Hydrobiol. Beih. Ergebn. Limnol. *14*:52–69.

Marker, A.F.H. and S. Jinks. 1982. The spectrophotometric analysis of chlorophyll a and phaeopigments in acetone, ethanol and methanol. Arch. Hydrobiol. Beih. Ergebn. Limnol. *16*:3–17.

McNabb, C.D. 1960. Enumeration of freshwater phytoplankton concentrated on the membrane filter. Limnol. Oceanogr. *5*:57–61.

Moore, J.K. 1963. Refinement of a method for filtering and preserving marine phytoplankton on a membrane filter. Limnol. Oceanogr. *8*:304–305.

Mullin, M.M., P.R. Sloan, and R.W. Eppley. 1966. Relationship between carbon content, cell volume, and area in phytoplankton. Limnol. Oceanogr. *11*:307–311.

Nusch, E.A. 1980. Comparison of different methods for chlorophyll and phaeopigment determination. Arch. Hydrobiol. Beih. Ergebn. Limnol. *14*:14–36.

Oliver, R., G. Ganf, S. Geary, J. Brookes, M. Fink, and M. Burch. 1996. Rapid measurement of algal biomass, species composition and physiological condition. *In*: R.J. Banens and R. Lehane, Editors. Riverine Environment Research Forum. Attwood Victoria Publ., Murray-Darling Basin Commission, Australia.

Oliver, R.L. and J. Whittington. 1997. Using measurements of variable chlorophyll-a fluorescence to investigate the influence of water movement on the photochemistry of phytoplankton. Physical Limnology, Coastal and Estuarine Studies, American Geophysical Union.

Olrik, K., P. Blomqvist, P. Brettum, G. Cronberg, and P. Eloranta. 1998. Methods for the quantitative assessment of phytoplankton in fresh waters. Part I. Rapport 4860, Naturvårdsverket Förlag, Uppsala, Sweden, 86 pp.

Olson, R.J., E.R. Zettler, and M.D. DuRand. 1993. Phytoplankton analyses using flow cytometry. pp. 175–186. *In*: P.F. Kemp, B.F. Sherr, E.B. Sherr, and J.J. Cole, Editors. Handbook of Methods in Aquatic Microbial Ecology. Lewis Publishers, Boca Raton.

Palmer, C.M. and T.E. Maloney. 1954. A new counting slide for nannoplankton. Spec. Publ. Amer. Soc. Limnol. Oceanogr. *21*. 6 pp.

Redalje, D.G. and E.A. Laws. 1981. A new method for estimating phytoplankton growth rates and carbon biomass. Mar. Biol. *62*:73–79.

Riaux-Gobin, C. and B. Klein. 1993. Microphytobenthic biomass measurement using HPLC and conventional pigment analysis. pp. 369–376. *In*: P.F. Kemp, B. Sherr, E. Sherr, and J.J. Cole, Editors. The Handbook of Methods in Aquatic Microbial Ecology. Lewis Publishers, Boca Raton.

Riemann, B. 1980. A note on the use of methanol as an extraction solvent for chlorophyll *a* determination. Arch. Hydrobiol. Beih. Ergebn. Limnol. *14*:70–78.

Reimann, B. 1982. Measurement of chlorophyll *a* and its degradation products: A comparison of methods. Arch. Hydrobiol. Beih. Ergebn. Limnol. *16*:19–24.

Rott, E. 1980. Spectrophotometric and chromatographic chlorophyll analysis: Comparison of results and discussion of the trichrometric method. Arch. Hydrobiol. Beih. Ergebn. Limnol. *14*:37–45.

Schanz, F. and H. Rai. 1988. Extract preparation and comparison of fluorometric, chromatographic (HPLC) and spectrophotometric determinations of chlorophyll-a. Arch. Hydrobiol. *112*:533–539.

Schmid, H., F. Bauer, and H.B. Stich. 1998. Determination of algal biomass with HPLC pigment analysis from lakes of different trophic state in comparison to microscopically measured biomass. J. Plankton Res. *20*:1651–1661.

Schröder, R. 1969. Ein summierender Wasserschöpfer. Arch. Hydrobiol. *66*:241–243.

Sicko-Goad, L., E.F. Stoermer, and B.G. Ladewski. 1977. A morphometric method for correcting phytoplankton cell volume estimates. Protoplasma *93*:147–163.

Stainton, M.P., M.J. Capel, and F.A.J. Armstrong. 1977. The Chemical Analysis of Fresh Water. 2nd Ed. Misc. Spec. Publ. Fish Environ. Canada *25*. 180 pp.

Sterman, N.T. 1988. Spectrophotometric and fluorometric chlorophyll analysis. pp. 35–45. *In*: C.S. Lobban, D.J. Chapman, and B.P. Kremer, Editors. Experimental Phycology: A Laboratory Manual. Cambridge Univ. Press, Cambridge.

Straškraba, M. and P. Javornický. 1973. Limnology of two re-regulation reservoirs in Czechoslovakia. Hydrobiol. Studies *2*:249–316.

Strathmann, R.R. 1967. Estimating the organic carbon content of phytoplankton from cell volume or plasma volume. Limnol. Oceanogr. *12*:411–418.

Strickland, J.D.H. and T.R. Parsons. 1972. A Practical Handbook of Seawater Analysis. 2nd Ed. Fisheries Research Board of Canada, Ottawa. 310 pp.

Turner, G.K. 1985. Measurement of light from chemical or biochemical reactions. pp. 43–78. *In*: K. Van Dyke, Editor. Bioluminescence and Chemiluminescence: Instruments and Applications.

Tyler, P.A. 1971. A simple and rapid technique for surveying size and shape variation in desmids and diatoms. Brit. Phycol. J. *6*:231–233.

Utermöhl, H. 1931. Neue Wege in der quantitativen Erfassung des Planktons. Verh. Int. Ver. Limnol. *5*:567–595.

Utermöhl, H. 1958. Zur Vervollkommnung der quantitativen Phytoplankton-Methodik. Mitt. Int. Ver. Limnol. *9*. 38 pp.

Wetzel, R.G. 1983. Limnology, 2nd Ed. Saunders Coll., Philadelphia. 860 pp.

Wiltshire, K.H., S. Harsdorf, B. Smidt, G. Blöcker, R. Reuter, and F. Schroeder. 1998. The determination of algal biomass (as chlorophyll) in suspended matter from the Elbe estuary and the German Bight: A comparison of high-performance liquid chromatography, delayed fluorescence and prompt fluorescence methods. J. Exp. Mar. Biol. Ecol. *222*:113–131.

Yentsch, C.S. and D.W. Menzel. 1963. A method for the determination of phytoplankton chlorophyll and phaeophytin by fluorescence. Deep-Sea Res. *10*:221–231.

Collection, Enumeration, and Biomass of Zooplankton

Zooplankton usually are distributed throughout a lake or reservoir. Most forms are motile, and thus their distribution both vertically and horizontally may be quite variable. Species have different habitat preferences that further accentuate spatial and temporal heterogeneity. Zooplankton are major herbivores as well as important predators in aquatic ecosystems. Therefore, to understand lake metabolism it is necessary to evaluate the biomass and the role of zooplankton in the ecosystem.

Quantitative collection of aquatic organisms within their natural habitats usually is difficult. The foremost problem is that only relatively small subsamples can be obtained from a population, which is often mobile, changing in size, and distributed in patches.

If zooplankton were distributed unevenly within a lake ecosystem, then discrete samples would grossly overestimate or underestimate the true population size. Thus characterization of the distribution of individuals is a prerequisite for quantitative sampling. Individuals may be dispersed randomly, contagiously, or uniformly (or various combinations of these) throughout the volume of a lake or reservoir (Fig. 11.1).

Errors associated with enumeration may be categorized into two types: (1) error associated with counting the individuals in a sample; and (2) errors in obtaining a representative sample of the population from a site within a lake, from the entire lake, or both. The purpose of this exercise is to examine these problems while, at the same time, attempting to characterize a zooplankton population for a lake.

Apparently random or uniform distributions are rather uncommon in zooplankton populations as distributions tend to be patchy (Tonolli, 1971; Levin and Segal, 1976). However, various scales of vertical and horizontal distance must be considered. A chi-square test may be used to test for contagion. This test is done by determining the significance of the difference between the observed distribution of individuals in unit volumes and a theoretical Poisson distribution for the same number of individuals and volumes (see Appendix 2). When the distribution is contagious, the degree of contagion can be estimated by the ratio of the variance to the mean. Ratios of 1 indicate randomness, >1 indicate contagion, and <1 indicate evenness.

The objectives of this exercise are

1. Evaluate counting procedures.
2. Compare various collection devices.
3. Characterize the population size, distribution, and biomass for a site and for an entire lake.

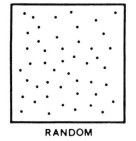
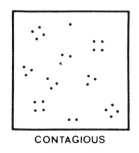
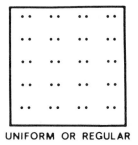

RANDOM   CONTAGIOUS   UNIFORM OR REGULAR

Figure 11.1. Random, contagious, and uniform distributions of individuals in populations of aquatic ecosystems.

SAMPLE COLLECTION

Various types of nets and traps have been used to concentrate (filter) zooplankton from large volumes of water [cf., Juday (1916), Welch (1948), Edmondson and Winberg (1971), and deBernardi (1984)]. Some of these devices are much more effective than others, depending on the habitat and the species one wants to catch.

In all cases a fine mesh is used to filter the organisms from the water. Choosing an appropriate mesh size is an important consideration in sampling zooplankton (Table 11.1). The mesh openings of new dry silk bolting cloth may decrease by 15 to 50% when wet. Somewhat less change is observed upon wetting of used dry silk and little or no change is observed for synthetic fabrics [cf., Tranter and Fraser (1968)]. It is important to determine the actual mean mesh size on wet fabric before beginning a study of zooplankton (see the section on microscopic aids in Exercise 10).

Obviously small mesh openings will clog more readily than larger ones, but, likewise, small organisms will pass readily through larger openings. In nets having the same filtering area, the volume that can be filtered before clogging becomes serious appears to be a function of the square of the mesh width (Tranter and Fraser, 1968). When the net clogs or is pulled through the water too fast, the ratio of the volume filtered to the volume swept by sampler mouth may fall far below 100%. Filteration efficiency (F) may be calculated as follows:

$$F = \frac{W}{(A)(D)}$$

Table 11.1. Average aperture size of standard grade silk bolting cloth (Defour) and nylon monofilament screen cloth (Nitex).[a]

Silk number	Mesh opening (μm)	Nitex® number	Mesh opening (μm)
000	1024	1050	1050
0	569	571	571
5	282	253	253
10	158	130	130
15	94	86	86
20	76	75	75
25	64	67	67

[a] ®Nitex is a registered trademark of Tobler, Ernst and Traber, Inc., 71 Murray St., New York, NY.

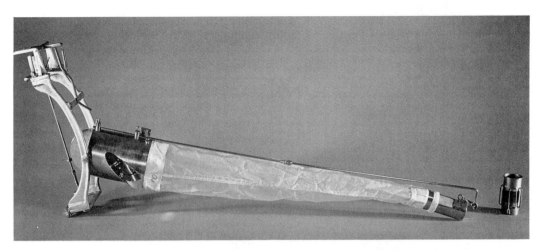

Figure 11.2. The Clarke-Bumpus zooplankton sampler with orifice door open, as would be the case while towing underwater. The door may be opened at the desired depth of sampling by a messenger and closed by another messenger when towing is completed. The housing contains a metered propeller from which the volume of water passing through the sampler can be calibrated. Note the large surface area of netting in relation to orifice size.

where W = volume of water filtered; A = area of mouth of sampler; and D = distance towed.

Towed, metered apparatus, such as the Clarke-Bumpus Plankton sampler (Fig. 11.2) and the Isaac Kidd sampler (Fig. 11.3), are used commonly to obtain samples from large volumes of water. However, as the meshes clog with organisms and debris, the metering may become unreliable. For this reason, McNaught (1971) recommended that mesh openings smaller than 363 μm (#2 silk) should not be used in very productive waters. In unproductive waters, a #10 silk net (158-μm mesh) is appropriate on metered devices.

Alternatively, volumetric devices such as the Schindler trap (Fig. 11.4) or the Likens-Gilbert water bottle and funnel technique (Fig. 11.5) have been used. With these devices, a relatively small volume of water (2 to 100 l) is collected and passed through fine mesh fabrics (<50-μm openings). The small volume may be a severe

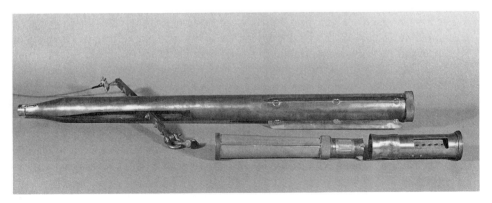

Figure 11.3. The Isaac Kidd zooplankton sampler.

Figure 11.4. The Schindler plankton trap, made of transparent Plexiglas with self-closing doors that seal when lowering of the device stops at the desired depth. (Photograph courtesy of D.W. Schindler.) See Schindler (1969) for construction details.

limitation when the population has a strongly contagious distribution, and motile forms may sense and evade these slowly moving devices [e.g., Tonolli (1971), Clutter and Anraku (1968), and Szlauer (1964)]. Constructing the trap of Plexiglas or other transparent plastic material helps to alleviate the latter problem. On the other hand, the use of fine mesh fabric with these devices gives a major advantage, especially when trying to collect small organisms such as rotifers [e.g., Schindler (1969) and Makarewicz (1974)]. In a study of a small oligotrophic lake (Mirror Lake) in New Hampshire, Makarewicz (1974) found that the Likens-Gilbert technique captured all of the species of the zooplankton community and collected as many or more individuals of each species, including the large motile Cladocera and Copepoda, than did the towed, metered devices. Large numbers of rotifers and nauplii passed through the nets of the towed devices.

"Integrated" vertical samples may be obtained by pulling a plankton net from near the bottom of a lake to the surface. In some cases a long, large-diameter

Figure 11.5. Funnel for filtering zooplankton from water samples obtained with a *clear* plastic Van Dorn water sampler (cf., Fig. 7.1). The three windows are covered with Nitex nylon screening, attached to the inside with methylene chloride. All dimensions are in centimeters. [From Likens and Gilbert (1970).]

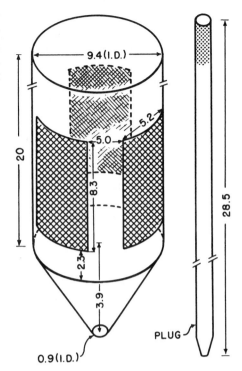

(10 to 20 cm) tube has been used to obtain an instantaneous sample from the entire water column. Pennak (1962) has suggested that such a device is useful in littoral regions with abundant rooted vegetation. With nonmetered tow nets or tubes the theoretical volume of water filtered is given by:

$$V = (A)(D)$$

Horizontal integration of samples can be achieved better with a towed device, such as the Clarke-Bumpus plankton sampler.

SAMPLE PRESERVATION

Zooplankton samples are usually preserved with enough neutralized formalin (40% formaldehyde) to produce a final concentration of about 4%. Delicate forms, such as ciliated protozoa and some rotifers, may disintegrate in formalin solutions. These forms must be studied while alive or after preservation with Lugol's solution (see Exercise 10). Details of zooplankton fixation and preservation are given in the monograph of Steedman (1976).

QUANTITATIVE ENUMERATION OF INDIVIDUALS

Once the sample is obtained from the lake and is concentrated, the organisms must be counted. When the number of individuals is relatively small, it is best to settle the entire sample and count all organisms with an inverted microscope (Utermöhl, 1958; Nauwerck, 1963; see Exercise 10).

When the number of organisms in the sample is very large, it will be necessary to take subsamples to facilitate counting. This subsampling introduces a manipulative error in addition to the error involved in obtaining a representative sample from the lake. However, the number of zooplankton per unit volume of lake water can be estimated within statistical limits by counting the number of individuals in several subsamples [see McCauley (1984)]. This technique is outlined as follows:

1. Depending on the concentration, dilute the concentrated sample of zooplankton from the lake to exactly 50 ml, 100 ml, or 200 ml with tap water. It is convenient to use a graduated cylinder.
2. Mix the organisms thoroughly within the graduated cylinder (vortex mixer or magnetic stirrer) and immediately obtain a subsample. Work rapidly to minimize the effect of settling of the organisms.
3. Obtain a subsample of 1 ml with an automatic, volumetric pipet (use a wide-mouth pipet so as not to restrict uptake of large zooplankton).
4. Add this subsample to a Sedgwick-Rafter (S-R) cell and cover with a coverglass. The S-R cell used in counting plankton is 50-mm long by 20-mm wide by 1-mm deep (Fig. 11.6). The cell is covered by a relatively thick coverslip and is calibrated to contain exactly 1.0 ml. The dimensions of each S-R cell should be checked carefully with calipers or a micrometer to determine the exact volume.

   With the coverslip at a slight diagonal to permit air displacement, the S-R cell is filled as illustrated in Fig. 11.6. Air spaces should not be permitted to develop during lengthy examination; replace evaporated water loss with a small drop of distilled water, fed into the cell by capillary action. The configuration of the S-R cell does not permit use of high-power microscope objectives, so that the identification of organisms smaller than 10 to 15 μm is either difficult or impossible. Thus, the use of the S-R plankton cell is limited to examination of only the larger forms of zooplankton.
5. Place the filled cell on a white background with grid lines under a dissecting microscope. Count all of the organisms in five subsamples. Use the grid lines for reference and use hand tallies to facilitate the counting procedure of different organisms. Record these data for each species and for the total zooplankton (see Table 11.2).

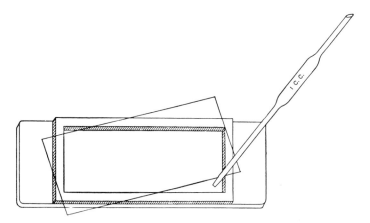

Figure 11.6. The Sedgwick-Rafter plankton counting cell and the method of filling. [From Whipple et al. (1927).]

Table 11.2. Exemplary table useful for recording data when using Sedgwick-Rafter cells.

Total volume of zooplankton settled (mm)	Volume of sample in ml = V_s	Counts from five 1-ml aliquots	N = average count per cell ± standard deviation of the mean	V_f = volume of lake water filtered in 1	Number of organisms per 1 $n = NV_s/V_f$	$n \times 1000$ equals Number of organisms/m^3

When the density of zooplankton is too great to count the entire S-R cell volume, horizontal strips may be counted along the length of the cell. Usually two to four strips are counted at the highest practical magnification. Knowing the width of the field using a Whipple ocular grid, calibrated for the particular set of oculars and objectives (see Exercise 10), the strip constitutes a volume (V_1) the length of the cell (ca. 50 mm), 1-mm deep, and the width (in mm) of the Whipple field (W):

$$V_1 = (50)(1)(W)$$
$$= mm^3$$

The plankton counts per strip then are determined by multiplying the actual count by a factor representing the counted portion of the whole S-R cell volume. Then:

$$No./ml = \frac{(C)(1000\ mm^3)}{(L)(D)(W)(S)}$$

where C = number of organisms counted; L = length of each strip in mm (usually length of the S-R cell); D = depth of the strip in mm (S-R cell depth); W = width of the strips in mm (Whipple grid image width); and S = number of strips counted.

MICROFLAGELLATES AND PROTOZOA

It is increasingly recognized that microzooplankton, particularly the heterotrophic protistan microflagellates or nanoflagellates (2–20 μm in size) and microprotozoa and larger flagellates (mostly >20–200 μm), are extraordinarily important as both consumers and decomposers of organic matter in the "microbial loop" of aquatic ecosystems as well as a potential food source for larger zooplankton (cf., detailed discussions in Wetzel, 1999). Because of their small size and relative fragility, different methods are needed for collection, preservation, and examination than is the case with larger zooplankton. Because they are very small and have relatively rapid rates of growth and reproduction, the metabolism and roles of microflagellates in inland waters can exceed those of the larger zooplankton. The following exercise is based upon methods of Caron (1983), Carrick et al. (1991), and others.

1. Water samples are preserved separately:
   a. Nanoflagellates: A final preservative concentration of 1% with 10% glutaraldehyde prepared in distilled water with 0.1 M sodium cacodylate buffer (pH 7.0).
   b. Ciliates: 1% final concentration of Lugol's acidic iodine preservative (see Exercise 10).

2. Filter 10 to 50 ml of preserved samples for nanoflagellates onto 0.8-μm pore size Nuclepore filters prestained with irgalan black (purchased or by Hobbie et al., 1977) with a vacuum differential of <0.3 atm (<5 kPa). A 0.5-μm pore size Millipore cellulose acetate filter is placed beneath the Nuclepore filter to promote even dispersion of the sample on the Nuclepore filter.

3. Rinse the filter twice with two 1-ml portions of a rinse solution of distilled water with 0.1 M Trizma-hydrochloride (pH 4). Flood the filter with a fluorochrome primulin (Direct Yellow 59, Color Index 49000, Aldrich Chemical Co.) solution at a concentration of 250 μg/ml. Allow the stain to react in the filtration tower for 15 min without vacuum. Then apply the minimal vacuum. Rinse further with two 2-ml portions of the 0.1 M Trizma-hydrochloride acidic solution.

4. Remove the Nuclepore black filter and place *onto* a thin film of immersion oil (Cargille type A) on a glass slide, specimen side up. Add one drop of oil on the center of the filter, followed by a cover slip.

5. Examine slides by epifluorescent microscopy with a Neofluar 100× objective (Palanachromat and Planapochromat lenses will not work) and a 10× objective. Filter sets for primulin fluorescence include, for example, Zeiss G365 exciter filter, FT420 chromatic beam splitter, and LP418 barrier filter; and for chlorophyll *a* Zeiss BP450 exciter filter, FT510 chromatic beam splitter, and LP520 barrier filter. View first for primulin fluorescence to locate nanoflagellate cells, and then for chlorophyll *a* to determine which of the cells are pigmented.

6. Convert the average number of cells per field to organisms per ml from the sample volume (×0.9 to compensate for dilution by preservatives), the area of the field of view, and the area of the filter covered by sample by:

$$\text{No./ml} = \frac{(\text{no. cells/field})(\text{funnel area/field area})}{(0.9)(\text{sample volume})}$$

7. Enumerate microprotozoan community composition with the Utermöhl sedimentation technique with 25 to 50-ml samples and an inverted microscope at 400× (see Exercise 10). Although the Utermöhl technique underestimates nanoflagellate concentrations greatly (Davis and Sieburth, 1982), the method is satisfactory for larger protozoa.

8. Estimate biovolume of nanoflagellates and protozoans from measurements and approximate geometric shapes (cf., Exercise 10). From these estimates, approximate values for cellular organic carbon (μg C/l) can be calculated from the viovolume (μm^3/l or μg/l) × the factor below:

Nanoflagellates   0.11
Microprotozoa   0.15

Estimates of organic carbon content of rotifers and crustaceans can be approximated by 0.5 × dry weight, directly from determinations (see the following discussion and Table 11.3).

EVALUATION OF BIOMASS

The mass of an organism can be estimated from its volume or it can be determined directly by weighing (Table 11.3). Many workers estimate average dry weight biomass of zooplankters from estimates of average length and regressions of length versus weight. This subject, developed by Dumont et al. (1975), has been reviewed

Table 11.3. Approximate volume and dry weight of some zooplankton genera.[a]

	Volume (μm^3)	Dry weight[b] (μg)
Protozoa		
small ciliates	2500	nd[c]
large ciliates	5×10^5	nd
Dileptus	3×10^6	nd
Bursaria	3×10^7	nd
Trichodina	5000	nd
Rotifers		
Brachionus	6×10^5	nd
Keratella	$0.5–1.0 \times 10^5$	0.05–0.1
Kellicottia	1×10^5	0.07–0.1
Trichocerca	$1.0–1.5 \times 10^5$	nd
Gastropus	5.5×10^5	nd
Ascomorpha	3×10^5	nd
Asplanchna	$3.0 \times 10^7–10^8$	0.2
Synchaeta	$1–2 \times 10^6$	nd
Polyarthra	$5–6 \times 10^5$	0.06
Conochilus	$4–6 \times 10^5$	0.08
Cladocera		
Daphnia	$10^7–19^9$	2–35
Ceriodaphnia	5×10^7	2–4
Bosmina	$4–7 \times 10^7$	1–3
Chydorus	1×10^7	2
Leptodora	1×10^{10}	nd
Copepods		
small nauplii	5×10^4	nd
large nauplii	1.5×10^6	nd
Cyclops	$10^7–10^8$	9–22
Diaptomus	$10^7–10^8$	5–7
Mesocyclops	3×10^7	1–3
Tropocyclops	nd	1–3

[a] After Nauwerck (1963), Makarwicz (1974), and Hall et al. (1970).
[b] Dry weights might be approximated by multiplying the volume by a specific gravity value of 1.025 (Hall et al., 1970).
[c] nd = not determined.

critically and in great detail by McCauley (1984). Volumetric determinations are difficult to make for small organisms of irregular shape. Dimensions may be measured (see Exercise 10) and volume calculated, assuming the organism approximates a cube, sphere, cylinder, or some other geometrical form [cf., Ruttner-Kolisko, 1977a, 1977b]. Alternatively, volume displacement of water may be used as an approximation. Since most zooplankton organisms are very small, it would be necessary to add numerous individuals to the water volume of a small volumetric chamber with fine graduations to obtain a reliable estimate of the mean volume for the organisms.

Before obtaining the weight of zooplankton they should be thoroughly dried. Freeze-dry or dry at 60°C until a constant weight is obtained (usually one to two days). Transfer from the drying oven to a desiccator and allow to cool for at least 1 h before weighing. When available, freeze-drying should be used to dry the zooplankton. Drafts and humidity fluctuations while weighing small organisms can produce serious errors in these measurements; appropriate precautions should be taken.

In most cases it will be necessary to weigh several individuals (a few to several hundred) at once and determine the mean weight. This measurement can be done by adding several animals to tared small aluminum foil pans or to a small cover-glass. A sensitive electrobalance (e.g., Cahn Gram electrobalance) is useful for determining very small weights.

Volume and weight of zooplanktonic species may vary by an order of magnitude or more depending on habitat and season (Dumont et al., 1975). For example, a relatively small change in length or width may result in a large change in volume for an individual.

For further details on sampling, enumeration, mass, and chemical determinations of zooplankton, see Edmondson and Winberg (1971), Winberg (1972), Tranter and Fraser (1968), de Bernardi (1984), and McCauley (1984).

EXERCISES

This exercise is designed for two or more field trips to a small lake or reservoir.

OPTION 1

A. Comparison of Collection Apparatus

1. Using at least one of the metered-type collectors (e.g., Clarke-Bumpus sampler) and one of the volumetric-type collectors (e.g., Schindler trap or Likens-Gilbert technique), obtain three replicate samples for each device from a depth of 3 m near the center of the lake.

   The Clarke-Bumpus sampler is towed from a boat at a constant speed at a specified depth. The speed should be maintained between 0.9 and 3.2 km/h so that the meter undergoes 40 to 140 rev/min (McNaught, 1971). This rate can be determined from the total number of revolutions and the elapsed time of the tow. At this speed the sampler is calibrated to pass about 4.4 l/rev. At slower or faster rates, the meter is inaccurate and samples should be discarded. With experience, the appropriate speed can be determined readily by the boat operator. The depth, z, of the sampler can be determined from the angle of the cable from: $z = (L) (\cos \alpha)$, where L = length of cable from surface of water to sampler; and α = angle subtended by the cable to the vertical. The angle can be measured readily with a clinometer. A useful "rule-of-thumb" when a clinometer is not available: an angle of 45° angle can be approximated. The cosine of a 45° angle is 0.7, so the cable should be extended 1.4 times the desired depth. The sampler is lowered to the appropriate depth and is opened by a messenger from the surface. A weight or hydrodynamically designed depressor plate (Fig. 11.7) should be attached below the sampler and at the end of the line to facilitate sampling at depth. A convenient tow period is 3 to 5 min, but time will depend on the density and size of particles in the water. At the end of the tow a second messenger is lowered to close the sampler, and the sampler is retrieved to the boat. Care should be taken to prevent movement of the propellers of the meter by air currents, as this can give erroneous results. Promptly record the number of revolutions and the time of the tow, and wash the organisms from the net into the bucket. Reduce the volume of water in the bucket and transfer to a labeled sample bottle. Preserve with formalin.

   In the Likens-Gilbert technique, a *transparent*, Van Dorn–type water bottle (see Exercise 10) is lowered to the desired sample depth and closed immediately. Closing the water bottle immediately is imperative if a quantitative sample of the larger, more motile

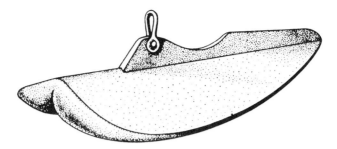

Figure 11.7. A depressor plate designed to keep devices being towed at depth. [From Ahlstrom et al. (1958).]

zooplankton is to be obtained (Smyly, 1968). The water bottle should hold 2 to 51 ml of water and be as short (0.5 m or less in length) as is practical. Upon hauling the bottle into the boat, the water is filtered through a plankton funnel fitted with a 48-μm or smaller mesh net (Fig. 11.5). Application of previously filtered water from a wash bottle to the net surfaces will facilitate filtration and reduction of sample volume. Transfer the sample to a labeled bottle and preserve with formalin.

2. Care must be taken in reducing the volume of the sample to about 100 ml. This concentration can be done by carefully draining excess water through the netting of the collection device. Patience and practice are required. Quantitatively transfer the sample to a sample bottle. A wash bottle filled with filtered water will be useful in washing the organisms from the collection device, particularly from the netting. Wash the net from the outside and from the top downward. Add a few ml of neutralized formalin to stop biological activity in the sample. Carefully label each sample bottle according to date, time, location, depth, volume filtered, collector's initials, and any other pertinent information.

3. Identify species and enumerate. For the purposes of this exercise, it is not essential to identify organisms exactly. Designation by letters or numbers will suffice as long as it is done consistently. Make sketches and notes as appropriate.

4. Determine dry weights for individuals of each species and calculate zooplankton biomass for each species on a volumetric and lake area basis (see Exercise 1).

5. Calculate the counting error for one of the samples as the standard error of the mean (see Appendix 2).

6. Compare statistically the results for the various collection devices. Which one was most efficient? Which collector provided the most representative sample?

B. Site Sampling Error

1. Using the collection device judged to be the best on the basis of the test above (part A), obtain five separate samples from a depth of 3 m at some location on the lake.

2. Identify species and enumerate.

3. Calculate the mean, standard error of the mean, and 95% confidence limits for the major species.

C. Horizontal Distribution and Variability

1. Using the collection device judged to be the best on the basis of the above test (part A), collect samples, from a depth of 3 m at five or more randomly located positions on the lake. Move in a random sequence to each of these positions, but collect the entire sequence of samples in as short a time as possible. Why? Alternatively, obtain samples from the same depth at prescribed location along a transect across the lake.

2. Identify species and enumerate.

3. Compute the mean, variance, and chi-square values for the major species collected at the various locations. Determine whether the population was distributed randomly, contagiously, or uniformly.
4. Based on the mean values, calculate the total number of zooplankton at the 3-m depth. Compute the 95% confidence limits. Would one station have been sufficient to quantitatively characterize the zooplankton at the 3-m depth?
5. Answer the questions and problems following Option 2.

OPTION 2

1. Use zooplankton samples provided by the instructor.
2. Assume that the samples were collected with a Clarke-Bumpus plankton sampler fitted with a #10 silk net. The device was towed through the water at a depth of 3 m for exactly 3 min. Calibration shows that 4.11 of water/rev pass through the sampling device under these conditions. Perform the calculations called for above in parts A-3, B, and C (Option 1).
3. Answer the following questions and problems.

Questions and Problems

1. Evaluate the various sources of error involved in collection and enumeration of zooplankton.
2. What is the effect of water movements such as Langmuir circulation on the collection efficiency of zooplankton?
3. What do you think causes the patchy nature of zooplankton in lakes?
4. Compare the advantages and disadvantages of a towed, metered-type sampler with a volumetric sampler.
5. Why should the Van Dorn sampler be closed immediately when sampling zooplankton at some depth?
6. What is "back pressure" in a plankton net?
7. How should sampling procedures vary from littoral to limnetic regions and throughout a 24-h period to obtain quantitative samples?
8. Why should plankton samples be dried at temperatures less than 105°C when obtaining dry weights? What are the disadvantages of drying at temperatures less than the boiling point of water?
9. Likens and Gilbert (1970) suggested that the pore size of plankton nets could seriously affect calculations of productivity in zooplankton. Why?
10. Why should nets not be towed too fast? Or too slow?
11. How does the predominance of species vary according to density and biomass? Which parameter (density or mass) is more important for the metabolism of the lake? Why?
12. What was the filtration efficiency for the samplers used?

Apparatus and Supplies

1. Zooplankton collection apparatus [e.g., Clarke-Bumpus plankton sampler (available from Wildco Instruments, Saginaw, MI, Isaacs-Kidd sampler (Isaacs and Kidd, 1953), 0.5-m diameter vertical townet (available from Wildco Instruments), Schindler trap (Schindler, 1969), Likens-Gilbert water bottle and plantkon funnel (Likens and Gilbert, 1970)].
2. Clinometer (available from Wildco Instruments).
3. Automatic volumetric pipets.
4. Graduated cylinders (50-, 100-, and 200-ml).
5. Sedgwick-Rafter cells (1-ml).
6. Hand tallies.

7. Wash bottles and filtered water.
8. Dissecting needles or fine insect needles with heads removed and inserted into matchsticks.
9. Buffered formalin (5%): Dilute 40% formaldehyde with water and add sufficient sodium acetate to yield a final pH of 7.5 to 8.0 in the preserved sample (see Exercise 10).
10. Compound microscopes (high and low power objectives and mechanical stage).
11. Stage micrometer (see Exercise 10).
12. Whipple disc (see Exercise 10).
13. Plastic rulers.
14. Wide mouth bottles, 8-oz, with screw-top caps.
15. Winch, meter wheel register, line, weight or depressor (available from Wildco Instruments) for zooplankton collection apparatus.
16. Small boat with motor (or untiring persons with oars!).
17. Hand tools (e.g., screwdriver, pliers, and adjustable end wrench).

References

Ahlstrom, E.H., J.D. Isaacs, J.R. Thrakill, and L.W. Kidd. 1958. High-speed plankton sampler. Fish Bull. Fish Wildl. Serv. *58*(132):187–214.

de Bernardi, R. 1984. Methods for the estimation of zooplankton abundance. pp. 59–86. *In*: J.A. Downing and F.H. Rigler, Editors. A Manual on Methods for the Assessment of Secondary Productivity in Fresh Waters. Blackwell, Oxford.

Caron, D.A. 1983. Technique for enumeration of heterotrophic and phototrophic nanoplankton, using epifluorescence microscopy, and comparison with other procedures. Appl. Environ. Microbiol. *46*:491–498.

Carrick, H.J., G.L. Fahnenstiel, E.F. Stoermer, and R.G. Wetzel. 1991. The importance of zooplankton-protozoan trophic couplings in Lake Michigan. Limnol. Oceanogr. *36*:1335–1345.

Clutter, R.I. and M. Anraku. 1968. Avoidance of samplers. pp. 57–76. *In*: D.J. Tranter and J.H. Fraser, Editors. Zooplankton Sampling. UNESCO Press, Paris.

Dumont, H.J., I. Van de Velde, and S. Dumont. 1975. The dry weight estimate of biomass in a selection of Cladocera, Copepoda and Rotifera from the plankton, periphyton and benthos of continental waters. Oecologia *19*:75–97.

Edmondson, W.T. and G.G. Winberg (eds). 1971. A Manual on Methods for the Assessment of Secondary Productivity in Fresh Waters. IBP Handbook No. 17. Blackwell, Oxford. 358 pp.

Hall, D.J., W.E. Cooper, and E.E. Werner. 1970. An experimental approach to the production dynamics and structure of freshwater animal communities. Limnol. Oceanogr. *15*:839–928.

Hobbie, J.E., R.J. Daley, and S. Jasper. 1977. Use of Nuclepore filters for counting bacteria by fluorescent microscopy. Appl. Environ. Microbiol. *33*:1225–1228.

Isaacs, J.D. and L.W. Kidd. 1953. Isaacs-Kidd midwater trawl. Final report. Scripps Inst. Oceanogr. (Ref. 53-3, 1-18. Oceanographic Equipment Rept. No. 1).

Juday, C. 1916. Limnological apparatus. Trans. Wisc. Acad. Sci. Arts Lett. *18*(Pt. 2):566–592.

Levin, S.A. and L.A. Segel. 1976. Hypothesis for origin of planktonic patchiness. Nature *259*:659.

Likens, G.E. and J.J. Gilbert. 1970. Notes on quantitative sampling of natural populations of planktonic rotifers. Limnol. Oceanogr. *15*:816–820.

Makarewicz, J.C. 1974. The community of zooplankton and its production in Mirror Lake. Ph.D. Thesis. Cornell Univ., Ithaca. 206 pp.

McCauley, E. 1984. The estimation of the abundance and biomass of zooplankton in samples. pp. 228–265. *In*: J.A. Downing and F.H. Rigler, Editors. A Manual on Methods for the Assessment of Secondary Productivity in Fresh Waters. Blackwell, Oxford.

McNaught, D.G. 1971. Appendix to Clarke-Bumpus plankton sampler. pp. 11–12. *In*: W.T. Edmondson and G.G. Winberg, Editors. Secondary Productivity in Fresh Waters. Blackwell, Oxford.

Nauwerck, A. 1963. Die Beziehungen zwischen Zooplankton und Phytoplankton im See Erken. Symbol. Bot. Upsaliensis *17*(5):1–163.

Pennak, R.W. 1962. Quantitative zooplankton sampling in littoral vegetation areas. Limnol. Oceanogr. 7:487–489.

Ruttner-Kolisko, A. 1977a. Comparison of various sampling techniques, and results of repeated sampling of planktonic rotifers. Arch. Hydrobiol. Beih. Ergebn. Limnol. *8*:13–18.

Ruttner-Kolisko, A. 1977b. Suggestions for biomass calculation of planktonic rotifers. Arch. Hydrobiol. Beih. Ergebn. Limnol. *8*:71–76.

Schindler, D.W. 1969. Two useful devices for vertical plankton and water sampling. J. Fish. Res. Bd. Canada *26*:1948–1955.

Smyly, W.J.P. 1968. Some observations on the effect of sampling technique under different conditions on numbers of some freshwater planktonic Entomostroca and Rotifera caught by a water bottle. J. Nat. Hist. *2*:569–575.

Steedman, H.F. (ed.) 1976. Zooplankton Fixation and Preservation. UNESCO Press, Paris. 350 pp.

Szlauer, L. 1964. Reaction of *Daphnia pulex* de Geer to the approach of different objects. Pol. Arch. Hydrobiol. *12*:15–16.

Tonolli, V. 1971. Zooplankton. pp. 1–14. *In*: W.T. Edmondson and G.G. Winberg, Editors. A Manual on Methods for the Assessment of Secondary Productivity in Fresh Waters. IBP Handbook No. 17. Blackwell, Oxford.

Tranter, D.J. and J.H. Fraser (eds). 1968. Zooplankton Sampling. UNESCO Press, Paris. 174 pp.

Utermöhl, H. 1958. Zur Vervollkommnung der quantitativen Phytoplankton-Methodik. Mitt. Int. Ver. Limnol. *9*. 38 pp.

Welch, P.S. 1948. Limnological Methods. Blakiston, Philaldelphia. 381 pp.

Wetzel, R.G. 1999. Limnology: Lake and River Ecosystems. 3rd Ed. Academic Press, New York (in press).

Whipple, G.C. 1927. The Microscopy of Drinking Water. 4th Ed. Wiley & Sons, New York. 586 pp.

Winberg, G.G. (ed). 1972. Methods for the Estimation of Production of Aquatic Animals. Academic Press, New York. 175 pp.

Benthic Fauna of Lakes

The animals living on and in the sediments and large plants of lakes are usually highly diverse. Much emphasis in the study of benthic fauna has been given to the immature stages of insects that often make up the dominant part of the total animal biomass of these habitats. Nearly all insect orders are represented. Some orders of insects are entirely aquatic; others inhabit fresh waters only during certain life stages. Segmented worms (oligochaetes and leeches), microcrustacea (ostracods), and macrocrustacea (mysids, isopods, decapods, and amphipods) often form major components of benthic fauna of fresh waters.

The sediment composition and characteristics of the water adjacent to the sediments are also highly variable. Attached microflora (bacteria, fungi, and algae) may occur in great abundance on sediments. These organisms and associated detrital organic matter often provide the predominant energy sources for the benthic fauna. Consequently, a variety of types of feeding and of reproductive dynamics occur among the benthic fauna. It is, therefore, essential in limnological analyses to obtain reliable quantitative estimates of the sizes and distributions of populations constituting these communities.

Quantitative estimations of benthic fauna require effective sampling procedures, separation of the organisms from the substratum, identification, and evaluation of biomass of species and of their life history stages. Methods to accomplish these tasks are not totally satisfactory, and the taxonomy of many immature stages of benthic invertebrates is difficult and incompletely understood. Nonetheless, in spite of these problems, often it is important to analyze the response of population growth and survival and physiological characteristics (e.g., respiration, excretion, and assimilation) of the benthic fauna to environmental variations, such as temperature, oxygen concentrations, and food quantity and quality.

The living biomass (g/m^2) of a population of animals at an instant in time and in a given habitat represents the net result of reproduction and growth and the opposing processes of loss (respiration, predation, mortality, emigration, and so on). Although the measurement of biomass gives an estimate of the extent of population development, from these data alone nothing can be said about the growth and reproduction of the organisms. More information is needed to evaluate production rates (g/m^2/time). This information can be obtained only by detailed analyses of reproductive and physiological characteristics or survivorship and growth.

There are some general direct correlations between the overall productivity of fresh waters and benthic animal productivity [cf., summary of Wetzel (1999)]. Moreover, some general insight into benthic faunal composition and distribution in relation to lake characteristics can be obtained from simple quantitative analyses. The following exercise is directed toward the understanding of problems of sampling, sorting, and quantitative estimates of population size. Obviously, much more detailed investigations would be needed to determine secondary productivity and its control in research efforts. The monographs of Edmond-son and Winberg (1971), Winberg (1971), Brinkhurst (1974), Waters (1977), Benke (1984), and Rigler and Downing (1984) are excellent critical compilations on methods and are highly recommended sources for further study.

SAMPLING

A number of basic requirements must be met to capture a quantitative sample of the populations living on and within the sediments. The sampler must penetrate into the sediments to a sufficient depth to capture all of the organisms inhabiting a defined area. The device should enclose the same area of sediment each time. As the sampling device is lowered, care should be taken not to disturb the sediments or give certain organisms an opportunity to escape prior to collection. The sampler should close completely so that sediment and organisms are not lost during retrieval. The efficacy of the many benthic samplers varies greatly; few fulfill all of these requirements adequately. Fewer samples are needed for a given precision when the population density is high and when samples are taken with a large sampler (Downing, 1979). However, the amount of sediment that must be examined increases with increasing size of the sampler. Hence, small-area samplers are most efficient. A sampler should not be prone to sediment loss upon retrieval or pressure-wave effects upon entry into the sediments. The sampler should be sufficiently large to avoid many zero values of organisms among replicates but should be as small as possible to minimize the time required to separate the organisms from the sediments.

Many insects emerge from the water as winged, flying adults. Accurate analyses of population dynamics of these aquatic organisms require a quantitative evaluation of such emigration. Emergence traps are discussed briefly in Exercise 13 and in extensive detail by Davies (1984).

Benthic samplers may be grouped into four general categories: (1) metal boxes closed by jaws at the bottom; (2) hinged grabs in which sediments are scraped toward the center, each side bucket forming a sector of a horizontal cylinder; (3) coring devices; and (4) specialized box- and core-type samplers for use in substrata that are difficult to sample, such as hard, stony sediments and sediments with dense macrovegetation. Several of the more commonly used devices and their operation will be discussed briefly here; a detailed treatment is found in Edmondson and Winberg (1971), Holme and McIntyre (1971), Brinkhurst (1974), Elliott and Drake (1981), and particularly Downing (1984).

Ekman-Type Grab Sampler

The Ekman grab consists of a square or rectangular box (15.2×15.2 cm), usually made of brass, with a pair of spring-operated jaws that close tightly when released to enclose a specific area of the sediments (Fig. 12.1). During operation, a strong metered line is passed through the tripping mechanism and knotted *securely* below the underlying plate. The jaws are cocked in the open position. *Caution*: Accidental closure can cause injuries; keep hands clear. Attach the dead end of the line to the boat. The sampler then is lowered into the water until it rests on the sediments. Usually its own weight is adequate to penetrate soft sediments for all or most of its

Figure 12.1. The Ekman grab. *Left*: In open form, as during descent through the water; upper doors would be open to permit water to flow through the sampler. *Right*: After the messenger has tripped the release mechanism and the jaws have closed.

height. The messenger is then sent down the line to close the jaws. The sampler is retrieved, and a container is placed beneath the sampler just as it breaks the surface of the water. The jaws then are opened and all of the sediments are washed into the container. If the jaws were not closed completely, the sample should be discarded.

The Ekman grab sampler is best suited for use in soft, finely divided sediments. The grab will not function well on sand substrates or at all where hard objects (sticks, leaves, stones, or mollusc shells) are common. In extremely flocculent, loosely dispersed organic sediments, the grab may penetrate below the surface of the sediments and a portion of the surface sediments may be excluded. Use of a high-form Ekman grab [cf., Kajak (1971)] will minimize this problem. This modified Ekman sampler is similar in every respect to that shown in Fig. 12.1 except that the box portion is about twice as high.

The Ekman grab may be modified by attaching a pole for controlled sampling in shallow waters (<2 m) where the sediments are very soft or where the grab must be forced into compacted sediments. The closure system may be replaced by levers for closing the jaws in compacted, sandy sediments.

The Ekman grab departs from the ideal sampling criteria in that it can create a pressure wave before entering the sediment, thus "blowing" away lightbodied organisms on the sediment surface. Also, mud and organisms may be lost out of the top of the sampler as it is being brought to the water surface. In addition, the bottom of the sampler is arc-shaped, so the entire sampler does not penetrate to a uniform depth; this is an important problem if the number of organisms were to change markedly with depth. Although the Ekman grab takes large samples, which is statistically advantageous, the sieving and sorting time is long [cf., Downing (1984)]. Nevertheless, Flannagan (1970) has demonstrated that total organism abundance is estimated as well with a standard Ekman grab as it is with a diver-operated hand core (which, theoretically, takes an unbiased sample), although the Ekman grab tends to underestimate oligochaete density. Burton and Flannagan (1973) have designed a greatly improved Ekman grab in which the problems of pressure wave and loss of organisms during retrieval are eliminated. Tests have shown that this device collects significantly more chironomids and oligochaetes than does the standard Ekman grab.

Petersen-Type Grab Samplers

The Petersen grab consists of two hinged, pincerlike buckets that are lowered in the open position to the sediments (Fig. 12.2). As the line slackens, the release mecha-

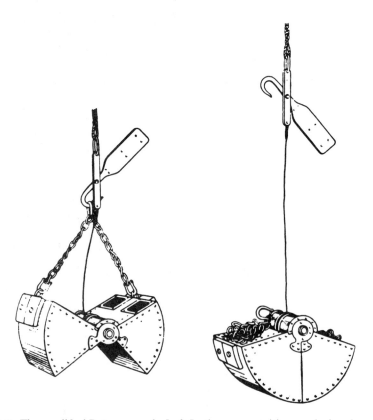

Figure 12.2. The modified Petersen grab. *Left*: In the open position, as during descent to the sediments. Screened ports on the upper surface attempt to reduce the pressure waves created during descent. *Right*: In the closed position. [From Kajak (1971).]

nism is actuated. Upon retrieval, the two buckets come together and enclose a semi-circular section of the sediments.

The advantage of the Petersen grab is its effective penetration into firm sediments, such as sand, because of its heavy construction. The area commonly sampled is somewhat greater than 1/20th of a m^2. As with any jawed device, stones, shells, and similar objects can prevent closure and result in washout losses on retrieval. The most severe disadvantage of this sampler is that its closed or nearly closed design creates pressure waves that disturb the sediments and benthic fauna as the sampler approaches the sediments. A modified grab, known as the Ponar grab, has been altered from the Petersen construction so that the upper portions are covered with screen to reduce the pressure waves (Powers and Robertson, 1967). It has been demonstrated that a portion of the sediment within the area sampled is lost and that sampling is deeper at the edges of the bite than in the center (Gallardo, 1965). The Ponar and Ekman grabs have been found to sample with acceptable efficiency in soft sediments up to a particle size of fine gravel (Elliott and Drake, 1981). The Ponar grab was found to be adequate for sampling a substratum of small stones; neither were adequate for sampling among larger stones (>16-mm diameter). Efficiencies of sampling always should be evaluated; for example, benthic biomass from box core samplers obtained with SCUBA were 1.7 times greater than in samples obtained with the Ponar grab (Nalepa et al., 1988).

Coring Samplers

A large number of core-type samplers have been devised. All consist of a tube that is pushed into the sediment, either by free fall, by force exerted from the surface, or by means of a piston system within the tube. A valve that closes upon retrieval usually is included at the top of the tube to prevent the sediments from washing out. It is most important that the closing device does not impede water flow through the tubes during descent. When the water outflow is impeded, even slightly, the tube and water pressure will push aside loose sediment at the sediment-water interface. Even with a closing device at the top of the tube, very soft and flocculent sediments can fall out during retrieval. Some core-type samplers have closing devices at the bottom to prevent loss of sediments.

In shallow sediments, plastic or metal tubes simply may be pushed into the sediment and then stoppered or sealed with a manually operated valve [e.g., Gillespie et al. (1985)]. Such cores also have been used at greater depths with SCUBA diving apparatus. In most cases, however, sampling from the surface is necessary. Coring devices, such as the Kajak-Brinkhurst corer (Brinkhurst, 1974), are allowed to fall freely to the sediments. The valve closes upon retrieval.

The relatively small area (10 to 50 cm^2) that corers sample is a drawback when organisms are distributed sparsely. It is mandatory to take a number of replicates at each depth. A multiple corer, such as that illustrated in Fig. 12.3, is excellent and permits several cores to be taken simultaneously. In this device, a messenger dropped along the line from the surface releases the soft half-balls that seal the top ends of the tubes (Hamilton et al., 1970). In a comparison with diver-operated cores and several other samplers, the abundance of organisms taken with this multiple corer showed no significant difference from the diver-operated cores, although it did slightly underestimate chironomid density. In addition, the multiple corer has the following advantages over the Ekman grab:

Figure 12.3. A multiple corer with ball-type closures that are released by a messenger after the tubes have penetrated into the sediments.

1. It is designed to create a minimal pressure wave.
2. Surface sediment is not lost during retrieval.
3. Variance within a sampling station can be determined since four cores are taken at once over a very small area.
4. Sieving and sorting time is reduced since the four cores combined produce a smaller volume than one Ekman grab sample.
5. It can be easily modified to perform experiments on benthic metabolism (e.g., with plastic core liners).
6. When the coring tubes are made of transparent plastic, the sediment profile, including the sediment-water interface, is readily available for visual inspection.

Neither the Ekman grab nor the multiple corer will take totally quantitative samples in coarse gravel or organic debris [see Elliott and Drake (1981) for comparisons and recommendations]. The use of SCUBA may facilitate quantitative sampling under many circumstances.

SEPARATION OF ANIMALS FROM SEDIMENTS

Separation of benthic organisms from sediments and debris is difficult to accomplish quantitatively. A number of methods have been developed to assist with this task; all are tedious and require patience and care. Often initial separation of the organisms from the sediments is done in the field by screening through sieves that remove some of the sediment and reduce bulk. When such separation cannot be accomplished within a few hours, the samples should be preserved by the addition of a formaldehyde solution (40%) to give a final formalin concentration of about 5% (a 5% formalin solution =2% formaldehyde).

When screening with sieves, a decision must be made as to what size of mesh to use. The smaller the mesh size, the more organisms that are retained but also the more sediment retained, and so the time required for sieving and sorting increases greatly. Compromise is needed. In general, as small a mash size as practical should be used. A mesh size of 0.40 mm will retain only large and older stages of benthic organisms. The better mesh size is 0.20 mm, which represents the practical lower size limit for general study of benthic animals. Even with this smaller mesh size, some small animals will be lost [e.g., see Nalepa and Robertson (1981), Storey and Pinder (1985), and Strayer (1985)]. In some instances where sediments are large-grained or contain fibrous organic materials, sieving with even 1-mm screens may require extraordinary amounts of time or prove practically impossible. The objectives of the studies should be kept in mind. When small stages of benthic organisms are required for analyses of growth and productivity the smallest mesh size is mandatory. On the other hand, when measurement of biomass of the composite benthic populations is the goal the instantaneous biomass will be associated with the larger organisms, and over 90% of the total biomass often will be retained using a coarser sieve [cf., Reish, (1959)].

When small core subsamples are taken simultaneously from surficial sediments (to a depth of 4 to 5 cm), estimates also can be made of the numbers and biomass of young stages of the benthic animals from the materials of the macrosample. The sieved materials should be washed carefully into collection bottles and preserved and the samples diluted with water to make the task more pleasant. The samples can be transferred to 90% alcohol, but this treatment will cause much shrinkage and confound biomass determinations (see the discussion below).

Small amounts of sediment samples can be placed into shallow, white enamel trays and sorted manually. When much organic debris is present, the addition of a small amount of eosin or rose bengal dye assists in seeing the organisms, since they tend to stain more intensely than debris. The organisms are even more conspicuous when stained with a very small amount of fluorescent dye such as rhodamine B and viewed under long-wave ultraviolet light (Hamilton, 1969a).

Flotation methods are used widely to assist in the separation of benthic organisms from sediments. Samples are placed into a salt or sugar solution of high density relative to the organisms. The organisms float, and sediment particles, more dense than the solution, sink. When the samples contain much plant material, the flotation method does not work well, as much plant debris also floats.

Although a number of solutions have been used (NaCl, CaCl$_2$, MgSO$_4$, and so on), sugar (sucrose) of a concentration sufficient to yield 1.12 to 1.13 g/ml density after mixing with the sediment sample works very well [see Anderson (1959) and Kajak et al. (1969)]. Samples are mixed with sugar or salt solution in a volumetric ratio of about 1:5 or 1:10, stirred vigorously, and the floating organisms removed *quickly* with forceps or a small strainer. Stirring is repeated several times. The remaining sediment is then sorted by hand because many organisms will not float, e.g., molluscs and caddisfly larvae in cases. Small larvae, such as those of chironomids, are very difficult to sort from sediments, and tedious efforts are required to do this quantitatively.

Many invertebrates can be separated from substrata by elutriation techniques, in which a flow of water is passed rapidly from below through the sediments. The invertebrates are less dense than sand and gravel and are separated from the sediments and pass with the flow to collecting sieves. Light-bodied organisms (e.g., copepods, chironomids, nematodes, odonates, and emphemeropterans) are separated most effectively [e.g., Whitman et al. (1983)]. Dense animals, such as molluscs, are recovered very poorly.

ESTIMATION OF BIOMASS

Biomass estimations of benthic organisms usually are made on the basis of dry weight and of organic weight (ash-free dry weight). Because of the large amount of inorganic materials that is part of many benthic organisms, e.g., shells of snails and clams, it is important that dry weight take these components into account. Dry weight is determined at 105°C. A small percentage of organic materials will be lost at this temperature; hence, for very exact work, lyophilization (freeze-drying) is necessary [cf., Downing (1984) and Leuven et al. (1985)]. The samples then are heated to 550°C in tared crucibles for 4 h to remove organic matter, cooled under desiccation, and reweighed.

When organisms have been preserved in alcohol, as much as 25% of the biomass (dry weight) will be lost by leaching (Holme, 1964; Stanford, 1973; Leuven, et al., 1985). A slight increase in weight occurs in some organisms when preserved in formalin. In both cases, evaluation of the changes is necessary, and appropriate correction factors must be applied.

Shrinkage may also occur upon preservation. The extent of shrinkage must be evaluated when making size measurements for the determination of growth and instar calculations.

Estimates of biomass can be extremely time-consuming. Rather than weighing all animals from field samples, it is common practice to develop independent length-weight relationships, so that the length of individuals can be measured from samples and converted into biomass. The following relationship is used:

$$\text{Dry wt} = (a) \cdot (\text{length}^b) \quad \text{or} \quad \log \text{dry wt} = \log a + b \log \text{length}$$

where b is usually $\cong 3$ [cf., Smock (1980)].

PRODUCTIVITY OF BENTHIC FAUNAL POPULATIONS

Secondary production is the formation of heterotrophic biomass through time (Benke, 1993). That formation is coupled bioenergetically to food ingestion and utilization, where:

$$I = A + E$$

Where I = ingestion, A = assimilation, and E = egestion. Assimilation $A = P + R + U$, where P = production, R = respiration, and U = excretion. Each of these terms are fluxes of energy or carbon, with units of energy per area per unit time. For an individual organism, P represents growth, whereas for a population it represents the composite growth of all individuals. Growth is in turn dependent upon assimilation efficiency (A/I) and net production efficiency (P/A). Assimilation efficiency can vary from very low values (<5%) among detritivores to much higher values for carnivores. Net production efficiency is less variable and is often close to 50% among benthic invertebrates.

Animal production methods usually consider populations. Estimates of rates of production of benthic animal populations have been made using four basic approaches [cf., Waters (1977), Rigler and Downing (1984), and Benke (1984, 1993)]. The first three methods often are referred to as "cohort" methods, in which changes in a population cohort are followed by sampling through time. The fourth method using size-frequency evaluates changes in average, pooled populations and their population frequencies in different size categories with time.

Removal-Summation Method. A simple life history of a single-species population is determined and a series of density and biomass estimates is made throughout the life of a single cohort or generation. The mortality, in terms of biomass, can be calculated between successive samples of the series, taking into account the size of the organisms during the period of loss. The sum of the observed mortalities over the entire life cycle then is equivalent to the total production of the cohort. When the species is univoltine, i.e., has only one reproductive generation per year, the cohort production then is equivalent to the annual production for this species [see Waters (1977), Benke (1984), and Rigler and Downing (1984)].

The removal- (= mortality) summation method evaluates production (P) as the sum of mean individual weight ($\bar{w}$, as dry weight in mg/individual) times the change in numbers (N) for each sample interval or cohort interval ($P = \Sigma\bar{w}\Delta N$). This removal-summation method essentially is identical to the growth increment summation method, based on the same population statistics, where production is equal to the sum of mean numbers times the change in mean individual biomass (weight) for each sample interval ($P = \Sigma\bar{N}\Delta w$).

Instantaneous Growth Method. The production rate for a given interval of time is estimated by the product of the instantaneous rate of growth and the mean biomass (standing stock) during the time interval:

$$P = G(B)$$

where P = production in biomass/area (e.g., g/m^2) over a given interval of time; G = instantaneous rate of growth during this period, computed as the natural logarithm of the ratio of the mean individual weight at the end of the time interval to the mean weight at the beginning of the interval; and B = mean biomass/area (e.g., g/m^2) during the time interval, computed as the average of the biomass at the beginning and end of the time period.

It should be emphasized that this simple exponential growth model assumes continuous growth. Many organisms usually exhibit discontinuous growth in which birth, death, immigration, and emigration vary and cause differing discrete changes in the population size and growth. It is important that sampling frequency take these

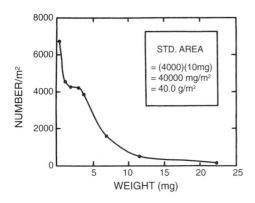

Figure 12.4. An Allen curve of numbers versus biomass for mean individual weight (wet) of *Ephemerella subvaria* over one year. A standard area of known dimensions is included for planimetry and calculations. [Drawn from data of Waters and Crawford (1973).]

factors into consideration in relation to rate of growth. While monthly sampling may be quite adequate for many benthic animals, much more frequent sampling may be needed for some forms. In general, the shorter the sampling interval, the better will be the estimations of rates of production. See Edmondson and Winberg (1971), Winberg (1971), Waters (1977), Benke (1984, 1993), and Rigler and Downing (1984) for detailed analyses.

The instantaneous growth method can be used for production estimates of asynchronous populations also when independent estimates of G from growth studies can be obtained. Furthermore, if there were size-specific differences in growth rates for a population, the best estimate of production would be obtained by $P = G_1B_1 + G_2B_2 + \ldots +$ (all sizes).

Allen Curve Method. The Allen curve method extends the instantaneous growth method and the removal-summation method to a graphical treatment, in which a curve is constructed from the numbers/weight relationship of a cohort over its life cycle (Gillespie and Benke, 1979). The method is basically a comparison of the survivorship of a cohort and the weights of individuals at selected points along the survivorship curve. Density in numbers per unit area is plotted against mean individual weight. The area under the resulting curve is an expression of cohort production [cf., Allen (1951)]. A simple example is given in Fig. 12.4, from which the area beneath the curve was determined by planimetry and was compared to a standard area within the graph. Area also could be estimated by electronic digitizing or counting of squares within the curve and comparison by proportion to a known area on the same scale.

Comparison of the Methods

Methods for estimates of secondary production must be coupled to life-history characteristics, particularly voltinism and length of the aquatic phase. Many insect species develop synchronously where newly hatched larvae all grow at about the same rate and are similar in size at any point in time (Benke, 1984). At the time of final instar or pupal stages, they all emerge synchronously. The actual cohorts can be followed by frequent quantitative sampling of the populations as they change over time (Fig. 12.5).

The methods for estimating production can be compared by reference to a plot of density against biomass of individuals (Fig. 12.6). The curve respresents the

Figure 12.5. Individual growth and survivorship curves for an aquatic insect population with synchronous cohort development. [Modified from Benke (1984).]

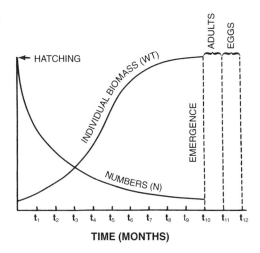

decline in density (numbers) and the increase in individual biomass (weight/individual) during sampling intervals (Δt). The standing biomass for any sampling date is equal to the area of the rectangle defined by that datum (or $N_t \times w_t$) [cf., Benke (1984)]. Production for the entire cohort is equivalent to the total area under the curve; the methods discussed above simply are different ways to calculate this area. The Allen curve is determined by either an exponential curve fitted to the data or by connecting the data points with a smoothed curve and then calculating the area under the curve.

Successive N and w values can also be used to calculate production or losses between sampling dates, rather than drawing a continuous curve across all samples. When time intervals are reasonably short in relation to cohort changes, linear approximations are usually sufficient and easier to calculate (Gillespie and Benke, 1979; Benke, 1984). The production lost during a sampling interval, such as by mortality from predation, is approximated by the area of the horizontal trapezoid ($X +$ Y of Fig. 12.6), calculated as $\bar{w}\Delta N$. Thus, the total cohort production by the removal-summation method can be determined by summing all of the horizontal trapezoids. The incremental-summation method sums the vertical trapezoids in a similar

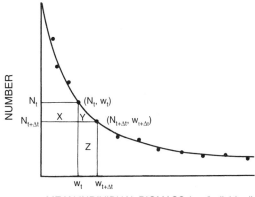

Figure 12.6. Plot of number (density) versus mean individual biomass (weight) for different means used in computation of cohort production (see text). [Modified from Gillespie and Benke (1979) and Benke (1984).]

fashion ($Y + Z$ in Fig. 12.6) with a production calculation as $\bar{N}\Delta w$, the amount of production during the change in time (Δt).

Size-Frequency Method (Average Cohort). The three methods discussed thus far usually can be applied only to synchronously developing populations in which cohorts can be followed through time. These methods require accurate measurement of the dynamics of benthic faunal communities and extensive knowledge of the life cycle of the species. Such information very often is not available for most species and can be acquired only from lengthy, detailed investigations. Until such time as this information is known for most species, a general method for estimating production of the composite species populations has been proposed by Hynes (1961; Hynes and Coleman, 1968) and variously corrected and refined (Hamilton, 1969b; Benke, 1979; Benke et al., 1984). The latter papers should be consulted for a detailed treatment of this method and its limitations. The size-frequency method has been used more than any other production method for macroinvertebrates (Benke, 1993).

The size-frequency method sums the losses between successive size classes, treating single species or small groups of species, rather than between successive sampling intervals. The procedural steps are best seen by way of example (Table 12.1). The field data are presented in columns 1 and 2, in which size groups are separated by the numbers (N/m^2; that is the average cohort) in each group. The loss of numbers between each pair of size classes is calculated in column 3. The median length of each size group is converted to volume by cubing (column 4) to obtain biomass in volumetric units (mm^3). The mean volume between each size class (column 6) is then multiplied by the loss at each stage (column 3) to yield the production loss (column 7).

It is assumed that the population(s) is univoltine, that individuals remain in each of the length classes for the same amount of time, and that they all can grow to the maximum length in this length-frequency table. Consequently, an animal growing through 10 length classes would be expected to remain in each of the length classes for 1/10 of the year, assuming its life expectancy is one year. The mean number in each length class then is multiplied by 10, in this case, to obtain an estimate of the number of individuals that grew to the mean length of that length class during the course of the year [cf., Hamilton (1969b)]. Thus, i of column 8 is the number of length classes that a population grows through to reach its maximum length. The production in volume units (column 9) then is the product of columns 7 and 8.

It is important to note that the last column is the algebraic sum of both positive and negative results, as production losses are summed between size classes. The total for all of the size groups yields a reasonable estimate of the annual production rate for the population. Where production rates of a given species have been determined comparatively by all four techniques, the size-frequency method compares well with the others (A. Benke, personal communication).

In natural heterogeneous populations, the size-frequency distribution can be regarded as a first estimate of an "average cohort" when the number of average cohorts equals the number of size classes through which the organisms grow (Hamilton, 1969b). Growth is assumed to be linear within the size classes. Under these conditions, numerical differences can be attributed to mortality. When all size classes are considered together, the effect of nonlinear growth on the estimate of annual production is small. When most of the organisms are not univoltine, however, a serious error is introduced.

Table 12.1. Calculation of production from insect fauna in 13 sets of samples from Afon Hirnant, Wales, 1955–1956.[a]

1	2	3	4[b]	5[c]	6[b]	7	8	9
Length of each size group (mm)	Density (Average No./m²)	Loss at each stage	Conversion factor to volume	Biomass in vol. units	Production conversion factor $\left(\dfrac{L_j + L_{j+1}}{2}\right)^3$	Loss × conversion factor	Number of times loss occurs	Production in volume units
L	N	$\bar{n}_j - \bar{n}_{j+1}$	L^3	Col. 2 × Col. 4		Col. 3 × Col. 6	i	Col. 7 × Col. 8
1	2,682.9	235.9	1	2,682.9	3.4	802.06	10	8,020.6
2	2,447.0	1,579.1	8	19,576.0	15.6	24,633.96	10	246,339.6
3	867.9	523.9	27	23,433.3	42.9	22,475.31	10	224,753.1
4	344.0	195.3	64	22,016.0	91.1	17,791.83	10	177,918.3
5	148.7	104.7	125	18,587.5	166.4	17,422.08	10	174,220.8
6	44.0	34.2	216	9,504.0	274.6	9,391.32	10	93,913.2
7	9.8	7.7	343	3,361.4	421.9	3,248.63	10	32,486.3
8	2.1	0.8	512	1,075.2	614.1	491.28	10	4,912.3
9	1.3	0.4	729	947.7	857.4	342.96	10	3,429.6
10	0.9	0.9	1,000	900.0	1,157.6	1,041.84	10	10,418.4
				$B_v = 102,084$				$P_v = 976,413$

[a] Modified from data of Hynes (1961) as treated by Hamilton (1969b).
[b] Columns 4 and 6 are estimates of mean individual volume.
[c] Column 5 is biomass in volumetric units, i.e.,

$$\frac{\text{Col.2}}{(\text{N/m}^2)} \times \frac{\text{Col.4}}{(\text{vol./individual})} = \frac{\text{Col.5}}{(\text{vol./m}^2)}$$

Estimates of rates of annual production of multivoltine invertebrates by the size-frequency method sometimes are made by multiplying the production value by the average number of generations per year (Hamilton, 1969b). When egg, pupal, or adult stages of aquatic insects constitute a significant portion of the total generation time, this procedure underestimates productivity (Benke, 1979). If reproduction were to occur before the final size class was attained, as in crustaceans, this procedure would overestimate productivity. These errors can be reduced by determining the average development time for the aquatic stages from hatching to final size, the *cohort production interval* (CPI), which applies only to the aquatic stages throughout which growth and production are occurring (Benke, 1979). The production value from the size-frequency table should be multiplied by 365/CPI, which yields annual production. Thus the average time from hatching to pupation or final size must be known, and this average time can be obtained from temporal patterns of size frequencies or growth studies.

It should be noted that weight units also can be used directly in construction of the size-frequency table. The mean weight for each size group replaces volume (columns 4 and 6) and production is calculated directly in g/m^2 over the year. When volume calculations are used, conversion to weight can be estimated by assuming that average organisms are cylinders five times as long as broad with a specific gravity of 1.05 (Hamilton, 1969b). Then, using a formula, $\pi \cdot r^2 \cdot L \cdot \rho$ each 1-mm unit of these cylinders weights, in grams *wet* weight:

$$\frac{\pi(0.1)^2(L)(1.05)}{1000}$$

A conversion factor is needed to correct the values from wet to dry weight, a more meaningful value. This factor is variable among different taxa and should be determined (approximately 80% water or a factor of 0.2).

Dry mass (W, in mg) can be estimated for each size class from body length (L, in mm) according to specimens in different insect orders by the regression equation constants given in Table 12.2.

P/B RATIOS

Biomass (B) is a measurement of mass for a population present at one point in time in units of mass per unit area (e.g., g/m^2) or volume, and is therefore a temporary storage of mass or energy. Production (P) is a flow or flux of mass (or energy) per area per time (e.g., g/m^2/yr). The ratio (P/B) of production (P) divided by mean biomass (B) is a rate with units of inverse time (e.g., 1/year). P/B is essentially a weighted mean value of biomass growth rates of all individuals in a population (Benke, 1993).

P/B ratios can be useful as an estimate of the turnover of a population and can be useful for comparative purposes among populations and growth in response to environmental conditions and perturbations. In the case of a cohort, the cohort P/B ratio is equal to the cohort production divided by the mean cohort biomass (Waters, 1969, 1977; Benke, 1984). The annual P/B ratio is the annual production divided by the mean biomass of the entire 12-month period, even though the generation under study may have been present for less than a year. Mean annual biomass is simply the mean of all monthly average biomass values.

Table 12.2. Constants ln *a* and *b* obtained form the log-transformed regression equation ln *W* = ln *a* + *b* ln *L*, where *W* is dry mass (mg) and *L* is body length (mm) according to insect order (Smock, 1980).

Order	Regression constants		Correlation coefficient
	ln *a*	*b*	*r*
Coleoptera	−1.878	2.18	0.94
Diptera	−5.221	2.43	0.96
Ephemeroptera	−5.021	2.88	0.94
Hemiptera	−3.461	2.40	0.93
Megaloptera	−5.843	2.75	0.95
Odonata	−4.269	2.78	0.94
Plecoptera	−6.075	3.39	0.95
Trichoptera	−6.266	3.12	0.83

For a univoltine species, the annual production rate is equal to that of the cohort. The cohort $P/\overline{B}$ ratio refers to the estimate of production rate and biomass over the duration of the generation under study (i.e., the CPI). Cohort $P/\overline{B}$ ratios are commonly ca. 5 (ranging between 2 and 8), regardless of voltinism (Waters, 1969, 1987; Benke, 1984, 1993). Benthic animals fall into two major groups: (1) animals that live their entire life history in the aquatic habitat (e.g., crustaceans), exhibit an Allen curve that is not truncated, and have an expected cohort $P/\overline{B}$ ratio of ca. 5 to 7; and (2) animals that terminate their CPI in pupation/emergence, exhibit an Allen curve that is truncated, and have a cohort $P/\overline{B}$ ratio of about 2 to 5 (Waters, 1987). Annual $P/\overline{B}$ varies approximately with voltinism (the CPI). Although most annual P/B ratios of benthic invertebrates are within 1 to 10, much higher values (50 to >100) occur in some dipterans and mayflies (Benke, 1984, 1993). The higher the annual P/B, the shorter the development time. For a bivoltine species, annual production is equal to that of the first generation plus the cohort production of the second generation. Cohort $P/\overline{B}$ ratios of the two generations are usually reported separately. The units of annual $P/\overline{B}$ are inverse time (year^{-1}) and the reciprocal equals turnover time, i.e., the amount of time needed to replace the biomass of the population.

EXERCISES

OPTION 1. FIELD ANALYSES (ALTERNATIVE A)

1. In a lake or reservoir, locate a representative transect of the basin extending from the littoral to the profundal zone of the open water.
2. At regular intervals of depth along the transect, collect duplicate or multiple quantitative samples of the sediment with two or more different samplers (e.g., an Ekman and corers).
3. Record the depth, vegetation (if any), type of sediment, and any other information that is germane to the distribution of the organisms (e.g., wave action in the area).
4. Determine the temperature and dissolved oxygen profiles (Exercises 1 and 6).
5. Sieve the samples with the finest sized mesh screens that time will permit to remove a major portion of the sediments.

6. In the laboratory, sort the samples. A comparison of manual-visual sorting to a flotation method on samples from the same depth should be made. Use flotation methods and dyes, as discussed earlier, to facilitate quantitative sorting.

7. Identify the organisms to the most specific level possible under the conditions of time and experience. Enumerate the organisms of each category.

8. Determine the biomass of the benthic organisms by dry weight and organic weight analyses.

9. Analyze your data in relation to gradients of depth, temperature, dissolved oxygen concentration, sediment composition, and other factors.

10. Answer the questions following Option 5.

OPTION 2. FIELD ANALYSES (ALTERNATIVE B)

1. In a lake or reservoir, locate one point at the same depth and collect five replicated samples of the sediment in a $10 \times 10\,$m area with two or more different sampling devices (e.g., an Ekman grab and multiple corer).

2. Proceed with steps 3 through 8 of Option 1 above.

3. By comparison of your results, test the variability of sampling and heterogeneity of distribution of benthic fauna.

4. Answer the questions following Option 5, emphasizing particularly Question 3.

OPTION 3. LABORATORY ANALYSES

1. Using the samples provided by your instructor, sieve the samples with the finest sized mesh screens that time will permit.

2. With replicate samples and a large homogeneous sediment sample, sieve the samples with (a) a screen of about 0.4-mm opening and (b) a screen of about 0.2-mm.

3. Sort the samples to remove the organisms from the sediments. Use flotation methods and dyes, as discussed previously, to facilitate quantitative sorting.

4. Identify the organisms collected to the most specific level possible under the conditions of time and experience. Enumerate the organisms of each category.

5. Determine the biomass of the benthic organisms by dry and organic weight analyses.

6. Analyze your data in relation to gradients of depth, temperature, dissolved oxygen concentration, sediment composition, and other factors as given to you by your instructor.

7. Answer the questions following Option 5.

OPTION 4. PRODUCTIVITY CALCULATIONS

1. Using the data given in Tables 12.3 and 12.4, calculate the annual production rate of the particular species of aquatic insect by the instantaneous growth method and the size-frequency method. Compare these results with that value obtained by the Allen curve method using Fig. 12.4.

2. What are the annual P/B ratios of this species by the different methods? What is the value of P/B ratio? [See Waters (1969), (1977) and (1987) and Benke (1984) and (1993)].

3. Discuss in detail the advantages and disadvantages of each of the methods and compare your conclusions to those of other workers [e.g., Cushman et al., (1978) and Walter (1985)].

4. Answer the questions following Option 5.

OPTION 5. SAMPLING METHODOLOGY

1. Using coring samplers constructed of metal and of clear plastic, collect at least five replicated samples of the sediment within a 10-m^2 area at the same depth.

Table 12.3. Population characteristics of the mayfly *Ephemerella subvaria* in a small stream of Minnesota.[a]

Date	Number/m^2	Mean weight (mg)	Biomass (g/m^2)
Sept. 26	6350	0.23	1.46
Oct. 22	4432	1.05	4.65
Nov. 27	4082	1.64	6.69
Dec. 27	4053	2.67	10.82
Jan. 30	3660	3.56	13.03
Mar. 12	1587	6.38	10.13
Apr. 17	230	11.61	2.67
Jun. 06	44	21.59	0.95

[a] From data of Waters and Crawford (1973).

2. Proceed with steps 3 through 8 of Option 1 above, concentrating on the numbers and biomass of the midges (*Chironomus*) and phantom midges (*Chaoborus*).
3. Analyze the data obtained to learn if some of the organisms were able to escape from opaque devices as they approached and penetrated the sediments.
4. Answer the following questions.

Questions

1. What types of benthic fauna predominate in shallower, littoral sediments? In the profundal sediments?
2. Where is the numerical diversity greater in relation to depth of water? What causal factors could contribute to the observed distribution? Where would you expect fish predation to be most intense?
3. Using simple statistics (see Appendix 2), compare the sampling precision (numbers and biomass) between replicated samples at the same depth in the shallow sediments and in presumably more uniform deep-water sediments. Where were confidence limits higher? How many samples would be required to get 95% confidence limits ±25% of the mean? Plot the total number of species found (in percent) against the number of replicated samples sorted. How many replicate samples would be needed to obtain a population estimate of 95% of the number of individuals of a species?

Table 12.4. Eleven samples of the mayfly *Ephemerella subvaria* collected throughout the year, separated into size classes.[a]

Size group length (mm)	Number/m^2	Mean weight (mg)	Biomass (g/m^2)
0–1	217	0.06	0.01
1–2	541	0.24	0.13
2–3	688	1.0	0.69
3–4	472	2.9	1.37
4–5	186	5.6	1.04
5–6	78	9.1	0.71
6–7	17	13.1	0.22
7–8	18	17.6	0.32
8–9	4	22.5	0.09
9–10	1	29.0	0.03

[a] After data of Waters and Crawford (1973).

4. If you were given unlimited resources for sampling procedures in an investigation of the benthic animal production rates and had the choices between random sampling, stratified random sampling, and systematic or transect sampling designs, which design would you choose? Why? If your resources were limited, which sampling design would provide the maximum information per unit sampling effort? [See Cummins (1962) and Resh (1975).]

5. If many aquatic macroinvértebrates have restricted distributions in the first and second larval instars, as is commonly the case, transect sampling would enhace collection of these organisms. Why?

6. If a number of the benthic animals were to emerge as adults and leave the water, how would this affect calculations of productivity by the different methods? Why? For detailed analyses of life cycles and identification, it is often necessary to capture emerging insects. How might one do this? [See Mundie (1971) and Davies (1984).]

7. How might an internal seiche affect the distribution of benthic organisms?

8. Why do organisms shrink when placed into alcohol? Why do some organisms gain weight in formalin?

9. What are the advantages to organisms of living in a benthic habitat? Disadvantages?

Apparatus and Supplies

1. Samplers (several types preferred for comparative analyses):
   a. Ekman grab(s), metered lines, messenger(s).
   b. Petersen grab or, preferably, a Ponar grab, metered cable and winch.
   c. Coring samplers, single, e.g., Kajak-Brinkhurst type [cf., Brinkhurst (1974)], corers, e.g., Freshwater Institute ball-closure type.
2. Formaldehyde.
3. Screening sieve boxes of two sizes, e.g., 0.4-mm and 0.2-mm mesh sizes; a stacked set of brass soil sieves works very well.
4. White enamel or plastic pans of various sizes, forceps, rhodamine B or similar fluorescent dye and a simple long-wave ultraviolet light source (used with water when fresh samples are being sorted; with 70 or 95% ethanol on formalin-preserved samples).
5. Organic dyes (e.g., rose bengal or eosin).
6. Sugar (sucrose); hydrometer for assaying density in the 1.12 to 1.13 g/ml range.
7. Dissecting microscopes; light sources.
8. Planimeter or electronic digitizer.

References

Allen, K.R. 1951. The Horokiwi Stream. A study of a trout population. Bull. N. Z. Mar. Dep. Fish. *10*:238 pp.

Anderson, R.O. 1959. A modified flotation technique for sorting bottom fauna samples, Limnol. Oceanogr. *4*:223–225.

Benke, A.C. 1979. A modification of the Hynes method for estimating secondary production with particular significance for multivoltine populations. Limnol. Oceanogr. *24*:168–171.

Benke, A.C. 1984. Secondary production in aquatic insects. pp. 289–322. *In*: V.H. Resh and D.M. Rosenberg, Editors. The Ecology of Aquatic Insects. Praeger, New York.

Benke, A.C. 1993. Concepts and patterns of invertebrate production in running waters. Verhand. Internat. Verein. Limnol. *25*:15–38.

Benke, A.C., T.C. van Arsdall, Jr., and D.M. Gillespie. 1984. Invertebrate productivity in a subtropical blackwater river: The importance of habitat and life history. Ecol. Monogr. *54*:25–63.

Brinkhurst, R.O. 1974. The Benthos of Lakes. St. Martin's Press, New York. 190 pp.

Burton, W. and J.F. Flannagan. 1973. An improved Ekman-type grab. J. Fish. Res. Bd. Canada *30*:287–290.

Cummins, K.W. 1962. An evaluation of some techniques for the collection and analysis of benthic samples with special emphasis on lotic waters. Amer. Midland Nat. *67*:477–504.

Cushman, R.M., H.H. Shugart, Jr., S.H. Hildebrand, and J.W. Elwood. 1978. The effect of growth curve and sampling regime on instantaneous-growth, removal-summation, and Hynes/Hamilton estimates of aquatic insect production: A computer simulation. Limnol Oceanogr. *23*:184–189.

Davies, I.J. 1984. Sampling aquatic insect emergence. pp. 161–227. *In*: J.A. Downing and F.H. Rigler, Editors. A Manual on Methods for the Assessment of Secondary Productivity. 2nd Ed. Blackwell, Oxford.

Downing, J.A. 1979. Aggregation, transformation, and the design of benthos sampling programs. J. Fish. Res. Bd. Canada *36*:1434–1463.

Downing, J.A. 1984. Sampling the benthos of standing waters. pp. 87–130. *In*: J.A. Downing and F.H. Rigler, Editors. A Manual on Methods for the Assessment of Secondary Productivity. 2nd Ed. Blackwell, Oxford.

Edmondson, W.T. and G.G. Winberg (eds). 1971. A Manual on Methods for the Assessment of Secondary Productivity in Fresh Waters. IBP Handbook No. 17. Blackwell, Oxford. 358 pp.

Elliott, J.M. and C.M. Drake. 1981. A comparative study of seven grabs used for sampling benthic macroinvertebrates in rivers. Freshwat. Biol. *11*:99–120.

Flannagan, J.F. 1970. Efficiencies of various grabs and corers in sampling freshwater benthos. J. Fish. Bd. Canada *27*:1691–1700.

Gallardo, V.A. 1965. Observations on the biting profiles of three $0.1\,m^2$ bottom samplers. Ophelia *2*:319–322.

Gillespie, D.M. and A.C. Benke. 1979. Methods of calculating cohort production from field data—some relationships. Limnol. Oceanogr. *24*:171–176.

Gillespie, D.M., D.L. Stites, and A.C. Benke. 1985. An inexpensive core sampler for use in sandy substrata. Freshwat. Invertebr. Biol. *5*:147–151.

Hamilton, A.L. 1969a. A method of separating invertebrates from sediments using long wave ultraviolet light and fluorescent dyes. J. Fish. Res. Bd. Canada *22*:1667–1672.

Hamilton, A.L. 1969b. On estimating annual production. Limnol. Oceanogr. *14*:771–782.

Hamilton, A.L., W. Burton, and J.F. Flannagan. 1970. A multiple corer for sampling profundal benthos. J. Fish. Res. Bd. Canada *27*:1867–1869.

Holme, N.A. 1964. Methods of sampling the benthos. Adv. Mar. Biol. *2*:171–260.

Holme, N.A. and A.D. McIntyre (eds). 1971. Methods for the Study of Marine Benthos. IBP Handbook No. 16. Blackwell, Oxford. 334 pp.

Hynes, H.B.N. 1961. The invertebrate fauna of a Welsh mountain stream. Arch. Hydrobiol. *57*:344–388.

Hynes, H.B.N. and M.J. Coleman. 1968. A simple method of assessing the annual production of stream benthos. Limnol. Oceanogr. *13*:569–573.

Kajak, Z. 1971. Benthos of standing water. pp. 25–65. *In*: W.T. Edmondson and G.G. Winberg, Editors. A Manual on Methods for the Assessment of Secondary Productivity in Fresh Waters. IBP Handbook No. 17. Blackwell, Oxford.

Kajak, Z., K. Dusoge, and A. Prejs. 1968. Application of the flotation technique to assessment of absolute numbers of benthos. Ekol. Polska. Ser. A, *16*:607–620.

Leuven, R.S.E.W., T.C.M. Brock, and H.A.M. van Druten. 1985. Effects of preservation on dry- and ash-free dry weight biomass of some common aquatic macro-invertebrates. Hydrobiologia *127*:151–159.

Mundie, J.H. 1971. Insect emergence traps. pp. 80–108. *In*: W.T. Edmondson and G.G. Winberg, Editors. A Manual on Methods for the Assessment of Secondary Productivity in Fresh Waters. IBP Handbook No. 17. Blackwell, Oxford.

Nalepa, T.F. and A. Robertson. 1981. Screen mesh size affects estimates of macro- and meiobenthos abundance and biomass in the Great Lakes. Can. J. Fish. Aquatic Sci. *38*:1027–1038.

Nalepa, T.F., M.A. Quigley, and R.W. Ziegler. 1988. Sampling efficiency of the Ponar grab in two different benthic environments. J. Great Lakes Res. *14*:89–93.

Powers, C.F. and A. Robertson. 1967. Design and evaluation of an all-purpose benthos sampler. Spec. Rept. Great Lakes Res. Div. Univ. Mich. *30*:126–131.

Reish, D.J. 1959. A discussion of the importance of the screen size in washing quantitative marine bottom samples. Ecology *40*:307–309.

Resh, V.H. 1975. The use of transect sampling in estimating single species production of aquatic insects. Verh. Int. Ver. Limnol. *19*:3089–3094.

Rigler, F.H. and J.A. Downing. 1984. The calculation of secondary productivity. pp. 19–58. *In*: J.A. Downing and F.H. Rigler, Editors. A Manual on Methods for the Assessment of Secondary Productivity. 2nd Ed. Blackwell, Oxford.

Smock, L.A. 1980. Relationship between body size and biomass of aquatic insects. Freshwat. Biol. *10*:375–383.

Stanford, J.A. 1973. A centrifuge method for determining live weights of aquatic insect larvae, with a note on weight loss in preservative. Ecology *54*:449–451.

Storey, A.W. and L.C.V. Pinder. 1985. Mesh size and efficiency of sampling of larval *Chironomidae*. Hydrobiologia *124*:193–197.

Strayer. D.L. 1985. The benthic micrometazoans of Mirror Lake, New Hampshire. Arch. Hydrobiol./Suppl. *72*:287–426.

Walter, R.A. 1985. Production and limiting factors: Benthic macroinvertebrates. pp. 280–288. *In*: G.E. Likens, Editor. An Ecosystem Approach to Aquatic Ecology: Mirror Lake and its Environment. Springer-Verlag, New York.

Waters, T.F. 1969. The turnover ratio in production ecology of freshwater invertebrates. Amer. Nat. *103*:173–185.

Waters, T.F. 1977. Secondary production in inland waters. Adv. Ecol. Res. *10*:91–164.

Waters, T.F. 1987. The effect of growth and survival patterns upon the cohort P/B ratio. J. N. Amer. Benthol. Soc. *6*:223–229.

Waters, T.F. and G.W. Crawford. 1973. Annual production of a stream mayfly population: A comparison of methods. Limnol. Oceanogr. *18*:286–296.

Wetzel, R.G. 1999. Limnology: Lake and River Ecosystems. 3rd Ed. Academic Press, San Diego (in press).

Whitman, R.L., J.M. Inglis, W.J. Clark, and R.W. Clary. 1983. An inexpensive and simple elutriation device for separation of invertebrates from sand and gravel. Freshwat. Invertebr. Biol. *2*:159–163.

Winberg, G.G. (ed). 1971. Methods for the Estimation of Production of Aquatic Animals. Academic Press, New York. 175 pp.

Benthic Fauna of Streams

Stream invertebrates are well adapted to the running water environment. The dominant taxa in headwater streams include the immature stages of the insect orders Ephemeroptera (mayflies), Plecoptera (stoneflies), Trichoptera (caddisflies), Megaloptera, Coleoptera (beetles), and certain Diptera. Many other types of invertebrates, attached and planktonic algae, rooted vascular plants, and various vertebrates also are common in streams of higher order. All of these organisms have certain traits that enable them to maintain their position and survive in rapidly flowing waters. Some mayflies (e.g., Heptageniidae) have become dorsoventrally flattened. Although they inhabit very rapid water, these mayflies live close to the substrate where the water velocity is nearly zero (Fig. 5.3). Many caddisflies (e.g., Limnephilidae) build elaborate cases that not only protect them from predators but also serve as ballast against the current or as attachment points on rocks. Others (Hydropsychidae) build intricate nets on submersed rocks and logs to catch food particles that are being transported downstream. Certain dipteran larvae (e.g., Blepharoceridae) have specialized suckers that are used to attach to the substrate. The common black fly larvae (Simuliidae) spin silk pads and attach themselves to rock surfaces by this means. [See Hynes (1970) for extensive treatment of this subject.]

The wide variety of microhabitats available to stream organisms will become apparent as you investigate stream ecosystems. Rock surfaces, plant surfaces, leaf debris, logs, backwaters, silty or sandy sediments, crevices in gravel, organic debris dams, and other spaces in the stream all provide special habitats for different organisms. The patchy distribution and abundance of organisms in a given stretch of stream is partially dependent on the availability of these microhabitats. Thus, any sampling of the benthic fauna must take this spatial heterogeneity into account. This patchiness adds both to the fascination as well as to the difficulty of research on streams.

Stream insects play a role in the processing of organic matter [see Cummins (1974) and Cummins et al. (1984)]. Streams in forested regions depend on allochthonous inputs of organic matter for much of the consumer productivity [see Fisher and Likens (1973)]. Many of the insects mentioned above are well adapted to utilize terrestrial organic matter as food. Ingested detrital particles are either assimilated by the organisms or egested and utilized as a food source by other consumers. Bacterial colonization of these organic particles may increase the food value of the detritus by increasing the nitrogen content. Many of the invertebrates themselves become sources of food for other carnivorous invertebrates or vertebrates.

Macroinvertebrates include invertebrate fauna retained by a 500-μm pore sieve or net. Because many of the early life stages important to life histories and production analyses are smaller than this size demarcation, collection methods often employ finer-meshed devices (e.g., 125–250 μm). Meiofauna are benthic animals that pass through a 500-μm pore sieve but are retained on a 40-μm pore sieve. Their composition is dominated by rotifers, copepods, ostracods, nematodes, and young stages of chironomid dipterans and oligochaete worms, as well as other animals such as gastrotrichs, tardigrades, and turbellarians. Because of their small size, the meiofauna has been studied rarely, even though these benthic animal communities dominate in terms of numbers and species diversity, and likely often are more important to ecosystem energetics than are the macroinvertebrates.

Meiofauna, like most macroinvertebrates, occur in all types of standing and running waters, and live on surfaces, including plants, debris, and sediments, as well as within the interstitial spaces of sediments and grains of sand. Because the availability of dissolved oxygen is important to the distribution of meiofauna within the interstitial water of sediments, meiofauna tend to be more abundant within the hyporheic zone of running waters than in organic dominated sediments of lakes.

The types and distributions of benthic macroinvertebrates also have been used widely as indicators quality. The distribution of certain macroinvertebrates and microorganisms can be quite specific because many organisms often have narrow physiological tolerance ranges [e.g., Sládeček, (1973)]. Colonization of many different artificial substrata has been used in comparative analyses of water quality among and within water bodies [e.g., Rosenberg and Resh (1982), Flannagan and Rosenberg (1982), and Lamberti and Resh (1985)].

This exercise briefly introduces some of the sampling equipment and problems of obtaining quantitative estimates of the distribution and abundance of stream invertebrates.

SAMPLING PROCEDURES

It is exceedingly difficult to obtain quantitative samples of the benthic invertebrates in a stream. Some of the major factors to consider are: (1) heterogeneity of habitat type and distribution of organisms; (2) depth of organisms in the sediments; (3) stage in the life cycle of the organisms (many insects may emerge as winged adults); (4) variations in current velocity; (5) variations in discharge, environmental conditions (e.g., ice), or both; and (6) movement or transport of organisms. Various types of procedures and devices have been used to obtain samples, e.g., kicking the bottom upstream of a net (Frost et al., 1971), Hess sampler (Hess, 1941), and Ekman grab samples (Usinger and Needham, 1956). The ponar and corer samplers, discussed in Exercise 12, are also used in stream ecosystems, particularly when sampling sandy substrata. Probably the most popular procedure, at least in the United States, has been to dislodge manually the organisms from the sediments and allow the current to transport them downstream where they can be filtered from the water with a net. The Surber sampler (Surber, 1937) has been used widely in this way (Fig. 13.1). However, various authors have shown this sampling procedure to be seriously deficient for obtaining representative and quantitative samples of stream invertebrates [e.g., Hess (1941) and Macan (1958), and Chutter (1972)]. Other devices such as the modified Hess sampler (Waters and Knapp, 1961) and so-called pot-samplers (Coleman and Hynes, 1970) represent improvements, but small (Fig. 13.2) and large emergence traps (Needham, 1908; Sprules, 1947; Munie 1957), covering the entire stream (Illies, 1971; Hall et al., 1980) (see Fig. 13.3), have the greatest possible potential for quantitative sampling of the dominant stream insects that leave the water as flying adults. Some of the advantages of large emergence traps are: (1) integration of estimates from heterogeneous habitats; (2) tedious sorting from debris and sediments is avoided; and (3) most adult stages can be identified to species. Except for the confounding influence of strong water currents, the general procedures for sampling and analysis of benthic invertebrates are essentially the same as those described in Exercise 12. The sampling procedures for stream benthos and the effi-

Figure 13.1. The Surber bottom sampler (*left*) and the Hess sampler (*right*) being used in a shallow stream. [From Usinger (1956), by permission.]

cacy of different techniques are discussed in detail by Ide (1940), Macan (1958), Cummins (1962), Ulfstrand (1967), Radford and Hartland-Rowe (1971), Mason et al. (1973), Peckarsky (1984), Davies (1984), and Hauer and Resh (1996). Quantitative sampling of the meiofauna is most frequently done by small corers of PVC or

Figure 13.2. Conical emergence traps in place over large rocks in a small stream. The white cups at the top of the cone contain a killing agent.

Figure 13.3. Emergence "house" spanning an entire stream. Insects emerging into this structure can be collected with a vacuum cleaner or by hand.

clear acrylic tubing (e.g., a cut-off 30-ml syringe) if the substratum can be penetrated (cf. Palmer and Strayer, 1996). Working downstream from the sampling site, the tubing is sealed at both ends and retrieved.

Separation of the macroinvertebrates from sediments of stream ecosystems is similar to those discussed in detail in Exercise 12. Cores of meiofauna samples are transferred to a container and preserved with several milliliters of 10% buffered Rose Bengal-formalin solution (1 g/l of 10% formalin). With coarse substratum, animals are extracted by a swirl-and-decant procedure with filtration through appropriate sieves (Palmer and Strayer, 1996). These procedures are repeated several times with collection in water. The organisms are suspended in the water and subsamples removed quantitatively for examination under a microscope.

EXERCISES

OPTION 1

We shall compare the results of a Surber sampler with those from a sediment core in a riffle and/or a pool area of a stream.

Procedures

1. Collect three samples of sediment to a depth of 15 cm with a coring device in a riffle and/or a pool area of a shallow stream (see Exercises 5 and 12).

2. Sieve the samples with the finest-sized mesh screen that time will permit to remove a major portion of the sediments. Transfer the remaining sample into strong plastic bags, add 70% alcohol, and transport to the laboratory.

3. In the laboratory, sort and collect the organisms from the samples. [See also Mason and Yevish (1967).] A comparison of manual-visual sorting to a flotation method on samples should be made. Empty each sample into a white porcelain or plastic pan and sort individual organisms and debris into small aluminum weighing pans. For samples containing organisms with case, e.g., caddisfly larvae and snails, the cases or shells must be removed before weighing or measuring.

4. Identify the organisms to the most specific level possible under the conditions of time and experience. It should be possible to separate organisms at least according to taxonomic order. Enumerate the organisms in each category.

5. Determine the biomass of the benthic organisms by dry weight and organic weight analyses.

6. In the same riffle and/or pool, if possible, collect five samples using a Surber sampler. The Surber sampler is designed for shallow flowing streams. Ideally the water depth should be <45 cm deep. The sampler consists of a silk or nylon net (approximately 60 cm long) supported by a folding solid-brass framework having sampling and net inlet areas of 30.5 × 30.5 cm.

   The frame should be positioned securely on the substrate. Hold the sampler in position in the current by applying pressure with one or both knees. Pick up all large stones from inside the frame area and wash all attached organisms into the net. After the larger stones have been removed, picked, washed and discarded, the remaining substrate should be churned gently with the fingers to release all burrowing forms. When finished, pick up the sampler and allow water to consolidate all of the organisms and debris at the end of the net. Carefully invert the net and transfer the sample to a plastic bag (Whirl-a-pac), add 70% alcohol to cover the sample, and transport to the laboratory.

7. Repeat items 3 through 5 above.

8. Using your results, test the variability between types of samplers and among collections made with each sampler.

9. How many species did you find in the first 100 organisms examined with each sampler? Prepare a frequency distribution for the number of individuals collected per taxonomic group. [See Williams (1964).]

10. Answer the questions following Option 2.

OPTION 2

In this option we shall examine the downstream transport of organisms and organic debris by water currents in a small stream. This downstream movement is referred to as drift. Significant quantities of benthic organisms may drift downstream [e.g., Müller (1974) and Smock (1996)] and show diurnal periodicity [(Waters, 1962, 1972, and Wiley and Kohler (1984)].

Procedures

1. Position three drift nets in the center of a shallow riffle area of the stream. Install the nets so that they will extend above the water surface even if the depth of the water should change during the study (Fig. 13.4). Nets should be left in place for a period (e.g., 30 to 60 min) sufficiently long to collect a representative sample of the benthos but not so long for flow through the nets to be impaired significantly by clogging from retained particulate matter. Alternatively,

Figure 13.4. Drift nets in a small stream. Stream flow is from left to right.

    a. Several nets could be placed across the stream channel to measure spatial variations in drift.

    b. Nets could be placed at the downstream end of a riffle and just below a pool.

    c. Diurnal variations could be examined by installing drift nets and collecting organisms at about 1 h before sunset and sunrise.

2. Transfer the organisms and debris from the drift nets to small plastic bags (Whirl-a-pacs), add 70% alcohol, and transport to the laboratory.

3. Sort and identify the organisms in the drift samples.

4. Determine the weight of the organisms and organic debris by dry weight and organic weight analyses.

5. Knowing the numbers of organisms drifting into the nets per unit time and the stream discharge rate (see Exercise 5), calculate the *drift rate* [numbers per time from the (number/cm^3 × cm^3/time)] for the whole stream. From the amount of stream volume passing through the net(s), calculate the *drift density* [as numbers per m^3, calculated from the area of the net opening (cm^2), velocity (cm/sec), and time; thus, volume = cm^2 × cm/sec × sec = cm^3]. Drift density is usually expressed as the number of macroinvertebrates drifting per 100 m^3 of water.

6. Using your results, test the variability of sampling and the spatial and temporal heterogeneity of distribution of drifting organisms.

7. Answer the following questions.

Questions

1. What are the advantages and disadvantages of a Surber or Hess sampler for collecting samples of benthic invertebrates in streams?

2. What are the advantages and disadvantages of emergence traps in collecting benthic invertebrates in streams?

3. How would you design a better sampler? [See Merritt and Cummins (1996).]
4. Which is richer in stream invertebrates, a pool or a riffle? Compare numbers and biomass. How do you explain the results?
5. What types of benthic fauna predominate in riffles? In pools?
6. What are the effects and interrelationships of current and sediment type on distribution and abundance of stream invertebrates?
7. How far into the sediments or into the banks do you think stream invertebrates may be found? [See Hynes (1974) and Williams (1984).]
8. What was the relationship between drift of organisms and drift of organic debris in your studies? Did this change diurnally? How do you explain this?
9. What differences would you expect in benthic invertebrates between upstream and downstream locations? Would problems of sampling differ? How?
10. Why would the placement of nets in the center of a riffle tend to overestimate rates of invertebrate drift for the entire stream? What would be the result of placing drift nets near the shore line?
11. What percentage of the total benthic fauna might move downstream as drift? [See Waters (1972).]
12. Also, see questions for Exercise 12.

Apparatus and Supplies*

1. Modified Hess sampler.
2. Surber sampler (Wildco Instruments, Saginaw, MI).
3. Drift net (Wildco Instruments, Saginaw, MI).
4. White enameled or plastic pans.
5. Forceps.
6. Balance.
7. Whirl-a-pac plastic bags.
8. Boots or waders.
9. Alcohol (70%).

Sample Data Sheet

Name of stream	_____	Recorder	_____
Type of sampler	_____	Collection date	_____
Sample no. _____		Time	_____
Stream width	_____	Stream depth	_____
Bottom type	_____	Current velocity	_____

Order	Numbers/m²	Biomass (dry g/m²)	Percent of total	
			Number basis	Weight basis
Trichoptera				
Ephemeroptera				
Diptera				
Etc.				

---

*See Exercise 12 for additional supplies.

References

Chutter, F.M. 1972. A reappraisal of Needham and Usinger's data on the variability of a stream fauna when sampled with a Surber sampler. Limnol. Oceanogr. *17*:139–141.

Coleman, M.J. and H.B.N. Hynes. 1970. The vertical distribution of the invertebrate fauna in the bed of a stream. Limnol. Oceanogr. *15*:31–40.

Cummins, K.W. 1962. An evaluation of some techniques for the collection and analysis of benthic samples with special emphasis on lotic waters. Amer. Midland Nat. *67*:477–503.

Cummins, K.W. 1974. Structure and function of stream ecosystems. BioScience *24*:631–641.

Cummins, K.W., R.W. Merritt, and T.M. Burton. 1984. The role of aquatic insects in the processing and cycling of nutrients. pp. 134–163. *In*: V.H. Resh and D.M. Rosenberg, Editors. The Ecology of Aquatic Insects. Prager, New York.

Davies, I.J. 1984. Sampling aquatic insect emergence. pp. 161–227. *In*: J.A. Downing and F.H. Rigler, Editors. A Manual on Methods for the Assessment of Secondary Productivity in Fresh Waters. 2nd Ed. Blackwell, Oxford.

Dickson, K.L., J. Cairns, Jr., and J.C. Arnold. 1971. An evaluation of the use of basket-type artificial substrate for sampling macroinvertebrate organisms. Trans. Amer. Fish. Soc. *100*:553–559.

Elliott, J.M. and P.A. Tullett. 1978. A bibliography of samplers for benthic invertebrates. Occas. Publ. Freshw. Biol. Assoc. U.K. *4*. 61 pp.

Fisher, S.G. and G.E. Likens. 1973. Energy flow in Bear Brook, New Hampshire: An integrative approach to stream ecosystem metabolism. Ecol. Monogr. *43*:421–439.

Flannagan, J.F. and D.M. Rosenberg. 1982. Types of artificial substrates used for sampling freshwater benthic macroinvertebrates. pp. 237–266. *In*: J. Cairns, Jr., Editor. Artificial Substrates. Ann Arbor Science Publs. Inc., Ann Arbor.

Frost, S., A. Huni, and W.E. Kershaw. 1971. Evaluation of a kicking technique for sampling stream bottom fauna. Can. J. Zool. *49*:167–173.

Hall, R.J., G.E. Likens, S.B. Fiance, and G.R. Hendrey. 1980. Experimental acidification of a stream in the Hubbard Brook Experimental Forest, New Hampshire. Ecology *61*:976–989.

Hauer, F.R. and V.H. Resh. 1996. Benthic macroinvertebrates. pp. 339–369. *In*: F.R. Hauer and G.A. Lamberti, Editors. Methods in Stream Ecology. Academic Press, San Diego.

Hess, A.D. 1941. New limnological sampling equipment. Limnol. Soc. Amer. Spec. Publ. *5*. 5 pp.

Hynes, H.B.N. 1970. The Ecology of Running Waters. Univ. of Toronto Press. 555 pp.

Hynes, H.B.N. 1974. Further studies on the distribution of stream animals within the substratum. Limnol. Oceanogr. *19*:92–99.

Ide, F.P. 1940. Quantitative determination of the insect fauna of rapid water, Publ. Ontario Fish. Res. Lab. *59*:1–20.

Illies, J. 1971. Emergenz 1969 in Breitenbach. Schlitzer Produktions biologische Studien (1). Arch. Hydrobiol. *69*:14–59.

Lamberti, G.A. and V.H. Resh. 1985. Comparability of introduced tiles and natural substrates for sampling lotic bacteria, algae and macroinvertebrates. Freshwat. Biol. *15*:21–30.

Macan, T.T. 1958. Methods of sampling the bottom fauna in stony streams. Mitt. Int. Ver. Limnol. *8*. 21 pp.

Mason, W.T., Jr., and P.P. Yevish. 1967. The use of phloxine B and rose bengal stains to facilitate sorting benthic samples. Trans. Amer. Microsc. Soc. *86*:221–222.

Mason, W.T., Jr., C.I. Weber, P.A. Lewis, and E.C. Julian. 1973. Factors affecting the performance of basket and multiplate macroinvertebrate samplers. Freshwat. Biol. *3*:409–436.

Merritt, R.W. and K.W. Cummins (Eds). 1996. An Introduction to the Aquatic Insects of North America. 3rd Ed. Kendall Hunt, Dubuque, IA.

Müller, K. 1974. Stream drift as a chronobiological phenomenon in running water ecosystems. Ann. Rev. Ecol. Syst. *5*:309–323.

Mundie, J.H. 1956. Emergence traps for aquatic insects. Mitt. Int. Ver. Limnol. 7. 13 pp.

Needham, J.G. 1908. Report of the entomological field station conducted at Old Forge, N.Y., in the summer of 1905. Bull. N.Y. State Mus. *124*:167–168.

Palmer, M.A. and D.L. Strayer. 1996. Meiofauna. pp. 315–337. *In*: F.R. Hauer and G.A. Lamberti, Editors. Methods in Stream Ecology. Academic Press, San Diego.

Peckarsky, B.L. 1984. Sampling the stream benthos. pp. 131–160. *In*: J.A. Downing and F.H. Rigler, Editors. A Manual on Methods for the Assessment of Secondary Productivity in Fresh Waters. 2nd Ed. Blackwell, Oxford.

Radford, D.W. and R. Hartland-Rowe. 1971. Subsurface and surface sampling of benthic invertebrates in two streams. Limnol. Oceanogr. *16*:114–119.

Rosenberg, D.M. and V.H. Resh. 1982. The use of artificial substrates in the study of fresh-water benthic macroinvertebrates. pp. 175–235. *In*: J. Cairns, Jr., Editor. Artificial Substrates. Ann Arbor Science Publs. Inc., Ann Arbor.

Sládeček, V. 1973. System of water quality from the biological point of view. Ergebnisse Limnol. Arch. Hydrobiol. 7. 218 pp.

Smock, L.A. 1996. Macroinvertebrate movements: Drift, colonization, and emergence. pp. 371–390. *In*: F.R. Hauer and G.A. Lamberti, Editors. Methods in Stream Ecology. Academic Press, San Diego.

Sprules, W.M. 1947. An ecological investigation of stream insects in Algonquin Park, Ontario. Univ. Toronto Stud., Biol. Ser. No. 56, Publ. Ontario Fisheries Res. Lab., No. 69.

Surber, E.W. 1937. Rainbow trout and bottom fauna production in one mile of stream. Trans. Amer. Fish. Soc. *66*:193–202.

Ulfstrand, S. 1967. Microdistribution of benthic species (Ephemeroptera, Plecoptera, Trichoptera, Diptera: Simuliidae) in Lapland streams. Oikos *18*:293–310.

Usinger, R.L. (ed). 1956. Aquatic Insects of California. Univ. of California Press, Berkeley. 508 pp.

Usinger, R.L. and P.R. Needham. 1956. A drag-type riffle-bottom sampler. Prog. Fish-Cult. *18*:42–44.

Waters, T.F. 1962. Diurnal periodicity in the drift of stream invertebrates. Ecology *43*:316–320.

Waters, T.F. 1972. The drift of stream insects. Ann. Rev. Ent. *17*:253–272.

Waters, T.F. and R.J. Knapp. 1961. An improved stream bottom fauna sampler. Trans. Amer. Fish. Soc. *90*:225–226.

Wiley, M.T. and S.L. Kohler. 1984. Behavioral adaptations of aquatic insects. pp. 101–133. *In*: V.H. Resh and D.M. Rosenberg, Editors. The Ecology of Aquatic Insects. Prager, New York.

Williams, C.B. 1964. Patterns in the Balance of Nature. Academic, New York. 324 pp.

Williams, D.D. 1984. The hyporheic zone as a habitat for aquatic insects and associated arthropods. pp. 430–455. *In*: V.H. Resh and D.M. Rosenberg, Editors. The Ecology of Aquatic Insects. Praeger, New York.

Primary Productivity of Phytoplankton

Although appreciable quantities of organic matter synthesized by terrestrial plants within the drainage basin can be transported to freshwater ecosystems in either dissolved or particulate forms (allochthonous primary productivity), much of the organic matter of lakes is produced within the lake by phytoplanktonic algae, by littoral macrophytic vegetation, and by sessile algae (autochthonous primary productivity). In situ rates of photosynthesis by phytoplankton are the subject of this exercise; those of other plant forms will be treated separately in Exercise 22.

It is a distinct advantage to measure rates of metabolism directly in situ since extrapolation of laboratory results to natural conditions is usually difficult. When the rates of synthesis of organic matter and changes in primary production can be measured over time, efforts can be directed to the experimental evaluation of causal mechanisms regulating the synthesis, utilization, and loss of the organic matter.

The complex biochemical reactions of photosynthesis can be summarized by the general redox reaction:

$$6CO_2 + 12H_2O \xrightarrow[\substack{\text{pigment} \\ \text{receptor}}]{\text{light}} C_6H_{12}O_6 + 6O_2 + 6H_2O$$

Cyanobacteria, prochlorophytes, eukaryotic algae, and higher plants use light energy to oxidize water to molecular oxygen, hydrogen ions, and electrons. The light reaction occurs in photosystem II located in the thylakoid membranes. Mitochondrial ("dark") respiration occurs both in the light as well as in darkness although at different rates, during which ATP and reductants are formed. Respiration associated with cell synthesis increases with temperature (Q_{10} ca. 1.7–2.0). Photorespiration, a light-dependent oxidation of ribulose bisphosphate produces glycolate that is excreted, oxidized, or used to synthesize amino acids, is an energy dissipating process that reduces the light-saturated rate of photosynthesis. Photorespiration increases with higher ratios of dissolved oxygen concentration to carbon dioxide concentration.

Techniques for measuring rates of photosynthesis are based on the stoichiometry of this reaction, e.g., rates of oxygen production, rates of utilization of CO_2 or water, or changes in the concentration of organic matter. The objective is to modify the natural community as little as possible during assays of in situ rates of photosynthesis. Variations in the metabolic state of the phytoplankton can be large; measurements of primary productivity may reflect the rates of certain species rather accurately, but for other species rates are estimated poorly. Nonetheless, with reasonable precautions and guarded interpretations, the methods provide useful estimates of in situ rates of phytoplanktonic photosynthesis.

Algae suspended in water are circulated within the epilimnion of stratified lakes and exposed to bright light at the surface and then moved in the circulation to low light habitats. In situ incubation is static and can suppress turbulent circulation of water containing algae for several hours during the incubation. Dynamic incubation methods were developed to move incubating samples within the mixed layer to derive an integral of primary production within the zone (e.g., Gervais et al., 1997). In general, however, for short-term incubations, agreement is relatively close between the results from static and dynamic incubations.

Although in situ incubation is most desirable, it is not always practical when studying large lakes where simultaneous measurements at distant stations are necessary. Therefore, samples of natural phytoplankton can be incubated under controlled light and temperature conditions in laboratory incubators that simulate underwater conditions. From the intensity and spectral distribution of light, temperature, and phytoplankton in relation to depth, the relationship between light and photosynthesis can be used to estimate primary production in the natural environment (cf., Strickland, 1960; Saunders et al., 1962; Fee, 1969, 1971, 1973; Parsons et al., 1984; Walsby, 1997a).

Details of the integration of phytoplankton photosynthesis through time and depth in a water column can be calculated by numerical analysis based on the photosynthesis irradiance curve of the phytoplankton, the vertical distribution of the community, and details of the underwater light field (Walsby, 1997b). Variations in the light field are calculated from continuous recordings of surface irradiance and measurements of vertical light attenuation, with corrections for losses by reflection at the water surface that depend on the sun's elevation and wave action. Effects of changes in phytoplankton distribution, light attenuation, photoinhibition, and water temperatures can be modeled for reasonable estimates of primary productivity within the euphotic depth integrated over 24 hours.

Alternatively, measurements of in situ photosynthesis can be estimated indirectly in the natural environment from changes in environmental parameters affected by photosynthesis, such as changes in CO_2 and oxygen concentrations, pH, or specific conductance. The resultant evaluation is a measure of community metabolism, and a number of critical assumptions must be made when assessing which components of the overall biological communities are causing the observed changes in environmental parameters over short periods of time. This approach will be undertaken in Exercise 24.

Finally, autotrophic productivity of lake ecosystems that are sufficiently large to stratify thermally can be estimated indirectly by measuring long-term changes in biomass, reductions in certain nutrients, or hypolimnetic oxygen deficits or accumulations of CO_2. This systems approach will be treated separately in Exercise 29.

IN SITU SAMPLING

To demonstrate the methodology with in situ sampling and its limitations, it is recommended that the primary productivity of a single vertical profile be determined for the deepest part of a representative lake(s). Research applications, in which accurate evaluations of phytoplanktonic productivity are necessary, would normally require some horizontal analysis of the spatial heterogeneity in phytoplankton distribution and productivity.

The water within a vertical profile should be collected, beginning at the surface and working downward, with an opaque, nonmetallic water sampler. The Van Dorn sampler discussed earlier (p. 86) is well suited. Sufficient metal ions leach from metallic samplers, even upon brief exposure to the sample, to affect rates of photosynthesis significantly in subsequent incubations.

The objective is to measure variations in the depth profiles of photosynthetic activity per unit volume of water. From the depth profile, the photosynthetic activity below a given area of lake surface can be calculated by integration. The number of samples required is governed by the gradient and depth of the euphotic zone. Photosynthesis is generally, but not always, reduced to zero at the depth where ambient light intensity is reduced to ca. 0.5% of the surface value. When possible, replicate samples should be taken at each meter of depth in low to moderately productive waters, and at 0.5-m intervals in shallow, productive waters where the light is attenuated rapidly. The narrower the depth intervals, the more accurate will be the profile of productivity. A minimum of five depths should be sampled through the euphotic zone. Bottles then are filled as rapidly as possible under shaded conditions and then held in a light-proof box until samples from all depths have been collected.

CHANGES IN DISSOLVED OXYGEN

Water samples from the various depths are enclosed both in transparent ("light") and in completely opaque ("dark") bottles. The bottles should be of high-quality glass, e.g., Pyrex, with the tips of the ground-glass stoppers tapered to permit sealing without including any air bubbles. While quartz bottles are preferred because they do not absorb short-wave radiation as normal silica glass does (Findenegg, 1966), quartz bottles are simply too expensive for most purposes. Ultraviolet light usually is absorbed rapidly in water (Exercise 2), and therefore its inhibitory effect would be most pronounced in the upper layers (1 to 2 m) of the lake. Dark bottles can be made by covering them with a double layer of black plastic electrician's tape to completely exclude light.

The bottles (usually duplicate light bottles and one dark bottle) are filled as rapidly as possible. The tube from the water sampler is inserted to the bottom of the bottle, and the bottle is flushed continuously for three times as long as it takes to fill the bottle initially. The bottles then are immediately stoppered and stored in a light-proof box until all of the samples have been collected. Duplicate oxygen bottles should be filled at each depth and immediately fixed chemically for measurement of the initial concentration of dissolved oxygen.

Using a metered suspension line and starting with the samples from the greatest depth, clip the bottles to the bottle spreaders (metal rods with small eyebolts) so that the duplicate light bottles are at the ends and the dark bottle is in the center. Better suspension alternatives are transparent cylinders with keyhole slots that permit rapid insertion and suspension at depths (Fig. 14.1). Cover the top of each dark bottle with aluminum foil to make certain that no light enters around the ground-glass joint. Continue to work in the shade and lower all of the bottles to the depth from which they were collected. Time zero of the incubation is recorded when half of the bottles have been lowered. Attach the suspension line securely to an anchored buoy and make certain that the buoy does not shade any of the suspended bottles. To alleviate the problem of shading, the suspension line may be attached to the center of a rod, approximately 2 to 3 m in length, which is floated by styrofoam blocks at each end. Such a buoy is easy and inexpensive to make.

The length of the incubation period should be governed by estimates of the intensity of photosynthetic activity. Sufficient time must be allowed for measurable changes in concentrations of dissolved oxygen to occur. In very unproductive waters the changes may never be sufficiently large for accurate measurements. At the other

Figure 14.1. A transparent cylindrical holder that is suspended on a line at depths of incubation of clear and opaqued Pyrex bottles. The keyhole slots permit rapid insertion and removal of bottles.

extreme, under "pea soup" conditions of very high algal biomass, an hour of incubation could result in supersaturated conditions of dissolved oxygen within the light bottles. Then, upon opening the bottles for chemical fixation, bubbles of oxygen likely would be lost and the productivity underestimated. Under conditions of moderate algal productivity, an incubation time of 2 to 4h should be adequate. The incubation is terminated by immediate chemical fixation of dissolved oxygen as the bottles are retrieved.

All traces of iodine, acid, and nutrients must be removed from the bottles prior to reuse. Bottles and stoppers must be cleaned thoroughly with acid, scrubbing, and copious rinsing with redistilled water.

Calculations

During the incubation of the samples of the phytoplankton, the initial concentration of dissolved oxygen (= initial bottle, IB) at given depth would be expected to decrease to a lower concentration (= dark bottle value, DB) in the opaque bottles from respiration. Conversely, the initial concentration (IB) would change (usually it would increase) to another concentration (= light bottle value, LB) in the transparent bottles as a result of the difference between photosynthetic production and respiratory consumption of oxygen.

When the influences on oxygen concentrations, other than algal photosynthesis and respiration, are small and can be neglected (see discussion below),

$IB - DB$ = respiratory activity per unit volume per time interval
$LB - IB$ = *net* photosynthetic activity per unit volume per time interval
$(LB - IB) + (IB - DB)$ = *gross* photosynthetic activity

The initial bottle values cancel each other in this equation, and it is possible to estimate gross photosynthetic activity directly from: $LB - DB$. The method then estimates: Gross photosynthesis = net O_2 evolved + O_2 used in respiration.

Gross photosynthesis refers to the gross true synthesis of organic matter resulting from exposure to light. *Net* photosynthesis refers to the net formation of organic matter after losses from respiration, extracellular release of soluble organic matter, and other losses (e.g., death) that occur from the metabolic activities of algae simultaneously with the photosynthetic processes.

It may be desirable to express the changes in oxygen concentration in terms of carbon, since carbon is both the initial material and the end product of synthesis and of respiration. The photosynthetic quotient (PQ) and respiratory quotient (RQ) are dimensionless numbers indicating the relative amounts of oxygen and carbon involved in the processes of photosynthesis and respiration:

$$PQ = \frac{+\Delta O_2}{-\Delta CO_2} = \frac{\text{molecules of oxygen liberated during photosynthesis}}{\text{molecules of } CO_2 \text{ assimilated}}$$

$$RQ = \frac{+\Delta CO_2}{-\Delta O_2} = \frac{\text{molecules of } CO_2 \text{ liberated during respiration}}{\text{molecules of oxygen consumed}}$$

The PQ and RQ values vary greatly among different algae, their chemical composition, and environmental conditions [cf., discussion of Strickland (1960)]. With "normal" algal populations exposed to moderate light intensities, a PQ of 1.2 and an RQ of 1.0 are typical. To convert from mass of oxygen to mass of carbon, the values of oxygen production and consumption must be multiplied by the ratio of moles of carbon to moles of oxygen (12 mg C/32 mg O_2 = 0.375). Then,

$$\text{Gross photosynthesis } (\text{mg C}/\text{m}^3/\text{h}) = \frac{[(O_2, LB) - (O_2, DB)](1000)(0.375)}{(PQ)(t)}$$

where t = hours of incubation and O_2 = oxygen in mg/l;

$$\text{Net photosynthesis } (\text{mg C}/\text{m}^3/\text{h}) = \frac{[(O_2, LB) - (O_2, LB)](1000)(0.375)}{(PQ)(t)}$$

$$\text{Respiration } (\text{mg C}/\text{m}^3/\text{h}) = \frac{[(O_2, IB) - (O_2, DB)](RQ)(1000)(0.375)}{t}$$

To estimate the rate of photosynthetic productivity through the water column of the euphotic zone below one square meter of water surface, the values of mg C/m³/h are plotted against depth. The area of the curve is integrated with an electronic digitizer or by planimetry (see Exercise 1, p. 11) and then compared to a known area on the same graph of known mg C/m³/h versus known depth (as in Fig. 1.5).

Assumptions and Errors

It must be remembered that the oxygen change method measures community metabolism. The respiration measured is not only that of the phytoplankton, but also includes respiration of bacteria and zooplankton in the sample. Filtration of the sample through nets to remove larger zooplankton is not recommended because the phytoplankton community may be altered in the process.

It is further assumed in this calculation that respiration is not affected by illumination; this assumption is not valid. Respiration results from mitochondrial activity, which has been shown to be altered by light, and of photorespiration, i.e., CO_2 generated from glycolate metabolism, which is influenced by and proportional to dissolved oxygen concentration, light intensity, and temperature. These factors may change markedly, and their variations are not taken into account by this method.

The rates of photosynthesis and of community respiration fluctuate throughout the course of a day. Rates of net photosynthesis often have been observed to be greater in the early hours of daylight and then to decrease markedly in the afternoon period of high light, increased concentrations of dissolved oxygen, and higher pH values. Incubations are made commonly from mid-morning to mid-afternoon (e.g., from 1000 to 1400 hours) to compensate to some extent for these variations and to obtain an average value. Expansion of the productivity values to the daylight period can be approximated by introducing a diurnal factor, whereby the proportion of insolation received during the incubation period is expanded to a total for the whole day (see calculations of the ^{14}C method, p. 229). This transformation assumes that photosynthetic rates are proportional to light intensity i.e., that photosynthesis is not subject to light saturation. Although photosynthetic rates at most depths of a water column are undersaturated with light because of the exponential attenuation of light with increasing depth, saturation with light may occur near the surface.

Under ideal conditions of sampling, fixation, and titration, dissolved oxygen determinations can be as precise as ± 0.02 mg/l (Exercise 6). However, these conditions are met rarely in experimental situations, and precision deteriorates as a result. Assuming a PQ of 1.2 and a statistical probability limit of 0.05, the smallest amount of productivity detectable is in the range of 20 mg C/m^3 (ca. 15 mg C/m^3 when duplicate titrations are averaged) (Strickland, 1960). Depending on the conditions, populations, and duration of the experiments, the lower limit of detection may be reduced to ca. 3 mg C/m^3/h, although values below 10 mg C/m^3/h obtained by the oxygen method must be viewed with caution. The method is applicable within a range of ca. 3 to 2000 mg C/m^3/h (Strickland and Parsons, 1972).

[14]C UPTAKE

The ability of phytoplankton to take up and incorporate tracer amounts of radioactive isotopes into organic matter during photosynthesis permits measurement of the in situ rates of primary production. Although tritiated water (^3H_2O) alone can be used (McKinley and Wetzel, 1977), dilution of the tracer by greater amounts of nonlabeled water necessitates addition of large amounts of high specific activity isotope in order to follow uptake and incorporation into the algae. Application of the ^3H_2O method is limited largely to specialized cases where use of other isotopes is complicated by various interferences.

The most commonly used method is to add a tracer amount of $^{14}CO_2$ as labeled bicarbonate (e.g., NaH$^{14}CO_3$) to the water sample. When the total CO_2 content of the water is known and the ^{14}C content of the phytoplankton is measured after a period of incubation, the total amount of carbon assimilated can be calculated by the proportional relationship (Steemann Nielsen, 1951, 1952):

$$\frac{^{14}C \text{ available}}{^{14}C \text{ assimilated}} = \frac{^{12}C \text{ available}}{^{12}C \text{ assimilated}}$$

In practice, a known amount of radioactive bicarbonate, $H^{14}CO_3^-$, is added to samples of known total DIC content. The amount of DIC added along with the tracer is usually negligible in relation to the total in the water. After photosynthesis by the phytoplankton has proceeded for a suitable period of time, the algae are filtered onto a membrane filter, treated, and assayed for the amount of radioactivity incorporated. The uptake of tracer carbon, as a fraction of the initial whole, is assumed to measure the assimilation of total DIC inorganic carbon, as a fraction of the whole, over the time period. The technique has excellent sensitivity. A number of assumptions must be made, however, and various correction factors are introduced to compensate for certain of these conditions. The values obtained are estimates of rates of primary production in $mg\,C/m^3/time$, which can be expanded to $mg\,C/m^3/day$ for the daylight hours. Knowing the productivity of the phytoplankton at various depths through a profile of the water column of the euphotic zone, the vertical profile can be integrated to estimate $mg\,C/m^2/day$.

Procedures

1. Collect water samples as discussed in the section on in situ sampling (p. 86) with a nonmetallic water sampler.
2. Shading the bottles, first fill one dark and then two light 130-ml Pyrex groundglass stoppered bottles with water from each depth. Immediately place these bottles into a light-proof box, taking care to maintain them in order of collection. Labeled partitions within the box are very useful.
3. Next, fill a screw-cap, amber polyethylene bottle to the brim for alkalinity and pH measurements. Avoid direct sunlight and store in an ice chest. Measurements of alkalinity and pH should be performed as soon as possible (see Exercise 8). Particularly in nutrient-poor, soft waters, direct analysis of DIC will be necessary (see pp. 120 and 124).
4. After samples have been collected from all depths, open each sample bottle in the darkened box. Inoculate each bottle with 1.0 ml of $NaH^{14}CO_3$ [74 to 185 kilobecquerels (kBq)/ml = 2 to 5 microcuries (μCi)/ml] with a spring-operated repeating syringe by the following procedure. Score the neck of the ^{14}C-ampoule (or use prescored ampoules) and snap-break the ampoule open. Fill the syringe to remove all air bubbles. The syringe should be equipped with a 10- to 15-cm hypodermic needle or cannula. Transfer 1.0 ml of the tracer solution *slowly into the bottom* of the sample bottles. Immediately stopper the bottles as soon as the tracer has been injected to avoid double inoculation or omission. Time zero of the incubation is taken to the nearest minute at midpoint of inoculation of the series of bottles.

   Caution: Although the amount of radioactivity employed is very small, the half-life of ^{14}C is very long (5760 yr), and the isotope must not be introduced into the body. All handling of the isotope must be done with extreme care. Disposable gloves are recommended, and all excess solutions and the glassware must be disposed of properly and cleaned per the directions of the instructor.

   The amount of ^{14}C required varies with growth rates, population density, light, temperature, length of incubation, amount of sample filtered, and other factors. Under many conditions of moderate productivity, 37 to 111 kBq (= 1 to 3 μCi) per 130-ml sample, a 4-h incubation, and filtration of a 50-ml subsample yields sufficient radioactivity in the algae for good statistical radioassay without excessive layering of the phytoplankton on the filter.

5. Make certain that each bottle is sealed tightly. Mix each gently but thoroughly by repeated inversion and cap the dark bottles with aluminum foil to assure that no light leaks through the stopper joint. In a shaded area, rapidly attach the bottles to bottle spreaders with the dark bottle first in the center and then light bottles on the ends. Lower the bottles to the depth from which they were collected. Incubate for 3 to 4 h (not exceeding 6 h) between 0900 and 1500 h of the day. Make certain that the bouy does not shade the upper bottles.

6. Perform analyses of pH, alkalinity, and/or DIC as soon as possible. Water temperatures should be taken at each sampling depth.

7. At the end of the incubation, retrieve the bottles and place them into a light-proof box as rapidly as possible. The time of the end of the incubation is the midpoint of the retrieval operation.

8. Aliquots from each bottle, starting with the surface samples (usually most active), are transferred by pipetting (no mouth pipetting) as rapidly as possible to the filtration apparatus. Use membrane filters with a pore size of $0.45\,\mu m$ (e.g., Millipore HA or equivalent) and vacuum not exceeding 0.5 atm (380 mm Hg) to reduce the possibility of rupturing more fragile cells as they aggregate on the filter. Filtration should be performed in a semidarkened area to minimize further photosynthesis. The vacuum should be released immediately after the water has passed the filter to avoid rapid air desiccation.

   To estimate extracellular release of dissolved organic carbon by the phytoplankton, obtain samples of the filtrate at this point (see discussion below, p. 232).

   The amount of water to be filtered is governed by the observed density of the phytoplankton. In oligotrophic waters 100-ml subsamples should be filtered; in mesotrophic waters filtration of 50 ml usually is adequate. Because the β-radiation of ^{14}C is absorbed readily by cellular materials, no more algae should be collected than is necessary to minimize self-absorption. Therefore, in eutrophic waters, 10 ml or less may provide sufficient radioactivity while simultaneously minimizing self-absorption problems.

9. The filters are removed while wet from the filtration apparatus with forceps and placed onto clean planchets. A ring weight that does not come in contact with the filtered materials is then placed on the edges of the filters to prevent curling. The filters are then placed under the desiccator at room temperature for a brief (6- to 24-h) period. Expose dry filters to fumes of HCl *under a hood* for 10 min to remove residual inorganic ^{14}C and carbonates that may have precipitated as a result of photosynthetic activity [cf., Wetzel (1965)]. The samples then are desiccated again before radioassay.

Radioassay of the Filters

The detection of the random emission of β-radiation over time follows a Poisson distribution (see Appendix 2), and hence a minimum of approximately 1500 counts is necessary for statistical significance at the 5% level. Reasonably "hot" radioactive samples, e.g., light bottles from shallow depths, may yield 5000 to 10,000 counts in a reasonable time. Samples from dark bottles are relatively inactive and may require longer time periods to yield 500 to 1000 counts. Modern automatic instruments reduce the time required for the assay, thus eliminating the need to separate active from inactive samples.

Radioassay by liquid scintillation counting has the advantages of greatly increased counting efficiencies (50 to 95%), estimations of counting efficiency for each sample counted, and determination of the activity of stock ^{14}C solutions, membrane filters, and filtrates by the same instrument.

Filters and other substances to be radioassayed are immersed in organic mixtures (fluors) consisting of solvents (e.g., toluene, dioxane) and organic scintillators. Development and improvement of fluors are proceeding rapidly. Some of the techniques that have been used are discussed in detail by Schindler et al. (1974; Schindler, 1966); see also manufacturer's recommendations. Algal pigments can cause severe interferences and losses in efficiency of counting; this quenching must be carefully evaluated. (*Caution*: Many fluor systems are highly flammable and some are suspected to be carcinogenic; adequate ventilation must be used at all times.)

Place the filters flat on the bottom of standard 20-ml scintillation vials with the algal coating on the upper surface, avoiding any contact with the plankton on the filter surface. Add the appropriate scintillation mixtures. Count each filter twice for at least 10 min. An internal standardization procedure employing ^{14}C-toluene or -hexadecane may be used to determine counting efficiency (Schindler and Holmgren, 1971). The amount of radioactivity on each filter then can be expressed as disintegrations per minute (dpm).

Activity of the ^{14}C Transfer Inoculum Solution

The radioactivity of the inoculum added to the sample bottles must be known with accuracy. Standardized ^{14}C solutions developed specifically for productivity analyses are available from the Danish ^{14}C Agency (Hørsholm, Denmark) but are expensive if purchased in small quantities. When ^{14}C solutions are purchased, diluted, ampoulated, and sterilized [cf., Strickland and Parsons (1968)], their activity must be determined exactly by comparison with National Bureau of Standards samples of the same compound. Liquid scintillation radioassay is used most commonly with fluors that accept appreciable quantities of water without severe quenching problems.

Calculations

1. $(x) = {}^{12}C$ assimilated
    = unknown quantity sought, in $mg\,C/m^3/time$
2. $(a) = {}^{12}C$ available (see Exercise 8, pp. 117 to 124)
    = (total alkalinity) × (pH factor from Table 8.2)
    or
    Alternative 1 = (total alkalinity-phenolphthalein alkalinity, if present) × (0.240)
    = $mg^{12}C/l$)
    Alternative 2 { = [In soft waters, total DIC should be obtained (see Exercise 8).]

3. $(b) = {}^{14}C$ assimilated
    = [(total filter counts per second × k_1)—(background × k_1)]
    × (isotropic effect, 1.06)

The total filter counts refer to those of filters obtained from the light bottles. Background values refer to those counts from the dark bottles, i.e., dark fixation, absorptive values, cosmic radiation.

Since ^{14}C has a slightly greater mass than ^{12}C, the radiocarbon is assimilated approximately 6% more slowly than is ^{12}C. The theoretical value is close to 6% and has been confirmed experimentally (Steemann Nielsen, 1955; Sorokin, 1959). Hence, a factor of 1.06 is used to account for the isotopic effect.

k_1 = volume correction factor for aliquots filtered and sample bottles. Although high-quality Pyrex bottles are machine-made, variation in volume occurs up to ±10% of the stated volume. It is therefore necessary to determine individually the volume of each bottle (by water displacement), since deviations affect the dilution of the ^{14}C added. If, for example, 50 ml were filtered from a bottle containing 133 ml, and 1 ml of ^{14}C was added, then this factor would be $(133 - 1)/50 = 2.64$.

4. (c) = ^{14}C available
     = ^{13}C activity added

$$= \frac{(\mu Ci \; C^{14} \text{added}) \times (\text{disinitegrations of } ^{14}C/\text{sec})}{\text{efficiency factor of scintillation counter}}$$

As ^{14}C distegrates, β particles are emitted from the source at a constant energy value. The rate of disintegration per microcurie (μCi) is constant at 3.7×10^4 disintegrations per second (dps) or, in Système Internationale terms, 1 Bq = 1 dps, where 1 Ci = 37 GBq or $1 \mu Ci = 3.7 \times 10^4$ Bq.

Most modern scintillation counters now automatically compensate for reductions in radioassay efficiencies from 100% by referencing against internal standards. As a result, counts per minute or per second are automatically corrected to disintegrations per minute or second. Thus, no efficiency factor for the scintillation counter is needed.

5. (d) = a dimensional factor to convert mg/l to mg/m^3, i.e., 1000.
6. Combining the relationships:

$$(x)(c) = (a)(b)(d)$$

For example, given the following data:
1.0 ml of NaH $^{14}CO_3$ added per bottle
Activity of NaH $^{14}CO_3$ = 4.36 μCi/ml
Counter efficiency = 100% or a factor of 1.0

Total activity	Disintegrations per second (dps)	Bottle volume
LB_1	75.2	126 ml
LB_2	74.1	132 ml
DB	1.3	123 ml

50-ml samples filtered
mg ^{12}C/l available = 12.32
Constant isotopic effect = 1.06.

Then,

$$(x)(4.36)(37,000) = (12.32)\frac{(75.2)(125/50)+(74.1)(131/50)}{2} - [(1.3)(122/50)](1000)(1.06)$$

$$(x) = \frac{(12.32)(187.90)(1000)(1.06)}{(4.36)(37,000)}$$

$$(x) = 15.2 \, mg\,C/m^3/\text{time of incubation}$$

Conversion of Volumetric Measurements to Areal Estimates of Productivity

In comparing aquatic ecosystems, it often is desirable to convert the measured volumetric productivity values in $mg\,C/m^3$/time at each depth to total productivity per square meter for a water column extending from the surface to the compensation depth. This conversion can be done by plotting the values in $mg\,C/m^3$/time against depth and then integrating the profile generated. Although the integration can be accomplished by computer analysis, it also can be done with an electronic digitizer or relatively simply by planimetry (see Exercise 1 and Fig. 1.5). The planimetered area of the productivity curve is compared proportionally with the planimetered area of a known standard area of mg/m^3/time versus depth drawn to the same scale on the graph. The resulting estimate of productivity is in $mg\,C/m^2$ of water column/time of incubation. This value may be converted to $mg\,C/m^2/h$ by dividing by the exact time in hours of incubation or, more meaningfully, converted to mg C/m^2/day by one of the following methods.

Expansion of Incubation Period to Daily Values

Incubation periods should be limited to 4 to 6 h for a number of reasons [cf., Vollenweider and Nauwerck (1961) and Vollenweider (1965)]. This period is long enough to allow measurement of ^{14}C assimilated but is also sufficiently brief to minimize errors incurred from longer exposure as a result of bacterial growth on the walls of the bottles, respiratory recycling of CO_2, nutrient depletion, and so on. To be meaningful, however, the results should be expanded to a daily value by correction factors. Several means are available; all possess disadvantages.

Ideally, productivity measurements should be taken in the euphotic zone at about 4-h increments from darkness to darkness each day for an entire year. Since this rigor is obviously impractical, estimates can be calculated from data that have been obtained. One method is to determine periodically (e.g., biweekly) the complete daily productivity by a series of 4-h measurements from dawn to dusk, e.g., 0500 to 0900; 0900 to 1300; 1300 to 1700; and 1700 to 2100. The productivity of each of these profiles at the different times of the day is converted to $mg\,C/m^3$/time interval. The value of each interval is plotted at the midpoint of its respective time period (Fig. 14.2A). The ends of the curve are the times of dawn and dusk as determined from a curve of solar radiation for the day. The area of the entire curve is integrated with a digitizer or by planimetry and compared to that area of one of the "typical" 4-h measurements (e.g., 0900 to 1300 h; shaded area in Fig. 14.2A). The ratio of the area of the fraction to the whole provides a factor by which the 4-h incubation could be expanded to a daily value.

Obviously the photosynthetic pattern of the daily curve will change from day to day with variations in meteorological conditions and other factors. Large errors could occur in such an extrapolation system from biweekly diurnal curves.

In an attempt to circumvent daily variations, productivity could be assumed to be proportional directly to light, i.e., algal photosynthesis is not light saturated or light limited below some threshold intensity. Certainly this often is not the case for phytoplankton in surface water but is true for algae throughout much of the water column under most circumstances. Also, there is some evidence that photosynthetic rates are somewhat greater in the morning hours than in the afternoon; perhaps this fact is related to increased photorespiration rates under brighter light conditions or reduced availability of nutrients or inorganic carbon as the pH increases. A midday

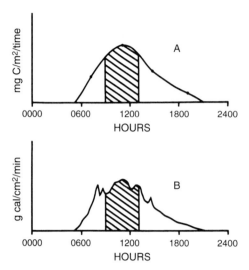

Figure 14.2. Estimates of a diurnal expansion factor by comparison of the area of the second incubation period with that of the whole day (A) and by comparison of the area of the insolation during the incubation period with that of the whole day (B).

incubation would tend to average out some of these variations. In this method, the relationship between the total daily insolation and that during the incubation period is used as the scaling factor (Fig. 14.2B).

Vollenweider (1965) demonstrated, with mathematical models of photosynthesis-depth curves and experimental data, that midday incubation results in very small errors. Day-rate estimates would be facilitated if the duration of experiments were chosen to be proportional to the day length factor, if the light day were divided into five equal parts, and if the ^{14}C exposures were performed during one or both of the central periods. In addition, it was shown that the specific shape of the photosynthesis-depth curve is of much less importance in regard to day-rate estimates than has been thought generally. The strongest influence is exerted by the drop of the photosynthetic activity during the course of the day, whereas neither the subsurface light inhibition nor the degree of adaptation to low light intensity will produce serious errors. The best period for obtaining experimental data from a day divided into five equal periods is generally the second one; the production rate of that period can be expected to be of the order of 30% of the total day rate, provided that nutrient depletion, or its effect on photosynthesis, remains within moderate limits.

When, instead, no nutrient depletion occurs at all (giving a symmetrical trend of surface rates), about 25% of the total day rate will be measured during the second, and 30% during the third, period. In any case, during both the second and third periods about 55 to 60% of the total day rate is produced. Accordingly, the error introduced in estimating day rate integrals of photosynthesis from exposures during the second and third periods will be of the order of ±10%, or less.

An extension of the latter method may be used to estimate the productivity on days between closely measured values. If the productivity were measured, for example, at 5-day intervals, a fair estimate of the productivity of the intervening days would be derived from the actual measurements, the light energy on the days of productivity measurements, and light values for days between the actual measurements.

For example, at Lake Hypothetical, the measurement of productivity on August 3, 1990, was 753 mg C/m^3/day, and the total light value was 638 gcal/cm^2/day. The next productivity measurement was not made until August 8, 1990, but light was measured continuously at the lake with a recording pyrheliometer. Hence, the estimated values must be viewed as approximate, and extrapolation for more than a few days is probably not justifiable because of changing densities, species composition, and other factors.

Date	Productivity ($mg\,C/m^3$/day)	Light ($gcal/cm^2$/day)	Remarks
Aug. 3	753	638	Measured productivity
Aug. 4	662	561	Estimated from productivity of Aug. 3
Aug. 5	851	721	Estimated from productivity of Aug. 3
Aug. 6	388	321	Estimated from productivity of Aug. 8
Aug. 7	152	126	Estimated from productivity of Aug. 8
Aug. 8	521	431	Measured productivity

Expression on an Annual Basis

The daily productivity values then are plotted on an annual scale (Fig. 14.3), and this annual curve is integrated by planimetry and compared to a standard area of the graph (e.g., $300\,mg\,C/m^3$/day versus 30 days). From these data, an annual mean productivity (mean $mg\,C/m^3$/day) may be calculated by dividing the total productivity ($g\,C/m^3$/year) by 365. The annual mean value is an excellent basis for comparison of productivity between lake ecosystems [see Wetzel (1983)].

An areal estimate of the primary productivity by phytoplankton for the entire pelagial zone of a lake may be made by several methods. The most accurate method is done by determining the annual productivity at each depth interval (0 to 0.5 m, 0.5 to 1.5 m, 1.5 to 2.5 m, and so on) and multiplying these values by the actual volume of water in that layer (Wetzel, 1964). The annual productivity of each of these strata, weighted for volume differences, then is summed and divided by 365 to give the $g\,C$/lake/day (Fig. 14.4).

It should be emphasized that estimates of the primary productivity of a lake ecosystem based only on the metabolic activity of phytoplankton can be grossly in error. The productivity of littoral macrophytes and algae attached to various substrates, as well inputs of organic matter from the drainage basin, often can exceed greatly the organic matter synthesized by the phytoplankton. In shallow lakes, the

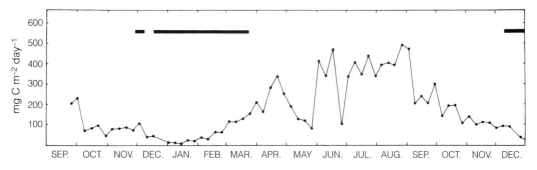

Figure 14.3. Integrated primary productivity per square meter of the pelagial zone of Lawrence Lake, MI, 1967–68. Bars indicate periods of ice cover. [From Wetzel (1983).]

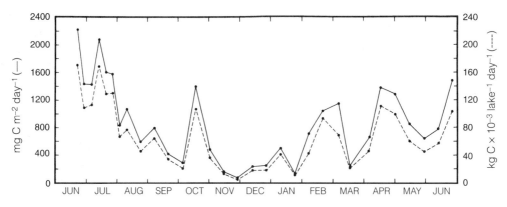

Figure 14.4. Integrated primary productivity of the phytoplankton of hypereutrophic Wintergreen Lake, MI, 1971–72, per square meter of the trophogenic zone (—) and for the lake (------) compensating for volumetric differences with depth. [From Wetzel (1983).]

phytoplanktonic productivity usually is quite subordinate to that of the rooted plants and associated sessile algae [cf., Wetzel (1990, 1999)].

Extracellular Release of Dissolved Organic Carbon

An estimate of extracellular release of dissolved organic carbon can be determined by obtaining a sample of filtrate, removing the labeled inorganic carbon, and then assaying the radioactivity in dissolved organic compounds. The filtrate from each sample can be collected in test tubes placed under the effluent tube of the filter support apparatus within the filtration flask. A 10-ml subsample of the filtrate is acidified to pH 3 with 3% H_3PO_4 in scintillation vials in an efficient hood and sparged with pure CO_2 for a time sufficient to remove the inorganic CO_2 [usually 4 min per 10 ml is adequate; see Hough and Filbin (1978)]. The vials then can be treated in either of two ways. A scintillation fluorescing compound that will accept appreciable quantities of water without severe quenching can be added in equal quantities (10 ml) to the sparged sample and then radioassayed. Alternatively, the vials containing 10-ml of acidified and CO_2-sparged sample can be frozen immediately and freeze-dried to concentrate the radioactivity of the organic compounds (McKinley et al., 1977). After the samples are dry, the material is resuspended in 1 to 5 ml of water and 15 to 19 ml of scintillation fluid (e.g., Insta-gel, Packard Instr. Co.).

The rate of organic carbon released extracellularly in dissolved form then can be calculated as $mg\,C/m^3/day$ in similar fashion to that outlined above. These values then can be compared for each depth to evaluate the percentage of dissolved organic carbon released to that fixed in particulate organic matter.

Sensitivity

Theoretically there is no upper limit of sensitivity for the ^{14}C method, but in practice errors of time required for manipulations increase as photosynthetic rates become exceedingly high. The lower limit is in the range of $0.01\,mg\,C/m^3/h$, some three orders of magnitude lower than that of the oxygen method.

Most evidence of comparative studies indicates that the ^{14}C method measures photosynthetic rates closer to net photosynthesis than to gross, especially when the rate of dissolved organic matter released is added to that carbon fixed in particulate organic matter. Organic carbon budgets for cultures of lake phytoplankton indicated that ^{14}C uptake overestimates net carbon production. The overestimate was slight (10 to 20%) for rapidly increasing cultures but was higher for nutrient-poor, slowly growing cultures (Peterson, 1978).

EXERCISES

OPTION 1. FIELD TRIPS

1. At the central depression of a lake, make a systematic comparison of the primary productivity of the phytoplankton using the oxygen and ^{14}C light and dark techniques. Collect water samples, as discussed, in teams.
2. One team should obtain replicated samples for light, dark, and initial oxygen concentration analysis at meter intervals (half-meter intervals near the surface, if productive). Part of the team should fix the initial bottles immediately. The dissolved oxygen concentrations can be determined by titration during the incubation period of the light and dark samples.
3. A second team should work with the instructor to collect water samples for replicated light and dark bottles for the ^{14}C method at the same depths as done for the oxygen technique. Samples for alkalinity, and pH, and DIC measurements should be taken simultaneously; these analyses can be made during the incubation period.
4. A third team should determine the temperature profile at the station and the underwater light distribution. When instruments are available, measure the incident and reflected light at each depth interval and, using specific filters, the distribution of spectral components (see Exercise 2).
5. When possible, another team could collect samples of the phytoplankton at each depth to determine simultaneously algal biomass by analyses of pigment concentrations (see Exercise 10).
6. Proceed with all of the analyses as discussed earlier. Determine the productivity of the phytoplankton by both the oxygen and ^{14}C methodologies and compare your results.
7. Compare the productivity measurements with the vertical stratification patterns of temperature and with the attenuation patterns of underwater light.
8. If phytoplanktonic biomass was determined, compare its distribution with the productivity values.
9. Make depth distribution graphical comparisons and calculations of all parameters with the data.

OPTION 2. LABORATORY ANALYSES WITH CULTURES

1. Using the photoplankton populations provided by your instructor, compare the productivity of the algae under decreasing light conditions by simultaneous application of the oxygen and ^{14}C light and dark methods in the laboratory as discussed earlier.
2. Perform titrations of the dissolved oxygen concentrations of the initial bottles and of the alkalinity of the water, and determine the pH and DIC of each series during incubation.
3. Determine the light intensities under the different regimes of incubation.
4. When possible, expose replicate populations to light at different intensities. The light should be filtered selectively by various absorption materials, e.g., colored celluloid filters

as used in theaters. (*Caution*: Many of such filter materials are water soluble.) Determine the intensity and spectral characteristics of light under these conditions. When instrumentation for direct measurements of the light is not available, pieces of the material can be analyzed in a spectrophotometer (scanning or at a series of wavelengths).

OPTION 3. LABORATORY ANALYSES WITH DATA

1. Using the data provided in Table 14.1, calculate the phytoplanktonic productivity for the given day by the different methods discussed earlier. *Note*: The oxygen and ^{14}C uptake data are not from the same day and therefore should not be compared to each other.
   a. Gross and net community productivity and respiration by the oxygen difference technique.
   b. Primary productivity by the ^{14}C method.
   c. Plot the data in mg $C/m^3/day$, integrate each curve with a digitizer or by planimetry, and determine the productivity in mg $C/m^2/day$ for the water column.
2. Compare the productivity with depth to the temperature profile and to the distribution of underwater light (Table 14.2).
3. Compare the productivity with depth to the observed distribution of pigment concentrations (Table 14.3).
4. Using the series of productivity values provided to you in Table 14.4, plot the annual productivity curve of mg $C/m^2/day$ versus time. Integrate the curve by planimetry and determine the annual mean productivity for this lake in $\bar{x}$ mg $C/mg^2/day$.

Questions

1. Assuming a photosynthetic quotient of 1.2, how do the gross and net photosynthetic estimates determined by the oxygen technique compare with the productivity estimate obtained from the ^{14}C method? Which method do you feel gives the most realistic results? Support your answer.
2. Under your particular set of conditions in the lake or laboratory, how is the productivity correlated with light? If surface photoinhibition was observed, how might the population overcome this problem in a circulating epilimnion?
3. Is the relationship between productivity rates and algal biomass direct? What may underlie any observed disparities?
4. Using microautoradiography, how might the relationship between total observed productivity of the composite algal community, species activity and growth turnover rates be clarified?
5. Is the mean rate of photosynthesis measured by the light and dark bottle techniques the same as the true rate of photosynthesis? Why? Support your answer.
   Note: When the true rate of photosynthesis is measured by the differential dp/dt, where p_t is the amount of fixed carbon at time t and p_0 the amount at a zero reference time, then the rate as measured experimentally is (Strickland, 1960)

$$\frac{p_t - p_0}{t} = \frac{D}{t}$$

   where D is the increase in carbon found in the light bottle. When plant growth is approximately exponential, the true rate at time t_0 is given by:

$$\frac{dp}{dt} = \left(\frac{p_0}{t}\right)\left(2.3 \log \frac{D + p_0}{p_0}\right)$$

   This expression requires the value of p_0, which is rarely known.
   How would duration of incubation affect this relationship?

Table 14.1. Field data from primary productivity assays with the oxygen and the ^{14}C light and dark bottle techniques. Note that these assays are from different times of the year and a direct comparison should not be attempted.

	Oxygen method in July					^{14}C method in October								
	Initial bottles (mg O$_2$/l)		Light bottles (mg O$_2$/l)		Dark bottles (mg O$_2$/l)	Light bottles				Dark bottles		Total alkalinity (mg/l)	pH	Temperature °C
						(a)		(b)						
Depth (m)	(a)	(b)	(a)	(b)		dps	Volume (ml)	dps	Volume (ml)	dps	Volume (ml)			
0	8.41	8.31	8.59	8.66	8.29	4.85	125	4.58	121	0.63	124	48.6	7.80	23.0
0.5	8.36	8.50	8.81	8.72	8.05	5.68	124	5.21	119	0.68	126	48.3	7.80	23.0
1	8.43	8.48	8.91	8.99	8.09	8.36	129	8.91	128	0.71	121	48.5	7.81	23.0
2	9.12	9.06	10.36	10.48	8.01	14.10	125	14.39	131	1.21	129	48.4	7.90	22.8
3	9.26	9.16	10.51	10.63	8.20	13.91	122	13.62	128	0.91	125	48.9	7.92	21.9
4	8.51	8.49	9.47	9.55	8.15	9.63	126	9.52	123	0.71	125	51.6	7.81	20.0
5	8.31	8.39	9.03	9.10	8.01	6.13	127	5.91	121	0.49	124	56.1	7.75	17.3
6	8.00	8.09	8.41	8.36	7.71	3.43	126	3.89	131	0.41	126	56.8	7.72	14.4
7	7.12	7.19	7.25	7.33	6.81	2.10	129	2.01	127	0.36	127	57.9	7.62	10.6
10	5.82	5.71	5.91	5.83	5.00	1.05	128	0.98	129	0.21	121	60.2	7.45	8.1
12	2.36	2.47	2.31	2.50	1.95	0.51	129	0.86	127	0.75	125	62.1	7.40	7.8

Oxygen Method
Incubation: 4h 35min
Diurnal expansion factor: 2.62

^{14}C Method
^{14}C added/bottle: 1.0ml
μCi/ml: 3.86
Counter efficiency: 100% by internal referencing
ml filtered: 50ml
Time of incubation: 0932 to 1348h
Diurnal expansion factor: 2.51

Table 14.2. Distribution of underwater light.[a]

Depth (m)	A Neutral filter	Red (640–720 nm)	Yellow (580–590 nm)	Green (500–580 nm)	Blue (460–500 nm)	B Upward-facing hemisphere	Downward-facing hemisphere
Surface (air)	100	100	100	100	100	100	100
0.1	75.0	62.9	90.6	79.4	81.3	65.0	97.8
1	45.5	42.9	40.4	50.5	46.0	47.5	93.6
2	30.4	25.9	28.9	32.4	28.4	37.5	91.7
3	17.8	18.5	21.9	24.0	21.0	30.0	80.0
4	13.5	13.2	16.6	17.6	11.1	23.0	61.9
5	9.6	9.7	11.9	12.3	7.9	17.3	50.0
6	6.8	4.4	9.1	7.0	5.6	13.5	36.9
7	4.9	3.1	6.8	5.0	4.0	10.3	27.4
8	3.4	2.2	3.8	2.6	3.0	8.0	22.6
9	1.8	1.6	2.8	2.3	2.1	6.0	19.0
10	1.2	1.1	2.1	2.0	1.6	4.8	17.9
11	0.8	0.8	1.6	1.4	1.1	3.5	20.2
12	0.5	0.6	1.0	0.8	0.8	1.3	21.4

[a] A: Underway photometer equipped with neutral and restricted spectral filters. B: Photometer that measures incident-scattered light from the upper hemisphere and reflected-scattered light from the lower hemisphere. All values in percent of the surface value.

Note: When a fractional increase in the size of a population of algae does not exceed about 20% during the measurement, the true rate at zero time, dp/dt, is equal to D/t to within a few percent (Strickland and Parsons, 1968). The true rate at a time $(t - t_0)/2$, however, is still close to the value for D/t with D as much as p_0 (i.e., when the population has doubled). Depending upon the average doubling time for algae, the experimental incubation time should not exceed 2 to 4 h in order to find dp/dt at t_0. In order to estimate dp/dt at time $(t - t_0)/2$, incubation times would have to be much longer (e.g., 12 h or more); other problems, such as bacterial respiration and proliferation, and supersaturation of dissolved gases, become severe.

6. What causes different values for PQ and RQ?
7. Researchers often observe an increase in dissolved oxygen concentration in dark bottles. What could be the explanation for this?
8. Why should you work in the shade and use light-proof storage boxes when measuring phytoplankton productivity?

Table 14.3. Concentrations of phytoplanktonic pigments.

Depth (m)	Chlorophyll a (mg/m³)	Chlorophyll b (mg/m³)	Chlorophyll c (mg/m³)	Plant carotenoids (mg/m³)	Phaeopigments (mg/m³)
0	0.38	0.17	0.00	0.38	0.13
1	1.08	0.23	0.57	1.08	0.00
2	2.00	0.37	0.52	1.82	1.05
3	1.86	0.16	0.72	1.60	2.06
4	4.17	0.39	2.05	3.29	2.15
5	6.07	0.13	2.52	6.73	4.18
7	5.18	0.42	2.27	4.18	5.23
10	1.11	0.49	0.88	1.26	3.98
12	0.75	0.13	1.35	3.47	3.56

Table 14.4. Daily values for primary productivity of phyto-plankton over an annual period (Lawrence Lake, MI, 1968).[a]

Date	mg C/m^2/day	Date	mg C/m^2/day
Jan. 2	14.3	July 3	335.8
Jan. 9	12.0	July 10	412.9
Jan. 15	7.8	July 17	349.3
Jan. 22	24.5	July 24	442.9
Jan. 29	21.3	July 31	337.7
Feb. 5	36.3	Aug. 6	396.2
Feb. 12	29.7	Aug. 13	417.1
Feb. 19	66.3	Aug. 20	393.7
Feb. 26	64.0	Aug. 27	496.1
Mar. 4	116.1	Sept. 3	473.1
Mar. 11	115.8	Sept. 9	203.4
Mar. 18	129.3	Sept. 16	243.2
Mar. 26	157.2	Sept. 23	206.7
Apr. 1	213.2	Sept. 30	304.1
Apr. 8	165.5	Oct. 7	142.3
Apr. 15	283.3	Oct. 14	194.6
Apr. 22	341.7	Oct. 21	197.8
Apr. 29	255.2	Oct. 28	108.5
May 6	197.1	Nov. 4	145.0
May 13	129.6	Nov. 11	98.7
May 20	121.9	Nov. 18	112.4
May 27	83.4	Nov. 25	109.9
June 3	415.8	Dec. 2	81.6
June 10	338.6	Dec. 9	94.1
June 17	474.3	Dec. 16	91.4
June 26	100.2	Dec. 27	35.9

[a] From Wetzel (unpublished data).

9. Does photorespiration occur in dark bottles?
10. How would respiration be measured in a ^{14}C light/dark bottle experiment?
11. If phytoplankton such as dinoflagellates were capable of extensive vertical migration during the day, how would this affect in situ measurements of primary productivity? [See Larson (1978) and Taylor et al. (1988).]
12. Calculations of primary productivity with the ^{14}C uptake method by extrapolation of the productivity estimates from the incubation period to day length (daylight periods) estimates of carbon fixation neglect respiration. What is the effect on the results? These measurements also neglect nighttime respiration. What is the composite effect on the results? How might one correct for respiration to obtain estimates of daily (24-h) productivity?
13. Do you think there would be differences between in situ incubation and laboratory incubations for primary production measurements? Why?

Apparatus and Supplies

1. Boats, anchors, Van Dorn or similar opaque, nonmetallic water samplers, thermistor for temperature measurements.
2. Oxygen change method:
   a. Labeled 250-ml Pyrex bottles with tips of ground-glass stoppers ground at an angle, transparent and opaqued with two layers of plastic electrician's tape; sheets of heavy aluminum foil.
   b. Buoy, suspension line with clips at 0.5- or 1-m intervals; bottle spreaders.

    c. Winkler reagents for fixation of dissolved oxygen (see Exercise 6, p. 82).

    d. Burettes and titration apparatus and reagents for determining dissolved oxygen concentrations (see Exercise 6, p. 84).

3. ^{14}C uptake method.

    a. Clear and opaqued (two layers of black plastic electrician's tape) 130-ml Pyrex ground-glass stoppered bottles, labeled and outfitted with clips wired to the neck of each; sheets of heavy aluminum foil.

    b. Buoy, suspension line with clip rings at 0.5- or 1-m intervals; rod bottle spreaders with eye-bolts at ends and suspension clips in the center.

    c. Apparatus and reagents for analyses of alkalinity, pH, and DIC (see Exercise 8).

    d. Precision syringe (spring-loaded) with 10- to 15-cm hypodermic needle or cannula; ampoules of ^{14}C of known specific activity, scoring file if needed, radioactive waste containers for liquid and solids.

    e. Pipets, membrane filters, filter forceps, multiple unit filtration apparatus, rings to weight edges of filters, desiccator, acidification chamber to fume *dry* filters with fumes of concentrated HCl (a glass-covered all-glass aquarium or battery jar with an internal glass rack to keep filters away from the HCl liquid at the bottom works satisfactorily).

    f. Radioassay liquid scintillation counter; appropriate reagents and vials.

4. Underwater photometers.

References

Fee, E.J. 1969. A numerical model for the estimation of photosynthetic production, integrated over time and depth, in natural waters. Limnol. Oceanogr. *14*:906–911.

Fee, E.J. 1971. Digital computer programs for estimating primary production, integrated over depth and time, in water bodies, Spec. Rept. *14*, Center Great Lakes Studies, Univ. Wisc. 42 pp.

Fee, E.J. 1973. A numerical model for determining integral primary production and its application to Lake Michigan. J. Fish. Res. Bd. Canada *30*:1447–1468.

Findenegg, I. 1966. Die Bedeutung kurzwelliger Strahlung für die planktische Primärproduktion in den Seen. Verh. Int. Ver. Limnol. *16*:314–320.

Gervais, F., D. Opitz, and H. Behrendt. 1997. Influence of small scale turbulence and large scale mixing on phytoplankton primary production. Hydrobiologia *342/343*:95–105.

Hough, R.A. and G.J. Filbin. 1978. Factors affecting the removal of ^{14}C from water. Verh. Int. Ver. Limnol. *20*:49–53.

Larson, D.W. 1978. Possible misestimates of lake primary productivity due to vertical migrations by dinoflagellates. Arch. Hydrobiol. *81*:296–303.

McKinley, K.R. and R.G. Wetzel. 1977. Tritium oxide uptake by algae: An independent measure of phytoplankton photosynthesis. Limnol. Oceanogr. *22*:377–380.

McKinley, K.R., A.K. Ward, and R.G. Wetzel. 1977. A method for obtaining more precise measures of excreted organic carbon. Limnol. Oceanogr. *22*:570–573.

Parsons, T.R., Y. Maita, and C.M. Lalli. 1984. A Manual of Chemical and Biological Methods for Seawater Analysis. Pergamon, Elmsford, NY. 173 pp.

Peterson, B.J. 1978. Radiocarbon uptake: Its relation to net particulate carbon production. Limnol. Oceanogr. *23*:179–184.

Saunders, G.W., F.B. Trama, and R.W. Bachmann. 1962. Evaluation of a modified ^{14}C technique for shipboard estimation of photosynthesis in large lakes. Publ. Great Lakes Res. Div., Univ. Mich. *8*. 61 pp.

Schindler, D.W. 1966. A liquid scintillation method for measuring carbon-14 uptake in photosynthesis. Nature *211*:844–845.

Schindler, D.W. and S.K. Holmgren. 1971. Primary productivity and phytoplankton in the Experimental Lakes Area, northwestern Ontario and other low carbonate waters, and a liquid scintillation method for determining ^{14}C activity in photosynthesis. J. Fish. Res. Bd. Canada *28*:189–201.

Schindler, D.W., J. Moore, and R.A. Vollenweider. 1974. Liquid scintillation techniques. pp. 76–80. *In*: R.A. Vollenweider, Editor. A Manual on Methods for Measuring Primary Production in Aquatic Environments. IBP Handbook No. 12. 2nd Edition. Blackwell, Sci. Publish. Oxford.

Sorokin, Yu.I. 1959. Opredeleniye velichin izotopicheskogo effekta pri fotosinteze v kulturakh *Scenedesmus quadricauda*. (Determination of the isotopic discrimination by photosynthesis in cultures of *Scenedesmus quadricauda*.) Bull. Inst. Biol. Vodokhranilisch *4*:7–9.

Steemann Nielsen, E. 1951. Measurement of the production of organic matter in the sea by means of carbon-14. Nature *167*:846–685.

Steemann Nielsen, E. 1952. The use of radioactive carbon (^{14}C) for measuring organic production in the sea. J. Cons. Int. Expl. Mer *18*:117–140.

Steemann Nielsen, E. 1955. The interaction of photosynthesis and respiration and its importance for the determination of ^{14}C-discrimination in photosynthesis. Physiol. Plant. *8*:945–953.

Strickland, J.D.H. 1960. Measuring the production of marine phytoplankton. Bull. Fish. Res. Bd. Canada *122*. 172 pp.

Strickland, J.D.H. and T.R. Parsons. 1972. A Practical Handbook of Seawater Analysis. 2nd Ed. Bull. Fish. Res. Bd. Canada *167*. 311 pp.

Taylor, W.D., J.W. Barko, and W.F. James. 1988. Contrasting dual patterns of vertical migration in the dinoflagellate *Ceratium hirundinella* in relation to phosphorus supply in a north temperature reservoir. Can. J. Fish. Aquat. Sci. *45*:1093–1098.

Vollenweider, R.A. 1965. Calculation models of photosynthesis-depth curves and some implications regarding day rate estimates in primary production measurements. Mem. Ist. Ital. Idrobiol. *18* Suppl.:425–457.

Vollenweider, R.A. and A. Nauwerck. 1961. Some observations on the ^{14}C method for measuring primary production. Verh. Int. Ver. Limnol. *14*:134–139.

Walsby, A.E. 1997a. Modelling the daily integral of photosynthesis by phytoplankton: Its dependence on the mean depth of the population. Hydrobiologia *349*:65–74.

Walsby, A.E. 1997b. Numerical integration of phytoplankton photosynthesis through time and depth in a water column. New Phytol. *136*:189–209.

Wetzel, R.G. 1964. A comparative study of the primary productivity of higher aquatic plants, periphyton, and phytoplankton in a large, shallow lake. Int. Rev. ges. Hydrobiol. *49*:1–61.

Wetzel, R.G. 1965. Necessity for decontamination of filters in ^{14}C measured rates of photosynthesis in fresh waters. Ecology *46*:540–542.

Wetzel, R.G. 1983. Limnology. 2nd Ed. Saunders Coll., Philadelphia. 860 pp.

Wetzel, R.G. 1990. Land-water interfaces: Metabolic and limnological regulators. Verhand. Internat. Verein. Limnol. *24*:6–24.

Wetzel, R.G. 1999. Limnology: Lake and River Ecosystems. 3rd Ed. Academic Press, San Diego (in press).

Feeding Rates by Protists and Larger Zooplankton

Direct consumption by zoolankton can have appreciable effects upon phytoplankton and bacterioplankton populations. By means of selective grazing, zooplankton can influence the seasonal succession of the phytoplankton [cf., Porter (1977)].

Most cladocerans and copepods remove particulate organic matter from the water by filtration and concentration of particles by water movements toward the mouth area. The size of the particles that can be cleared from the water is a function of the morphology of the setae on the moving appendages or of entrapment as the movement of the animal brings particle-laden water to the setae. The maximum rate at which energy is gained from food is a function of the combined rates of filtering and ingestion, and of the abundance, size, and digestibility of food [cf., Lehman (1976)].

Clearance rate is generally defined as the volume of water cleared of suspended particles per unit time (hour or day). This term, usually synonymous with filtration capacity, should not imply, however, that the volume of water passed over the filtration appendages is known, that all particles of a given type have been removed from the water, or that all particles retained by the filtration apparatus have been consumed (Rigler, 1971). The contemporary view of zooplanktonic feeding behavior is that the animals do not filter the water in the sense of sieving. Instead, high speed photography has shown that particles are captured as parcels of water are moved within the feeding structures (Peters, 1984; Omori and Ikeda, 1984). Although the term *filtering rate* is still used widely, the terms *clearance* or *clearance rate* are more representative of the actual process being measured.

In contrast, *feeding rate* is a measure of the quantity of food ingested by an animal in a given time (measured as number, volume, dry weight, or chemical content of cells, or other components of the ingested food). Measurements of feeding rate are made by observing changes in the number of particles removed over time by grazing, or by measuring the rate of removal of food particles labeled radioactively or fluorescently. These methods measure food actually taken into the gut. Loss of filtered particles may occur either through active rejection or during the maceration process. It should be kept in mind that ingestion rates do not equal assimilation rates, which can be highly variable depending on the type and concentrations of food particles and their chemical content.

This exercise evaluates several methods of measuring clearance and grazing rates in protists and in larger zooplankton. Quantification of the rates of clearance and feeding rates allows an evaluation of the effects of various environmental factors on ingestion.

FEEDING RATES OF LARGE ZOOPLANKTON

Particle Removal

1. Fill small breakers with algal cells or yeast uniformly mixed in a known volume of water.
2. Determine the initial cell concentrations (see Exercise 10).
3. Add a known number of zooplankton to each container and incubate at room temperature under constant light for a known period of time. Maintain a control to which no zooplankton is added.
4. Manipulate food types, food particle concentrations, zooplankton types and concentrations, and environmental parameters as discussed below.
5. Determine the final concentrations of food cells.
6. Calculations are based on the following relationships:
   a. The original number of cells (N_0) is reduced at any given time t to the number of cells remaining (N_t):

   $$N_t = N_0 e^{-kt}$$

   where k = coefficient of grazing.
   b. Using logarithms to determine k:

   $$k = \frac{\ln N_0 - \ln N_t}{t}$$

   where t = time of the experiment in hours.
   c. The volume (V) of water filtered under the experimental conditions is

   $$V = vk$$

   where v = the volume of water available in the vessels for each animal.

   This calculation assumes that no significant changes in food particle concentrations have occurred in the controls during the experiment. If long incubation times are required, however, some growth or death can occur in the controls. Computations of grazing must then correct for these changes in the controls [see Frost (1972) and Peters (1984)].

Removal of Food Particles That Have Been Labeled with Radioactivity

1. Radioactively label a culture or several cultures of algae by allowing photosynthetic uptake of $H^{14}CO_3^-$. The cultures should be well mixed, and a known volume distributed into each of several small flasks (ca. 500 ml). Alternatively, algae or yeast labeled with $^{32}PO_4$ can be used. *Caution*: Strict adherence to regulations for handling and disposal of radioactive materials, as specified by the instructor, is of the utmost importance.
2. Add a known number of zooplankton to each container and incubate under the experimental conditions (see below) for a known period of time. Since the radioassay methods are very sensitive, incubation periods can be very short (the time for passage of food particles through the gut of *Daphnia* is about 10 min). Maintain controls to which no zooplankton is added.
3. Manipulate food types and concentrations, zooplankton types and concentrations, and environmental parameters as discussed below.

4. At the beginning of the experiment, remove an aliquot (e.g., 1 ml) of the algal or yeast cell suspension for radioassay. Filter onto a membrane filter for radioassay with scintillation procedures (see Exercise 14 and below).

5. At the end of the incubation period, kill the zooplankton immediately by bringing the culture solutions to a boil (use a vented hood). Then remove the zooplankton from the media by filtration through fine-meshed sieves as instructed (see Exercise 11). Wash the zooplankton rapidly with a small volume of water and place them with fine forceps and dissecting needles into containers suitable for counting the radioactive disintegrations. Transfer the zooplankton directly to scintillation vials and solubilize with an appropriate solubilizer reagent (see "Apparatus and Supplies," p. 247).

6. The radioactivities of the zooplankton and of the algal suspension are then determined, correcting for counting efficiencies and for self-absorption. If ^{14}C radioassay is used, the weak β-radiation is partially absorbed by the chitinous carapace of the zooplankton. Approximate correction factors are given in Table 15.1 if scintillation solutions are used directly without prior treatment with tissue solubilizers. Such correction factors are not needed for radioassay of ^{32}P radiation. Why?

7. From the radioactivity accumulated in the organisms, the radioactivity of the food (algal suspension), and the duration of the feeding period, the clearance rate can be calculated as follows (Haney, 1973):

$$\text{Clearance rate} \atop \text{(ml/individual/day)} = \left(\frac{\text{counts/min/individual}}{\text{counts/min/ml of suspension}}\right)\left(\frac{1440\ \text{min/day}}{\text{minutes feeding time}}\right)$$

The grazing rate (%/day) can be calculated by considering the volume of water filtered per unit volume of suspension:

$$\text{Grazing rate} \atop \text{(%/day)} = \left(\frac{\text{clearance rate}}{\text{in ml/individual/day}}\right)\left(\frac{100}{\text{ml containing zooplankton}}\right)$$

Table 15.1. Average self-absorption coefficients of ^{14}C radiation in several types of zooplankton.[a]

Organisms	Self-absorption coefficients	Organisms	Self-absorption coefficients
Rotifera		Copepoda	
Asplanchna priodonta	1.17	*Diaptomus gracilis*	1.88
A. herrichi	1.17	*D. graciloides*	1.88
		D. oregonensis	1.23
Cladocera		Ostracoda	
Polyphemus pediculus	1.56	*Darwinula stevensoni*	1.85
Bosmina longirostris	1.16	*Cyclocypris ampla*	5.03
Daphnia longispina	1.51	*Cypria turneri*	2.26
D. pulex	1.66	*Physocypria pustulosa*	2.38
Ceriodaphnia quadroangula	1.18	*Canodona rawsoni*	3.23
Simocephalus vetulus	1.60	*C. inopinata* (♀)	3.35
S. exospinosus	1.86	*C. inopinata* (♂)	2.86
Cydorus sphaericus	1.14		

[a] From McGregor and Wetzel (1968) and references cited therein.

Clearance rate and grazing rate as calculated above are not two different measurements, but rather are simply different ways of expressing the rate of clearance of particles from a suspension.

EXPERIMENTAL MANIPULATIONS

Zooplankton

Evaluate the effects of several of the following variables on the clearance and grazing rates by either or both of the food removal techniques. Use control flasks without zooplankton to determine changes in food levels by natural growth and reproduction.

1. Differences in concentrations of zooplankton, e.g., replicates of zooplankton, at 2, 4, 8, 16, and 32 zooplankters/l.
2. Differences in concentrations of food particles, e.g., 1000, 10,000 and 100,000 cells/ml.
3. Different species of zooplankton, e.g., two or more species of the same genus such as *Daphnia*; comparison of a mixture of rotifers, cladocerans, and/or copepods (separation can be done manually at the end of the experiments prior to radioassay).
4. Different food particles (separate controls for each type):
   a. Different species of algae of similar size.
   b. Different-sized algae and/or bacteria of similar geometry, e.g., spherical.
   c. Differently shaped algae, such as a green alga versus cyanobacterial species of filamentous or colonial morphology.
   d. Dissect out the guts of several zooplankters and examine them microscopically to ascertain different apparent digestion rates of different algae.
5. Varying temperatures, e.g., 10°C indicative of hypolimnetic conditions and 25°C similar to summer epilimnetic temperatures.
6. Clearance rate under bright light versus that under total darkness.
7. Using the radioactive food method, allow introduced zooplankters to feed only 5 to 7 min. Rapidly remove the animals by sieving. Transfer about half of the live animals to nonradioactive food of the same type and concentration and allow them to feed for 15 min. Transfer the other half to scintillation vials (enumerate these animals). After allowing them to feed for 15 min on the nonradioactive food, remove by sieving the live animals that have now cleared their digestive tracks of radioactive material and radioassay (enumerate these organisms). Compare the amount of radioactivity that was assimilated with the amount ingested.
8. Graph all data and compare the results. Answer the following questions as related to your results.

Questions

1. What does the rate of ingestion of food measured by these methods tell about the assimilation of food materials? Why?
2. If the "assimilation rate" is determined by procedure 7 above, is this a true rate of assimilation? What assumptions are made?
3. Among zooplankton that migrate vertically, such as many species of *Daphnia*, clearance rates have been observed to increase (double or more) at night when they migrate toward the epilimnion. Discuss in terms of (a) differences in food

concentrations, (b) differences in food quality at different times of the day, (c) differences in temperature, (d) varying predation pressure at night versus daylight conditions (see Exercise 23).

4. How might the concentrations of dissolved oxygen in a stratified lake affect the rate of clearance? [See Haney (1973).]

5. How might you design an apparatus to measure the rates of clearance and grazing in situ with natural phytoplankton and zooplankton populations at several depths in a vertical profile? [See Haney (1971).]

6. Compare your observed clearance rates at different food and zooplankton concentrations to the observed levels of both in one or more lakes or ponds studied in Exercises 10 and 11. Assuming that the zooplankton were effective in their grazing, what impact would you anticipate they would have on phytoplankton or bacterial growth? Phytoplankton species succession? Why?

7. Zooplankton feed on particles. During maceration and digestion of food particles, much of the food is lost as fragments, leaching of soluble organic matter, and incompletely digested food material. What is the fate of this partially utilized food in the lake ecosystem? Describe in detail.

8. How might inorganic nutrients (such as phosphorus) of the food, released during the ingestion process, affect the phytoplankton and/or bacterioplankton?

9. How effective is the grazing by rotifers compared to that of cladocerans? Copepods to cladocerans?

10. How would you expect clearance rates of the same species to change with increasing body length? Increasing environmental temperatures?

11. How would you anticipate clearance efficiency to change with size of food particles? With increasing food concentrations? [See Lehman (1976).]

PROTISTAN FEEDING RATES

Rates of feeding by protists, such as ciliates and heterotrophic flagellates, on bacteria or small phytoplankton in natural waters may be estimated by evaluating the average rate of particle ingestion by specific components of the protists in a water sample, and then multiplying the cell-specific ingestion rate by the total number of protists in each component (size fractions or identified morphological types). Fluorescently labeled prey particles can be used, either bacteria or algae, or inert plastic microspheres coated with protein as analogs of live prey (Pace and Bailiff, 1987), in order to determine cell-specific uptake rates. The use of inert plastic microspheres allows short-term, cell-specific uptake rates to be determined by quantifying the average number of particles within protistan cells over a time course of 10 to 60 minutes. Because algae and bacteria are digested by the protists, their rate of disappearance can be followed over longer times, e.g., 12 to 24h or more. The numerous advantages and disadvantages of each approach are discussed by Sherr and Sherr (1993). The methods employ an epifluorescent microscope outfitted with both ultraviolet light (DAPI) and blue light (AO, FITC, DTAF, chlorophyll autofluorescence) filter sets.

Ciliate Feeding

1. Obtain a bacterial enrichment by adding filtered lake water and a grain of rice to a culture flask. After a few days of growth, inoculate this culture with ciliated protozoa and allow several additional days for growth of the ciliates.

2. Pour the culture through Nitex netting of 30- to 100-μm mesh in accordance with the size of the protozoa. This procedure simply removes large particles and clumps of bacteria.

3. Take a preliminary bacterial sample and enumerate the microorganisms (see Exercise 19). Make a fluorescent microsphere standard by adding a drop of the stock solution of microspheres (recommended sizes between 0.5 and 1.0 μm) to 10 ml of filtered lake water. Sonicate the standard to disperse the microspheres, if possible, and take a subsample for enumerating by the same procedure as was used for bacteria except omitting the use of acridine orange stain. Add fluorescent microspheres from the standard to the culture at a concentration of about 10% of the bacterial density (Pace and Bailiff, 1987).

4. Immediately take a 5- to 10-ml subsample and preserve with formalin solution to yield a final concentration of 1 to 2%.

5. Continue taking subsamples at 10-min intervals for 1 h. Keep careful track of the time of the experiment and the actual time that the subsamples were killed to stop feeding.

6. Filter subsamples with gentle vacuum (<100 mm Hg) through 2- or 5-μm pore size Irgalan black-stained polycarbonate filters. Sandwich each filter between drops of nonfluorescent immersion oil (e.g., Cargille Type A) and cover with a cover slip.

7. Examine the filters under blue light with an epifluorescent microscope. Ciliates will autofluoresce a light apple-green color. Shift to high power and enumerate the number of microspheres per ciliate. Count 50 to 100 cells on each filter.

8. Plot the number of microspheres per ciliate versus time.

9. Calculate the feeding rate from the slope of the line for microspheres ingested versus time multiplied by the ratio of bacterial cells to fluorescent microspheres.

Heteroflagellate Feeding

1. Harvest bacteria and/or small (<10 μm) phytoplankton during log-phase growth by centrifugation (e.g., 25 ml in 50-ml tube at 22,000 $\times$ g for 10 to 20 min for bacteria, 800 $\times$ g for 10 min for phytoplankton).

2. Suspend pellets in 10 ml phosphate buffer, pH 9. Add 2 mg of yellow-green fluorescing dye DTAF [5-(4,6-dichlorotriazin-2-yl) aminofluorescein] that binds to protein. Incubate in 60°C water bath for 2 h. Centrifuge, decant the DTAF solution, and wash and centrifuge three times with phosphate buffer.

3. Resuspend cells in 20 ml of buffer, vortex, and distribute in 1- or 2-ml aliquots in plastic vials. Store frozen.

4. Determine natural concentrations of bacteria and small algae in natural samples collected from the field in order to determine amount of fluorescently labeled particles to add. The range recommended (Sherr and Sherr, 1993) is ca. 5% of total becterioplankton for ciliate uptake (but not less than $1 \times 10^5 \, ml^{-1}$) and ca. 30% of total bacterioplankton for flagellate uptake (but not less than $1 \times 10^6 \, ml^{-1}$). The volume of the diluted stock solution added should be less than 1% of the total volume of the sample.

5. Add the particles and gently create a uniform suspension. At selected time intervals, withdraw 10- or 20-ml subsamples and preserve with alkaline Lugol's solution to achieve a 0.5% final concentration followed immediately by 2% borate buffered formalin and a drop of 3% sodium thiosulfate to clear the Lugol iodine color.

6. Filter subsamples separately onto 0.2-μm pore size plain membrane filters and enumerate the fluorescently labeled particles to determine the amount added.
7. Stain interval subsamples with concentrated DAPI (1 mg/ml; see Exercise 19), filter onto 0.2-μm pore size black membrane filters, and examine via epifluorescent microscopy. Locate protistan cells with the UV filter set, and then switch to the blue light filter set to locate and enumerate ingested particles within cells.
8. Determine ingested particles per protistan cell over time. Divide the particle uptake rate (particles/cell/h) by the concentration of particles per volume to obtain an estimate of the hourly per cell clearance rate of protists for the type of prey. Multiply the per cell clearance rates by the total number of protists per ml to obtain an assemblage clearance rate.

Questions

1. The rate of ingestion of microspheres is typically linear for the first several time points and then declines. Why?
2. What is the purpose of the time zero sample? Were any microspheres "ingested" at time zero?
3. How would you test the idea that the fluorescent microspheres provide an accurate estimate of feeding on bacteria?
4. Calculate the clearance rate from your estimate of ingestion. How does this rate compare with that measured for the large zooplankton in the previous experiments? Now divide the clearance rates of the zooplankter and the ciliate by their respective weights or volumes. Weights can be estimated from size by assuming that the animals conform to simple geometric shapes (see Exercise 10) and then by converting volume estimates to weight assuming a specific gravity of 1. The estimate of size corrected feeding rate is called the specific rate. Which rate is greater and what does this imply about the relative significance of a unit of protozoan biomass versus a unit of zooplankton biomass?

Apparatus and Supplies

Zooplankton

1. Cultures of algal or bacterial food supplies.
   a. Cultures of spherical algae of a size between 10 and 20 μm (e.g., *Chlamydomonas*) and between 30 and 50 μm.
   b. Cultures of large algae, preferably a cyanobacterium such as *Anabaena*.
   c. If available, cultures of bacteria of large size (e.g., *Pseudomonas*) or yeast (e.g., *Rhodotorula*).
2. Media for the culturing of algae [e.g., Stein (1973)] and microflora [e.g., Rodina (1972)].
3. Flasks or beakers, ca. 500 ml; graduated cylinders for volume measurements.
4. Zooplankton, preferably several species in pure or nearly pure cultures.
5. Counting cells to evaluate cell numbers of food organisms (see Exercise 10).
6. Algae or bacteria labeled with $H^{14}CO_3$ or $^{32}PO_4$.
7. Safety apparatus for handling isotopes and labeled materials (e.g., waterproofed absorbent pads, gloves, cleaning agents, and disposal facilities).
8. Scintillation vials. Accurate volumetric pipe (e.g., Eppendorf) for small volumes. Appropriate biological solubilizers (e.g., Beckman NCS or BTS-450) and water-

accepting scintillation reagents (e.g., Packard Instagel). Scintillation radioassay systems.

9. Membrane filtration apparatus for filtration of labeled food sources for subsequent radioassay (see Exercise 14).

10. Fine-mesh sieves (cylinders of plastic about 10cm in diameter and 5cm in length, to which Nitex netting is glued over one end, work very well). The mesh size of the netting should be sufficiently small to retain all zooplankton and yet pass all food particles (see Exercise 11).

Protistans

1. Cultures of bacteria and ciliates. Monospecific cultures of bacterial-feeding ciliates such as *Paramecium* and *Cyclidium* can be obtained from culture supply houses. Alternatively ciliates can be cultured from hay infusions.

2. Fluorescently labeled microspheres impregnated with a fluorescent stain (e.g., Polysciences, Inc., Warrington, PA). Microspheres 0.5-μm YG (yellow-green fluorescence) in size are recommended as an analog for suspended bacteria, and 1.0-, 2.0-, 3.0-, and 6.0-μm diameter labeled particles for small size classes of phytoplankton.

3. Disposable culture tubes for subsamples.

4. Nitex netting, 30- to 100-μm mesh size in relation to protozoan size.

5. Filtration apparatus with vacuum source.

6. Irgalan black-stained polycarbonate filters: 0.2-μm pore size for making preparations for counting bacteria and microspheres, and 2.0- or 5.0-μm pore size for making preparations for counting microspheres ingested by ciliates.

7. Epifluorescent microscope with appropriate filter sets for UV and blue illumination (see Exercises 10 and 19).

8. Formalin and Lugol's solutions.

9. Sonicator (optional).

10. Microscope slides, coverslips, and nonfluorescent immersion oil.

References

Frost, B.W. 1972. Effects of size and concentration of food particles on the feeding behavior of the marine planktonic copepod *Calanus pacificus*. Limnol. Oceanogr. *17*:805–815.

Haney, J.F. 1971. An *in situ* method for the measurement of zooplankton grazing rates. Limnol. Oceanogr. *16*:970–977.

Haney, J.F. 1973. An *in situ* examination of the grazing activities of natural zooplankton communities. Arch. Hydrobiol. *72*:87–132.

Lehman, J.T. 1976. The filter-feeder as an optimal forager, and the predicted shapes of feeding curves. Limnol. Oceanogr. *21*:501–516.

McGregor, D.L. and R.G. Wetzel. 1968. Self-absorption of ^{14}C radiation in freshwater ostracods. Ecology *49*:352–355.

Omori, M. and T. Ikeda. 1984. Methods in marine zooplankton ecology. Wiley, New York.

Pace, M.L. and M.D. Bailiff. 1987. Evaluation of a fluorescent microsphere technique for measuring grazing rates of phagotrophic microorganisms. Mar. Ecol. Progr. Ser. *40*:185–193.

Peters, R.H. 1984. Methods for the study of feeding, filtering and assimilation of zooplankton. pp. 336–412. *In*: J.A. Downing and F.H. Rigler, Editors. A Manual for the Assessment of Secondary Production in Fresh Waters. 2nd Ed. Blackwell, Oxford.

Porter, K.G. 1977. The plant-animal interface in freshwater ecosystems. Amer. Sci. *65*:159–170.

Rigler, F.H. 1971. Feeding rates: Zooplankton. pp. 228–255. *In*: W.T. Edmondson and G.G. Winberg, Editors. A Manual on Methods for the Assessment of Secondary Productivity in Fresh Waters. IBP Handbook 17. Blackwell, Oxford.

Rodina, A.G. 1972. Methods in Aquatic Microbiology. (Translated and revised by R.R. Colwell and M.S. Zambruski.) University Park Press, Baltimore. 461 pp.

Sherr, E.B. and B.F. Sherr. 1993. Protistan grazing rates via uptake of fluorescently labeled prey. pp. 695–701. *In*: P.F. Kemp, B.F. Sherr, E.B. Sherr, and J.J. Cole, Editors. Handbook of Methods in Aquatic Microbial Ecology. Lewis Publishers, Boca Raton.

Stein, J.R. (ed). 1973. Handbook of Phycological Methods. Culture Methods and Growth Measurements. Cambridge Univ. Press, Cambridge. 448 pp.

Zooplankton Production

Most zooplankton, and some benthic animals, reproduce continuously. As a population changes by addition and growth over a given time interval, a demographic turnover occurs [see reviews of Edmondson (1974) and Rigler and Downing (1984)]. Few, if any, of the individuals present at the peak of an exponentially developing population were alive at the beginning of the exponential phase. An individual also undergoes a biochemical turnover during its lifetime so that, upon completing a mean lifespan, it will have assimilated several times its final mass.

Production, in the context of a population, then, is growth and is only one factor in the material or energy budget for the whole population (Edmondson, 1974):

$$P = C - F - U - R - G_r$$

where P = production; C = consumption or feeding; F = egestion (feces and regurgitation); U = excretion; R = respiration; and G_r = gametes.

Assimilation (A) is the difference between ingestion and egestion ($A = C - F$).

With continuous reproduction, the cohorts of the population overlap, so that it is either difficult or impossible to observe changes in abundance over time. To analyze the production of populations with continuous reproduction, it is necessary to use methods that do not require complete evaluation of cohort differences.

A number of models of zooplankton production have been developed, but they fall into two general classes: (1) direct models based on time-dependent parameters of the zooplankton species [e.g., Edmondson and Winberg (1971) and Rigler and Downing (1984)], and (2) indirect models based on inferred rates of zooplankton filtering, assimilation, and consumption by fish and other predators [e.g., Winberg (1971)]. Both discrete time-interval and instantaneous models are used to estimate production.

The change in numbers (N) in a population over a time (t) interval is given by:

$$\frac{\Delta N}{\Delta t} = \text{birth} + \text{growth} - \text{mortality}$$

The population at time t, then, is

$$N_t = N_0 + \text{birth} + \text{growth} - \text{mortality}$$

Mortality includes both natural death and losses by predation.

Since it is necessary to measure recruitment or birth rate, and with continuous birth and death one cannot identify distinct cohorts, production (P) can be calculated from either the rate of growth (i.e., the duration) of different size classes or of life history stages (Edmondson, 1974):

$$P = \frac{N_1 \Delta w_1}{T_1} + \frac{N_2 \Delta w_2}{T_2} + \frac{N_3 \Delta w_3}{T_3} + \dots \frac{N_n \Delta w_n}{T_n}$$

where the subscripts $1,2,3 \dots n$ = stages or size classes; w = the weight of the class; T = the duration of the stage, in days; and Δw = the weight increment of the class (birth + growth − mortality).

This relationship implies that, when a stage lasts two days, one-half of the individuals will pass from that stage on one day and the other half on the following day, i.e., the number of individuals in a given stage will be inversely proportional to the duration of that stage. It is assumed that the animals have a uniform age distribution, which is not always the case.

The objective of this exercise is to construct a realistic, logical model of zooplankton production. The model then can be used to estimate the production of a sample zooplankton species over a given time interval. By this means, the future population size of the species can be predicted on the basis of its present population structure and size, observed size-specific production of eggs, and certain assumed or known information on biomass and survival. This procedure will be demonstrated using *Daphnia*. For precise results, detailed information on birth, growth rates, and mortality would be required for each species under investigation.

PROCEDURES

1. From a large reservoir of a very well-mixed, nearly monospecific population of *Daphnia*, collect a sample of known volume (approximately 1 l). Pass the sample through a stacked series of screens of known mesh size to separate the zooplankton into four or more size categories [e.g., large (>2.0 mm in length), medium (<2.0 to >1.5 mm), small (<1.5 to >1.0 mm), and very small (<1.0 mm)]. Carefully wash all zooplankton from the screens into labeled petri dishes. Each student or pair of students should obtain size-fractionated samples in this manner.

2. The more live animals, the better the results that are obtained; disregard any dead (inactive) animals. When time does not permit immediate analysis, preserve with a cold (ca. 6°C) mixture of 40 g/l sucrose and 4% formalin to prevent carapace distortion and loss of eggs (Haney and Hall, 1973; Prepas, 1978).

3. Enumerate the *Daphnia* in each of the size categories. When your sample contains more than 100 individuals in a size group, count the number of *Daphnia* in several subsamples (see Exercise 11). Calculate the number per liter in each size category for the total sample volume.

4. Determine the average length of the carapace (head to base of anal spine) of each size category and estimate the biomass by:

$$\text{Dry wt(mg)} = (0.0052)(\text{mm})^{3.012}$$

Alternatively, use the estimates given in Table 16.1.

5. Count the number of eggs per female in each size category; primarily the larger *Daphnia* will be carrying eggs (Table 16.2).

CALCULATIONS AND CONSTRUCTION OF A POPULATION MODEL

1. Knowing the biomass of your total population at $t = 0$, predict the net change in biomass of this population of *Daphnia* at $t = 14$ days. Estimates of natality (birth rate) (B), growth (G), and mortality (death rate) (D) are required. These population parameters are functions of many environmental and physiological factors, e.g., age, food availability and quality, temperature, predation, and other variables. While age is not determined easily in zooplankton, size and age are related closely. For purposes of this exercise, we can assume that all influencing factors,

Table 16.1. Length-weight, age structure, and survival rates of two species of *Daphnia* at 20°C.[a]

Size category	*D. pulex*		*D. magna*	Median age of category from hatching (birth) (days)	Survival rate per day (%)
	Length (mm)	Dry weight (μg)[b]	Dry weight (μg)		
Very large	—	—	300		
Large	2.13	50.7	100	14	50
Medium	1.85	33.2	36	10	60
Small	1.25	10.2	18	6	80
Very small	0.91	4.0	—	3	95
Eggs	—	3	3	2[c]	60

[a] From D.J. Hall (personal communication).
[b] Dry weight (mg) = $(0.0052)(\text{mm})^{3.012}$.
[c] I.e., development time in days to hatching.

Table 16.2. Matrix of biomass in size classes versus days of population development (numbers × biomass per size class).

Time (days)	Very large	Large	Medium	Very small	Eggs (in very large)	Eggs (in large)	Eggs (in medium)	Biomass (mg)
0								
1								
2								
3								
4								
5								
6								
7								
8								
9								
10								
11								
12								
13								
14								
Total								Total

except age, are constant, and that B, G, and D are related to the size of the animals.

2. Calculate natality (B) per female per day for each size class by:

$$B = \frac{E}{D}$$

where E = number of eggs/female and D = development time (given as 2 days in Table 16.1).

3. Calculate the initial biomass ($t = 0$) for each size category and for the total population sample. Estimate the size-specific mortalities for each size category on the basis of survival rates given in Table 16.1.

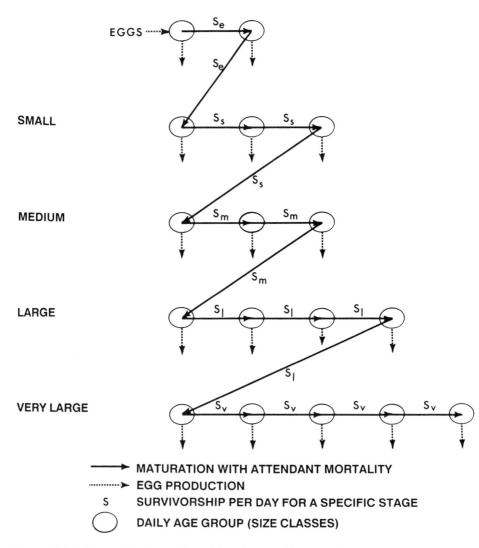

Figure 16.1. Discrete time interval model pathways with assumed directions of survivorship and mortality. S_e = egg survivorship; S_s = small survivorship; S_m = medium survivorship; S_l = large survivorship; and S_v = very large survivorship per day.

4. Using your data on the biomass of your total *Daphnia* population at $t = 0$, predict the net change in biomass of this population of *Daphnia* at $t = 14$ days (see Fig. 16.1). Use the calculations for a discrete time interval model, as discussed above, with the assumptions made below. Then,

$$\text{Number in the size class at } t+1 = [(\text{survival rate})(\text{number in size class at } t \text{ or in egg class})]$$

$$
\begin{array}{ccc}
-(\text{number that} & & (\text{number of survivors} \\
\text{advances to the} & + & \text{recruited in from} \\
\text{next size class}) & & \text{smaller size class})
\end{array}
$$

$$
\begin{array}{ccc}
\text{Biomass for size} & & (\text{numbers in size class} \qquad (\text{biomass of that} \\
\text{class at } t+1 & = & \text{at a given day}) \qquad \times \qquad \text{size class})
\end{array}
$$

Then the sum of all biomass of the different size classes = zooplankton biomass on day_n.

Assumptions for the model are
a. Includes the biomass (dry weight) lost as mortality.
b. Assumes a constant temperature of 20°C.
c. Assumes constant mortality and egg production and no emigration of immigration throughout the period.
d. Assumes one-half of eggs found were 0 to 1 day old and one-half were 1 to 2 days old.
e. Uses one-day time intervals.
f. Assumes that the eggs produced on a given day occurred before mortality occurred.

Then, from your model, make the following calculations:
a. For each day calculate the egg development in selected females, mortality of the day class, hatched eggs (i.e., 2 days old), and advance the *Daphnia* through each day class.
b. Determine the biomass at $t = 14$ days. Add the cumulative mortality to the t_{14} biomass for total production.
c. Determine the net change in biomass $t_{14} - t_0$.
d. Graph biomass (mg) and mortality (mg/day) versus time in days over the 14-day period.

QUESTIONS

1. We assume that your population consists only of females. Why? What is the typical life cycle of *Daphnia*?
2. Survival rates decrease with increasing age (e.g., Table 16.1). What would you expect to be dominant causes of mortality in differing age classes?
3. How might survival rates change for a population of *Daphnia* in the open water versus one in a littoral zone of dense emergent macrophytes?
4. Using a time interval larger than 1 day would simplify computations in a discrete model. What would result? How realistic would such production values be?
5. What would be the potential effects of daily vertical migration of *Daphnia* in a thermally stratified lake on:
   a. Growth rates?
   b. Predation?
   c. Production values calculated by a discrete time interval model?
6. Was your population expanding or contracting by $t = 14$ days? Why?

7. Was the rate of increase (or decrease) in production occurring at a constant rate? Why or why not?
8. What would be the effect of an increase in fecundity on population production? A decrease in mortality?
9. Is a discrete time interval model more accurate than an instantaneous model? Why? [See Edmondson and Winberg (1971).]
10. What is the effect of biasing the egg count through faulty sampling or enumeration? [See Likens and Gilbert (1970).]
11. Among rotifers, production has been calculated by the doubling time method:

$$P = \frac{(\overline{N})(\overline{w})}{T_{e+p}}$$

where $\overline{N}$ = average numbers/time interval; $\overline{w}$ = average weight increment; and T_{e+p} = development time from hatching to hatching.

How does this method differ from that outlined in this exercise? What is assumed?

APPARATUS AND SUPPLIES

1. Culture of nearly monospecific *Daphnia*. (A large fiberglass or similar tank containing 200 to 500 l of water may be inoculated with a *Daphnia* population collected from a particular layer of a stratified lake several weeks in advance of this exercise. The *Daphnia* may be fed adequately by daily additions of a water slurry of dog food.)
2. Screening sieves of differing mesh sizes (Plexiglas cylinders ca. 10 cm in diameter and 6 cm high with the differing sizes of Nitex mesh glued over one end work very well).
3. Dissecting microscopes.
4. Petri dishes, probes (insect needles with heads removed, inserted into wooden match sticks), and mechanical counters.

References

Edmondson, W.T. 1974. Secondary production. Mitt. Int. Ver. Limnol. *20*:229–272.

Edmondson, W.T. and G.G. Winberg (eds). 1971. A Manual on Methods for the Assessment of Secondary Productivity in Fresh Waters. IBP Handbook 17. Blackwell, Oxford. 358 pp.

Haney, J.F. and D.J. Hall. 1973. Sugar-coated *Daphnia*: A preservation technique for Cladocera. Limnol. Oceanogr. *18*:331–333.

Likens, G.E. and J.J. Gilbert. 1970. Notes on quantitative sampling of natural populations of planktonic rotifers. Limnol. Oceanogr. *15*:816–820.

Prepas, E. 1978. Sugar-frosted *Daphnia*: An improved fixation technique for Cladocera. Limnol. Oceanogr. *23*:557–559.

Rigler, F.H. and J.A. Downing. 1984. The calculation of secondary productivity. pp. 19–58. *In*: J.A. Downing, and F.H. Rigler, Editors. A Manual on Methods for the Assessment of Secondary Productivity in Fresh Waters. 2nd Ed. Blackwell, Oxford.

Winberg, G.G. (ed.) 1972. Methods for the Estimation of Production of Aquatic Animals. Academic Press, New York. 175 pp.

Predator–Prey Interactions

Predator–prey interactions have been among the most intensively studied areas of aquatic biology during the past several decades. Investigations have focused particularly on theories of "optimal foraging," which seeks to describe predator behavior [e.g., Charnov (1976), Werner and Hall (1974), and Pyke (1984)], and "predator mediated community structure" (Hrbacek, 1962; Brooks and Dodson, 1965; Hall et al., 1976; Zaret, 1980; Werner and Gilliam, 1984; Sih et al., 1985; Kerfoot and Sih, 1987; Lampert and Sommer, 1997), which interprets community structure in relation to predatory activities. Many of these hypotheses still are speculative, although supporting evidence for some is growing. These concepts form a useful basis for the study of predator–prey relationships. The literature on this subject is extremely large; a few summary articles relative to limnology are cited in this exercise.

In evaluating predator–prey interactions, both the predators and prey have physiological and behavioral characteristics that must be considered. Major *predator* characteristics that can be evaluated include: (1) visual, mechanical, or chemical detection of prey; (2) how much energy is required during searching for and attacking prey; and (3) energy and time expended in handling, and total or partial consumption of prey. Evaluation of the ways in which *prey* respond to predation requires considerations of: (1) behavior and energy expended in escape responses, often by refuge in space and/or time; and (2) means by which prey adapt and coexist with predators. The latter coexistence can be accomplished by camouflage, differences in size, release of repulsive chemical compounds, frightening displays, evasive movements, morphological structures that impede or prevent consumption, and aggregation in large groups.

Predators and prey respond continuously to each other's adaptations, which results in constant, although slow, coevolution and changing interactions. The extent of interactive couplings can be evaluated by exposing prey to predators under controlled experimental conditions [e.g., Thompson (1978) and Werner (1974)] or by the introduction of new predators or prey into established ecosystems [e.g., Langeland (1981) and Morgan et al. (1978)]. The latter approach should not be done without considerable forethought and understanding of ecosystem properties. The former experimental approach will be used in this exercise to gain insight into some basic predator–prey interactions.

OPTION 1

Procedures

The objective of this exercise is to compare the feeding characteristics of two predators, a common fish (bluegill, *Lepomis macrochirus*) and a dragonfly nymph

(Corduliid odonate, *Tetragoneuria* spp.) under different conditions of light and prey densities. The prey will be *Daphnia magna*, a relatively large zooplanktonic cladoceran.

Effects of Light on Predator Success

Procedures

1. Place one bluegill and one odonate into separate containers filled with equal volumes of water at room temperature. Allow the organisms to acclimate for 15 min without disturbance (such as loud noise or startling moves), particularly around the fish. These predators have been starved and should feed readily.
2. Select 30 *Daphnia* of similar size from the available culture. Carefully add 15 *Daphnia*, one at a time with a wide-bore pipet, to each container and observe the feeding for 3 min. Remove the predators and count the prey remaining in each container. Record the number of prey consumed by each predator. Place each fed predator in a designated repository to avoid accidental reuse in subsequent experiments.
3. Repeat steps 1 and 2 above, but place the containers into total darkness immediately after adding the prey to each. Speed is essential when adding prey and placing the containers into darkness and when removing the predators after the 3-min feeding period.
4. Determine instantaneous prey mortality rates (M) for prey under different conditions, where:

$$M = (\ln N_0 - \ln N_1)/t$$

and N_0 = initial prey density, N_1 = final density of live prey, and t = time units (Dodson, 1975). The units of M are prey mortality per predator per time, usually a day.

Questions

1. Based on your observations and pooled class data on the feeding in light, compare predation by the fish and by the odonate.
   a. How does each predator detect prey?
   b. How does the prey organism detect the predator? What defenses do the prey appear to have?
   c. What predation strategy (searching or ambushing) is employed by each predator?
   d. What energy considerations are associated with each strategy?
   e. How are the prey consumed by each predator, and how does this affect feeding?
   f. What additional information would be useful in making a more complete evaluation of the effects of light on these predators?
   g. What environmental (habitat) factors favor each predator?
   h. What sources of error are possible in these experiments? How would they affect the results?
2. Why should exotic predators or prey not be introduced into natural ecosystems without "considerable forethought"?

The Effect of Prey Density on Predator Success, A Functional Response

Procedures

1. Place one bluegill into each of three containers and an odonate in each of three additional containers. The containers should be filled with an equal volume (same as in the above) of water at room temperature before adding the predators.
2. Select 320 *Daphnia* of similar size from the available culture. Separate these prey organisms into two sets of 30 prey, two sets of 50 prey, and two sets of 80 prey.
3. Pair the six containers with predators such that each of the three pairs has a fish and an odonate. Add 30 prey to each container of one pair, 50 to each container of another pair, and 80 to each container of the third pair. All trials should not be attempted simultaneously. Each laboratory team should be responsible for one or two prey densities, with each predator tested separately. While feeding is in progress, carefully observe the behavior of each predator. Allow 3 min for feeding, remove the predators, and count the number of prey remaining. At high prey densities, record predator strikes and successful captures; confusion of the predator can occur at high prey densities with an increase in strikes but a decline in capture, particularly by the fish. Record the number of prey consumed by each predator at each prey density.
4. Graph the relationship between prey density and prey consumed for each predator type. Include the results of the trials done in the light (15 prey per container), above, with the data from this part.
5. Optional experimental variations:
   a. Feed predators live and dead prey and observe the differences.
   b. Compare fish attacks on clear test tubes containing different sizes of *Daphnia* prey.
   c. Place small red and blue dots on a flask or the aquaria and observe the fish feeding behavior.
   d. Try repeating fish predation experiments with different backgrounds (paper or boxes) around beakers, or while shining differently colored lights on the beakers.
   e. Place the *Daphnia* in one of several suspensions of food color for ca. 15 min and repeat fish predation experiments. What effect does color have?

Questions

1. Evaluate the plots of feeding data for each predator.
   a. What does the shape of each curve suggest about the predator?
   b. How do the curves compare?
   c. What factors appear to be responsible for the observed responses?
   d. How reliable are these data for extrapolation to natural environments?
2. Compare the results statistically with a *t*-test:

$$t = \frac{\bar{x}_1 - \bar{x}_2}{\sqrt{(n_1 - 1)s_1^2 + (n_2 - 1)s_2^2}} \sqrt{\frac{n_1 n_2 (n_1 + n_2 - 2)}{n_1 + n_2}}$$

where x_1 = initial arithmetic mean; x_2 = final arithmetic mean; n_1 = initial number of sampling units; n_2 = final number of sampling units; and s_1^2 and s_2^2 = initial and final variance, i.e., mean of the squares of the deviations (see Appendix 2).

Comparison[a]	t value	Probability of obtaining t value by chance
B_{15L} versus B_{15D}		
O_{15L} versus O_{15D}		
B_{15L} versus O_{15L}		
B_{15D} versus O_{15D}		
etc.		

[a] B = bluegill; O = odonate; L = light; D = dark.

OPTION 2

With the data provided in Table 17.1, perform all of the analyses discussed in Option 1 and answer the questions.

EQUIPMENT AND SUPPLIES

1. Beakers, 50-ml and 500-ml.
2. Wide-bore pipets.

Table 17.1. Predation of a bluegill sunfish and a dragonfly nymph on *Daphnia*.

	Prey density per 500 ml									
	Bluegill as predator					Odonate as predator				
Trial	15L	15D	30L	50L	80L	15L	15D	30L	50L	80L
1	10	0	22	20	9	8	7	4	12	6
2	5	5	20	39	10	4	1	2	4	16
3	15	0	19	38	3	2	3	0	8	7
4	13	6	23	15	12	9	7	9	23	3
5	11	1	29	16	3	4	9	9	10	19
6	14	0	0	16	15	5	2	12	4	7
7	15	10				0	9			
8	15	3				0	5			
9	13	4				2	12			
10	4	8				4	5			
11	14	11				2	5			
12	15	4				3	5			
13	15	4				6	1			
14	12	1				5	6			
15	7	5				0	9			
16	3	4				9	8			
17	11	5				2	1			
18	0	1				0	0			
19	15	14				3	4			
20	2	1				7	6			
Mean	10.5	4.4	18.8	24.0	8.7	3.8	5.3	6.0	10.2	9.7
S	5.0	3.4	9.9	4.9	4.8	2.9	3.3	4.7	7.1	6.3
CV (%)	48	77	53	20	55	76	62	78	70	65
SE	1.1	0.8	4.0	2.0	2.0	0.6	0.7	1.9	2.9	2.6

3. Finger bowls.
4. Test tubes with caps.
5. Wide, flexible forceps.
6. Fine meshed net, <80-μm mesh size.
7. Stream or nonchlorinated well water.
8. Dispensing flasks, 4-1.
9. Stopwatch.
10. Nitex net/filter apparatus to separate predators from prey at the end of the feeding period and for counting remaining prey.
11. Darkroom or darkened chamber.
12. Organisms:
    a. Bluegill, or equivalent, of similar size.
    b. Odonate nymphs (e.g., *Tetragoneuria* or *Anax*).
    c. *Daphnia* (intermediate to large size).

References

Arts, M.T., E.J. Maly, and M. Pasitschniak. 1981. The influence of *Acilius* (Dytiscidae) predation on *Daphnia* in a small pond. Limnol. Oceanogr. *26*:1172–1175.

Brooks, J.L. 1968. The effects of prey size selection by lake planktivores. Syst. Zool. *17*:273–291.

Brooks, J.L. and S.I. Dodson. 1965. Predation, body size, and composition of plankton. Science *150*:28–35.

Charnov, E.L. 1976. Optimal foraging theory: The marginal value theorem. Theor. Pop. Biol. *9*:129–136.

Clepper, H. and R.H. Stroud, (eds). 1979. Predator-Prey Systems in Fisheries Management. Sport Fishing Institute, Washington, DC.

Dodson, S.I. 1975. Predation rates of zooplankton in arctic ponds. Limnol. Oceanogr. *20*:426–433.

Hall, J.D., S.T. Threlkeld, C.W. Burns, and P.H. Crowley. 1976. The size–efficiency hypothesis and the size structure of zooplankton communities. Ann. Rev. Ecol. Syst. 7:177–208.

Hassell, M.P. 1977. Sigmoid functional responses by invertebrate predators and parasitoids. J. Anim. Ecol. *46*:249–262.

Holling C.S. 1965a. The functional response of predators to prey density, and its role in mimicry and population regulation. Mem. Ent. Soc. Can. *45*:1–60.

Holling, C.S. 1965b. The functional response of invertebrate predators to prey density. Mem. Ent. Soc. Can. *48*:1–86.

Hrbacek, J. 1962. Species composition and the amount of the zooplankton in relation to the fish stock. Rozpravy Ceskosl. Akad. Ved. Rada Matem. Prir. Ved. *72*:1–114.

Iwasa, Y., et al. 1981. Prey distribution as a factor determining the choice of optimal foraging strategy. Amer. Nat. *117*:710–723.

Jacobs, J. 1965. Significance of morphology and physiology of *Daphnia* for its survival in predator-prey experiments. Naturwissenschaften *52*:141.

Kerfoot, W.C. and A. Sih (eds.) 1987. Predation: Direct and Indirect Impacts on Aquatic Communities. Univ. Press of New England, Hanover, NH. 386 pp.

Lampert, W. and U. Sommer. 1997. Limnoecology: The Ecology of Lakes and Streams. Oxford Univ. Press, New York. 382 pp.

Langeland, A. 1981. Decreased zooplankton density in two Norwegian lakes caused by predation of recently introduced *Mysis relicta*. Verh. Int. Verein. Limnol. *21*:926–937.

Morgan, M.D., S.T. Threlked, and C.R. Goldman. 1978. Impact of the introduction of Kokanee (*Oncorphynchus nerka*) and the oppossum shrimp (*Mysis relicta*) on a subalpine lake. J. Fish. Res. Bd. Can. *35*:1572–1579.

Murdoch, W.W. 1973. The functional response of predators. J. Appl. Ecol. *10*:335–342.

Pyke, G.H. 1984. Optimal foraging theory: A critical review. Ann. Rev. Ecol. Syst. *15*:523–575.

Oaten, A. and W.W. Murdoch. 1975. Switching, functional response and stability in predator-prey systems. Amer. Nat. *109*:299–318.

Schoener, T.W. 1971. Theory of feeding strategies. Ann. Rev. Ecol. Syst. *2*:369–404.

Sih, A., P. Crowley, M. McPeek, J. Petranka, and K. Strohmeier. 1985. Predation, competition, and prey communities: A review of field experiments. Ann. Rev. Ecol. Syst. *16*:269–311.

Taylor, R.J. 1981. Ambush predation as a destabilizing influence upon prey populations. Amer. Nat. *118*:102–109.

Thompson, D.J. 1978. Towards a realistic predator-prey model: The effects of temperature on the functional response and life history of larvae of the damselfy *Ishnura elegans*. J. Anim. Ecol. *47*:757–767.

Werner, E.E. 1974. The fish size, prey size handling time relation in several sunfishes and some implications. J Fish. Res. Bd. Can. *31*:1531–1536.

Werner, E.E. and D.J. Hall. 1974. Optimal foraging and the size selection of prey by the bluegill sunfish (*Lepomis macrochirus*). Ecology *55*:1042–1052.

Werner, E.E. and J.F. Gilliam. 1984. The ontogenetic niche and species interactions in size-structured populations. Ann. Rev. Ecol. Syst. *15*:393–425.

Zaret, T.M. 1980. Predation and Freshwater Communities. Yale Univ. Press, New Haven, CT.

Enumeration of Fish or Other Aquatic Animals

There are various ways of estimating the size of a natural population of large, mobile animals in a freshwater habitat. Obviously, the most accurate method would involve catching or, in some way, counting the entire population. However, this approach is usually either impossible to do in practice or, at least, destructive to the natural populations. Instead, a mark-and-recapture procedure frequently is used to obtain a statistical estimate of the size of the natural population.

The mark-and-recapture technique is based on the premise that recognizable (marked) organisms released to the population will be recaught in numbers proportional to their abundance in that population. The size of the natural population can be estimated from the proportion of marked to unmarked organisms in random samples obtained from the entire population.

The basis for this technique apparently was first published in 1662 by John Graunt in an article on human demography in London (Ricker, 1975). Later, Petersen (1896) applied the method to fish populations, and others have used it in various ways since, e.g., the Lincoln (1930) index for birds. According to this procedure:

$$\hat{N} = \frac{SM}{R}$$

where $\hat{N}$ = estimate of total number in the population; S = total number of organisms in a sample from the population; M = total number of marked organisms in the population; and R = number of marked organisms in the sample.

However, certain basic assumptions must be made for the mark-and-recapture technique to be valid:

1. There can be no difference in mortality or emigration between marked and unmarked organisms.
2. Tags or other marks must remain recognizable and must not be lost. All marks on recaptures must be reported.
3. There must not be a difference in catchability between marked and unmarked organisms.
4. Marked organisms must be mixed randomly within the entire population.
5. There can be no unknown recruitment or immigration to the population.

There are many modifications of the basic Petersen relationship [cf., Ricker (1975)], but only the two commonly used for fish or other aquatic vertebrates will be presented here.

THE SCHNABEL METHOD

The Schnabel method weights individual Petersen averages throughout the study period:

$$\hat{N} = \frac{\sum SM}{\sum R}$$

Note that marked organisms must be returned to the population for possible recapture a second time, or more. In addition to all assumptions applying to the Petersen method, this method assumes no mortality. Table 18.1 may be useful in executing the Schnabel method.

SCHUMACHER-ESCHMEYER METHOD

If a population were to be sampled indefinitely using a mark-and-recapture technique, eventually all of the fish in the population would become marked. Keeping this in mind, and using the proportion of fish marked in a sample as an estimate of the proportion of fish marked in the population, the total number of fish in a population (i.e., the number of fish marked when 100% of the population is marked) can be estimated (after several trials). This approach is done by constructing a Schumacher-Eschmeyer plot (Fig. 18.1). Note that, in this plot, any factor that changes the slope of the regression line will change the population estimate. Thus, an increase in slope will result in a *lower* population estimate, while a decrease in slope will result in a *higher* estimate. In the case of "trap-happy" animals, a greater proportion of marked animals are seen in the sample. This problem would *increase* the slope of the line and thus *decrease* the population estimate.

RELIABILITY OF POPULATION ESTIMATES

The question may be raised as to what proportion of the population must be marked before a reliable estimate of the total can be obtained. One criterion that can be used is the point at which running estimates (or the Petersen and Schnabel types)

Table 18.1. A useful table for recording results from the Schnabel method.

Reference, trial number, and date	Total number captured (S)	Number marked and returned	Total number marked in population (M^a)	$(S)(M)$	$\Sigma(SM)$	Number of marked recaptures (R)	$\Sigma(R)$	$\dfrac{\sum(SM)}{\sum(R)} = \hat{N}$

[a] At time of sample.

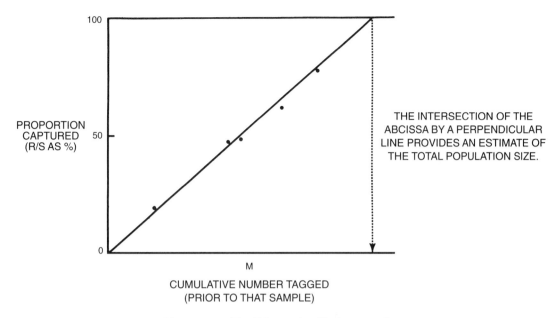

Figure 18.1. The Schumacher-Eschmeyer plot.

appear to stabilize around some particular value. Two methods can be used to determine the location of this point:

1. Plot the estimates versus trial number. In those trials in which the estimate appears to stabilize (Fig. 18.2), the proportion marked in the population can be estimated as the proportion marked in the samples (R/S).
2. Plot the estimate versus the proportion marked in the sample (Fig. 18.3). Again, the desired proportion of the population marked to obtain a stable estimate can be approximated from the proportion of the sample that is marked.

CONFIDENCE INTERVALS ABOUT THE ESTIMATED POPULATION SIZE

Petersen Method

If a sample of predetermined size were taken from a population in which some individuals have been marked, the confidence interval would be calculated by first calculating the variance $[\hat{V}(\hat{N})]$ of the population estimate $[\hat{N}]$:

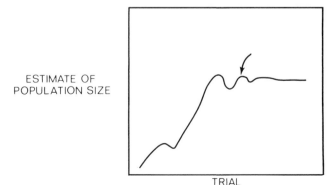

Figure 18.2. Population estimates versus trial number.

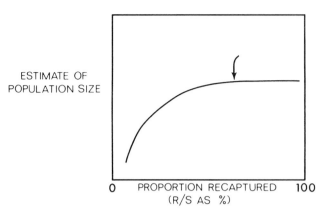

Figure 18.3. Population size versus proportion marked in the sample.

ESTIMATE OF
POPULATION SIZE

0   PROPORTION RECAPTURED   100
            (R/S AS %)

$$\hat{V}(\hat{N}) = \hat{N}^2 = \frac{(\hat{N} - M)(\hat{N} - S)}{(M)(S)(\hat{N} - 1)}$$

The confidence interval on the estimate of N is

$$\hat{N} \pm t\sqrt{\hat{V}(\hat{N})}$$

where t is the value of the t statistic at the appropriate level of significance. For large samples (>120), the t value at the 95% level of probability equals 1.96. For the 99% level, it is 2.58. The Petersen estimate tends to average less than the true population size. This bias is most pronounced with small sample sizes but drops to more acceptable levels when $(M)(S) > 4(\hat{N})$. Charts are available for determining the necessary sample size to obtain a desired level of accuracy (Everhart et al., 1975).

Schnabel Method

The confidence interval for the Schnabel estimate follows a similar procedure, except that the reciprocal of N is used for all calculations because this value is more normally distributed. The variance is

$$\hat{V}\left(\frac{1}{\hat{N}}\right) = \frac{\sum R}{\left(\sum S \cdot M\right)^2}$$

The reciprocal confidence interval is

$$\frac{1}{\hat{N}} \pm t\sqrt{\hat{V}(1/\hat{N})}$$

and the regular confidence interval must be calculated from:

$$\hat{N} \pm \frac{\hat{N}}{1 \pm \hat{N}t\sqrt{V\left(\frac{1}{\hat{N}}\right)}}$$

Notice that the use of the reciprocal for estimation of confidence intervals produces an interval unbalanced about the mean. This compensates for the fact that the Schnabel estimate requires "sampling with replacement," that is, recaptured fish are returned to the population for possible recapture a second time or more.

EXERCISES

OPTION 1

Select a small pond, reservoir, or stream with a readily observable population of fish or salamanders. On several consecutive sampling days (at least four or five), collect as many individuals as possible, count and mark them, and then return them to the lake or stream. If this were a class project, various groups would be responsible for sampling on different days, but be certain that identical procedures are followed.

1. A variety of methods have been used to mark fish and other animals so that individuals would be recognizable at a later date. The most common method of marking fish or salamanders in short-term studies is by mutilation of a fin or other appendage. Clipping part of a fin or foot provides an easily recognizable mark, but regeneration usually obliterates the mark within a few months. Alternatively, small holes (or other shapes) can be punched into the opercula or fins of some fish and are identified easily during short-term studies. Branding, tags, tattooing, radioisotopes, and other markings all have been used to identify organisms in mark-and-recapture studies. See Everhart et al. (1975) for additional information on marking fish.
2. For each sampling date or trial, record the following information:
   a. Total number of individuals captured.
   b. Number of marked individuals captured.
   c. Number of marked individuals returned to habitat.
   d. Time spend in collection.
   e. Type of equipment used.
   f. Location of sample area.
   g. General observations.
3. Estimate the population of individuals by:
   a. The Petersen method.
   b. The Schnabel method.
   c. The Schumacher-Eschmeyer method.
4. Calculate the confidence interval for estimates made by both the Petersen and the Schnabel methods.

OPTION 2

This option can be done as a demonstration or laboratory exercise.

Procedure: Work in Pairs!

1. Set up a data sheet with headings according to the Schnabel method.
2. Place a few handfuls of dried beans (the "fish") in a can (the "lake").
3. With the eyes closed to facilitate taking a random sample, remove several seeds with your hand.
4. Count the seeds; mark each with a waterproof marking pen; record the numbers in the proper places on the data sheet (Table 18.1); and return the seeds to the can and mix thoroughly by stirring or shaking vigorously.

5. With eyes closed again, take another random sample; count; record the marked and unmarked ones; mark the unmarked ones and return them all to the can; mix; and calculate the total number of seeds in the can.
6. Repeat step 5 until estimates begin to approximate one another.
7. When estimates become more or less constant, verify the last estimate obtained (presumably the best of the series) by actually counting all of the seeds in the can. While counting, separate the marked and unmarked seeds in order to verify your previous counts of the marked ones. Discard the marked beans and return all of the seeds to the original container.
8. Use the data from the last sample for a final estimate of the total population with the Petersen method.
9. Use the data from the series for an estimate by the Schumacher-Eschmeyer method.
10. Calculate the confidence interval for estimates made by both the Petersen and the Schnabel methods.

OPTION 3

Using the data provided in Table 18.2, estimate the population of fish in Farce Lake by the:

1. Petersen method.
2. Schnabel method.
3. Schumacher-Eschmeyer method.

Calculate the confidence interval for estimates made by the Petersen and the Schnabel methods.

Problems and Questions

1. Estimate the number of "fish" by the Petersen method at each step of the Schnabel method. Plot the "running estimates" for each method.
2. State the error of the estimate in terms of percent and explain the deviations by relating them to the requirements for studies of a fish population by the mark-and-recapture method described above.
3. Basing your conclusion on the results of all three methods, what percentage of the total population must be marked to obtain valid population estimates?
4. Propose means by which the mark-and-recapture method could be verified.
5. Summarize your information on the usefulness and limitations of these methods for estimating the population size of aquatic animals.

Table 18.2. Population data collected from Farce Lake during July 1960.

Date 1960	Number of fish caught	Number of marked fish recaptured	Total number of marked fish returned to lake
July 3	30	0	30
July 4	13	1	6
July 5	13	3	13
July 6	68	9	29
July 7	41	19	9
July 8	47	10	27
July 9	20	4	17
July 10	51	22	21

Apparatus and Supplies

Field

1. Nets or other collecting equipment. (Don't forget to obtain a permit from the State Fish and Game Department or other appropriate authority!)
2. Anesthesia. A variety of anesthetics have been used to facilitate the handling of fish and other organisms [see McFarland (1960) and McFarland and Klontz (1969) for reviews of this subject]. MS-222 (methane tricainesulfonate) is used widely and is effective at a concentration of about 0.025 g/4l of water.
3. Surgical scissors, sturdy paper punch, or tags.
4. Boots, boats, and ropes.

Laboratory

1. Dried beans or peas.
2. Small containers (coffee cans or plastic freezer containers with lids).
3. Waterproof markers.
4. Hand tally.
5. Graph paper.

References

Bennett, G.W. 1962. Management of Artificial Lakes and Ponds. Reinhold, New York.

Carlander, K.D. 1953. Handbook of Freshwater Fishery Biology. William C. Brown, Dubuque, IA. 429 pp.

Carlander, K.D. 1977. Handbook of Freshwater Fishery Biology. Vol. 2. Iowa State Univ. Press, Ames, IA. 431 pp.

Cooper, G.P. 1952. Estimation of fish populations in Michigan lakes. Trans. Amer. Fish. Soc. *81*:4–16.

Everhart, W.H., A.W. Eipper, and W.D. Youngs. 1975. Principles of Fishery Science. Cornell Univ. Press, Ithaca, NY. 288 pp.

Lagler, K.F. 1956. Freshwater Fishery Biology. 2nd Ed. William C. Brown, Dubuque, IA. 421 pp.

Lincoln, F.C. 1930. Calculating waterfowl abundance on the basis of banding returns. U.S.D.A. Circ. *118*:1–4.

McFarland, W.N. 1960. The use of anesthetics for the handling and the transport of fishes. California Fish and Game *46*:407–431.

McFarland, W.N. and G.W. Klontz. 1969. Anesthesia in fishes. Federation Proc. *28*:1535–1540.

Petersen, C.G.J. 1896. The yearly immigration of young plaice into the Limfjord from the German Sea, etc. Rep. Dan. Biol. Sta. *6*:1–48. [As cited in Ricker (1975).]

Ricker, W.E. 1968. Methods for Assessment of Fish Production in Fresh Waters. Blackwell Scientific Publ., Oxford. 313 pp.

Ricker, W.E. 1975. Computation and interpretation of biological statistics of fish populations. Bull. Fish. Res. Board Can. *191*. 382 pp.

Rounsefell, G.A. and H.W. Everhart. 1953. Fishery Science: Its Methods and Applications. Wiley, New York. 444 pp.

Schnabel, Z.E. 1938. The estimation of the total fish populations of a lake. Amer. Math. Monthly *45*:348–352.

Southwood, T.R.E. 1978. Ecological Methods: With Particular Reference to the Study of Insect Populations. 2nd Ed. Chapman and Hall, London. 524 pp.

Bacterial Growth and Productivity

Earlier discussions emphasized that dead organic matter, called detritus, exists as a spectrum from dissolved organic compounds, organic colloids, and larger particles of organic matter. Dissolved organic matter is in much greater abundance, by about five to ten times, than is particulate organic matter. All microflora must degrade particulate organic matter enzymatically to the dissolved form prior to assimilation for further metabolic breakdown and eventual mineralization to inorganic solutes and gases.

The rates of growth by bacteria and their productivity are known poorly from aquatic ecosystems. Much of this lack of progress resulted from the slow application of laboratory methodology to heterogeneous natural communities of bacteria. Often methods were developed for isolated strains at high laboratory concentrations. A number of methods have been developed recently that permit examination of growth and productivity of in situ communities of bacteria. These methods are evolving as more is learned about metabolic constraints and variations.

Natural variance is high in microbial communities, and many organisms are metabolically inactive or dormant. The methods used to examine bacteria are labor intensive. We describe methods here in considerable detail because they require rigorous application in order to obtain meaningful results. Conversion of rates of biomass change or genetic replication to actual rates of carbon flux in bacterial secondary productivity is difficult and is undergoing intensive study by many researchers. A spectrum of conversion factors is given for different environmental conditions in order to indicate the range of probable in situ values.

We stress the importance of these evaluations of in situ metabolism and growth rates of bacteria. Much of the understanding about biogeochemical cycling in aquatic ecosystem has been limited severely by poor or inadequate techniques for evaluation of bacterial metabolism and its limitations by environmental parameters. The methods presented here are not perfect, and they are laborious. Certainly they will improve with time. However, these methods offer means to examine in situ rates of bacterial growth and productivity and, importantly, means to evaluate in situ effects of experimental manipulations of environmental conditions.

BACTERIAL ENUMERATION AND BIOMASS DETERMINATION

Evaluation of the species composition and relative numbers of heterotrophic bacteria by direct enumeration or by plating on culture media may give useful information about the heterotrophic potential of the microbial community but yields little insight into their in situ rate processes. All plating media are highly selective, and none will give "total viable counts"[cf. reviews of Caldwell (1977), Austin (1988), and numerous papers in Kemp, et al. (1993)]. Direct counting of aquatic bacteria has been greatly facilitated by staining the bacteria with fluorescent dyes, filtration onto polycarbonate (Nuclepore) filters stained black with irgalan dye, and examination with epi-illuminated fluorescence microscopy (Francisco et al., 1973; Hobbie et al., 1977; Bowden, 1977; Bührer, 1977).

Direct counting of bacteria and other natural microflora is considered the most reliable method for the evaluation of community dynamics [e.g., Daley (1979) and Sorokin and Kadota (1972)]. Estimates of biomass from cell numbers and average cell dimensions improve further the estimates of microbial dynamics, particularly when coupled to direct measures of community bacterial productivity. The ensuing exercise demonstrates one approach to bacterial community analyses by enumeration. Populations, cell size, DNA, and biomass distributions of pelagic bacteria can be obtained by flow cytometry from cell frequency, forward scatter, and fluorescence intensity of cells stained with dyes specific for nucleic acids (Button and Robertson, 1993). Among the many advantages of these methods are the speed with which large numbers can be evaluated with resulting much greater statistical accuracy than can be obtained by manual enumeration. The instrumentation required is expensive and not routinely available.

Detection, enumeration, and size evaluations of small bacteria and cyanophytes are difficult. The binding of organic dyes to the cells, however, can produce fluorescence when excited with light of appropriate wavelengths (Bratbak, 1993). As a result, cells can be differentiated from other particles at the lower limits of detection for light microscopy.

Numerous fluorescing dyes are available. Acridine orange binds with DNA and RNA of living cells. When excited with light (436 or 490 nm), the DNA complex fluoresces green and the RNA complex red. Unfortunately, nonliving particles may also become stained and fluoresce. In waters containing much suspended seston, other fluorescing stains can be used for epifluorescent enumerations. For example, DAPI [4'6-diamidino-2-phenylindole] is specific for DNA of cells (Porter and Feig, 1980). When excited with light at 365 nm, the DNA-DAPI complex fluoresces blue at or above 390 nm, while DAPI associated with non-DNA material may fluoresce a weak yellow.

Procedures

1. Obtain water samples with a clean Van Dorn or similar nonmetallic water sampler from several strata within a lake or stream ecosystem. Flush the sample bottles well. Preserve to achieve a 5% glutaraldehyde solution. Record the initial and final volumes to determine the dilution factor. Keep the samples on ice and in the dark.

2. Set up the filtration unit with a large-pore Millipore filter on the fritted glass base. Mount the black Nuclepore (see "Apparatus and Supplies," p. 284) of 0.2-μm pore size on the other filter and assemble the funnel unit. All filters should be moistened with distilled water.

3. Filter an appropriate water sample (e.g., 0.3 to 3 ml from productive waters; 5 ml from an oligotrophic water) with a low (<0.3 atm) vacuum. Release the vacuum immediately as filtration is completed. Rinse the funnel walls briefly with sterile, filtered (0.2-μm) distilled water.

4. Add 2.0 ml 0.01% acridine orange stain (2 ml stock plus 18 ml filtered water) to the funnel and expose for 2 min. (DAPI at a concentration of 50 μg/ml would be exposed similarly at 2 ml for 2 or more min.) After staining, draw the stain through the filter with modest vacuum and rinse with 2 to 3 ml of filtered sterile distilled water. Alternatively, the stain (0.2 ml of 0.05 to 0.10% acridine orange per 2-ml sample) can be added to the initial sample and exposed for 2 min before filtration and rinsing.

5. Remove the Nuclepore filter, blot for 30 sec on a cellulose acetate filter (e.g., Gelman), and place flat onto a drop of nonfluorescent immersion oil on a clean microscope slide.

6. Allow one drop of immersion oil *to drop onto* the center of the filter. Cover with a cover glass. After a brief period, press slightly to flatten and expel the excess oil. Retain the slide mount in the dark until examination.

7. With an appropriately matched epifluorescent compound microscope (see "Apparatus and Supplies"), observe samples in a semidarkened room. *Caution:* Do not look directly at the ultraviolet light sources of the microscope system. Use oil immersion at 1250 ×. Count the bacterial cells with a Whipple eyepiece grid (ca. 71 × 71 μm). An appropriate density of organisms filtered onto the filter would yield 10 to 30 organisms per grid field. Count 10 to 15 fields per slide to achieve at least 300 cells for each sample.

8. Calculate bacterial densities:

$$\text{Bacteria/ml} = \frac{(\text{membrane conversion factor} \times N)}{D}$$

where N = average number of bacteria per micrometer field by

$$\frac{(\text{total number bacteria enumerated})}{(\text{number of micrometer fields counted})}$$

D = dilution factor.

$$\text{Membrane conversion factor} = \frac{\text{filtration area} *}{\text{area of micrometer field}}$$

9. Using microscopic techniques discussed earlier (p. 153ff), determine the average length and width of the filamentous and rod cells, and the diameter of the coccoid cells. Categorize your counts into these groups and estimate biomass from simple geometric shapes (see pp. 161–162).

Alternatively, the volume of the bacteria can be calculated by (Bratbak, 1993):

$$V = (\pi/4)\, W^2 (L - W/3)$$

where V = volume, L = cell length, and W = cell width. Although this formula is based on the assumption that the bacteria are straight rods with hemispherical ends, it applies equally well for cocci.

As a further alternative, apply an allometric relationship of biomass to volume described by a power function: $m = CV^a$, where m = biomass, V = volume, C = conversion factor between biomass and volume for unity volume where

---

* Determined by using the internal diameter of a filtration funnel or the optical field as determined with a stage micrometer.

$V = 1$, and a = a scaling factor of 0.72 which suggests that smaller cells tend to have a higher biomass-to-volume ratio than larger ones (Norland, 1993).

10. Convert bacterial biovolumes to organic carbon. Carbon comprises approximately 50% of dry weight among bacteria. Carbon content per cell volume changes as cell sizes change (Psenner, 1990), but assume a carbon-to-volume ratio of $0.35 \, pg/\mu m^3$.

11. Graph the vertical distributions of bacterial numbers, biovolume, and carbon content against depth in the lake.

NUCLEIC ACID SYNTHESIS AS A MEASURE OF BACTERIAL GROWTH RATES

A number of methods have attempted to measure the biomass and metabolic activity of bacteria [e.g., van Es and Meyer-Reil (1983)]. The best expression of bacterial activity, however, is the rate of cell division. The method described here evaluates heterotrophic bacterial growth rates in the natural environment by measuring the rates of synthesis of nucleic acids, in particular DNA.

The principle of measurement of growth rates by synthesis of nucleic acids is the determination of the rate of incorporation of a radioactively labeled precursor into a macromolecule such as DNA. Thymidine generally is efficiently transported across the cell membrane and converted to thymidine monophosphate by the enzyme thymidine kinase (Moriarty, 1986). Cyanobacteria, eukaryotic algae, and fungi generally lack thymidine kinase of the transport mechanism needed to assimilate thymidine. This characteristic is important in that it restricts, in short incubation times, uptake to direct DNA synthesis by the heterotrophic bacteria. Adenine and adenine nucleotides, in contrast, are less frequently used as measures of bacterial growth rates.

In practice, natural bacterial communities are incubated at in situ conditions with a known quantity and specific activity of tritiated thymidine for a brief period (<45 min). The macromolecules then are separated chemically to evaluate synthesis into different constituents. The measures of changes in DNA then are converted to rates of bacterial productivity in carbon units.

Laboratory Preparation for Field Operations: Stock ^{3}H-Thymidine Solution

1. Disperse 1.05 ml of the purchased [methyl-^{3}H]thymidine (usually stored in 2% ethanol) solution into *clean* 2-ml ampoules. Seal the ampoules immediately; do *not* autoclave (Robarts and Zohary, 1993). Label well and store and 4°C.

2 On the day prior to the situ assays:
   a. Freeze one of the stock ampoules prepared above.
   b. Open the ampoule and carefully place the vial in a centrifuge tube; fasten the filter paper over the tube mouth with a rubber band. Place into a beaker and similarly cover with a filter paper. Place into a third container and similarly cover with a filter paper.
   c. Lyophilize according to freeze-dryer procedures to remove solvents (e.g., ethanol) and tritiated water. Thymidine should not be stored in the dried state for any appreciable time because this condition can accelerate self-decomposition.
   d. Remove the filter paper covers carefully with plastic gloves, as they may be slightly contaminated with radioactivity.

e. Add 4.2 ml of ultrapure water and transfer to a clean 5-ml ampoule (working stock solution).

f. Remove a 100-μl sample for calibration (see below).

g. Immediately seal the 5-ml ampoule of working stock solution and autoclave 15 min (SLOW EXHAUST). Store well labeled at 4°C.

3. Calibration procedure (needed for each series):

a. Remove a 100-μl sample from an ampoule containing the working stock solution and add it to 5.00 ml ultrapure water. After thorough mixing, remove two 100-μl aliquots and add to two 900-μl aliquots of water in liquid scintillation (LSC) vials. Prepare a third LSC vial with 1.00 ml of pure water for later internal standardization.

b. Add 9.0 ml of LSC cocktail for aqueous solution (e.g., Instagel or Safety-Solve) to each vial. Radioassay.

Field Procedures

1. The assay is better accomplished with two persons, and both are needed when inoculation and fixation are to be completed within 1 min.

2. Fill three 60-ml sample bottles with water from each depth (see "Apparatus and Supplies," p. 275ff), stopper, and place into sample bottle holders of a dedicated field box.

3 After all depths have been sampled and the bottles filled:

a. Remove 2 ml of water from each bottle.

b. Preserve one bottle from each depth (killed blank) with 1.6-ml 37% formaldehyde.

c. Wearing plastic gloves, and working with your partner to keep order, add 0.20 ml ^{3}H-thymidine to each of the active experimental bottles and then to the formaldehyde-fixed blank bottles to avoid contaminating the active bottles. Record the beginning and ending times of the injections to the nearest minute.

d. Immediately incubate the samples at the depths from which they were collected in situ. The incubation should be at least 30 min but should not exceed 45 min.

e. After the incubation period, retrieve and place the bottles in the box.

f. Wearing plastic gloves, inoculate each active bottle with 1.6 ml of 37% formaldehyde. Record the time to nearest minute.

Laboratory Treatment and Assay of Field Samples

1. Place the sample bottles on ice until the analyses are undertaken.

2. Soak the required number of filters in ultrapure water in a petri dish.

3. Wearing plastic gloves, use a 30-ml disposable plastic syringe with ca. 18 cm of 3-mm diameter Tygon tubing fitted to the syringe end to remove the samples.

a. Withdraw 5 ml of the sample and rinse the syringe; discard.

b. Withdraw another 5-ml sample and use it to remove the air bubbles from the syringe and tubing. Discard.

c. Withdraw exactly 20 ml and dispense into a labeled test tube. Discard the remainder in tubing.

d. Repeat steps a to c for all of the other samples.

e. Change contaminated gloves.

f. With an automatic transfer pipet, pipet 10 ml of ice-cold 15% TCA into each of the first 12 tubes. Mix thoroughly (vortex). Allow 10 min of exposure to TCA.

g. Prepare a 12-port filtration apparatus.

h. After 10 min of exposure, vortex each sample and filter at a pressure differential of 1 atm. Each sample will fill the well of the Millipore manifold filtration apparatus twice. Some wells filter more rapidly than others; stopper wells that have completed filtration while waiting for the others.

i. Allow the well to empty prior to rinsing. Rinse each tube with 5 ml of ice-cold 5% TCA; vortex and pour into well.

j. Add 5 ml of ice-cold 80% ethanol and filter in order to remove thymidine taken into the cells but not incorporated into the DNA (Robarts and Wicks, 1989).

k. Rinse each filter four additional times: (1) Add 1 ml of 5% TCA and filter; (2) repeat the first step twice; and (3) add 1 ml of 50% ethanol/algal mixture for marking the filtration area on the filters when the waters are oligotrophic and require visualization (see "Apparatus and Supplies," p. 283).

l. Carefully slide the filter off the filtering unit with forceps and place it in a labeled vial with the filtered surface up. Filter the remainder of the samples as above.

m. Excise the filter margins with a cork borer of the same diameter as the filtered area by placing the filters on cardboard. Remove the contaminated edge and discard it in radioactive waste. Place the filter into a vial with the filtered side up. Clean the cork borer and forceps (70% ethanol, or Isoclean, rinse, and dry) between sample cuttings.

n. Freeze the vials at −20°C.

o. Prepare five calibration vials as follows:
   i. One (1.00) ml of unfiltered water from the well-mixed surface thymidine sample bottle, plus 10 ml of scintillation cocktail for the liquid samples.
   ii. As above, but from a replicate sample bottle from the same depth.
   iii. As above, but from a replicate blank sample bottle from the same depth.
   iv. One (1.00) ml of formalin-fixed lake water plus 10 ml of scintillation cocktail.
   v. Internal standard vial with 1.0-ml formalin-fixed surface water plus 10 ml of scintillation cocktail. ^{3}H-hexadecane standard to be added later.

p. Waste disposal and decontamination:
   i. Rinse the bottles with deionized water. Soak the bottles and stoppers for 2 days minimally in 2 to 4% alkaline detergent (e.g., isoclean) plus distilled water.
   ii. Dispose of filtered waste in waste containers that can contain acids.

Acid Extraction of Thymidine Uptake Samples and Radioassay

1. Each sample is separated into two fractions: a perchloric acid soluble fraction for DNA and the residue on the filter. Samples can be processed in groups of ca. 18, which is a common capacity of a centrifuge rotor.

2. Perchloric acid soluble fraction:
   a. Prepare and label a new liquid scintillation (LS) cap for each sample. Start centrifuge refrigeration (4°C).
   b. Add 50 μl protein +DNA carrier solution (see "Apparatus and Supplies") to each filter in LS vials.

  c. Add 6.0 ml of 0.5 M $HClO_4$.
  d. Place into 70°C water bath and extract for 45 min; swirl to mix about every 15 min.
  e. Chill on ice.
  f. Decant liquid to labeled centrifuge tube and cap. Centrifuge for 20 min at 4°C and 10,000 rpm.
  g. Carefully pipette 4.50 ml of supernatant to an LS vial with a labeled cap. Add 4.50 ml of appropriate scintillant (e.g., Packard Instagel), cap, and mix thoroughly. Keep undisturbed for 12 to 24 h before radioassay.
  h. Include two background LS vials of 4.50 ml 0.5 M $HClO_4$ plus 4.50 ml scintillant.
  i. Clean-up: Vortex the centrifuge tubes and decant suspension into a separate perchloric waste container. Add a small amount of distilled water to each, vortex, and decant. Wash the tubes thoroughly before reusing.
3. Residue on filter:
  a. Make up two background samples with filters only.
  b. Add 1.0 ml of 1 N HCl and digest in an oven incubator at 90°C for 5 h. Retighten the caps after 1 h to correct for expansion and loosening.
  c. Add 2.0 ml of ethylacetate and wait several hours for dissolution of filters. Dissolution is accelerated by affixing the sample tray to a rotary shaker (ca. 175 rpm).
  d. Add 10.0 ml of colloidal scintillant (e.g., Safety-Solve) to each LS vial and shake thoroughly. After several hours, shake well again.
4. Radioassay:
  a. As the vials are organized for counting, wipe each externally with 70% ethanol. Use plastic gloves.
  b. Radioassay with an appropriate LS spectrometer with a standard tritium program (e.g., 20 min or 2%).
  c. Select six vials that have differing quench numbers for internal standardization (samples with >1500 cpm are preferred). Add an appropriate amount of a standard (e.g., ^{3}H-hexadecane or ^3H_2O) and radioassay (ca. 10 min or to high statistical efficiency, 99%).

Bacterial Productivity Calculations

1. Convert counts per minute (cpm) to disintegration per minute (dpm):

$$dpm = \frac{cpm - background\ cpm}{\%\ efficiency\ of\ radioassay}(100)$$

The efficiency is determined from standards of known activity within the LS spectrometer capabilities. Follow the instructions of the manufacturer.

2. The estimate of in situ rates of bacterial community productivity is calculated from the rates of thymidine incorporation:

$$\frac{moles\ thymidine}{(1)\ \ (h)}$$

$$= \frac{(dpm_{sample} - dpm_{blank})(ml\ per\ 1)(min\ per\ h)\left(\begin{array}{c} volume\ factor\ for \\ formalin\ added,\ 1.03 \end{array}\right)}{(dpm\ per\ Ci)\left(\begin{array}{c} specific\ activity \\ in\ Ci/mmol \end{array}\right)\left(\begin{array}{c} volume\ filtered, \\ ml \end{array}\right)\left(\begin{array}{c} incubation \\ time,\ min \end{array}\right)\left(\begin{array}{c} mmol \\ per\ mole \end{array}\right)}$$

3. Calculate pmol/l/h for the two fractions:
   a. Nucleic acids (DNA) of perchloric acid soluble fraction.
   b. Total thymidine incorporation, consisting of the nucleic acid fraction plus the residue on the filter that may be incorporated into proteins.
   Determine the percentage that each fraction is of the whole.
4. Plot the thymidine uptake (pmol/l/l) against depth in the lake.
5. The estimation of bacterial productivity or growth rates from thymidine uptake requires an estimation of the number of cells produced per amount of thymidine incorporation. Then, knowing the mean cell biomass from the cell volume of the bacteria from direct measurements and the average amount of cellular carbon, the amount of carbon produced for the measured thymidine uptake can be calculated:

$$\mu g\,C/liter/hour = (moles\ thymidine/l/h)(cells/mole)(carbon/cell)$$

Estimate bacterial secondary productivity in carbon units by the following factors:
   a. Approximately 2.0×10^{18} cells are produced for each mol of thymidine uptake based on empirical measurements (cf., Table 19.1). These empirically based conversion factors are considerably lower than the theoretical values [e.g., between 2.3×10^{17} and 8×10^{17} cells per mol (Moriarty, 1989; Coveney and Wetzel, 1988)].
   b. The mean cell volume of bacteria varies greatly in the range of 0.025 to 0.08 μm^3. Use your measured values or a value of $0.04\,\mu m^3$.
   c. The amount of carbon per cell varies considerably, and smaller cells generally have more carbon per unit biovolume. Empirically determined values are often higher than theoretically derived values (Bratbak and Dundas, 1984; Bratbak, 1985, 1993). A value of $2.2 \times 10^{-13}\,g\,C\,\mu m^{-3}$ is recommended for estimating the carbon content of bacteria in which the biovolume was estimated from living cells. A value of $3.5 \times 10^{-13}\,g\,C\,\mu m^{-3}$ [= $0.35\,pg\,C/\mu m^3$] is recommended to estimate the carbon content of bacteria where biovolume was estimated microscopically with preserved bacterial cells. An estimate of bacterial

Table 19.1. Empirically derived conversion factors for calculation of production of bacterial cells from [³H]thymidine incorporation.[a]

Conversion factor Range	Mean	System	Source
	1.4	Marine	Fuhrman and Azam (1982)
1.9–8.9	4.4	Salt marsh	Kirchman et al. (1982)
3.0–5.9		Freshwater pond	Kirchman et al. (1982)
1.9–2.2	2.0	Eutrophic lake	Bell et al. (1983)
5.8–8.7	6.9	Freshwater swamp	Murray and Hodson (1985)
1.6–7.3	2.2	Eutrophic lakes	Lovell and Konopka (1985)
2.8–6.2	4.0	Ocean	Ducklow and Hill (1985)
2.7–5.5		Eutrophic lakes	Riemann and Søndergaard (1985)
0.3–2.7	1.1	Coastal marine	Riemann et al. (1987)
5.0–25[b]		Lake Michigan	Scavia and Laird (1987)
1.7–4.0[b]	2.5	Estuary	Autio (1990)
—	2.0	Marine (median of 97)	Ducklow and Carlson (1992)

[a] Most factors are based on ³H incorporation into total macromolecules. Units are 10^{18} cells per mole thymidine.
From Moriarty (1989) and Coveney and Wetzel (1989).
[b] Range does not include some higher values rejected by the authors.

productivity using this example then might be (2.0×10^{-18}) (0.04) (3.5×10^{-13}) or $2.8 \times 10^4 \, gC$ per mol thymidine uptake.

BACTERIAL PRODUCTION OF PROTEIN

The method just discussed for estimating bacterial productivity by DNA synthesis determined by incorporation of tritiated thymidine measures cell multiplication. The rate of cell multiplication then must be converted into bacterial carbon production from a knowledge of the carbon content of the growing bacteria in natural communities. As was explained above, the content of cell carbon is difficult to determine. Bacterial production could be estimated also in terms of a cellular component that constitutes a large fraction of the bacterial biomass and is proportional to changes in cellular carbon and dry weight. Because the protein of the nuclear materials comprises a large fraction (ca. 60%) of the biomass of planktonic bacteria, the rate of protein synthesis could be used as a measure of bacterial biomass and carbon production.

Bacterial protein production has been estimated on the basis of the rates of tritiated leucine incorporation into bacterial protein (Kirchman et al., 1985, 1986; Kirchman, 1993). When added at nM concentrations, leucine is assimilated nearly exclusively by bacteria in the plankton. The method has been examined in detail and evaluated as a means for estimating directly bacterial carbon production (Simon and Azam, 1989). The latter workers determined the necessary parameters, such as intracellular dilution of the isotope by de novo synthesis of leucine, for converting the synthesis rate of ^{3}H-leucine incorporation into protein. The bacterial protein production technique was compared to the method of ^{3}H-thymidine incorporation into DNA; the two methods yielded comparable rates of bacterial production (Simon and Azam, 1989). The bacterial protein production method was, however, about ten times more sensitive than the thymidine method and yielded bacterial carbon production directly without the need to know the cell size of the bacterial community in the growth state.

Field Procedures

1. Fill three 60-ml sample bottles with water from each depth, stopper, and place into sample bottle holders of a dedicated field box.
2. After all of the depths have been sampled and the bottles filled:
    a. Remove 2 ml of water from each bottle.
    b. Preserve one bottle from each depth (killed blank) with 1.6 ml of 37% formaldehyde.
    c. Wearing plastic gloves, and working with your partner to keep order, add 0.20 ml of $[3,4,5^{-3}H]$-1-leucine to achieve ca. 10 nM final concentration to each of the active experimental bottles and then to the formaldehyde-fixed blank bottles to avoid contaminating the active bottles. Record the beginning and ending times of the injections to the nearest minute.
    d. Immediately incubate the samples at the depths from which they were collected in situ. Incubate for at least 30 min but do not exceed 60 min.
    e. After the incubation period, retrieve the bottles and place them in the box.
    f. Wearing gloves, inoculate each active bottle with 1.6 ml of 37% formaldehyde. Record time to the nearest minute.

Laboratory Treatment and Assay of Field Samples

1. Soak the required number of filters in ultrapure water in a petri dish.
2. Wearing protective gloves, use a 30-ml disposable plastic syringe with ca. 18 cm of 3-mm diameter Tygon tubing fitted to the syringe end to remove the samples.
   a. Withdraw 5 ml of the sample and rinse the syringe; then discard it.
   b. Withdraw another 5-ml sample and use it to remove air bubbles from the syringe and tubing; then discard it.
   c. Withdraw exactly 20 ml and dispense into a labeled test tube. Discard the remainder from the tubing.
   d. Repeat steps a to c for all of the samples.
   e. Change the contaminated gloves and discard them in an appropriate manner.
   f. With an automatic transfer pipet, pipet 6.5 ml of 15% TCA into each test tube to expose the bacteria to a 5% TCA solution. Mix thoroughly (vortex).
   g. Immediately lower the tubes in an appropriate rack into a water bath of 80–95°C; extract for 15–30 min.
   h. Prepare a 12-port filtration apparatus.
   i. After a 30-min exposure, vortex each sample and filter at a pressure differential of 1 atm onto 0.45-μm pore size cellulose membrane filters (Millipore or Sartorius). Each sample will fill the well of the Millipore manifold filtration apparatus twice. Some wells filter more rapidly than others: stopper the wells that have completed filtration while waiting for others.
   j. Allow the wells to empty prior to rinsing. Rinse each tube twice with *cold* 5% TCA (3 ml per rinse), vortex, and pour into the appropriate well. Rinse each filter with 2 ml of *cold* 80% ethanol (2 ml per rinse).
   k. Carefully slide the filter off of the filtering unit with forceps and place it in a labeled vial with the filtered surface up. Filter the remainder of the samples as above.
   l. Dissolve the filter by adding 1.0 ml of ethylacetate to each vial.
   m. Prepare five calibration vials as was described in the tritiated thymidine incorporation procedures (see p. 275).
3. Radioassay:
   a. Add 10 ml of an appropriate scintillant fluor to each vial and shake thoroughly.
   b. As the vials are organized for counting, wipe each externally with 70% ethanol.
   c. Radioassay with an appropriate spectrometer with a standard tritium program (e.g., 20 min)
   d. Select six vials that have differing quench numbers for internal standardization (samples with >1500 cpm are preferred). Add an appropriate amount of a standard (e.g., ^{3}H-hexadecane or ^{3}H$_2$O) and radioassay (ca. 10 min).

Calculation of Bacterial Protein Production

1. Convert counts per minute (cpm) to disintegrations per minute (dpm):

$$dpm = \frac{cpm - background\ cpm}{\%\ efficiency\ of\ radioassay}(100)$$

The efficiency is determined from standards of known activity within the LS spectrometer capabilities. Follow the instructions of the manufacturer.
2. Bacterial protein production (BPP) from ^{3}H-leucine incorporation is calculated as follows (Simon and Azam, 1989; Kirchman, 1993):

Biomass production (g C/liter/hour) = (mol leu_{inc}) $(7.3/100)^{-1}$ (131.2) (C/protein) (ID) Where:

mol leu_{inc} = rate of leucine incorporation (mol/liter/hour)

$$= \frac{(dpm_{sample} - dpm_{blank})(\text{ml per 1})(\text{min per h})\left(\begin{array}{c}\text{volume factor for}\\\text{formalin added, 1.03}\end{array}\right)}{(\text{dpm per Ci})\left(\begin{array}{c}\text{specific activity}\\\text{in Ci/mmol}\end{array}\right)\left(\begin{array}{c}\text{volume filtered,}\\\text{ml}\end{array}\right)\left(\begin{array}{c}\text{incubation}\\\text{time, min}\end{array}\right)\left(\begin{array}{c}\text{mmol}\\\text{per mole}\end{array}\right)}$$

131.2   = gram molecular weight of leucine

$(7.3/100)^{-1}$ = mole % of leucine in protein, hence 0.073

C/protein   = ratio of cellular carbon to protein; best estimate = 0.86

ID   = intracellular isotope dilution of ^{3}H-leucine; best estimate = 2

If the isotope dilution is not known, the theoretical minimum estimate of BPP would assume no isotope dilution, i.e., all leucine of protein would be derived from exogenous sources. Studies of this problem using independent methods demonstrated isotopic dilution to be twofold (Simon and Azam, 1989). When these best estimates are used, the resulting conversion factor is 3.1 kg C/mol, which is multiplied times the leucine incorporation rate to obtain rates of bacterial biomass production.

Alternatively, BPP can be estimated directly from intracellular pool specific activity (SA):

$$BPP(g) = (dpm_{inc})(SA)(131.2)$$

The specific activity of intracellular ^{3}H-leucine (Ci/mmol) reached 42% of extracellular specific activity within 10 min and a maximum of 61% of extracellular specific activity within 30 min of incubation (Simon and Azam, 1989).

The ratios of protein : dry weight and of carbon : dry weight were found to be quite constant. Thus, BPP measurements for all cell sizes can be converted to rates of dry weight production simply by multiplying by 1.6, and into bacterial carbon production by multiplying by 0.86.

EXERCISES

OPTION 1. FIELD ANALYSES

1. Collect water samples at regular intervals from the central depression of a lake or reservoir with a clean Van Dorn or similar nonmetallic water sampler. Dispense into two replicate experimental bottles and a blank bottle from each depth for the productivity analyses. Also fill a small bottle for bacterial enumeration; preserve with glutaraldehyde (see "Apparatus and Supplies," p. 284).

2. Perform the ^{3}H-thymidine incorporation assay of bacterial productivity and/or the measures of bacterial production of protein as detailed in this exercise with in situ incubations at the depth from which the samples were collected.

3. If possible, collect other vertical series of water samples for comparison from (a) within a littoral zone among dense stands of higher aquatic plants, and (b) in the open water near the mouth of an inlet stream.

4. If possible, make comparative collections and analyses of bacteria and bacterial productivity from vertical profiles in (a) a relatively unproductive oligotrophic lake; (b) a pro-

ductive, eutrophic lake or reservoir; (c) the open water of a bog; and (d) a stream at several points along its drainage (e.g., in the central stream and in backwaters from downstream and upstream stations and directly below the outlet from a lake source to the stream). In stratified waters, determine the temperature and dissolved oxygen profiles.

5. In the laboratory, determine the incorporation rates into nucleic acids or protein in order to estimate the bacterial productivity. Estimate the biovolume of bacteria from different depths and sites by microscopic examination.

6. Answer the questions following option 2.

OPTION 2. LABORATORY EXERCISES

1. Using water samples provided by your instructor, determine the bacterial productivity by the thymidine incorporation method or by bacterial production of protein with incubations in the laboratory.

2. With samples of natural bacterial communities provided by your instructor, determine the biovolume of the bacteria by microscopic examination and enumeration.

3. Answer the following questions.

Questions

1. Numbers, biomass, and productivity of planktonic bacteria generally increase with increasing photosynthetic productivity of the phytoplankton. Studies have indicated that often between 20 and 80% of the extracellular organic carbon released by phytoplankton is utilized rapidly by planktonic bacteria [e.g., Coveney and Wetzel (1989)]. If 30 to 50% of the photosynthetic productivity were released either extracellularly or during autolysis as dissolved organic substrates, what would this release mean for higher trophic levels?

2. What is the common lower limit for the removal of bacteria through ingestion by cladoceran zooplankton? By rotifers? By protozoans?

3. What is meant by the "microbial loop"? What effects would such a cycling among algae, bacteria and protozoans/microflagellates have on nutrient cycling via mineralization? [See Pomeroy (1974).]

4. If many of the algae and bacteria were not ingested by animals, what would become of them?

5. Carbon flux through planktonic bacteria is usually less than 50% of the net phytoplankton production. Where is the rest of the phytoplanktonic production utilized or stored?

6. As the hypolimnion of a productive lake becomes anaerobic during seasonal stratification, how might these conditions alter the metabolism and total bacterial productivity?

7. If inorganic turbidity in a lake or reservoir were to increase markedly, such as following a rainstorm, how would this particulate loading influence bacterial productivity?

8. When a lake contains a major development of littoral plants and attached algae, how might the productivity and development of planktonic bacteria be altered?

9. Fresh water draining from forested and extensive wetland areas often contains high concentrations of dissolved organic matter. How might these dissolved organic compounds influence the productivity of bacterioplankton? Of phytoplankton? Of zooplankton?

Apparatus and Supplies

1. Boats, anchors, a Van Dorn or similar nonmetallic water sampler, and suspension buoy for the bottles.

Thymidine-DNA Change Method

1. Thymidine-DNA change method:
   a. A sturdy box for thymidine uptake assay bottles that has holders for each bottle separately and with a firm lid to secure the bottles even if the box were dropped or inverted.

    b. Ca. 20 60-ml ground-glass-stoppered reagent bottles with spring clips at the necks. Label three bottles for each depth (two experimental measures and one control).

    c. A calibrated line with clip rings at appropriate depths.

    d. A thymidine kit for field, consisting of:

        i. Formaldehyde autopipet, adjusted to 1.6 ml.

        ii. Water withdrawal autopipet, adjusted to 2.0 ml.

        iii. ^{3}H-thymidine ampoule (see procedure) in a capped vial with padding.

        iv. Thymidine styringe, adjusted to 0.20 ml.

        v. 50 ml of 37% formaldehyde.

        vi. Six pairs of disposable vinyl gloves.

        vii. Absorbent laboratory paper wipes.

        viii. Sealable plastic bags for radioactive solid wastes.

2. ^{3}H-Thymidine stock preparation:

    a. [Methyl-^{3}H]thymidine of highest purity and high specific activity (e.g., Amersham TRK.686) at ca. 80 Ci/mmol. The stock solutions should be prepared (see stock preparation example) to yield a 10 nM solution when adding 200 μl thymidine to a 63-ml sample in the field. Because ca. 4 ml of working solution are needed to inoculate the 20 samples (200 μl each), ca. 1 ml, depending upon the specific concentrations, of the stock solution should be diluted to ca. 4.2 ml for the working field solution.

    b. Whatman filter paper, fine (e.g., No. 42 or 50).

    c. Lyophilizer.

    d. Polycarbonate centrifuge tube; glass jar.

    e. Clean 2-ml and 5-ml ampoules, combusted at 500°C, capped with aluminum foil, and autoclaved.

    f. Laboratory radioisotope-absorbent bench pads.

    g. On the day of the in situ assay:

        i. Millex GV 0.22-μm filter unit (25-mm diameter) or equivalent.

        ii. A 5-ml disposable syringe and two $1\frac{1}{2}$-in., 22-gauge needles.

        iii. Lyophilized stock ^{3}H-thymidine vial.

        iv. A clean 5-ml ampoule, combusted at 500°C; glass annealing torch.

        v. Disposable vinyl gloves.

        vi. Disposable radioisotope-absorbent bench pads.

3. Laboratory assays:

    a. Volumetric and Erlenmeyer flasks, graduated cylinders, syringes, and pipets as noted.

    b. A 10-ml automatic transfer pipet with a 250-ml base flask, if available.

    c. A multiple filtration manifold unit (e.g., Millipore 1225) to permit simultaneous vacuum filtration of up to 12 samples, 25-mm diameter, if available; 12 stoppers (#6).

    d. Polypropylene test tubes, 50 ml.

    e. Cellulose acetate membrane filters (e.g., Millipore), 25-mm diameter, 0.22-μm pore size.

    f. Cork borer, size #12 (20.2-mm diameter).

    g. Vortex mixer.

    h. Trichloroacetic acid (TCA), (*Caution: TCA is caustic, adheres to skin, and is very corrosive*):

        i. Measure 30.0 g of TCA into 200-ml volumetric flask, bring up to mark with pure water, and mix thoroughly. Carefully transfer to a 250-ml automatic transfer pipet. label 15% TCA and store at 4°C.

        ii. Measure 10.0 g of TCA into a 200-ml volumetric flask, bring to mark with pure water, and mix thoroughly. Carefully transfer to a 250-ml glass-stoppered Erlenmeyer flask, label 5% TCA and store at 4°C.

    i. Algal marker, if needed:

        i. An aliquot of dense algal culture is harvested by centrifugation and then extracted in 50% ethanol and recentrifuged until the supernatant is clear.

        ii. The pellet of algal cell fragments then is suspended in ca. 25 ml of 50% ethanol and retained for dilution and resuspension for marking filtration areas in samples from oligotrophic waters (see procedures).

4. Acid extraction of thymidine uptake samples:
   a. Water bath, 70°C, with rack for ca. 20 LS vials.
   b. Ice bath with rack for LS vials.
   c. 0.5 M HClO$_4$; keep refrigerated, discard after one week.
   d. Refrigerated centrifuge (4°C), chilled rotor; polypropylene centrifuge tubes (15-ml) with caps.
   e. Transfer pipets, automatic pipets, vortex mixer, Pasteur pipets, LS vials and caps, and ice.
   f. Oven incubator, 90°C.
   g. Protein-DNA carrier solution:
      i. Transfer 50 mg of deoxyribonucleic acid (e.g., DNA, sodium salt, Sigma D-1501 from calf thymus) to a 50-ml, flask.
      ii. Add 50 mg of purified bovine serum albumin.
      iii. Add 50 ml of 0.01 M NaOH; cover and stir with a magnetic stir bar until dissolved.
      iv. Store refrigerated in tightly capped vials. Stable for at least six months.

Bacterial Production of Protein Method

1. Sturdy box for leucine uptake assay bottles with holders for each bottle separately and with a firm lid to secure bottles even if the box were dropped or inverted.
2. Ca. 20 60-ml ground-glass-stoppered reagent bottles with spring clips at the necks. Label three bottles for each depth (two experimental measures and one control).
3. A calibrated line with clip rings at appropriate depths.
4. Leucine kit for field (identical to the thymidine kit discussed above; see p. 283).
5. ^{3}H-Leucine stock preparation:
   a. [3,4,5-^{3}H]-1-leucine of highest purity and high specific activity at ca. 140 Ci/mmol. The stock solutions should be prepared to yield a 10-nM solution in final concentration when adding 200 μl of leucine to a 63-ml sample in the field. Because ca. 4 ml of working solution are needed to inoculate the 20 samples (200 μl each), ca. 1 ml, depending on the specific concentrations of the stock solution should be diluted to ca. 4.2 ml for the working field solution.
   b. Laboratory radioisotope-absorbent bench pads.
6. Laboratory assays.
   a. Volumetric and Erlenmeyer flasks, graduated cylinders, syringes, and pipets as noted.
   b. A multiple filtration manifold unit, if available, to permit simultaneous vacuum filtration of up to 12 samples, 25-mm diameter; 12 stoppers (#6).
   c. Cellulose acetate membrane filters (e.g., Millipore or Sartorius), 25-mm diameter, 0.22- or 0.45-μm pore size.
   d. Vortex mixer.
   e. Trichloroacetic acid (TCA). *Caution: TCA is caustic, adheres to skin, and is very corrosive.* Measure 30.0 g of TCA into a 200-ml volumetric flask, bring up to mark with pure water, and mix thoroughly. Carefully transfer to a 250-ml automatic transfer pipet. Label 15% TCA.
   f. Lake or stream water, filtered through 0.45-μm pore size filters.
   g. Polypropylene test tubes, 50 ml.
   h. Water bath or oven, 95 to 100°C, with rack for holding tubes during extraction.
   i. Ethylacetate and scintillation fluors.
   j. Ethanol, 80% v/v.

Bacterial Enumeration and Biomass

1. Glutaraldehyde [1,5-Pentanedial], added to field samples to preserve at a 5% solution. *Caution*: Glutaraldehyde at 50%, the maximum available commercially, is caustic. Use

of 25% is recommended as a stock solution. Use only in a well-ventilated area or in a hood.

2. Acridine orange stock solution: To obtain a 0.1% solution in 1% glutaraldehyde, add 4.0 ml of 25% glutaraldehyde (0.8-μm pore size filtered) to 100 mg acridine orange in a 100-ml volumetric flask and bring to volume with water.

3. Polycarbonate Nuclepore filters, 0.2-μm pore size, 25-mm diameter, stained black by manufacturer or by soaking 2 to 24 h in a solution of 2 g of irgalan black (Chemical Index, acid black 107) in 1 liter of 2% acetic acid. When stored in the dye, rinse several times in clear water and use immediately. Alternatively, filters can be dried on absorbent paper and stored (Francisco et al. 1973; Hobbie et al. 1977).

4. Millipore filters, 1.2-μm pore size, 25-mm diameter.

5. Filtration unit (25-mm diameter base, funnel, clamp, filtration flask) and vacuum pump.

6. Microscope slides, coverslips, and Cargille Type A nonfluorescent immersion oil.

7. Compound epifluorescent microscope (e.g. Zeiss Standard or Universal or Leitz Ortholux) with Xenon or other halogen lamp system. Appropriate beam splitters and filters to achieve excitation at 436 or 490 nm.

References

Austin, B. (ed.). 1988. Methods in Aquatic Bacteriology. Wiley, Chichester, 425 pp.

Autio, R.M. 1990. Bacterioplankton in filtered brackish water cultures: Some physical and chemical parameters affecting community properties. Arch. Hydrobiol. *117*:437–451.

Bell, R.T., G.M. Ahlgren, and I. Ahlgren. 1983. Estimating bacterioplankton production by the [³H]thymidine incorporation in a eutrophic Swedish lake. Appl. Environ. Microbiol. *45*:1709–1721.

Bowden, W.B. 1977. Comparison of two direct-count techniques for enumerating aquatic bacteria. Appl. Environ. Microbiol. *33*:1229–1232.

Bratbak, G. 1985. Bacterial biovolume and biomass estimations. Appl. Environ. Microbiol. *49*:1488–1493.

Bratbak, G. 1993. Microscope methods for measuring bacterial biovolume: Epifluorescence microscopy, scanning electron microscopy, and transmission electron microscopy. pp. 309–317. *In*: P.F. Kemp, B.F. Sherr, E.B. Sherr, and J.J. Cole, Editors. Handbook of Methods in Aquatic Microbial Ecology. Lewis Publishers, Boca Raton.

Bratbak, G. and I. Dundas. 1984. Bacterial dry matter content and biomass estimations. Appl. Environ. Microbiol. *48*:755–757.

Bührer, H. 1977. Verbesserte Acridinorangemethode zur Direktzählung von Bakterien aus Seesediment. Schweiz. Z. Hydrol. *39*:99–103.

Button, D.K. and B.R. Robertson. 1993. Use of high-resolution flow cytometry to determine the activity and distribution of aquatic bacteria. pp. 163–173. *In*: P.F. Kemp, B.F. Sherr, E.B. Sherr, and J.J. Cole, Editors. Handbook of Methods in Aquatic Microbial Ecology. Lewis Publishers, Boca Raton.

Caldwell, D.E. 1977. The planktonic microflora of lakes. CRC Critical Rev. Microbiol. *5*:305–370.

Coveney, M.F. and R.G. Wetzel. 1988. Experimental evaluation of conversion factors for the (³H)thymidine incorporation assay of bacterial secondary productivity. Appl. Environ. Microbiol. *54*:2018–2026.

Coveney, M.F. and R.G. Wetzel. 1989. Bacterial metabolism of algal extracellular carbon. Hydrobiologia *173*:141–149.

Daley, R.J. 1979. Direct epifluorescence enumeration of native aquatic bacteria: Uses, limitations, and comparative accuracy. pp 29–45. *In*: J.W. Costerton and R.R. Colwell, Editors. Native Aquatic Bacteria: Enumeration, Activity, and Ecology. ASTM STP695. American Society of Testing Materials, Washington, D.C.

Ducklow, H.W. and C.A. Carlson. 1992. Oceanic bacterial production. Adv. Microbial Ecol. *12*:113–181.

Ducklow, H.W. and S.M. Hill. 1985. Tritiated thymidine incorporation and the growth of heterotrophic bacteria in warm core rings. Limnol. Oceanogr. *30*:260–272.

van Es, F.B. and L.-A. Meyer-Reil. 1983. Biomass and metabolic activity of heterotrophic marine bacteria. Adv. Microbial Ecol. *6*:111–170.

Francisco, D.E., R.A. Mah, and A.C. Rabin. 1973. Acridine orange-epifluorescence technique for counting bacteria in natural waters. Trans. Amer. Microsc. Soc. *92*:416–421.

Fuhrman, J.A. and F. Azam. 1982. Thymidine incorporation as a measure of heterotrophic bacterioplankton production in marine surface waters: Evaluation and field results. Mar. Biol. *66*:109–120.

Hobbie, J.E., R.J. Daley, and S. Jasper. 1977. Use of Nuclepore filters for counting bacteria by fluorescence microscopy. Appl. Environ. Microbiol. *33*:1225–1228.

Kemp, P.F., B.F. Sherr, E.B. Sherr, and J.J. Cole (eds). 1993. Handbook of Methods in Aquatic Microbial Ecology. Lewis Publishers, Boca Raton. 777 pp.

Kirchman, D.L. 1993. Leucine incorporation as a measure of biomass production by heterotrophic bacteria. pp. 509–512. *In*: P.F. Kemp, B.F. Sherr, E.B. Sherr, and J.J. Cole, Editors. Handbook of Methods in Aquatic Microbial Ecology. Lewis Publishers, Boca Raton.

Kirchman, D., H. Ducklow, and R. Mitchell. 1982. Estimates of bacterial growth from changes in uptake rates and biomass. Appl. Environ. Microbiol. *44*:1296–1307.

Kirchman, D.L., E. K'Nees, and R. Hudson. 1985. Leucine incorporation and its potential as a measure of protein synthesis by bacteria in natural aquatic systems. Appl. Environ. Microbiol. *49*:599–607.

Kirchman, D.L., S.Y. Newell, and R.E. Hodson. 1986. Incorporation versus biosynthesis of leucine: Implications for measuring rates of protein synthesis and biomass production by bacteria in marine systems. Mar. Ecol. Prog. Ser. *32*:47–59.

Lovell, C.R. and A. Konopka. 1985. Seasonal bacterial production in a dimictic lake as measured by increases in cell numbers and thymidine incorporation. Appl. Environ. Microbiol. *49*:492–500.

Moriarty, D.J.W. 1986. Measurement of bacterial growth rates in aquatic systems using rates of nucleic acid synthesis. Adv. Aquatic Microbiol. *9*:245–292.

Moriarty, D.J.W. 1989. Accurate conversion factors for calculating bacterial growth rates from thymidine incorporation into DNA: Elusive or illusive? Ergebn. Limnol. Arch. Hydrobiol. (In press).

Murray, R.E. and R.E. Hodson. 1985. Annual cycle of bacterial secondary production in five aquatic habitats of the Okefenokee Swamp ecosystem. Appl. Environ. Microbiol. *49*:650–655.

Norland, S. 1993. The relationship between biomass and volume of bacteria. pp. 303–307. *In*: P.F. Kemp, B.F. Sherr, E.B. Sherr, and J.J. Cole, Editors. Handbook of Methods in Aquatic Microbial Ecology. Lewis Publishers, Boca Raton.

Pomeroy, L.R. 1974. The ocean's food web: A changing paradigm. BioScience *9*:499–504.

Porter, K.G. and Y.S. Feig. 1980. The use of DAPI for identifying and counting aquatic microflora. Limnol. Oceanogr. *25*:943–948.

Psenner, R. 1990. From image analysis to chemical analysis of bacteria: A long-term study? Limnol. Oceanogr. *35*:234–237.

Riemann, B. and M. Sjøndergaard. 1985. Regulation of bacterial secondary production in two eutrophic lakes and in experimental enclosures. J. Plankton Res. *8*:519–536.

Riemann, B., P.K. Bjørnsen, S. Newell, and R. Fallon. 1987. Calculation of cell production of coastal marine bacteria based on measured incorporation of [^{3}H]thymidine. Limnol. Oceanogr. *32*:471–476.

Robarts, R.D. and R.J. Wicks. 1989. [Methyl-^{3}H]thymidine macromolecular incorporation and lipid labeling: Their significance to DNA labeling during measurements of aquatic bacterial growth rate. Limnol. Oceanogr. *34*:213–222.

Robarts, R.D. and T. Zohary. 1993. Fact or fiction—Bacterial growth rates and production as determined by [*methyl-^{3}H*]-thymidine? Adv. Microbial Ecol. *13*:371–425.

Scavia, D. and G.A. Laird. 1987. Bacterioplankton in Lake Michigan: Dynamics, controls, and significance to carbon flux. Limnol. Oceanogr. *32*:1017–1033.

Simon, M. and F. Azam. 1989. Protein content and protein synthesis rates of planktonic marine bacteria. Mar. Ecol. Prog. Ser. *51*:201–213.

Sorokin, Y.I. and H. Kadota (eds). 1972. Techniques for the Assessment of Microbial Production and Decomposition in Fresh Waters. Blackwell, Oxford.

Decomposition: Relative Bacterial Heterotrophic Activity on Soluble Organic Matter

The decomposition of organic matter in aquatic ecosystems by microorganisms generally involves two processes: (1) the hydrolytic degradation of high molecular weight organic polymers into compounds of low molecular weight, such as glucose, cellobiose, and amino acids; and (2) the nonhydrolytic oxidative mineralization of low molecular weight organic compounds to inorganic compounds, especially CO_2, H_2S, NH_4^+, and PO_4^{-3}. Measurement of the rates of decomposition and of mineralization of organic matter in natural waters is difficult and has been approached in a number of ways.

To analyze the rates of reactions involved in situ decomposition and mineralization of organic matter, both biochemical and geochemical approaches have been taken. Geochemical methods involve the chemical analysis of organic and inorganic compounds in water and sediments. Changes in the chemical composition often reflect the biochemical events that have occurred in the environment as a consequence of microbial activities.

Biochemical methods address more directly fundamental questions of the chemical nature, concentrations, and the rates at which substrates are utilized by bacteria for energy. In the previous exercise, we outlined current state-of-the-art methods used to estimate the rates of productivity of the heterogeneous communities of heterotrophic bacteria in situ. Alternative techniques have been used to measure a community response by following the rates of respiratory activity, of utilization of specifically labeled organic substrates, or of polymer degradation by enzymatic activity [e.g., Cunningham and Wetzel (1989)]. Heterotrophic bacterial communities consist of a heterogeneous composite of populations in various physiological states. Moreover, the dissolved organic substrates comprise a spectrum from very labile compounds that are readily reactive with the enzymes of the microflora, to highly recalcitrant substrates that are utilizable only slowly by highly specialized organisms.

Estimates of in situ rates of planktonic community respiration have been made by separating the larger photosynthetic organisms from the bacteria by filtration and then analyzing either the oxygen consumption or CO_2 production [see Sorokin and Kadota (1972)]. However, separation of the organisms by filtration does not give consistent results. As we have seen earlier, picophytoplankton ($<2\,\mu$m) can constitute a major portion of the phytoplankton community. Furthermore, under anaerobic conditions, decomposition occurs by multifaceted fermentation processes. Fermentation uses alternate electron acceptors to produce a variety of reduced metabolic end products, such as methane, volatile fatty acids, and alcohols, in addition to CO_2.

This exercise offers a simple approach to measure the relative heterotrophic activity of microbes in natural waters. In situ bacterial communities are analyzed by the uptake and mineralization of a ^{14}C-labeled organic substrate with Michaelis-Menten enzyme kinetics. This technique was proposed originally by Parsons and Strickland (1962) and later developed by Wright and Hobbie (1965, 1966) and Hobbie and Crawford (1969a). Labeled organic compounds are added to natural communities at serially increasing concentration (μg/l) in a closed system so that the $^{14}CO_2$ evolved by microbial degradation can be recovered. After a period of incubation at in situ temperatures, both the amount of substrate synthesized into cellular components and that portion respired as CO_2 are measured. The rate of turnover of the substrate then is evaluated by a model system of enzyme kinetics. The method evaluates the utilization of only one simple substrate at a time and is subject to a number of interpretative difficulties [e.g., Wright (1973) and Caldwell (1977)], some of which will be discussed below. In spite of the availability of other more direct methods of evaluating in situ bacterial productivity and limitations of estimates of relative heterotrophic activity, this method is still used to evaluate comparatively the relative in situ rates of utilization of specific dissolved organic substrates.

UPTAKE OF ORGANIC SOLUTES

1. Prepare 25-, 50-, or 125-ml modified serum reaction flasks (see Fig. 20.1, and "Apparatus and Supplies," p. 298), by first rigorously cleaning with a strong acid, copiously rinsing with organic matter-free redistilled water, and allowing to air-dry in a dust-free location. The glassware used in these analyses *must* be scrupulously clean. Cap with aluminum foil and heat sterilize just before use.

2. Prepare serum caps by inserting the assembled plastic cup attached to a support rod through the cap (Fig. 20.1). Fold small pieces (ca. 1.5×3 cm) of Whatman No. 1 filter paper accordian-style and place into the cups.

3. Using an automated micropipet (e.g., Eppendorf), pipet aliquots of the ^{14}C organic substrate solution (see p. 298) into at least duplicate, preferably into triplicate, flasks to obtain the following: 10, 50, 100, 200, 300, 400, and 500 μl, to yield substrate concentrations in the final samples of 10 to 500 μg/l.

4. To one flask of each concentration of substrate, add 0.5 ml of 2 N sulfuric acid for controls (killed blanks). Label the flasks appropriately.

5. Homogenize the water sample by *gentle* inversion. Add 10, 25, or 50 ml of sample as rapidly as possible to all of the flasks. Immediately seal the flasks with the serum cap-cup assemblies and note the starting time for each flask. Swirl each flask *gently* to mix the samples thoroughly, but avoid any contact with the suspended cup assemblies.

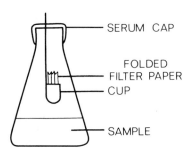

— SERUM CAP

— FOLDED
— FILTER PAPER
— CUP

— SAMPLE

Figure 20.1. Experimental flask for assaying the uptake and mineralization of organic substrates by planktonic bacteria.

6. Incubate the flasks in darkness at the temperature of the in situ sample. The time of incubation depends on the temperature and suspected rate of activity of the bacterial populations. In productive waters at warm (20 to 25°C) temperatures, 1 h of incubation should be adequate to give an appreciable amount of uptake and metabolism for reliable results from radioassay (ca. 500 cpm at the lowest level of substrate addition). In oligotrophic, colder waters, several hours of incubation may be needed. Swirl the samples *gently* several times during the incubation or place on a very slowly turning rotary shaker.

7. Terminate the incubation by killing the bacteria. Carefully inject 0.5 ml per 10 ml of sample of 2 N sulfuric acid solution into the sample through the serum stopper [or through the side-arm port when these modified flasks are used (see "Apparatus and Supplies," p. 298)]. Avoid contact with the cup assembly. Note the time of injection for each flask.

8. Allow the sealed flasks to stand for a few minutes with *gentle* shaking on a rotary shaker. Add 0.2 ml of monoethanolamine directly to the filter paper in the cup by injection through the serum cap. Incubate with shaking for 1 h.

9. *Carefully* remove the cup assembly and, with it poised over a labeled scintillation vial, clip the support rod (Fig. 20.1) and allow the cup and filter paper to fall into the vial. Add 15 ml of scintillation cocktail (PPO-bis-MSB-toluene-methanol, see p. 298) and seal the vial. Mix well.

10. Filter the entire water sample onto 0.45-μm pore size membrane filters (e.g., Millipore HA). Keep an accurate account of the order of the samples since the edges of the filters cannot be written on (interferes with radioassay). Rinse each flask with 10 ml of prefiltered (0.45-μm pore size) water sample and filter with the sample. The coating of the filtration funnels with a silicone compound (e.g., Beckman Desicote) prior to use prevents adhesion of cells to the walls.

11. Using forceps, place the wet filters directly into labeled scintillation vials, when Instagel or other dioxane-based scintillation cocktails are used (see p. 298). The filters must be dried under desiccation prior to radioassay when toluene-based cocktails are used [e.g., Hobbie and Crawford (1969a)].

12. Radioassay the samples with a liquid scintillation counter, correcting for quenching with the channels ratio method to yield the counting efficiency for each sample. Convert the counts per minute values to disintegrations per minute.

Calculations

1. The uptake of organic compounds should exhibit saturation kinetics. With saturation kinetics, the uptake velocity approaches a constant value in spite of increasing concentrations (Wright and Hobbie, 1965, 1966), and a plot of uptake velocity versus concentration demonstrates a hyperbolic relationship. Although the measured activity likely resulted from bacterial enzyme-mediated transport systems, the observed activity is a measure of a community response to the substrate.

2. The uptake can be analyzed by the well-known Michaelis-Menten kinetics for enzyme-substrate systems:

$$v = -\frac{(V_{max})(S)}{K_t + S} \tag{1}$$

where v = uptake velocity, V_{max} = theoretical maximum uptake velocity when all uptake sites are saturated, S = substrate concentration (natural plus added), and K_t = transport constant—by definition, the substrate concentration at which $v = V_{max}/2$. This equation can be transformed by inversion and multiplication of both sides of the equation by S to derive the Lineweaver-Burke equation:

$$\frac{S}{v} = \frac{K_t + S}{V_{max}} \qquad (2)$$

and then expanded to include the known and unknown (natural) substrate concentrations ($S = S_n + A$):

$$\frac{S_n + A}{v} = \frac{A}{V_{max}} + \frac{K_t + S_n}{V_{max}} \qquad (3)$$

where S_n = natural substrate concentration of a water sample and A = substrate concentration added to the sample in labeled and unlabeled forms (see below). Applying these relationships to the experimental assays of uptake velocity versus substrate concentration:

$$v = \frac{F(S_n + A)}{t} \qquad (4)$$

where v = velocity of substrate uptake ($\mu g/time$) and F = fraction of available labeled substrate taken up in time t, in hours.

This equation (Eq. 4) is now solved for ($S_n + A/2$) of Eq. 3, resulting in the form of a slope-intercept equation*:

$$\frac{t}{F} = \frac{(1)(A)}{V_{max}} + \frac{K_t + S_n}{V_{max}} \qquad (5)$$

When t/F is plotted against A, the values for the slope and intercept may be determined (Fig. 20.2). Equation 5 circumvents the problem of not knowing the

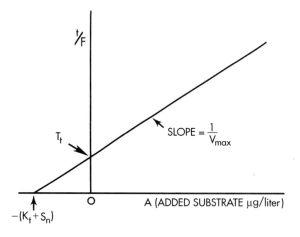

Figure 20.2. Graphic analysis of bacterial uptake at low organic substrate concentrations following Michaelis-Menten enzyme kinetics (see text).

---

*$y = mx + b$, where m = slope, b = y intercept.

natural substrate concentration (S_n). The measurement of S_n requires sophisticated techniques, since the natural concentrations of many simple substrates (e.g., carbohydrates and amino acids) are very low (usually $< 25\,\mu g/l$). When S_n of the compound under study can be measured independently, the natural uptake and mineralization rates can be calculated.

The plot of t/F against A (Fig. 20.2) permits the following calculations:

a. V_{max}, the maximum rate of uptake ($\mu g/l/h$), is the reciprocal of the slope, which is affected by temperature as well as by the density and activity of the composite bacterial population.

b. The quantity ($K_t + S_n$), the composite of the transport constant (K_t) and natural substrate concentration ($\mu g/l$), is the absolute value of the x intercept and approximates the natural substrate concentration when K_t is small, as often is, but not always, the case. Since S_n is seldom known, interpretation of the ($K_t + S_n$) value is difficult and provides only limited information about the bacterial community.

c. T_t, the turnover time in hours (the ordinate intercept), is the time required for complete removal of the natural substrate by the microflora, assuming a constant rate of replacement of the substrate.

3. The counts per minute (cpm) for each filter (assimilation) and corresponding filter paper and solution of the cup (mineralization to CO_2) and the blanks (killed control) are converted to disintegrations per minute (dpm) by correcting for quench and counting efficiency. A rigorous treatment would also include an analysis of the trapping efficiency of CO_2.

4. Determine the total dpm:

$$\sum dpm = (dpm_a - dpm_{ba}) + (dpm_m - dpm_{bm})$$

where dpm_a = mean dpm assimilated (corrected counts of filters), dpm_{ba} = dpm of killed control filter, dpm_m = mean dpm mineralized (corrected counts of filter paper and solution in the cup), and dpm_{bm} = dpm of killed control filter paper in the cup.

5. Plot v in $\mu g/h$ on the ordinate versus A, using Eq. 4 to determine whether a linear (uptake proportional to substrate concentration) or a hyperbolic (uptake velocity approaches a saturation level) relationship occurred. Use the following definitions:

a. dpm added to the sample is known (see p. 298).

b. F = total sample dpm/dpm added.

c. t = hours of incubation.

d. $A = \mu g$ of substrate added (see. p. 298).

e. S_n is assumed to be $5\,\mu g/l$ in this example.

6. Calculate t/F and plot on the ordinate against A. Draw the line of best fit (a least square fit is preferable), and determine (Eq. 5):

a. V_{max} from the slope ($1/V_{max}$) in $\mu g/l/h$.

b. ($K_t + S_n$) from the x intercept, which equals $-(K_t + S_n)$.

c. T_t from the y intercept in hours. Since $A = 0$ at this point, from Eq. 3, $T_t = K_t + S_n/V_{max}$.

7. Calculate the ratio of mineralization (M) of the substrate to assimilation (As):

$$M/As = \frac{(dpm_m - dpm_{bm})}{(dpm_a - dpm_{ba})}$$

and determine the percentage of the substrate mineralized to CO_2:

$$\%M = \frac{(\mathrm{dpm_m} - \mathrm{dpm_{bm}})}{\sum \mathrm{dpm}}(100)$$

RESPIRATION OF BACTERIA

Respiration generally occurs in mitochondria while respiratory enzymes occur in or on cellular membranes. Measurement of the final limiting step of the electron transport reactions of both aerobic and anaerobic microorganisms yields an estimate of the maximum potential respiratory rate of cells. During the degradation of organic compounds within the Krebs cycle, electrons are transported by NADH//NADPH and ATP is formed. Oxygen is consumed with the release of water and CO_2.

Dehydrogenase activity can be considered to be a good measure of microbial oxidative activity. An artificial final electron acceptor, such as a soluble iodonitrotetrazolium salt (INT), is reduced by microbial activity. Measurement of INT reduction allows evaluation of the rate respiration of either a community of microbes or, at a microscopic level, of individual cells (e.g., Bott and Kaplan, 1985; Johnson and Ward, 1993). Respiratory enzymes convert the INT to INT-formazan, a red-colored complex that can be evaluated spectrophotometrically or microscopically.

Procedures

1. Obtain water samples in triplicate with a clean Van Dorn or similar nonmetallic water sampler from several strata within a lake or stream ecosystem. Flush sample bottles well.
2. Filter as large a volume as possible (50–250 ml) onto a 0.2-μm pore size membrane filter with minimal vacuum differential (<0.3 atm). If not analyzed immediately, the filter samples can be stored frozen in tightly capped, small (e.g., 2 ml) plastic vials.
3. Place the filter sample in a 10-ml glass centrifuge tube with 3.0 ml of ice cold homogenization buffer (see Apparatus and Supplies, p. 299).
4. Disperse and homogenize with a sonicator at intermediate intensities with pulsed intervals (e.g., 50% time) for 3 min. *Important*: Keep sample tubes in an ice bath during homogenization to avoid heating by the sonication. Retain all samples on ice until homogenization is completed.
5. After homogenization, mix well with a vortex mixer and disperse 1.0 ml of homogenate in triplicate to 10-ml centrifuge tubes containing 3.0 ml of substrate buffer solution (see Apparatus and Supplies, p. 299). Additionally, prepare three different blanks: (a) Include a control blank tube for chromatophores containing the substrate buffer solution plus the homogenate; (b) a control blank containing the homogenate, substrate buffer solution, and 1.0 ml of the quench solution (see Apparatus and Supplies, p. 299), and (c) a control blank containing only the substrate buffer.
6. Place the reaction tubes in water baths at temperatures adjusted to simulate those in situ. Allow 15 to 30 minutes for temperature equilibration.
7. Add 1.0 ml of INT solution (see Apparatus and Supplies, p. 299) to each tube with a rapid, repeating pipet. Record time and gently mix.

8. After incubations for 15–60 minutes, depending upon the activity, add 1.0 ml of quench solution. Record times. Centrifuge for 15 minutes at ca. 5,000 rpm in a cooled centrifuge.
9. Measure absorbance spectrophotometrically with a 1-cm pathlength cell at 490 nm, corrected against 750 nm for turbidity from colloidal matter. Preferred absorbances are below 0.8 OD units.

Calculations

1. Calculate the electron transport respiratory activity from the following equations.
2. Determine the corrected spectrophotometric absorbance (A_{cor}):

$$A_{cor} = A_{sample} - A_{chromatophore} - A_{incubated\ buffer} - A_{killed\ buffer}$$

3. Determine electron transport respiratory activity (INT):

$$INT = \frac{(A_{cor})(V_{hom})(60)(V)(1000)(1.42_a)}{(1.42_b)(V_f)(t)}$$

where: INT = electron transport respiratory activity ($\mu g\ O_2/l/h$)
A_{cor} = optical absorbance corrected for blanks
V_{hom} = volume of homogenate (ml) (e.g., 3.0 ml)
60 = minutes per hour
V = final reaction volume (ml) (e.g., 6.0 ml)
1000 = ml per liter
1.42_a = conversion of O_2 volume to mass of O_2 (e.g., 31.8/22.4 = 1.4196)
1.42_b = extinction of 1 mM INT (15.9/mM/cm) – 1 mol INT
= 0.5 mol O_2 – (2 × 15.9) / mole volume (22.41 dm^3/mole)
= 31.8 / 22.41
= 1.42 (i.e., extinction per μl O_2 at 490 nm per cm)
V_f = filtered volume (ml)
t = time of incubation (min)

When the incubation is performed at 20°C or some other constant temperature, the activities should be corrected to in situ temperatures via the Arrhenius relationship:

$$INT_{in\ situ} = (INT_{in\ vitro})e^{[E_a(1/T_{in\ vitro} - 1/T_{in\ situ})/R]}$$

Where: $INT_{in\ situ}$ = corrected in situ activity for the temperature ($\mu g\ O_2/l/h$)
$INT_{in\ vitro}$ = activity at the incubation temperature
E_a = activation energy (15,800 cal/mol)
R = gas constant (1.987 cal/mol degree)
T = temperature, degrees Kelvin

EXERCISES

OPTION 1. VERTICAL DISTRIBUTION OF ACTIVITY

1. Collect water samples from the epi-, meta-, and hypolimnion of the central area of a lake or reservoir with a clean Van Dorn or similar nonmetallic water sampler. Place the samples

into clean 1-liter bottles in an ice chest and transport to the laboratory as rapidly as possible. Take the water temperatures at meter intervals throughout the water column.
2. Determine the relative heterotrophic bacterial assimilation and mineralization rates of an organic substrate such as glucose or acetate by the procedures outlined. Incubate the samples as close to the in situ temperatures as possible.
3. Calculate the assimilation and mineralization parameters as outlined earlier. Compare observed differences in the heterotrophic activity against depth.
4. Answer the questions following Option 6.

OPTION 2. SPATIAL DIFFERENCES IN ACTIVITY

1. Collect water samples from a depth of 2 m from the central area of a lake or reservoir, from the littoral zone among emergent macrophytes near the sediments, and from the mouth of an inlet stream. Place the samples into clean 1-liter bottles in an ice chest and transport to the laboratory as quickly as possible, after taking the water temperatures at the points of collection.
2. Determine the relative bacterial assimilation and mineralization rates of an organic substrate (e.g., glucose or acetate) by the procedures outlined. Incubate the samples as close as possible to in situ temperatures.
3. Calculate the assimilation-mineralization parameters as outlined. Compare the spatial differences in heterotrophic activity.
4. Answer the questions following Option 6.

OPTION 3. COMPARATIVE UTILIZATION OF SUBSTRATES

1. Collect a water sample of several liters from a depth of ca. 2 m from the central portion of a lake or reservoir. Measure the water temperature at the point of collection.
2. Determine the relative bacterial assimilation and mineralization rates of several organic substrates (e.g., glucose, acetate, or an amino acid) by the procedures outlined. Incubate the samples at the temperature as close to in situ as possible.
3. Calculate the assimilation-mineralization rates of each substrate and compare the differences.
4. Answer the questions following Option 6.

OPTION 4. RELATIVE HETEROTROPHIC ACTIVITY IN WATERS OF DIFFERING PRODUCTIVITY

1. Divide the class into teams and collect water samples from ca. 2 m or the epi- and hypolimnion of an oligotrophic lake and from a eutrophic lake, reservoir, or pond. Place the samples into clean 1-liter bottles in an ice chest and transport to the laboratory as rapidly as possible. Determine the water temperatures at the points of collection.
2. Determine the relative assimilation-mineralization rates of an organic substrate (e.g., glucose or acetate) by the procedures outlined. Incubate the samples at in situ temperatures.
3. Calculate the rates of substrate utilization and mineralization as outlined and compare the results for the lakes of differing productivity.
4. Answer the questions following Option 6.

OPTION 5. COMPARISON OF ACTIVITY OF WATER VERSUS SEDIMENTS

1. Collect water samples from the epi-, meta-, and hypolimnion at a central station of a lake or reservoir. In addition, collect a sample of surface sediments with an Ekman grab or

coring device (Exercise 12). Place the water samples into clean 1-liter bottles, remove about 100 ml of sediment slurry from the upper first centimeter of the sediments, and place into a clean container. Place all of the samples in an ice chest and transport to the laboratory as rapidly as possible, after measuring the water temperatures at the points of collection.

2. Determine the relative assimilation-mineralization rates of an organic substrate (e.g., glucose or acetate) by the bacteria of the water samples using the procedures outlined. Determine the mineralization rate of the same substrate using a 10-ml slurry of sediments with a 15-min incubation period. Incubate at temperatures as close to the environmental conditions as possible.

3. Calculate the rates of substrate utilization and mineralization as outlined and compare the results between the planktonic and sediment communities.

4. Answer the questions following Option 6.

OPTION 6. LABORATORY ANALYSES

1. Using water samples provided by the instructor, determine the relative bacterial assimilation and mineralization rates of one or more organic substrates (e.g., glucose, acetate, or amino acid) by the procedures outlined. Incubate at a constant temperature or incubate replicates at high (e.g., 25°C) and low (10°C) temperatures.

2. Calculate the assimilation-mineralization parameters as outlined and compare observed differences in heterotrophic activities.

3. Answer the following questions.

Questions

1. Kinetic analyses of heterotrophic activity assume that active transport is the rate-limiting step in the metabolism of the substrate and that the transport system is in steady-state equilibrium. Do you believe these assumptions are met in applying the method to natural communities? If not, how would the results be affected?

2. If the incorporated substrate does not remain in the cells or is liberated as some gas other than CO_2, how would the results be affected? Would it be possible to measure labeled substrates that were metabolized and released in soluble form?

3. The kinetic method assumes that there was an insignificant change in the size of the populations comprising the bacterial community during the incubation. Do you believe that this assumption is valid? Explain. Do you believe that the uptake rates measured over the incubation period estimate accurately the instantaneous initial uptake velocities? Why?

4. When your data do not follow a hyperbolic relationship where uptake velocity approaches a saturation level with increasing substrate concentrations, what do you think was happening? [See Wright (1973).]

5. The kinetic method further assumes that various bacteria in a sample of water have similar enzymatic transport constants (K_t) for the substrate being assayed. Do you believe that this assumption is valid? [See Hobbie and Wright (1965).]

6. What does the turnover time (T_t) value for a naturally occurring substrate mean? How might T_t be useful in comparing different lakes or as an index of eutrophication of lakes? [See Hobbie and Crawford (1969b) and Wetzel (1983).]

7. The kinetic method as presented here can analyze the utilization activity of only one simple organic compound at a time. What about the myriad other dissolved organic substrates that occur in natural waters that are being decomposed at varying rates? How might the utilization rates of these compounds be determined as well? Several substrates simultaneously?

8. How might this kinetic approach be applied to the bacterial populations utilizing organic substrates in the interstitial water of sediments? [See Harrison et al., (1971) and Mayer-

Reil (1978).] Would it be possible to evaluate the amount of substrate assimilated in microbial biomass in a sediment system? What problems would be encountered?

9. The kinetic method assumes that the removal of the organic substrate by algae is negligible in comparison to the active permease uptake system of bacteria. Do you believe that this assumption is true for natural algal and bacterial communities at naturally occurring substrate concentrations? Why?

10. Sometimes one organic substrate can inhibit competitively the uptake of another e.g., Burnison and Morita (1973). How could this potential problem be evaluated? What would be the effect of such competitive inhibition?

11. Under acidic conditions, some substrates, such as acetic acid and glycollic acid, are volatile and would be assayed as CO_2 mineralized. What effects would this event have on the blank (control) values? How would this complication affect the estimates of the various kinetic parameters?

12. How could specific rates of decomposition of dissolved organic matter be determined? [Cf., Cunningham and Wetzel (1989).]

Apparatus and Supplies

1. Reaction flasks (25-, 50-, or 100-ml) (e.g., K.882300, Kontes Glass Co., Vineland, NJ), with serum stopper caps and suspended plastic cups (e.g., Kontes Center Well K-882320-0000). Reaction flasks with a side-arm port (e.g., Kontes K-882360) facilitate addition of the killing agent but are not essential.

2. Filter paper, e.g., Whatman No. 1.

3. Micropipets, e.g., Eppendorf of 10-, 50-, 100-μl capacities.

4. 2N H_2SO_4 (50% v/v phosphoric acid can also be used as a killing agent).

5. Temperature control (bath or environmental chamber).

6. Rotary shaker.

7. Monoethanolamine. [Alternatively, β-phenethylamine can be used (Hobbie and Crawford, 1969a).]

8. Radioassay:
   a. Standard scintillation vials.
   b. Filter paper—cups—cocktail: Mix 15 g PPO (2,5-diphenyloxazole) and 1 g bis-MSB [p-bis-(o-methylstyryl)-benzene] in 1 liter of scintillation grade toluene. Mix well three parts of this solution to one part of spectral quality methanol. [Alternative cocktails are given in Hobbie and Crawford (1969a).] *Caution*: Some of these reagents are highly toxic and flammable.
   c. Bacteria on membrane filters: Place wet filters directly into Instagel (Packard Instrument Co., Downers Grove, Il) or similar scintillation solution.
   d. Liquid scintillation counter.

9. Labeled organic substrates:
   a. Uniformly carbon-14 labeled glucose, acetate, and/or glutamate or other amino acids, e.g., glycine, of highest specific activity available (e.g., Amersham/Searle, Des Plaines, Il).
   b. Dilute the isotop to activities of the range of 37 kBq/ml (=1 μCi/ml) with sterile water of the greatest purity available (deionized, organic-free, redistilled). Dispense into clean precombusted (see Exercise 9) ampoules and seal and freeze immediately. All glassware and ampoules must be cleaned scrupulously.
   c. From the specific activity of the isotope, calculate the amount of substrate present in μg/ml in the final dilution from the specific activity (usually given as mCi/mmol).
   d. Determine the radioactivity of the substrate(s) per ml by comparison to standards (see Exercise 14).

10. Additions of substrate:
   a. From Section 9, Part (d) above, determine the amount of "carrier" organic substrate that is added with the labeled substrate. This amount is less than that needed for the serially increasing substrate concentrations of the experiment.

   b. The substrate concentration of the radioactively labeled ("hot") material must be augmented with unlabeled ("cold") substrate of the same organic compound. (i) The nonlabeled substrate can be mixed with the labeled material in such a way that added μl amounts (Procedures, Section 3) contain final substrate concentrations of 10 to 500 μg/l when diluted with the water sample; or (ii) a constant amount of labeled substrate is added to each flask and supplemented with increasing concentrations of nonlabeled substrate. When the latter method is used, several stock solutions of increasing concentrations should be prepared so that a constant volume of substrate is added to each flask.

11. Respiration of bacteria:

   a. Homogenization buffer: Add 9 mg $MgSO_4$, 1.5 g polyvinylpyrrolidone (PVP), and 3 ml Triton X-100 detergent to 997 ml 0.1 M Na_2HPO_4 phosphate buffer. Store in small (e.g., 100 ml) aliquots frozen.

   b. Substrate buffer: Add 30 ml of homogenization buffer to 970 ml of 0.1 M Na_2HPO_4. Store frozen in small aliquots. Thaw 100-ml aliquot and add 70.9 mg NADH [NAD+, reduced form, disodium salt] and 20.8 mg NADPH [NADP+, reduced form, tetrasodium salt] (1.0 and 0.25 mmol/l, respectively); mix *gently*.

   c. INT-solution: Mix 20 mg INT [2-(*p*-iodophenyl)-3-(*p*-nitrophenyl)-5-phenyl tetrazolium chloride salt] for 30 minutes. Filter through a 0.2-μm pore size filter and store on ice.

   d. Quench solution: Dilute 13.45 ml (=23 g) of 85% H_3PO_4 to 180 ml with pure water. Separately mix 20 ml of 37% formic acid with 200 ml pure water. Mix the two solutions and adjust the volume to 400 ml with water. Store in glass.

12. Van Dorn water sampler.

13. Electrical underwater thermometer.

References

Bott, T.L. and L.A. Kaplan. 1985. Bacterial biomass, metabolic state, and activity in stream sediments: Relation to environmental variables and multiple assay comparisons. Appl. Environ. Microbiol. *50*:508–522.

Burnison, B.K. and R.Y. Morita. 1973. Competitive inhibition for amino acid uptake by the indigenous microflora of Upper Klamath Lake. Appl. Microbiol. *25*:103–106.

Caldwell, D.E. 1977. The planktonic microflora of lakes. CRC Critical Rev. Microbiol. *5*:305–370.

Cunningham, H.W. and R.G. Wetzel. 1989. Kinetic analysis of protein degradation by a freshwater wetland sediment community. Appl. Environ. Microbiol. *55*:1963–1967.

Harrison, M.J., R.T. Wright, and R.Y. Morita. 1971. Method for measuring mineralization in lake sediments. Appl. Microbiol. *21*:698–702.

Hobbie, J.E. and C.E. Crawford. 1969a. Respiration corrections for bacterial uptake of dissolved organic compounds in natural waters. Limnol. Oceanogr. *14*:528–532.

Hobbie, J.E. and C.C. Crawford. 1969b. Bacterial uptake of organic substrate: New methods of study and application to eutrophication. Verh. Int. Ver. Limnol. *17*:725–730.

Hobbie, J.E. and R.T. Wright. 1965. Bioassay with bacterial uptake kinetics: Glucose in freshwater. Limnol. Oceanogr. *10*:471–474.

Johnson, M.D. and A.K. Ward. 1993. A comparison of INT-formazan methods for determining bacterial activity in stream ecosystems. J.N. Amer. Benthol. Soc. *12*:168–173.

Meyer-Reil, L.A. 1978. Uptake of glucose by bacteria in the sediment. Mar. Biol. *44*:293–298.

Parsons, T.R. and J.D.H. Strickland. 1962. On the production of particulate organic carbon by heterotrophic processes in sea water. Deep-Sea Res. *8*:211–222.

Sorokin, Y.I. and H. Kadota (eds). 1972. Techniques for the Assessment of Microbial Production and Decomposition in Fresh Waters. IBP Handbook No. 23, Blackwell, Oxford. 112pp.

Wetzel, R.G. 1983. Limnology. 2nd Ed. Saunders Coll., Philadelphia. 860pp.

Wright, R.T. 1973. Some difficulties in using ^{14}C-organic solutes to measure heterotrophic bacterial activity. pp. 199–217. *In*: L.H. Stevenson and R.R. Colwell, Editors. Estuarine Microbial Ecology. Univ. South Carolina Press, Columbia.

Wright, R.T. and J.E. Hobbie. 1965. The uptake of organic solutes in lake water. Limnol. Oceanogr. *10*:22–28.

Wright, R.T. and J.E. Hobbie. 1966. The use of glucose and acetate by bacteria and algae in aquatic ecosystems. Ecology *47*:447–464.

Decomposition: Particulate Organic Matter

Decomposition completes the biogeochemical cycles that photosynthesis initiates. Thus, complete decomposition results in the conversion of the organic (reduced) products of photosynthesis back into the inorganic (generally oxidized) constituents used as the reactants for photosynthesis (see Exercise 14). The major biogeochemical cycles affected by decomposition are those of C, N, P, S, and O, although it is important to realize that all of the minor constituents of biomass (cations, trace metals, etc.) also are released (mineralized) by decomposition. When plants or animals senesce and die, both dissolved (DOM) and particulate (POM) organic matter are available for degradation. Leakage of DOM from dying cells and autolysis of the tissue increase during senescence and reach maximum levels soon after death. The DOM thus produced is leached readily from the tissue into the aquatic environment and can constitute as much as 30% of the total amount of combined particulate and dissolved organic matter in lake water. This detrital DOM is available as an energy source for microflora in the sediments and waters adjacent to particulate detritus. The rate of degradation is dependent on both the enzymatic capabilities of the microflora and the environmental conditions (see Exercise 20). Some compounds of the DOM are more stable than others, but warmer temperatures and increased availability of oxygen reduce their resistence to oxidation (recalcitrance) to some extent (Godshalk and Wetzel, 1978a).

Particulate detritus is colonized by various microflora. The rate of degradation depends on: (1) the chemical composition of specific substrates within the particles, (2) the ability of microbes to get at the tissue (e.g., particle size:surface area), and (3) the rate of microbial metabolism as governed by enzymatic capacities, temperature, and availability of electron acceptors (e.g., oxygen) and mineral nutrients (Alexander, 1965; Godshalk and Wetzel, 1978b; Webster and Benfield, 1986). A succession of types of microflora associated with the detrital particles occurs over time as a result of changes in substrate availability and environmental conditions caused by their metabolism. Total microbial metabolism and biomass often increase initially after colonization of the detritus as a result of increased concentrations of organic nitrogen relative to that of carbon. As the resistance of the residual detritus increases with continued decomposition, the degradability of the detritus decreases. Organic nitrogen concentration then decreases relative to that of carbon, resulting in high organic C:N ratios.

Measurement of the rates of decomposition of organic matter in situ is difficult. Commonly, changes in the dry weight of a known amount of POM are measured over a period of time. Such parameters as percent weight loss continually change through time and only yield information about the end result of decomposition. Decomposition is a continuous process, but the *rate* of decomposition varies through time depending on a number of substrate and environmental variables.

The relative rate of decomposition, k, can be viewed in a simple way in relation to the controlling variables (Godshalk and Wetzel, 1978c):

$$k \propto \frac{(T)(O)(N_u)}{(R_e)(S_p)}$$

where T = temperature (within biological limitations), O = dissolved oxygen or other electron acceptors, N_u = mineral nutrients required for microbial metabolism, R_e = initial tissue chemical composition and recalcitrance, and S_p = particle size (i.e., particle size : surface area).

For example, k will be low and conditions will not be conducive to rapid decomposition if relative values of T, O, and/or N_u are low. Since these factors interact with one another, high values of one variable will offset low values of another only to a limited extent. Mechanical fragmentation by water turbulence or by animals can cause lower values of S_p, which, in turn, theoretically will increase decay rates in spite of constant chemical recalcitrance or constant environmental conditions.

The relative rate of decomposition k is the amount of detrital carbon metabolized per unit time (e.g., POM→DOM, POM→microbial cells, DOM→CO_2, and so on). Commonly observed rates of decay through time may be separated into three phases (Fig. 21.1). The first phase (A) is a period of increasing weight loss from leaching, autolytic release, or both, of DOM. Much of this DOM is of simple composition and is decomposed readily. The quantity and composition of DOM persisting over time (days) are influenced greatly by temperature and oxygen or alternate electron acceptors. The rapid release and decay of DOM in phase A

may follow a logistic S curve (line a', Fig. 21.1) as, for example, in the release and decay of DOM from a phytoplankton cell. As larger organic particles begin to decompose under natural conditions, decay rates may be slow at first, followed by an increasing rate as more cells senesce and lyse, and then finally slow again as all cells complete senescence (line a).

Following the maximum rate of weight loss in phase A, the decay rate decreases during the phase of decomposition (B) of POM, when the interactions of the factors controlling degradation have their greatest influence. Utilization of the most readily available substrates usually occurs first (phase A), so that the relative chemical recalcitrance of the POM increases over time. Simultaneously, concentrations of dissolved oxygen and mineral nutrients decrease especially under nonturbulent conditions as is commonly the case.

As conditions become more productive, the increased loading of detrital organic matter causes deterioration of conditions conducive to rapid and complete mineralization. As a result, more organic matter persists in the ecosystem in reduced form [cf., Rich and Wetzel (1978) and Wetzel (1979, 1995)]. In the last phase (C of Fig. 21.1), the rate of decay of this increasingly recalcitrant POM approaches closely an asymptotic limit of zero. Decomposition in phase C can be altered or accelerated by changes in physical conditions or by replenishment of mineral nutrients or electron acceptors, as may occur during turbulent circulation. Much of the detritus of this phase is highly recalcitrant and subject to such slow rates of decay that it may be incorporated permanently into the sediments (see Exercise 27).

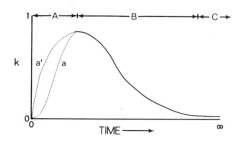

Figure 21.1. Generalized sequence of decay rates. [From Godshalk and Wetzel, (1978c).]

The following exercises examine the rates of decomposition and mineralization of phytoplankton, macrophytes, and leaf fall. Since rates of decomposition are slow (≈ 1 to 3%/day), it is necessary to use long-term experiments, or to use radioctively labeled organic matter to measure the rates of production of metabolic end products in short-term experiments.

EXERCISES

MINERALIZATION OF PHYTOPLANKTON

The method discussed here measures the mineralization rates of phytoplanktonic carbon (both particulate and dissolved fractions) in lakes. The rates of decomposition are determined by monitoring $^{14}CO_2$ evolved from the microbial respiration of naturally occurring particulate and dissolved phytoplankton detritus labeled with ^{14}C (Cole and Likens, 1979; Cole et al., 1984). The designs of the incubation bottle and of the $^{14}CO_2$ trapping system provide a method for measuring mineralization rates in intact microbial communities under in situ temperatures and dissolved gas concentrations without extreme manipulations.

Algal DOC can be mineralized rapidly by bacteria. For example, Cole (1985) found that only about 12% of the net algal primary productivity in Mirror Lake, a softwater lake in New Hampshire, accumulated in the sediments each year. Many studies of this subject have shown that from 75 to 95% of the phytoplanktonic organic carbon is decomposed before reaching the sediments (cf., Wetzel (1983)].

Procedures

1. Obtain large (ca. 3 l) volumes of water with natural phytoplankton populations from two or more depths within the epi- and metalimnion of a lake.
2. Place the samples into large, ground-glass-stoppered bottles (2.5 l) and add approximately 40 μCi NaH$^{14}CO_3$ to each. Mix well by *gentle* inversion several times. *Caution*: Strictly observe all precautionary measures for handling radioactive materials (see Exercise 14). Clip the bottles to a suspension line and return the bottles to the depth from which they were collected. Incubate the samples in situ for 3 days. This time of incubation should exceed the average turnover time of the algae and should be adequate to obtain reasonably uniform labeling of the algal cells.
3. After the incubation, divide the well-mixed sample into two parts. From one part, harvest cells by filtration from a known volume onto precombusted (500°C) glass fiber filters. Use a 2-l filtration flask and save a large portion of the filtrate. Remove the filters, subject them to *fumes* (*not* liquid) of HCl for 30 min *in a hood*, and dry under desiccation (see the simple methods for this operation discussed in Exercise 14). The treatment with fumes kills the algae and removes residual, adherent H$^{14}CO_3^-$.
4. From the second part of the sample, remove several 50-ml aliquots of the labeled algae, filter onto membrane filters (0.45-μm pore size), and treat the filters identically to the glass fiber filters (Step 3). Determine the amount of ^{14}C assimilated (primary productivity) by assaying the incorporated ^{14}C; use the methods discussed in Exercise 14 (liquid scintillation radioassay).
5. The filtrate from the labeled algae and detritus (Step 3) presumably contains phytoplankton exudates as well as solubilized phytoplankton components in various stages of decomposition. Bubble this filtrate vigorously *in a hood* with CO_2 or CO_2-enriched air for 12 h to remove residual $^{14}CO_2$. Filter the sparged filtrate through 0.45-μm pore size membrane filters. Slowly evaporate the filtrate to dryness at 50°C or lyophilize to dryness.

Reconstitute the dried residue with distilled water to a final volume of 50 to 100 ml and seal in glass ampoules or bottles. Use immediately or freeze until use.

6. Collect lake water from the same depths as the original samples were taken and fill four clean 300-ml opaque BOD (oxygen) bottles. Measure the temperature at these depths. Add materials to duplicate bottles as follows:

   a. Add one glass fiber filter containing radioactively labeled algae and detritus into duplicate bottles; seal and cover the top of each with aluminium foil to exclude all light; mix well.

   b. Remove 2.5 or 5.0 ml of water from two bottles; discard. Add 2.5 or 5.0 ml of the concentrated filtrate to each, seal, and cap with aluminium foil. Mix well.

   c. Incubate the bottles in situ or in the laboratory at ambient temperatures for between 12 and 72 h

   d. Control bottles, using lake water that has been autoclaved (1 h; cooled) or treated with formalin, are essential. Otherwise, treat identically to the experimental samples.

7. At the end of the incubation, remove 60 ml from each bottle in duplicate 30-ml aliquots for measurement of dissolved inorganic carbon (see Exercise 8) and to create a head space in the bottle necessary for the next step.

8. Place a simple sparging apparatus into each bottle, containing the remaining 240 ml of sample, as indicated in Fig. 21.2. Acidify the samples to about pH 2 by injection of H_2SO_4 through the serum stopper. *Slowly* bubble the solution with N_2 gas (10 ml/min) for 6 h, forcing the output through a CO_2 trap (2 ml of ethanolamine-methyl cellosolve or similar

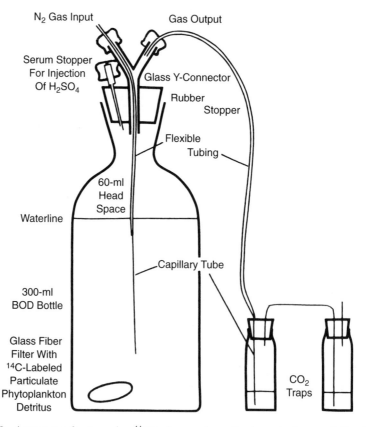

Figure 21.2. Apparatus for trapping $^{14}CO_2$ from mineralized particulate and dissolved phytoplanktonic detritus. [Modified from Cole and Likens (1979).]

organic material that absorbs CO_2 readily; see Exercise 20). Add 10 ml of Bray's solution (or other scintillation fluid; Exercise 20) and radioassay by liquid scintillation. The efficiency of trapping is usually >95%; a second trap positioned in series will remove virtually all of the CO_2 evolved.

9. After measuring the volumes and radioactivities of labeled particulate detritus and of dissolved detritus and the radioactivity of the CO_2 evolved during the incubation, calculate the percentage of the particulate and dissolved organic matter mineralized during the incubation period and per day.

10. Answer the questions on pp. 309–310.

DECOMPOSITION OF AQUATIC PLANTS

The productivity of aquatic macrophytes of the wetland and littoral land–water interface zones constitutes a major source of organic matter input for a majority of the lakes of the world. Much of the organic matter produced by these larger aquatic plants remains in the wetlands and littoral zone of lakes and undergoes decomposition. During senescence and after death of organisms, much of the organic matter is released as soluble compounds. The particulate components decompose at various rates depending on their location, composition, and environmental conditions, particularly those of temperature and oxygen availability. While floating-leaved and many submersed macrophytes decompose relatively rapidly, the tough structural components of emergent macrophytes have markedly reduced rates of decomposition [e.g., Godshalk and Wetzel (1978a, 1978b, 1978c) and Webster and Benfield (1986)]. When the decomposing tissue falls to the sediments in detrital masses, the environment of the aggregates rapidly becomes anaerobic. Under these reducing conditions, rates of decomposition are decreased greatly.

Experimental techniques are available to analyze in situ rates of degradation of aquatic macrophytes. These methods, however, require long-term labeling of the plants with radioisotopes, specialized incubation chambers, and instrumentation to analyze the metabolic end products of decomposition. Alternatively, controlled laboratory experiments related to macrophyte decomposition similarly are long term and complex, not lending themselves to classroom analyses.

A commonly used method of analyzing the decomposition of particulate organic matter of macrophytes or terrestrial leaf fall into lakes and streams is to incubate in situ known amounts of tissues in mesh containers. Changes in weight and chemical composition of the particulate plant detritus then are followed through time. Problems with the methodology and interpretation of data obtained are many and have been analyzed critically by many workers (e.g., Boulton and Boon, 1991). Although this approach neglects the important factor of dissolved detrital organic matter that is leached from the plants and enters the ecosystem, and provides only general information about the controlling mechanisms of microbial metabolism, analyses of gross changes in the particulate organic matter can be instructive.

Procedures

1. Locate a transect through a representative wetland and littoral zone of a lake. Extend the transect through emergent, floating-leaved, and submersed vegetation.

2. Collect the above-ground foliage (leaves and culm material) of representative plants of each group. Obtain additional samples for identification to species. Mark the sites of collection with permanent wooden stakes, which later will be used to support the plant samples.

3. Clip representative subsamples of each species into pieces about 15 to 30 cm in length. The submersed plants should be *gently* blotted (15 sec) between paper toweling to remove excess water. Place 100 to 500 g fresh weight onto labeled drying trays made of aluminium foil and freeze to kill the plants.

4. Weigh the thawed plants promptly to the nearest 0.1 g. Immediately transfer about half of the plant sample to another tared aluminium foil tray and dry at 105°C to constant weight. Determine the dry weight : wet (fresh) weight of the dried sample and calculate the dry weight of the fresh sample to be used in the next step.

5. Transfer the fresh (killed) tissue to nylon mesh bags, "litter bags" (ca. 2-mm mesh; ca. 15 × 30 cm), and close with nylon line (e.g., fishing line). After removal of the samples, dry the aluminium trays and obtain a tare weight for each. Make a sufficient number of litter bags to be able to place replicate samples under natural aerobic and anaerobic conditions and to be able to sacrifice one of the replicate samples at each of several time intervals.

6. Place the litter bags in the littoral areas in appropriate sites. Examples could be:
   a. Submersed and floating-leaved plant samples: Place bags (tethered to a wooden stake) on the surface of the sediments and suspend above the sediments (e.g., 50 cm) in well-aerated water.
   b. Emergent plants: Tether bags on the surface of the sediments among particulate detritus and suspend above the water exposed to the air. The later site simulates "standing dead" of many emergent macrophytes that remain long after death. [See, for example, Davis and van der Valk (1978).]

7. At weekly or biweekly intervals, remove the litter bags from each site. Carefully place the residual plant detritus into tared aluminium foil trays, dry to constant weight at 105°C, cool under desiccation, and weigh. Determine the dry weight and calculate the change in weight over the time intervals.

8. If possible, remove subsamples of the dry material and grind in an appropriate mill (e.g., Wiley mill, 40-mesh) or to a fine powder with a clean mortar and pestle. Determine the organic carbon, nitrogen, and phosphorus content by methods outlined in Exercises 7, 9, and 27. Determine changes through time.

9. Answer the questions on pp. 309–310.

DECOMPOSITION OF LEAF FALL IN STREAMS

Organic matter the enters aquatic ecosystems from allochthonous sources originates primarily from terrestrial primary productivity. This organic matter of plant origin occurs in both dissolved and particulate forms and is transformed variously by microbial degradation and animal utilization during transport to the receiving lake system. Much of the input of terrestrial organic matter to streams is in the form of soluble compounds either derived as leachates from the flora or carried in drainage water from the terrestrial ecosystem in various stages of fungal and bacterial decomposition.

Particulate organic matter can fall directly into streams from overhanging vegetation, be transported there by drainage water, or be windblown into the stream. The POM from trees and ground vegetation can provide highly significant amounts of organic matter to streams, both as POM and as leached DOM [e.g., Fisher and Likens (1973)]. The inputs can be seasonal, especially in autumn, to woodland streams passing through deciduous forests. Some of the large POM may become trapped in the stream channel, as, for example, in organic debris dams or aggregations of leaves [see Bilby and Likens (1980)]. Accumulated leaves undergo colonization by bacteria, fungi, and invertebrates in complex successional patterns within the various microhabitats [e.g., Suberkropp and Klug, (1976)]. As the resistant plant material is degraded, solubilized products of decomposition are utilized by bacteria living as stratified populations in the steep redox gradients of the compacted plant tissue (Suberkropp et al., 1976). The detrital POM and its associated microflora serves as a food source for numerous aquatic invertebrates [cf., review by Cummins (1973)]. The shredding, collecting, and grazing activities of aquatic insects can accelerate the reduction of the size of POM and subsequent microbial degradation. Much of the animal nutrition is obtained from the attached microflora rather than from the POM itself.

Procedures

1. Obtain a number of dead leaves from two types of trees, for example, "tough" leaves, such as oak or hickory, and more readily degradable leaves, such as aspen or willow. Conifer leaves (needles) also could be used as a resistant type (would require large-mesh litter bags as used in the macrophyte exercise).

2. Tether a known air dry weight of the leaves of each type in packs with monofilament nylon line to regular construction bricks. Label each brick appropriately. Make sufficient leaf "packs" of each species to permit the removal of replicates for analysis at three-week intervals for 12 to 15 weeks.

3. Place the leaf packs within a relatively uniform area of the stream substrata with the leaves facing upstream, to stimulate natural entrapment. Stake the shoreline area so that the samples can be found later. If the current is strong or the stream is subject to spates that could move the bricks, firmly anchor with rods penetrating through the holes of the bricks well into the substratum.

4. Enclose some of the leaf packs completely with fine mesh (ca. 1.0-mm) nylon screening to restrict access of large aquatic invertebrates. If large conifer needles are used, enclose the samples in large mesh (e.g., 5-mm) and small mesh (ca. 1.0-mm) litter bags.

5. At biweekly intervals, remove a replicate of each leaf species (exposed and those in enclosures). Carefully remove any aquatic invertebrates from the residual leaf material, and examine the groups and amounts of invertebrates that have colonized the leaf material. Then place the POM into tared trays made of aluminum foil and dry to constant weight at 105°C. Cool under desiccation and immediately weigh.

6. If possible, remove subsamples of the dry material and grind in an appropriate mill (e.g., a Wiley mill, 40-mesh) or to a fine powder with a clean mortar and pestle. Determine the organic carbon, nitrogen, and phosphorus content by the methods outlined in Exercises 7, 9, and 27.

7. Determine the changes in dry weight and, if possible, organic C, N, and P through time for the leaves of different species and exposed versus unexposed to the feeding activities of large invertebrates.

8. Answer the following questions.

FUNGAL BIOMASS AND GROWTH RATES DURING DECOMPOSITION OF PARTICULATE ORGANIC MATTER

The fungi are an extraordinarily large (10^6 species) and diverse group of microorganisms. An appreciable number of fungi have evolved biochemical pathways that allow decomposition of complex macromolecules of the structural tissues of higher plants. In aquatic ecosystems, fungi are critical to the degradation of the organic matter generated in wetland and littoral zones by emergent, floating-leaved, and submersed vascular plants. Many fungi degrade plant macromolecules by internal pervasion of mycelia within the solid organic substrates. Many of the mycelial hyphae are morphologically indistinct and very difficult to identify or distinguish between living and senescent cells. The fungal mycelial mass thus becomes a part of the decomposing plant mass, and as a result cannot be separated readily from the plant organic matter or examined quantitatively by direct microscopy.

An organic sterol, ergosterol, is highly specific to fungi and is located in the plasma membranes of active fungal cells. Dead hyphae without cytoplasm would contribute little to fungal ergosterol, and therefore estimates of the ergosterol content can provide an estimate of the living fungal biomass. Additionally, an estimate of fungal specific growth rate can be evaluated by the rate of synthesis of ergosterol by measurement of the flux of a sterol-precursor molecule acetate, labeled with ^{14}C, into ergosterol. The following exercise examines differences in the quantities and activities of fungal communities by estimating active fungal biomass by ergosterol analysis (methods derived from Newell et al., 1988; Gessner et al., 1991; Suberkropp and Weyers, 1996) and fungal growth by rates of acetate incorporation into ergosterol (methods modified from Newell and Fallon, 1991; Gessner and Chauvet, 1993; Newell, 1993; Suberkropp and Weyers, 1996).

Procedures

1. Collect samples of standing dead emergent aquatic plants and decaying aquatic plant or fallen leaves from wetland, lake, or stream ecosystems. Store samples as briefly as possible at conditions of temperature and humidity close to ambient. Remove subsamples of standard size or volume from each (e.g., cork-borer discs from leaves). Separate into two groups of several replicates each, one for organic mass analyses and the other for ergosterol procedures.

2. Dry organic-mass replicates in a drying oven at 105°C until fully dry, cool under desiccation, and weigh for dry mass. Heat samples in prefired and tared crucibles at 550°C for at least 4 h, cool under desiccation, and weigh. Subtract ash mass from dry mass to yield organic mass.

3. Under sterile conditions, as in a laminar-flow hood, place ergosterol replicates each into clean, sterile scintillation vials. Add 5.0 ml of bacteria-free water (0.2-μm pore size filtered) and incubate in darkness 1 h while avoiding excessive agitation of the samples. Add formaldehyde to yield a 2% final concentration to adsorption (killed) controls and cap tightly.

4. Add [1-^{14}C]-sodium acetate in bacteria-free solution with carrier acetate to a final concentration of 5 mM to both living and adsorption-control vials. Experimentation is required to determine the minimum acceptable levels of carrier acetate that is acceptable (see Newell and Fallon, 1991; Suberkropp and Weyers, 1996). For example, about 10^{8} Bq/mmol specific activity (1 Bq $\approx$ 27 pCi = 1 dpm) of acetate at 5 mM acetate concentration was adequate for approximately a 70-mg organic mass of a decaying leaf of several weeks.

5. Incubate for a defined period commensurate with temperature (2-4 h at 20°C; 3–4 h at 10°C).

6. End the incubation by adding formaldehyde to a 2% final concentration. Carefully remove litter or plant sample from the radiolabeled solution, rinse, and place in vials under 5 ml methanol (HPLC grade) at 4°C in darkness.

7. Prepare a series of concentrations of ergosterol standards in 5 ml methanol over the range expected for samples.

8. Transfer methanol and subsamples from living incubations, controls, and standards to reflux flasks (25 × 200-mm screw-capped culture tubes with the Teflon-lined caps loosely placed on top). Rinse sample vials with 2.5 ml methanol and add rinse to the reflux flasks. Add a measured additional quantity of methanol to dissolve the ergosterol fully. Reflux operations should be done at 80°C in a fume hood with the temperature of air flowing over the tube tops projecting from the block heater much lower (>10°C) than the boiling point of methanol (65°C).

9. Add 5 ml of ethanolic KOH solution (see Apparatus and Supplies) to each reflux flask. Add a Teflon boiling chip to each flask and reflux (80°C) for 30 min in a block heater. Remove tubes from block heater and allow to cool.

10. The boiling chip and plant material are then removed from each tube with a pair of forceps. Forceps should be rinsed with methanol between each sample.

11. Add 5 ml ultrapure water and 5 ml pentane (HPLC grade) to each tube and seal tightly. (*Caution*: Pentane is very flammable.) With the tubes held together tightly with two test-tube racks, invert gently 30 times to extract the ergosterol into the non-polar upper pentane fraction. Collect the lower polar fraction in a separate beaker. Transfer the pentane layer with a Pasteur pipet into a 15 × 125-mm screw-capped culture tube. Repeat the pentane extraction two additional times on the polar fraction with 5 ml of pentane with repeated agitation. Discard the lower, polar fraction into radioactive waste. Combine the second and third pentane fractions with the first.

12. With the block heater or dry-bath temperature at 30°C, fit vials into wells and immediately situate air drying manifold outlet above each vial. Caution: *Do not* allow pentane in vials to boil (36°C boiling point; 40°C flash point). Continue air flow into each vial until all pentane has evaporated.

13. Add 2.0 ml of methanol (HPLC grade) to each vial to redissolve neutral lipids. Tightly cap vials and partially submerge in a sonic cleaning bath and sonicate for 5–10 min. Push each redissolved sample through a Teflon filter cartridge into a 2-ml vial. Immediately cap securely to prevent evaporation. These samples can be kept frozen (−20°C) for later HPLC analysis.

14. Allow methanol eluent to flow and equilibrate in the HPLC system for 1 h at 282-nm wavelength. Rinse syringe and injector valve with methanol between all samples and inject methanol blank. Inject pure standard and remaining standards and samples. During passage of standards, record times of appearance and completion of ergosterol peaks. Use these times and recorder outputs to activate fraction collectors for collection of radiolabeled samples in scintillation vials at the sample outlet valve. Collect two standard ergosterol fractions for determinations of background levels of ^{14}C. A solid phase extraction technique offers an expedient alternative for concentration and purification prior to HPLC analysis (Gessner and Schmitt, 1996).

15. Add 10-ml scintillation fluor to each vial of the ergosterol fraction. Mix thoroughly; wipe external walls of vials, and radioassay in a scintillation counter.

Calculations

1. Determine the ergosterol concentration per sample by comparison of areas of sample and standard peaks. Divide these values by the organic mass to yield ergosterol per unit organic mass. To convert approximately ergosterol contents determined in samples to biomass values of aquatic hyphomycetes, multiply by 182 (Gessner and Chauvet, 1993).

2. Subtract dpm for adsorption controls from values for the incubations to yield dpm from fungal synthesis of ergosterol from the radiolabeled acetate.

3. Determine the growth rate (μ_c, per day) by:

$$\mu_c = (24)\,(dpm_c)/(Q)(SA)(I)(t)(E_c)$$

where:

24	= pg C per pmol acetate	
dpm_c	= corrected dpm	
Q	= correction for counting efficiency	
SA	= specific activity of radioacetate (dpm per pmole acetate)	
I	= decimal fraction of sample injected	
t	= time of incubation in days	
E_c	= amount of ergosterol carbon (pg) per sample	

Among several species of stream fungi examined, ratios of rates of biomass increase to rates of acetate incorporation into ergosterol were similar, with a mean of 19.3 μg of biomass per nmol of acetate incorporated (Suberkropp and Weyers, 1996).

Questions

1. In analyses of phytoplankton mineralization, what types of controls might be incorporated into the decomposition incubations? Since the filtrate was not acidified to remove $^{14}CO_2$, rather relied on extensive sparging with nonlabeled CO_2 for exchange and removal, would a small amount of background contamination be expected? How severe do you think this contamination would be? [See Cole and Likens (1979).]

2. Compare the estimated rates of mineralization to the estimated residence time (i.e., sedimentation rates) of phytoplankton in the water column. How might large thermal discontinuities in the metalimnion alter the rate of sedimentation?

3. If much of the phytoplankton particulate detritus were to reach the sediments, how might the rates of decomposition be altered in comparison to those in the water column? How might the mineralization method be modified to measure rates in the sediments?

4. How might the benthic mineralization differ between hard and soft waters? [See Fallon and Brock (1982) and Cole (1985).]

5. How might the rates of decomposition of algae differ among various species, temperatures, pH, and other factors? [See Mills and Alexander (1974) and Gunnison and Alexander (1975).]

6. How would the rates of mineralization of phytoplankton determined by the method outlined be altered if the water sampled were completely anaerobic? What gaseous products might be evolved? Soluble products?

7. It is easy to criticize methods that attack the difficult problem of measuring in situ rates of decomposition. Suggest alternative methods. [See Saunders (1972) and Storch and Saunders (1978).]

8. How might you calculate the theoretical turnover time for the POM of the macrophytes or stream leaf packs?

9. What is the fate of organic matter (both DOM and POM) of macrophytes decomposing in wetlands surrounding a lake basin? Of macrophytes in the littoral zone of the lake per se?

10. What are the possible fates of DOM leached from POM and decomposing POM, and POM detrital plant materials entering a stream? [See Fisher and Likens (1973) and Wetzel and Manny (1972).]

11. What chemical and metabolic pathways could be involved in the shifting of DOM of POM? [See Lush and Hynes (1973) and Lock and Hynes (1976).]

12. How do anaerobic conditions affect the rates of decomposition of macrophytes? Affect the methods of measuring in situ rates of decomposition by the analysis of gaseous metabolic end products?

13. Is the microhabitat within a leaf pack in a flowing stream necessarily aerobic? Why? [See Boulton and Boon (1991).]

14. What is the role of invertebrates in the decomposition of particulate organic matter?

15. Compare decomposition rates and processes in freshwater and marine environments. [See Capone and Kiene (1988).]

Apparatus and Supplies

1. Phytoplankton mineralization
   a. Nonmetallic water sampler; thermistor thermometer.
   b. 2.5-1 bottles (ground-glass-stoppered); suspension clips, line, and buoy; anchor for buoy.
   c. $NaH^{14}CO_3$ stock solution.
   d. Filtration apparatus; precombusted (500°C) glass fiber filters (e.g., Whatman GF/F); chamber for exposing filters to fumes of HCl (see Exercise 14); desiccator; membrane filters (0.45-μm pore size).
   e. CO_2 or CO_2-enriched air; N_2 gas.
   f. 50°C oven or block heater (or lyophilization apparatus).
   g. 300-ml ground-glass-stoppered BOD (Biological Oxygen Demand) bottles, made opaque with black electrician's tape or aluminum foil.
   h. Apparatus for measuring dissolved inorganic carbon (see Exercise 8).
   i. N_2-sparging apparatus (as illustrated in Fig. 21.2); ethanolamine-methyl cellosolve (or similar, Exercise 20) CO_2-trapping agent; scintillation solution (e.g., 2:1 toluene:2-methoxyethanol plus 6 g PPO/1); liquid scintillation radioassay instrumentation.
2. Decomposition of macrophytes
   a. Tape rule (ca. 30 m); clippers, scissors; wooden stakes (ca 1 m long).

  b. Balance; drying oven (105°C); aluminum foil; nylon monofilament line; nylon mesh (ca. 2-mm mesh); desiccator.

  c. Apparatus and supplies for analyses of organic carbon, nitrogen, and phosphorus (optional) (see Exercises 7, 9, and 27).

3. Decomposition of leaves in streams:

  a. Construction bricks (with holes); nylon monofilament line; metal stakes, if needed; nylon mesh (ca. 1.0–mm mesh).

  b. Balance; drying oven (105°C); aluminum foil; desiccator.

  c. Apparatus and supplies for analyses of organic carbon, nitrogen, and phosphorus (optional) (see Exercises 7, 9, and 27).

4. Fungal biomass and growth:

  a. Refluxing system of multiple glassware units.

  b. 80°C water bath.

  c. Dry bath heater, 35–40°C, aluminum wells and glass vials to fit.

  d. Air distribution drying manifold.

  e. Sonicating cleaning bath.

  f. HPLC system with reverse phase column, UV detector, and integrator/recorder.

  g. 500-μl syringe for HPLC injection.

  h. Liquid scintillation spectrometer.

  i. Inert boiling chips.

  j. Separatory funnels.

  k. Teflon filter cartridges (0.45-μm pore size).

  l. HPLC grade solvents: methanol, ethanol, pentane.

  m. Ergosterol.

  n. Potassium hydroxide.

  o. Sodium acetate.

  p. [1-^{14}C]sodium acetate.

  q. Hydroxide ethanol solution: Mix potassium hydroxide (40 mg/ml) in reagent ethanol at a ratio of 95:5, ethanol: water because KOH will not dissolve without the water. For example, 1.6 g KOH dissolved in 2 ml water and added to 38 ml methanol. Renew frequently as the solution is not stable.

References

Alexander, M. 1965. Biodegradation: Problems of molecular recalcitrance and microbial fallibility. Adv. Appl. Microbiol. *7*:35–80.

Bilby, R.E. and G.E. Likens. 1980. Importance of organic debris dams in the structure and function of stream ecosystems. Ecology *61*:1107–1113.

Boulton, A.J. and P.I. Boon. 1991. A review of methodology used to measure leaf litter decomposition in lotic environments: Time to turn over an old leaf? Aust. J. Mar. Freshwat. Res. *42*:1–43.

Capone, D.G. and R.P. Kiene. 1988. Comparison of microbial dynamics in marine and freshwater sediments. Contrasts in anaerobic carbon metabolism. Limnol. Oceanogr. *33*:725–749.

Cole, J.J. 1985. Decomposition. pp. 302–310. *In*: G.E. Likens, Editor. An Ecosystem Approach to Aquatic Ecology: Mirror Lake and Its Environment. Springer-Verlag, New York.

Cole, J.J. and G.E. Likens. 1979. Measurements of mineralization of phytoplankton detritus in an oligotrophic lake. Limnol. Oceanogr. *24*:541–547.

Cole, J.J., G.E. Likens, and J.E. Hobbie. 1984. Decomposition of phytoplankton in an oligotrophic lake. Oikos *42*:257–266.

Cummins, K.W. 1973. Trophic relations of aquatic insects. Ann. Rev. Entomol. *18*:183–206.

Davis, C.B. and A.G. van der Valk. 1978. The decomposition of standing and fallen litter of *Typha glauca* and *Scirpus fluviatilis*. Can. J. Bot. *56*:662–675.

Fallon, R.D. and T.D. Brock. 1982. Planktonic blue-green algae. Production, sedimentation and decomposition in Lake Mendota, Wisconsin. Limnol. Oceanogr. *25*:72–88.

Fisher, S.G. and G.E. Likens. 1973. Energy flows in Bear Brook, New Hampshire: An integrative approach to stream ecosystem metabolism. Ecol. Monogr. *43*:421–439.

Gessner, M.O., M.A. Bauchrowitz, and M. Escautier. 1991. Extraction and quantification of ergosterol as a measure of fungal biomass in leaf litter. Microb. Ecol. *22*:285–291.

Gessner, M.O. and E. Chauvet. 1993. Ergosterol-to-biomass conversion factors for aquatic hyphomycetes. Appl. Environ. Microbiol. *59*:502–507.

Gessner, M.O. and A.L. Schmitt. 1996. Use of solid-phase extraction to determine ergosterol concentrations in plant tissue colonized by fungi. Appl. Environ. Microbiol. *62*:415–419.

Godshalk, G.L. and R.G. Wetzel. 1978a. Decomposition of aquatic angiosperms. I. Dissolved components. Aquatic Bot. *5*:281–300.

Godshalk, G.L. and R.G. Wetzel. 1978b. Decomposition of aquatic angiosperms. II. Particulate components. Aquatic Bot. *5*:301–327.

Godshalk, G.L. and R.G. Wetzel. 1978c. Decomposition of aquatic angiosperms. III. *Zostera marina* L. and a conceptual model of decomposition. Aquatic Bot. *5*:329–354.

Gunninson, D. and M. Alexander. 1975. Basis for the resistance of several algae to microbial decomposition. Appl. Microbiol. *29*:729–738.

Lock, M.A. and H.B.N. Hynes. 1976. The fate of "dissolved" organic carbon derived from autumn-shed maple leaves (*Acer saccharum*) in a temperate hardwater stream. Limnol. Oceanogr. *21*:436–443.

Lush, D.L. and H.B.N. Hynes. 1973. The formation of particles in freshwater leachates of dead leaves. Limnol. Oceanogr. *18*:968–977.

Mills, A.L. and M. Alexander. 1974. Microbial decomposition of species of freshwater planktonic algae. J. Environ. Qual. *3*:423–428.

Newell, S.Y. 1993. Membrane-containing fungal mass and fungal specific growth rate in natural samples. pp. 579–586. In: P.F. Kemp, B.F. Sherr, E.B. Sherr, and J.J. Cole, Editors. Lewis Publishers, Boca Raton.

Newell, S.Y., T.L. Arsuffi, and R.D. Fallon. 1988. Fundamental procedures for determining ergosterol content of decaying plant material by liquid chromatography. Appl. Environ. Microbiol. *54*:1876–1879.

Newell, S.Y. and R.D. Fallon. 1991. Toward a method for measuring instantaneous fungal growth rates in field samples. Ecology 72:1547–1559.

Rich, P.H. and R.G. Wetzel. 1978. Detritus in the lake ecosystem. Amer. Nat. *112*:57–71.

Saunders, G.W., Jr. 1972. The kinetics of extracellular release of soluble organic matter by plankton. Verh. Int. Ver. Limnol. *18*:140–146.

Storch, T.A. and G.W. Saunders. 1978. Phytoplankton extracellular release and its relation to the seasonal cycle of dissolved organic carbon in a eutrophic lake. Limnol. Oceanogr. *23*:112–119.

Suberkropp, K. and M.J. Klug. 1976. Fungi and bacteria associated with leaves during processing in a woodland stream. Ecology *57*:707–719.

Suberkropp, K. and H. Weyers. 1996. Application of fungal and bacterial production methodologies to decomposing leaves in streams. Appl. Environ. Microbiol. *62*:1610–1615.

Suberkropp, K., G.L. Godshalk, and M.J. Klug. 1976. Changes in the chemical composition of leaves during processing in a woodland stream. Ecology *57*:720–727.

Webster, J.R. and E.F. Benfield. 1986. Vascular plant breakdown in aquatic environments. Ann. Rev. Ecol. Syst. *17*:567–594.

Wetzel, R.G. 1979. The role of the littoral zone and detritus in lake metabolism. pp. 145–161. *In*: G.E. Likens, W. Rodhe, and C. Serruya, Editors. Lake Metabolism and Lake Management. Arch. Hydrobiol. Beih. Ergebn. Limnol. *13*:145–161.

Wetzel, R.G. 1983. Limnology. 2nd Ed. Saunders Coll., Philadelphia. 860 pp.

Wetzel, R.G. 1995. Death, detritus, and energy flow in aquatic ecosystems. Freshwat. Biol. *33*:83–89.

Wetzel, R.G. and B.A. Manny. 1972. Decomposition of dissolved organic carbon and nitrogen compounds from leaves in an experimental hardwater stream. Limnol. Oceanogr. *17*:927–931.

The Littoral Zone

The littoral zone represents an interface between the land of the drainage basin and the open water of the lake, reservoir, or stream (Wetzel, 1990). The extent of the development of the littoral zone is highly variable and depends on both the geomorphology of the basin and the rate of sedimentation that has occurred since its formation.

The emergent macrophytes of littoral wetlands and the shore regions of lakes are among the most productive habitats of the biosphere [cf., Whittaker and Likens (1975)]. Productivity decreases with increasing distance from the lake, as one passes into terrestrial conditions, and decreases as one passes toward the lake through zones of floating-leaved macrophytes, submersed macrophytes, and phytoplankton. For a number of physiological reasons [cf., Wetzel (1983a, 1990)], the surface area of submersed vegetation is much greater than that for other types of littoral macrophytes. These surfaces are colonized densely by epiphytic algae which can contribute much to the total productivity of the lake. In addition, algae can grow profusely when attached to sediments or in loosely attached clumps among the littoral macrovegetation. All of this primary productivity in the littoral zone results in a large input of dissolved and particulate organic detritus to the lake system. Particulate organic detritus is deposited largely in the sediments of the littoral zone; a portion is transported to sediments of deeper water where degradation continues.

The high productivity of living organic matter and the detrital accumulations in sediments of the littoral zone provide an abundance of habitats and food resources for zooplankton, invertebrates, and vertebrates such as fishes, amphibians, birds, and muskrats. Many of the littoral animals are specialized in their adaptation to conditions in the littoral zone (cf. many articles in Jeppesen et al., 1997). As with the littoral algae, the diversity of littoral fauna is very high.

The spatial conditions within the littoral zone are complex. Extreme heterogeneity exists in the distributions and productivities of the macro- and microflora along the gradient from the emergent wetland vegetation to submersed conditions. The heterogeneity and abundance of microhabitats make quantitative sampling difficult. As a result, knowledge of the metabolism of the flora and fauna is poor in comparison with what is known of the pelagic biota. Nonetheless, it now is known that the littoral biota are a major component of the metabolism of the whole ecosystem in most lakes of the world (Wetzel, 1979, 1990). An understanding of the physiological relationship of the littoral biota to the ecology of the lake presents an imposing challenge to ecologists when evaluating the role of this interface zone in the metabolism of the whole ecosystem.

The following exercise is designed to introduce some aspects of the structure and function of littoral biota. The limitations of the various analytical techniques will become readily apparent. The problems of extreme heterogeneity in the distribution and metabolism of organisms also will be obvious. However, the interrelationships of the littoral organisms to each other, to environmental gradients, and to the rest of the lake system can be seen in overview.

LITTORAL MACROFLORA

Distribution and Biomass of Vegetation

The qualitative distribution of the macrovegetation may be analyzed by careful sampling within quadrats. Sampling sites can be selected from either a stratified random design or along a transect (see discussions in Exercise 11 and Appendix 2). The calculation of relative abundance (percent) of species within a given area is helpful in interpreting data obtained by intensive quantitative measurements of biomass. Although many marsh and aquatic plants can be identified to species from their vegetative characteristics, reproductive parts are frequently needed for positive identification [cf., Muenscher (1944), Fassett (1957), Mason (1957), Prescott (1969), Correll and Correll (1972), and Godfrey and Wooten (1979, 1981)].

As much as 90% of the biomass of aquatic plants can occur below ground. While the average percentage of biomass production below the sediments is higher among emergent macrophytes than in submersed plants, in all cases it is essential to sample biomass from both above and below ground in any quantitative study. Among annual plants which reproduce by seed, the biomass persisting from one year to the next is negligible. Because most aquatic plants are perennial, the above-ground biomass dies and enters the detrital pool annually; part of the below-ground biomass dies, but much of it remains alive and perenniates new shoots for several years (up to 15 yr in certain water lilies) (Eissenstat and Yanai, 1997). Analysis of growth, then, is complicated in that the quantity of roots/rhizomes often constitutes the cumulative result of several years' growth. Estimations of annual turnover rates of root material are possible but require rather elaborate experimental techniques.

The determination of the below-ground biomass of macrophytes requires care because it is difficult to differentiate which parts are living and which are dead. Often when the below-ground materials are separated, the color of the roots is a rough indicator of which are alive (lighter color, more firm in texture) and which are dead (darker browns, black; softer). Tetrazolium dyes, some of which become red when they react with respiratory products, can be helpful in determining viability.

Since the below-ground biomass is usually very dense, it is impractical to estimate this biomass from whole quadrats (e.g., $0.25\,m^2$) used to determine above-ground biomass. Therefore, replicate sediment cores including the below-ground biomass often are taken within the quadrats. The material from these cores then is washed with a vigorous stream of water over a straining sieve to catch any fragments. The living components are separated from the dead.

The dry weight of the biomass is determined after drying the material for 24 h at 105°C or by freeze-drying. Representative samples then are homogenized, and subsamples are combusted for at least 4 h at 550°C to remove organic matter from ash materials. Dry weight less ash weight yields ash-free dry (organic) weight. The average amount of carbon in the organic matter of aquatic macrophytes has been found to be 46.5% of ash-free dry weight (Westlake, 1965).

Productivity of Macrophytes

The primary productivity of macrophytes is most commonly evaluated by changes in biomass. In an annual plant, the initial biomass of seeds is negligible, and biomass changes follow a typical sigmoid growth curve (Fig. 22.1). Gross productivity

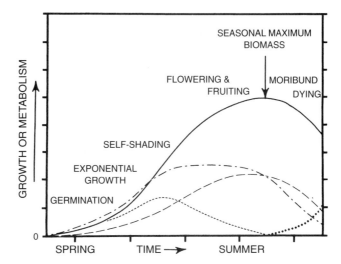

Figure 22.1. Generalized growth and metabolic patterns for a typical annual aquatic macrophute. —— = Biomass; -——- = current gross productivity; ---- = current net productivity; —— = current respiration rate; ⋯⋯ = death losses. [From Wetzel, (1983a) after Westlake (1965).]

decreases with time and becomes negative as respiration continues to increase with greater biomass. Maximum biomass, the maximum cumulative net production, is reached when the current daily net productivity becomes zero. This type of annual growth is characteristic of many plants in which the number of perenniating organs decreases to a minimum.

When much of the plant survives by means of below-ground tissue to the next growing season, the annual net production may be estimated from the difference between the final and initial biomass. Many aquatic plants lose a portion of the biomass produced during the growing season, e.g., sloughing of lower leaves as the main apical tissue grows. This complication requires that the rate of turnover of foliage be evaluated [cf., Westlake (1965), Mathews and Westlake (1969), Dickerman and Wetzel (1985), Dickerman et al. (1986), and Wetzel and Howe (1999)].

LITTORAL ALGAE

In comparison with the quantity of knowledge about phytoplankton, little is known of the physiology or ecology of algae attached to substrata, loosely aggregated in the littoral areas of lakes, or in shallow zones of streams. The littoral zones contain a substantial number of microhabitats associated with surfaces of submersed macrophytes, with particulate detritus, and with the sediments. Moreover, these habitats are changing constantly, e.g., as the larger plants grow, as sediments are redistributed, and so on. As a result, attached algae are distibuted heterogeneously and are exposed to more variable environmental conditions than are planktonic algae.

Because of the heterogeneity in distribution and growth of attached algae, approaches to quantitative analyses have included placing artificial substrata (plates of glass, blocks of styrofoam, and other materials) of defined surface area in the littoral zone. These substrata are left for a period of several weeks for colonization by the microflora [cf., reviews by Sládečková (1962), Wetzel (1964), and Wetzel

(1983b)]. The attached algae then are removed and analyzed quantitatively for biomass by enumeration, pigment analyses, or other parameters (cf., Exercise 10). By the incubation of colonized substrata in chambers, the oxygen difference and ^{14}C-uptake methods have been used to estimate rates of primary productivity [e.g., Allen (1971)]. The ^{14}C method also has been applied following undisturbed incubation in chambers of natural substrata [e.g., Wetzel (1964), and Burkholder and Wetzel (1989)].

Field analyses of the physical, chemical, and biological environmental factors regulating the growth and productivity of periphyton on natural substrata are difficult because of extreme heterogeneity of habitat and attached communities (Wetzel, 1996). In situ manipulations of factors that control the development of attached communities have most frequently involved changes to chemical or biotic characteristics in the water *overlying* the periphyton. Clearly recent evidence has demonstrated the importance of internal recycling of metabolic products and nutrients *within* the periphyton, and the importance of supplemental acquisition of nutrients and gases from the supporting substrata (Wetzel, 1996). The microbiota are metabolically interdependent and can recycle resources rapidly for maintenance, whereby newly acquired inorganic and organic nutrients can be directed to new growth.

Analyses of the effects of external environmental parameters on growth have involved both manipulations in the overlying water as well as alteration of substances diffusing from substrata that support the periphyton. None of these analyses is free of problems. For example, often the periphyton communities are enclosed for various periods of time during which changes in physical (e.g., light, temperature), chemical (e.g., nutrients), or biota (e.g., grazing) are varied. Alternatively, nutrient-diffusing substrata can be used to simulate release of nutrients from supporting materials into the overlying attached communities (Fairchild and Lowe, 1984; Fairchild et al., 1985; Pringle and Bowers, 1984; Tate, 1990; Pringle and Triska, 1996). Experimental demonstrations of these methods are set forth in Option 3 below to indicate some of the approaches that can be used to evaluate environmental effects on these communities. Many variations and modifications are possible dependent upon the questions being asked.

LITTORAL ZOOPLANKTON AND PHYTOPLANKTON

The zooplankton of the littoral zone of lakes generally are quite distinct from those of the open water. Many are adapted to living in close association with aquatic macrophytes. Some, such as *Chydorus*, attain maximum population density in the spring and decline to low numbers in the summer [e.g., Keen (1973)]. Other species are aestival, overwinter as ephippial eggs, and attain maximum densities in late summer and autumn (Goulden, 1971; Keen, 1973; Whiteside, 1974). High mortality in the summer has been associated with predation by other invertebrates and by small fish.

The close association of littoral zooplankton with the macrophytes makes quantitative sampling difficult. A long tube can be lowered into the water, stoppered tightly at the upper end, and the composite vertical sample withdrawn [cf., also Pennak (1962) and Exercise 15]. The sample then is passed through sieves to filter the zooplankton from the water. Fast-swimming zooplankton may avoid collection by this method.

The activity of littoral zooplankton apparently is greater at night (Whiteside and Williams, 1975), when presumably predation pressure is reduced. Many littoral zooplankton migrate upward at night, away from vegetation. This migration has been used to advantage in sampling, by placing a series of inverted funnels, connected to jars by tubing, over the vegetation (Szlauer, 1963; Whiteside and Williams, 1975; Brakke, 1976). Zooplankton swim into these traps, and few can find their way out. The samplers are left in place overnight, and zooplankton are collected in the early morning. Efficiency is especially good for chydorid cladocerans.

LITTORAL BENTHIC FAUNA

The diversity and productivity of benthic animals can be particularly great in the littoral zone. The presence of macrovegetation makes sampling difficult. A number of specialized devices has been developed to sample the vegetation and associated fauna quantitatively [see review of Kajak (1971) and Downing (1984)]. Once the plants and sediments of a known area of the littoral zone are sampled, it is necessary to separate the organisms from the extraneous materials collected. No simple method has been devised, and most are tedious. All of the criteria discussed earlier (Exercise 13) for quantitative sampling apply here as well, and certain of the sorting aides, such as staining, can make easier the task of separating the organisms from the plant material.

EXERCISES

OPTION 1. FIELD ANALYSES ALONG A TRANSECT GRADIENT

Divide into five teams; each team will analyze different components of the littoral zone.

1. Establish a transect perpendicular to the shore line in the littoral zone of a representative lake or reservoir. Extend the ends of the transect from the lower depth limit of submersed vegetation into the emergent vegetation, to a point where environmental conditions indicate that the sediments are not continually water-saturated.
2. One team should collect plants along the transect for species composition, relative abundance, and sediment characteristics (see Exercises 5 and 12).
   a. In the laboratory, these plants should be identified as specifically as experience and time permit. Use taxonomic keys of regional marsh and aquatic flora.
   b. Plot graphically the vegetation distribution and relative abundance along the transect gradient.
   c. Plot the relative changes in the sediment characteristics along the gradient.
3. The second team should evaluate the foliage and root/rhizome biomass of the macrophytes in four or more quadrats along the transect.
   a. Using 0.25- to 0.50-m^2 quadrats, collect all above-ground biomass within the sampling area.
   b. Take at least five random cores of the root/rhizome materials within the quadrat extending down through the root mass. Keep records of the sections of the root cores associated with different dominant plant populations.
   c. In the laboratory, determine the dry weight (105°C) biomass of the foliage. On representative subsamples, establish the ash by combustion at 550°C and calculate the ash-free organic weight of the plant materials.

    d. In white enamel or plastic pans, separate the viable roots and rhizomes from dead or partially decayed materials and sediments. Determine the dry weight (105°C) and the ash-free dry weight (dry wt–ash wt) of all of the materials from the combustion of representative samples at 550°C.

    e. Calculate the dry weight and ash-free dry weight per m² of foliage, rooting material, and the total combined biomass for dominant species of the quadrats. Repeat for other species as time permits.

    f. Calculate the estimated organic carbon of the materials from the ash-free dry weight data.

    g. Plot the data graphically and compare the differences along the transect.

    h. Assuming that the macrophytes are annual plants and that you sampled at maximum seasonal biomass, calculate and compare the primary productivities of the dominant species.

4. The third team should collect both qualitative and quantitative samples of algae attached to plant surfaces and the sediments.

    a. Using the artificial substrata, such as glass slides, that were suspended vertically about one month earlier among the macrophytes and adjacent to littoral sediments, analyze for the following algal biomass components:

      i. Carefully collect the substrata so as to minimize loss upon retrieval. Place into containers so that the surfaces do not rub against the substrata (e.g., wide-mouth bottles with rubber stoppers containing cuts extending 5 mm into the narrow end; one edge of a slide can be inserted into each slot of the stopper and then suspended into the bottle).

      ii. In the laboratory, carefully remove all attached algae from a known area (e.g., cork borer from a styrofoam substratum if these are used; or, using the edge of a clean slide, scrape all microflora from one or both sides of the colonized slide) into a known amount of filtered lake water.

      iii. Mix thoroughly and measure a known portion into a cell counting chamber (e.g., Palmer-Malony cell or sedimentation chamber—see Exercise 10).

      iv. Enumerate and identify algae.

      v. Filter a known volume of the homogeneous algae onto membrane filters and extract chlorophyll a with 90% acetone as described in Exercise 10. Chlorophyll a and phaeopigment concentrations can then be calculated as follows (Wetzel and Westlake, 1971):

$$\mu g \ Chl \ a \ per \ sample = 11.9 \ [2.43(OD_b - OD_a)](v/l)$$

where OD_b = optical density at 665 nm in basic acetone, corrected for turbidity at 750 nm and cell differences; OD_a = optical density at 665 nm after acidification, corrected for turbidity at 750 nm and cell differences; v = volume (ml) of solvent used to extract the sample; l = path length (cm) of spectrophotometric cell; and

$$\mu g \ phaeopigment \ per \ sample = [11.9(v/l)][(1.7 \, OD_a) - Chl \ a]$$

    b. Collect the attached algae from the surfaces of representative macrophytes for which the surface area can be estimated [e.g., submersed portion of a bulrush (*Scirpus*) plant where the area (A) can be estimated as the convex surface of the frustum of a cone: $A = h/2(c + c')$, where c and c' are circumferences ($c = 2\pi r$) of the bases and h is the slant height]. Determine the pigment concentrations per unit area and compare to those determined from artificial substrata located within the same area.

    c. Take cores (ca. 2 cm² in area) of the sediments and extract the pigments from the upper 2 cm, correcting for phaeopigment degradation products. Compare these concentrations with those found in algae attached to substrata "incubated" near the sediments.

    d. If possible, determine the in situ rates of primary productivity of algae attached to the stems and leaves of underwater portions of emergent macrophytes by modification of the oxygen difference of ^{14}C-uptake methods as outlined in Exercise 14.

     i. Place segments of the stems into light and dark bottles (inject ^{14}C) and incubate at the depth from which the plants were taken. Care must be taken not to disturb the epiphytic microflora.

     ii. Simultaneously, incubate control bottles containing only the littoral water. Why?

     iii. At the end of the incubation, fix the oxygen chemically or, when the ^{14}C-uptake method is used, remove the attached algae by scraping and filter onto membrane filters along with the water content of the bottles.

     iv. Determine the surface area of the macrophyte segment (see above).

     v. Evaluate the rate of primary productivity of the attached algae, correcting for that of littoral phytoplankton of the control bottles.

     vi. Calculate the primary productivity of the epiphytic algae per unit area per unit time (see Exercise 14). Using the data of other teams, estimate the productivity per square meter of the littoral zone.

5. A fourth team should collect phytoplankton and zooplankton samples in the littoral zone and make qualitative and quantitative comparisons of these communities to those found at a similar depth in the open water of the pelagic zone.

    a. Using a tube sampler, take composite vertical samples of a known volume along transects through the littoral zone and different vegetation.

    b. Put the sample into a large container, mix well, and withdraw a small sample for phytoplankton analysis. Preserve with Lugol's solution (see Exercise 10).

    c. Filter the whole sample through Nitex mesh screens of a mesh size of ca. $45\,\mu m$. Concentrate by washing and preserve in a sample bottle (see Exercise 11).

    d. Proceed to the open water of the lake and take several samples of the pelagial water to analogous depths by the same methods.

    e. In the laboratory, identify the dominant species of littoral phytoplankton and zooplankton. By the methods outlined in Exercise 10 and 11, enumerate the organisms and calculate the numbers per liter or cubic meter. Compare to those species and abundances found in the open water of the lake.

6. A fifth team should sample the littoral zone for benthic fauna.

    a. Because of the difficulties associated with the separation of the benthic fauna from the vegetation and sediments, it is suggested that three areas be sampled: one among the emergent vegetation with some standing water and two sites among submersed vegetation, one in shallow and one in deeper water.

    b. Using the large plastic cylinder (see "Apparatus and Supplies"), carefully but quickly lower the cylinder through the vegetation and work into the sediments to a depth of ca. 10 cm. With several persons holding the cylinder rigid, work a sharpened metal or plastic plate under the end penetrating the sediments. With the end sealed, bring the entire sample to the surface and deposit it into a large pail (see Exercise 5).

    c. In a laboratory, separate the vegetation from the sediments in large white enameled or plastic trays and remove all macrofauna.

    d. Sieve the sediments for benthic animals as outlined in Exercise 13.

    e. Identify and enumerate the organisms found and calculate the numbers found per area of littoral zone at the different sites.

    f. If possible, determine the biomass of the different dominant taxa found.

    g. Compare your results to those gathered for the open water sediments (Exercise 13).

7. Answer the questions following Option 3.

OPTION 2. ANALYSES OF LITTORAL HETEROGENEITY

1. Select two littoral sites of relatively homogeneous communities of aquatic macrophytes: one among emergent macrophytes in standing water and another among submersed vegetation in water of a depth of about 0.5 to 1 m.

2. Divide the class into five teams, each of which should analyze the different components, as discussed in Option 1 for each team.

3. Emphasis in this option should be on obtaining good quantitative, replicated samples. A minimum of three replicates should be taken at each site for each component sampled (macrophytes, attached algae, zooplankton, phytoplankton, and benthic animals). Analyze the samples as outlined in Option 1.

4. From the quantitative data, analyze each for the mean, range, standard deviation and standard error of the mean, and coefficient of variation (see Appendix 2). Determine the adequacy of sampling (cf., Appendix 2).

5. Answer the questions following Option 3.

OPTION 3. IN SITU NUTRIENT ENRICHMENTS

1. Construct nutrient-diffusing artificial substrata from unglazed clay flowerpot saucers, the rims of which are glued to Plexiglas plates. Fill with 2% agar solutions with experimental nutrient additions, each in triplicate. Plug drainage holes with neoprene stoppers. Examples follow, although many variations are possible in an attempt to simulate nutrient or chemical releases from substrata of different types and porosities/hydraulic conductivities.
   a. Control: No nutrient enrichments.
   b. P enrichment: Add 0.1 mol P/liter as KH_2PO_4 (ca. 3.1 g P/l).
   c. N enrichment: Add 0.5 mol N/liter as $NaNO_3$ (ca. 7.0 g N/l).
   d. P + N enrichment: Add 0.1 mol P/l as KH_2PO_4 and 0.5 mol N/l as $NaNO_3$.

2. Place replicated substrata (a) on the sediments facing upward and (b) attached to frames, facing upward, that are suspended from stakes at specific water depths along a gradient from shallow water to deeper water in a representative lake or reservoir. At each station, position substrata in a randomized block design.

3. Collect replicated sets of substrata after 1, 2, and 3 weeks of incubation. Carefully transport back to the laboratory in separate containers.

4. Remove replicated samples from a precisely determined flat area of each clay substratum. Carefully scrape an area delineated by a tool (e.g., a cork borer) and remove all attached microorganisms. A tubular brush designed specifically for quantitative removal from hard surfaces will greatly facilitate accurate removal (see Douglas, 1958) and can be adapted to remove quantitative samples in situ without removal of the substrata from the water (Loeb, 1981).

5. Suspend the attached community in filtered (0.45-μm pore size) lake water of known dilution. Mix gently but well. Remove samples for examination by sedimentation chambers or Palmer-Maloney chambers for microscopy (see Exercise 10) and for filtration and quantification of pigment concentrations (chlorophyll a and phaeopigments) discussed earlier (Option 1, Item 4).

6. Graphically analyze the growth responses of the attached communities and their algal-cyanobacterial composition changes over time and along the depth gradient.

7. A variation in the design would be to place a large number of replicated diffusing substrata with different nutrients and concentrations at one depth in a representative lake. The prevention of macroinvertebrate grazing from a portion of the replicates could be accomplished by enclosing the diffusion substrata with a nylon mesh (0.5- or 1.0-mm mesh size) well above (e.g., 2 to 4 cm) the surfaces. Because the mesh also functions as a neutral density filter of light, it would be necessary for additional controls over which the mesh was suspended but with spaces through which grazing invertebrates could have access to the periphyton.

Questions

1. How might you devise better methods to measure the living biomass of roots/rhizomes of aquatic macrophytes? Why is this biomass important to computations of productivity? [See, for example, Ervin and Wetzel (1997).]

2. How could one estimate turnover rates of organic matter produced by macrophytes:
   a. During the season when plant tissues are being lost by sloughing and cohort senescence? [See Dickerman and Wetzel (1985) and Wetzel and Howe (1999).]
   b. Over several years among perennial plants when a portion of one or more years' growth carries over to subsequent years?
3. When resources are limiting, does sampling along a transect or in a stratified random pattern yield more information? Which approach provides greater statistical accuracy?
4. Based on your biomass data, where is maximum productivity likely to occur among the three major plant zones (emergent, floating-leaved, or submersed)?
5. When macrophytes senesce, what is the fate of both dissolved and particulate organic detritus? Where would you anticipate the decomposition of each fraction to occur?
6. It has been demonstrated that submersed and floating-leaved macrophytes release significant quantities of dissolved organic compounds and nutrients, such as nitrogen and phosphorus, during active growth as well as during senescence. What are the implications of these findings in relation to the use of artificial substrata for studying epiphytic algal growth?
7. Among heterogeneous attached algal populations, the extrapolation of changes in biomass to estimates of productivity are complicated by many species of differing and variable generation times. How could one reasonably estimate the composite population turnover rates?
8. The epiphytic algae and bacteria are metabolically coupled to each other and to the macrophyte. How might this symbiotic relationship function and be advantageous to all biota included in the association?
9. As the large plants grow, new surfaces are available for epiphytic colonization. How would these changes affect epiphytic populations (positively and negatively) and measurement techniques?
10. Loosely aggregated sediments present special problems to epipelic algae. How might epipelic algae cope with disturbances of the substrata?
11. What advantages might epipelic algae accrue from living on sediment?
12. How do the littoral phytoplankton differ from those of the open water? What are the disadvantages for phytoplankton living in the littoral zone among macrophytes? How would light, temperature, and nutrient composition differ?
13. How do the zooplankton of the littoral zone differ qualitatively and quantitatively from those of the open water? How would feeding types differ in each zone? Predation pressure by fishes and invertebrates?
14. How might you devise improved sampling apparatus for sampling zooplankton among littoral macrovegetation?
15. Why does the species diversity of benthic fauna generally increase in the littoral zone as compared to that of the pelagial?
16. What feeding types among benthic fauna would you anticipate to dominate in the littoral zone? How significant would you expect direct feeding by macroinvertebrates on tissue of macrovegetation to be? What alternatives do these animals have?
17. How might surface seiches affect littoral organisms? Progressive waves?
18. How would you diagram the food web for the littoral zone?
19. How would you diagram the qualitative and estimated quantitative metabolic relationships (e.g., in terms of carbon pools and fluxes) for the littoral zone?

Apparatus and Supplies

1. Stakes (ca. 2 m in length), 100-m tape, and meter sticks.
2. Large plastic bags, permanent marking pens, and data sheets.
3. Herbarium presses and associated mounting supplies.
4. Weighted grappling hooks ("cat-o-nine tails"), plastic buckets, and snorkel or SCUBA equipment.

5. Quadrats (0.25- or 0.50-m²) of steel rod or heavy electrical wire.
6. Drying ovens (105°C), muffle furnace (550°C), and crucibles.
7. Large balances and a semimicro-analytical balance.
8. Coring tubes for root/rhizome sampling (ca. 30 cm by 2 cm diameter) and stoppers to fit.
9. Artificial substrata (e.g., glass slides) incubated for 2 to 4 weeks in the littoral areas prior to exercise [see Sládečková (1962) for a discussion of types of substrata and apparatus] and transport containers for substrata (see p. 318).
10. Algal counting cells (see Exercise 10).
11. Apparatus and reagents for pigment analyses (see Exercise 10).
12. Apparatus and supplies for oxygen or ^{14}C productivity analyses (see Exercise 14).
13. Vernier calipers to determine the diameters of emergent macrophytes.
14. Zooplankton counting cells and zooplankton sieving nets (see Exercise 13).
15. Large cylinder of rigid plastic (ca. 15 cm in diameter by 2 m in length) and a metal or plastic plate ca. 20 × 20 cm².
16. Large white enameled or plastic pans and sorting sieves of mesh size ca. 0.2 and 0.4 mm.
17. Clay pottery saucers, ca. 10 cm in diameter, unglazed; Plexiglas plates; 0.5- or 1.0-mm mesh nylon screening.
18. Agar; $NaNO_3$, KH_2PO_4.

References

Allen, H.L. 1971. Primary productivity, chemo-organotrophy, and nutritional interactions of epiphytic algae and bacteria on macrophytes in the littoral of a lake. Ecol. Monogr. *41*:97–127.

Brakke, D.F. 1976. Modification of the Whiteside-Williams pattern sampler. J. Fish. Res. Bd. Canada *33*:2861–2863.

Burkholder, J.A. and R.G. Wetzel. 1989. Epiphytic microalgae on a natural substratum in a hardwater lake: Seasonal dynamics of community structure, biomass and ATP content. Arch. Hydrobiol. *Suppl. 83*:1–56.

Correll, D.S. and H.B. Correll. 1972. Aquatic and Wetland Plants of Southwestern United States. U.S. Govt. Printing Office, Wat. Poll. Control Res. Series, 16030 DNL 01/72. Washington, D.C. 1777 pp.

Dickerman, J.A. and R.G. Wetzel. 1985. Clonal growth in *Typha latifolia*: Population dynamics and demography of the ramets. J. Ecol. *73*:535–552.

Dickerman, J.A., A.J. Stewart, and R.G. Wetzel. 1986. Estimates of net annual aboveground production: Sensitivity to sampling frequency. Ecology *67*:650–659.

Douglas, B. 1958. The ecology of the attached diatoms and other algae in a small stony stream. J. Ecol. *46*:295–322.

Downing, J.A. 1984. Sampling the benthos of standing waters. pp. 87–130. *In*: J.A. Downing and F.H. Rigler, Editors. A Manual on Methods for the Assessment of Secondary Productivity in Fresh Waters. 2nd Ed. Blackwell, Oxford.

Eissenstat, D.M. and R.D. Yanai. 1997. The ecology of root lifespan. Adv. Ecol. Res. *27*: 1–60.

Ervin, G.N. and R.G. Wetzel. 1997. Shoot: Root dynamics during growth stages of the rush *Juncus effusus* L. Aquat. Bot. *59*:63–73.

Fairchild, G.W. and R.L. Lowe. 1984. Artificial substrates which release nutrients: Effects on periphyton and invertebrate succession. Hydrobiologia *114*:29–37.

Fairchild, G.W., R.L. Lowe, and W.B. Richardson. 1985. Algal periphyton growth on nutrient-diffusing substrates: An in situ bioassay. Ecology *66*:465–472.

Fassett, N.C. 1957. A Manual of Aquatic Plants. Univ. Wisconsin Press, Madison. 405 pp.

Godfrey, R.K. and J.W. Wooten. 1979. Aquatic and Wetland Plants of Southeastern United States. Monocotyledons. Univ. Georgia Press, Athens. 712 pp.

Godfrey, R.K. and J.W. Wooten. 1981. Aquatic and Wetland Plants of Southeastern United States. Dicotyledons. Univ. Georgia Press, Athens. 933 pp.

Goulden, C.E. 1971. Environmental control of the abundance and distribution of the chydorid Cladocera. Limnol. Oceanogr. *16*:320–331.

Jeppesen, E., Ma. Søndergaard, Mo. Søndergaard, and K. Christoffersen (eds). 1997. The Structuring Role of Submerged Macrophytes in Lakes. Springer Verlag, New York. 423 pp.

Kajak, Z. 1971. Benthos of standing water. pp. 25–65. *In*: W.T. Edmondson and G.G. Winberg, Editors. A Manual on Methods for the Assessment of Secondary Productivity in Fresh Waters. IBP Handbook No. 17. Blackwell, Oxford.

Keen, R. 1973. A probabilistic approach to the dynamics of natural populations of the Chydoridae (Cladocera, Crustacea). Ecology *54*:524–534.

Loeb, S.L. 1981. An in situ method for measuring the primary productivity and standing crop of the epilithic periphyton community in lentic systems. Limnol. Oceanogr. *26*:394–399.

Mathews, C.P. and D.F. Westlake. 1969. Estimation of production by populations of higher plants subject to high mortality. Oikos *20*:156–160.

Mason, H.L. 1957. A Flora of the Marshes of California. Univ. California Press, Berkeley. 878 pp.

Muenscher, W.C. 1944. Aquatic Plants of the United States. Comstock, Ithaca, NY. 374 pp.

Pennak, R.W. 1962. Quantitative zooplankton sampling in littoral vegetation areas. Limnol. Oceanogr. *7*:487–489.

Prescott, G.W. 1969. How to Know the Aquatic Plants. Wm. C. Brown, Dubuque, IA. 171 pp.

Pringle, C.M. and J.A. Bowers. 1984. An in situ substratum fertilization technique: Diatom colonization on nutrient-enriched, sand substrata. Can. J. Fish. Aquat. Sci. *41*:1247–1251.

Pringle, C.M. and F.J. Triska. 1996. Effects of nutrient enrichment on periphyton. pp. 607–623. *In*: F.R. Hauer and G.A. Lamberti, Editors. Methods in Stream Ecology. Academic Press, San Diego.

Sládečková, A. 1962. Limnological investigation methods for the periphyton ("Aufwuchs") community. Bot. Rev. *28*:286–350.

Szlauer, L. 1963. Diurnal migrations of minute invertebrates inhabiting the zone of submerged hydrophytes in a lake. Schweiz. Z. Hydrol. *25*:56–64.

Tate, C.M. 1990. Patterns and controls of nitrogen in tallgrass prairie streams. Ecology *71*:2007–2018.

Westlake, D.F. 1965. Some basic data for investigations of the productivity of aquatic macrophytes. Mem. Ist. Ital. Idrobiol. *18*(Suppl.):229–248.

Wetzel, R.G. 1964. A comparative study of the primary productivity of higher aquatic plants, periphyton, and phytoplankton in a large, shallow lake. Int. Rev. ges. Hydrobiol. *49*:1–64.

Wetzel, R.G. 1979. The role of the littoral zone and detritus in lake metabolism. pp. 145–161. *In*: G.E. Likens, W. Rodhe, and C. Serruya, Editors. Symposium on Lake Metabolism and Lake Management. Ergebnisse Limnol., Arch. Hydrobiol. *13*:145–161.

Wetzel, R.G. 1983a. Limnology. 2nd Ed. Saunders Coll., Philadelphia. 860 pp.

Wetzel, R.G. 1983b. Attached algal-substrata interactions: Fact or myth? pp. 207–215. *In*: R.G. Wetzel, Editor. Periphyton of Freshwater Ecosystems. Developments in Hydrobiology *17*. Dr. W. Junk Publishers/Kluwer, Dordrecht.

Wetzel, R.G. 1990. Land-water interfaces: Metabolic and limnological regulators. Baldi Memorial Lecture. Verh. Int. Int. Ver. Limnol. *25*:6–24.

Wetzel, R.G. 1996. Benthic algae and nutrient cycling in standing freshwater ecosystems. pp. 641–667. *In*: R.J. Stevenson, M. Bothwell, and R. Lowe, Editors. Algal Ecology: Benthic Algae in Freshwater Ecosystems. Academic Press, New York.

Wetzel, R.G. and D.F. Westlake. 1971. Periphyton. pp. 42–50. *In*: R.A. Vollenweider, Editor. A Manual on Methods for Measuring Primary Production in Aquatic Environments. IBP Handbook No. 12. Blackwell, Oxford.

Wetzel, R.G. and M.J. Howe. 1999. Maximizing production in a herbaceous perennial aquatic plant by continuous growth and synchronizing population dynamics. Aquatic Botany *64*:111–129.

Whiteside, M.C. 1974. Chydorid (Cladocera) ecology: Seasonal patterns and abundance of populations in Elk Lake, Minnesota. Ecology *55*:538–550.

Whiteside, M.C. and J.B. Williams. 1975. A new sampling technique for aquatic ecologists. Verh. Int. Ver. Limnol. *19*:1534–1539.

Whittaker, R.H. and G.E. Likens. 1975. The biosphere and man. pp. 305–328. *In*: H. Lieth and R.H. Whittaker, Editors. Primary Productivity of the Biosphere. Springer-Verlag, New York.

Experimental Manipulation of Model Ecosystems

Ecosystems have been described as the "functional units of the landscape" (Odum, 1971), where organisms interact with their physical-chemical environment and with other organisms. For limnologists, lakes, ponds, streams, and rivers and their associated drainage basins represent the ecosystems of interest. However, these ecosystems are large and complex, and thus it is difficult to decipher the various interactions that occur within them.

Natural ecosystems are exposed continuously to changing environmental conditions, of both internal and external types. These conditions can be changed either by natural (for example, seasonally) or by unnatural means (e.g., through human intervention). Whatever the cause, the effects of any single change may be difficult to separate from those resulting from all of the other changes that are occurring simultaneously. A changing natural ecosystem can be likened to an experiment in which the investigator is trying to control and, at the same time, understand many known and unknown variables. For the creative scientist, this situation can present a real challenge. Nevertheless, the understanding of causation in the ecosystem rarely can be delineated without the benefit of some carefully designed and rigorously controlled experimental work.

The cost in both money and time of doing experimental work on whole ecosystems often is prohibitive. Moreover, while such studies have been done [see, for example, Hasler and Johnson (1954), Likens et al. (1970), and Schindler (1974)], we usually do not have the privilege of being able to alter or to disturb seriously an entire ecosystem for experimental purposes. A common solution is to recreate, in the laboratory, microecosystems or microcosms [see, for example, Warington (1851) and Beyers (1963)]. An ecosystem brought into the laboratory can mimic the natural ecosystem in some respects but will differ in others: a microcosm is a simplified ecosystem with discrete boundaries. The scales of events in both time and space are abbreviated. Succession to a new steady state takes place in weeks, rather than in years. Microcosms generally have fewer species than do natural ecosystems and have, in consequence, simpler communities of organisms. Some characteristics of microcosms make them valuable objects of study. Microcosms are expendable, and the experimenter has control over the environmental boundary conditions to a degree impossible to achieve in the field. Also, it generally is assumed that the investigator can establish reproducible or replicable units, thereby allowing statistical evaluation of the data obtained from experimental treatments and controls for each manipulation.

In this exercise, two different approaches, the chemostat and the microcosm, will be described for the study of ecosystems in the laboratory. Both of these approaches require several weeks for stabilization, manipulation, and evaluation. Thus students should select one of these approaches and should understand that appreciable time must be allotted for independent study of these systems. The chemostat requires somewhat more sophisticated equipment and procedures.

DEVELOPMENT OF AN EXPERIMENTAL DESIGN

Before beginning the experimental study, each student should prepare a research proposal that describes the objective and plan of the experiment. This proposal should be evaluated by the instructor and returned to the student. The experimental design should not be too complex. Manipulate one variable at a time and establish and evaluate adequate controls. For example, if 1 ml of inorganic phosphorus solution were added to an experimental ecosystem, then the control would get 1 ml of the same solution without the inorganic phosphorus. Evaluate all parameters pertinent to the questions being addressed. Analytical procedures should be rigorous, standardized, and consistent.

The research proposal should include:

1. A clear statement of the hypothesis.
2. A brief explanation of hypothesis development (justification of the study).
3. An explicit description of how the hypothesis is to be tested, including manipulation of parameters, how the manipulation is to be done, to what level, and how often.
4. A description of experimental and control ecosystems, number of each, types of materials to be used, and so forth.
5. A brief outline of the methods of measurement and evaluation techniques. Indicate a time schedule for sampling and for the analyses of samples. Identify any plans for statistical evaluation of the data.

You may want to use flow diagrams or compartmental models to clarify the experimental design.

A CHEMOSTAT APPROACH

The *chemostat* is a continuous culture apparatus most commonly used in physiological studies of algae and bacteria. Basically it consists of a culture vessel into which a fresh culture medium is fed continuously at a constant, metered rate. The chemostat can serve as a model of a natural aquatic ecosystem, and its relative simplicity will provide a starting point for understanding the more complicated interactions which occur between phytoplankton, bacteria, and their environments. [For more information, see Herbert et al. (1965), and Barlow et al. (1973).]

The Chemostat

Fresh culture medium is held in a 20-l Pyrex carboy and forced into the culture chamber by a metered gas flow (Fig. 23.1). The gas may be generated by electrolytic

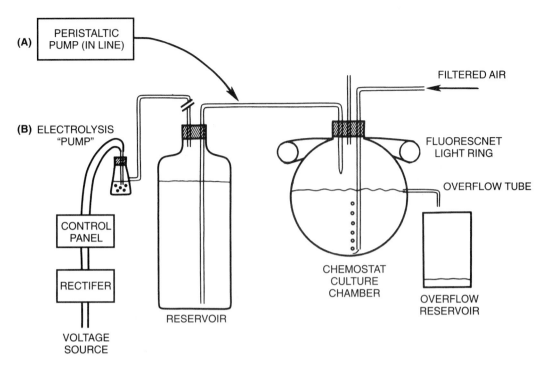

Figure 23.1. Chemostat continuous culture apparatus, in which the nutrient media of the reservoirs are supplied by a peristaltic pump (A) or by pressure of gas generated by electrolysis (B).

decomposition of water from a 250-ml Erlenmeyer flask and connected to the carboy, which in turn is connected to the culture chamber. Alternatively, a peristaltic pump could be used to transfer the culture medium to the culture flask. A 12-l Florence flask is used for the culture chamber. An overflow tube keeps the volume of the culture chamber constant. The initial culture should be 8l of natural water from a local pond or stream containing a full complement of the organisms present.

Operate the chemostat for this exercise for three weeks at a flow rate of 160 ml/h. The dilution rate, D (i.e., number of volumes of medium passed through the culture chamber in 1h), can be calculated by $D = R/V$, where R = flow rate in ml/h and V = volume (ml) of culture. Its reciprocal, $1/D$, is the mean residence time (i.e., average time a water molecule or suspended particle spends in the culture chamber).

When the chemostat is in operation, it is possible to calculate the theoretical concentration of the medium created in the culture chamber after any given number of hours:

$$C_t = C_0 + \left[\frac{C_i R}{R} - C_0\right]\left[1 - e^{-(t-t_0)R/V}\right] \qquad (1)$$

where C_t = outflow concentration at time t, C_0 = concentration at time zero (t_0), C_i = concentration of inflow, R = flow rate (ml/h), V = volume of culture chamber (ml), and t_0 = time 0; initiation of experiment.

Sample calculation: Given $C_0 = 0$, $t = 115\,\text{h}$, $R = 160\,\text{ml/h}$, $V = 8000\,\text{ml}$, and $C_i = 155\,\mu\text{g P/l}$:

$$C_t = 0 + \left[\frac{(155)(160)}{160} - 0\right]\left[1 - e^{-(115)(160/8000)}\right]$$

$$C_t = (155)(1 - e^{-2.3}), \quad \text{where } e^{-2.3} = 0.100259$$

$$C_t = 139.5\,\mu\text{g P/l created in chamber after 115 h}$$

The actual concentration of dissolved inorganic phosphorus (PO_4-P) will differ from this theoretical value owing to a number of factors, including uptake and assimilation by organisms in the chamber, precipitation or absorption after entering the chamber, and existing concentration in the chamber at time 0. With some of these factors measured, others can be calculated. For example, the uptake rate of a particular nutrient (N, P, K, Zn, and so on) by the organisms can be determined by comparing the theoretical concentration with the actual concentration, the difference being the amount taken up, assuming to precipitation or absorption.

The basic equation used in calculating the increase or decrease of phytoplankton cells from the chemostat is

$$N_t = N_0 e^{r t_d} \tag{2}$$

where N_t = number of organisms at some time t; N_0 = number of organisms at the start of the experiment, t_0; r = total rate of increase or decrease in number of organisms; and t_d = time in days.

Using this equation, the growth rate of the organisms and their washout rate can be calculated. If the concentration of organisms were to remain constant during operation of the chemostat, then the increase in the number of cells would balance exactly the number washed out. This special case is called a *steady state* and can be expressed mathematically as:

$$(r_{tot}) + (W) = 0 \tag{3}$$

where r_{tot} is the rate in increase in number of cells of the organism and W is the washout rate. W must be a negative value since it represents a rate of decrease. W is equal to the negative of the dilution rate or $(-D)$. Therefore, in the example given above,

$$W = -D = -R/V = -\frac{160\,\text{ml/h}}{8000\,\text{ml}} = -1/50\,\text{h}^{-1} \tag{4}$$

or −0.48/day. The rate of increase by the organism required to balance the washout rate is 0.48/day. For convenience, we shall call this rate of increase r_w. The number of divisions per day then can be calculated. This is another special case of Eq. 2 in which we want to calculate the time needed to double the original number of organisms, i.e., $N_t/N_0 = 2$. Using Eq. 2, replacing r with r_w, and solving for t_d, we get:

$$t_d = \left(\ln \frac{N_t}{N_0} \right) \frac{1}{r_w} \tag{5}$$

where in this case t_d is the time in days needed for the number of organisms to double and $1/t_d$ is the number of divisions per day. Thus, for the example,

$$t_d = (\ln 2) \frac{1}{0.48}$$

$$t_d = \frac{0.69315}{0.48}$$

$$t_d = 1.444 \quad \text{and} \quad 1/t_d = 0.6925 \text{ divisions } (d)/\text{day}$$

for the number of cells of the organism to remain constant.

Example. Beginning with 1000 *Chamydomonas* cells per ml in the chemostat, what would be the concentration after one day, assuming no growth?

$$N_t = N_0 e^{r t_d}, \quad \text{replace } r \text{ with the washout rate } (W)$$

$$N_t = N_0 e^{W t_d}$$

$$N_t = 1000 e^{(-0.48)(1)} = (1000)(0.61878)$$

$$N_t = 619 \text{ cells/ml after one day}$$

If an actual count were made after day 1 and 2000 cells/ml were recorded, calculate the growth rate and the division rate ($t_d = 1$):

$$r = \left(\ln \frac{N_t}{N_0} \right) \frac{1}{t_d} = \left(\ln \frac{2000}{1000} \right) \frac{1}{1} = 0.69315$$

and

$$d/\text{day} = 1/t_d = r \Big/ \left(\ln \frac{N_t}{N_0} \right) = 0.69315/(\ln 2) = 0.69315/0.69315 = 1$$

This would be the total growth rate of cells only if there were no loss from the chemostat. If there were loss from the chemostat, the above rate of increase would be only an apparent rate (r_{obs}). The total rate of increase (r_{tot}) equals the apparent rate (r_{obs}) plus the rate needed to balance the washout rate, which was earlier defined as r_w:

$$r_{tot} = r_{obs} + r_w = 0.69315 + 0.4800 = 1.173$$

The total number of divisions/day is also the sum of that observed plus the number of divisions needed to balance washout:

$$d/\text{day}_{tot} = d/\text{day}_{obs} + d/\text{day}_w = 1.0000 + 0.6925 = 1.6925$$

The formula presented earlier for calculating the doubling time and divisions/day (Eq. 5) may be used to calculate either r_{tot} or d/day_{tot} once the other is known.

Problem. Calculate N_t after 3 days, given $N_0 = 100$ *Chlamydomonas* cells/ml and $r = 0.4$. Calculate r if N_t were found to be 985 cells/ml by an actual count after 3 days. Calculate d/day in the latter case.

The growth rate could be positive, zero, or negative. If the growth rate were positive and greater than 0.6925 d/day, then the population would increase in the chamber. With a zero growth rate, the population would decline according to the formula for the washout rate. A negative growth rate would indicate that the organisms are disappearing faster than can be accounted for by washout alone. This decrease could be explained by losses from death, zooplankton feeding, or both.

The preceding discussion should provide some indication of the manipulations and calculations that are possible with the chemostat. Can you think of others?

The following is an example of an experiment that could be done as a class exercise (four weeks) or as an independent project using the chemostat.

NUTRIENT ENRICHMENT EXPERIMENT

Locate two chemostats in a constant temperature chamber or room for the three- or four-week duration of the study. Filter water from an oligotrophic pond or stream through a 50-μm mesh net to remove large zooplankton and pour the water into the culture chamber (up to the 8-l mark) of each chemostat. For the next 7 days, the culture medium will consist of filtered (0.45-μm pore size Millipore filter) water augmented by autoclaved water from the same source.

During the first session of the second week, add enough nitrogen, phosphorus, or both to one of the reservoirs (containing sterile water) to create an excess concentration of these elements over that already present in the water. This enriched medium will be pumped into one of the culture chambers for the next one to two weeks.

A number of parameters will be measured by each laboratory section or group during the three-week period. The following is a chronology of the sampling and analysis requirements. Analytical techniques are discussed elsewhere in the appropriate exercises. Because of the limited amount of water delivered from the chemostat, some parameters can be measured only on one water sample during a day. For these samples, it will be necessary to work as a group and to share the data.

First Week

1. Obtain nutrient-poor pond water, filter it through a 50-μm mesh net to remove larger zooplankton, and add 8l to the chamber of each chemostat. Start the chemostats. The chemostats (both culture chamber and medium reservoirs) should be labeled "control" or "enriched," as appropriate.

2. As a group, collect two separate 40-ml samples from the net-filtered water. Analyze immediately for N, P, and K concentrations (see Exercise 7) or freeze for analysis at a later time.

3. As a group, collect a 2-l sample for phytoplankton from the filtered water. Add Lugol's preservative and mix well. Distribute into 25- to 50-ml settling-counting chambers and label. Place the chambers in an area of the laboratory where they will remain undisturbed for one week.
4. Count phytoplankton in a sample of the filtered water (see Exercise 10).
5. Measure the pH of the filtered water.

Sections B, C, and D (Days 2, 4, and 6)

1. Form two groups; each group collects a 40-ml water sample from the outflow tube from one of the chemostats. Analyze these samples for N, P, and K content or freeze. Label all samples with the following information: collection data, time, chemostat type, your section, and your initials.
2. Each group collects a 200-ml phytoplankton sample from the outflow tube of one of the chemostats and processes it according to the directions given in Exercise 10.
3. Each group measures the pH and temperature of water sample from the outflow (about 30 ml). Record the temperature as the water is being collected to get a more reliable measurement.
4. Each group checks the "culture medium reservoir" and refills it as required with autoclaved water (about 8l removed every 50 h).

Second Week

Section A (Day 8)

1. Each group collects and chemically analyzes (or freezes) two separate 40-ml water samples from one of the chemostats. Collect the water from the outflow tube and label it carefully.
2. Each group measures the pH and temperature of a water sample from the outlet (about 30 ml).
3. Half of each group: (a) Collects a 200-ml sample from the outflow tube and processes it to settle phytoplankton. (b) Counts the phytoplankton sample that was collected and preserved during the previous week. Handle the sample *very carefully* so as not to resuspend any of the settled organisms.
4. Half of each group should perform pigment analyses on a chemostat sample following the procedure given in Exercise 10.
5. At the end of the laboratory period add the N- and/or P-enrichment solution to one of the medium reservoirs labeled "Enriched." Refill both of the reservoirs to the 20-l mark with autoclaved water prior to adding the nutrients.

Enrichment procedure: Using NH_4NO_3 and/or K_2HPO_4, enrich the culture medium concentrations of N and P by about tenfold.

Sections B, C, and D (Days 10, 12, and 14)

1. Complete Parts 1 through 4 given under the second week for Section A. In Part 3, you will be processing the 200-ml sample taken the previous week.

2. Check the nutrient reservoirs to see if they need refilling. If so, be careful to create the proper nutrient concentration.

Third Week

All Sections (2-Day Intervals)

1. Complete Parts 1 through 4 from the second week's schedule. Process the 200-ml sample, which your section (group) collected from the chemostat during the previous week.
2. Section D will turn off and clean the chemostat.

Fourth Week

All Sections

1. Run analyses for NH_4-N, NO_3-N, and PO_4-P on any frozen samples. Follow the procedures given in Exercise 7.
2. Complete all counts of phytoplankton samples.
3 Check all records and data for possible errors and omissions.

Data Analysis

1. Record all of your data on the sheet provided on p. 335. Also fill in the data collected by the other sections. Answer the questions in items 4 and 5 below.
2. Determine the growth rates for the two dominant species for the interval between the time of initiation of the experiment and the first collection, and thereafter between successive collection periods.
3. Calculate the theoretical nitrogen and phosphorus concentrations in the culture chamber for 2, 4, 6, 8, and 10 days after the beginning of enrichment. Compare these values with the actual concentrations as determined from the water samples. Determine the uptake of N and P per unit volume of Chl *a* or cells for each of the above days.
4. Interpreting the results of the chemostat:
   a. Compare and contrast the parameters before and after enrichment.
   b. What can be concluded about the relative dependence upon nitrogen and/or phosphorus for the enrichment of this system (i.e., which seemed to have the greatest effect on, or which was required the most by, the phytoplankton)? What can be said about the role of potassium?
   c. What can you conclude about the water originally used to operate this chemostat? Discuss the use of the chemostat in estimating the potential of a natural system (lake or stream) for studies of the eutrophication process.
5. Relating the chemostat to a natural system:
   a. Considering the parameters measured in the chemostat, how would you collect the same information in a lake? Would you sample more frequently and at more sites? It so, why? [See deNoyelles and O'Brien (1974).]

b. Considering the estimates that you made for growth rates and nutrient uptake rates, how would you make the same calculations from a lake study? Would there be different or additional parameters that must be measured, considering the relative complexity of the lake as compared to the chemostat? Discuss in detail any additional information that would have to be collected. Consider some particular reasons why these estimations would be more difficult to make for a lake study.

A MICROCOSM APPROACH

Aquatic microcosms may be established by bringing together collections of biota, sediment, and water. Microcosms may be established conveniently and economically in 1-gal glass jars and maintained in the laboratory. Ideally, it would be most informative to compare ecosystems constructed or collected from different kinds of environments (i.e., oligotrophic and eutrophic). It is critically important that all replicate microcosms for a given experiment be established at the same time from the same samples of water and sediment. Why?

At least four major habitats may be recognized in these model ecosystems: (1) the free-water column containing planktonic organisms; (2) the water surface, where duckweed (*Lemna*) and watermeal (*Wolffia*), both angiosperms, may grow in abundance; (3) the glass wall of the container where algae and bacteria may attach and grow; and (4) the sediments, containing an abundance of microorganisms, immature insects, and other invertebrates, and a reservoir of nutrients. Eutrophic microcosms may develop a large snail population. Planktonic cladocera and copepods may be replaced by a stable population of seed shrimp (ostracods).

The kinds of microcosms ultimately established will depend on the nature of the experimental design. It may be appropriate to exclude snails, zooplankton, or sediments from some microcosms. The composition of the microcosms can be modified according to the experimental design, but steps necessary to insure replication must be taken.

Collect large volumes of sediments and overlying water from a nearby lake or pond. Pass the sediment through one or two coarse-mesh sieves to remove snails and large animals as well as larger stones and organic debris. Collect this sieved sediment into another bucket. Sort the animals both by type and by size. After thoroughly mixing the sediment, add equal aliquots (about 300 ml) to the microcosm jars. Stir to allow even settling. Some of the extra animals may be maintained in stock microcosms for addition to the experimental microcosms at appropriate times. If they were added, use equal biomass and size distributions. Add equal amounts of water to each microcosm, so that the water level is near the shoulder of the jar. Each jar holds about 3.5 l.

Note that adult snails may begin to lay eggs within days after they are brought into the laboratory (they are hermaphroditic). To avoid unwanted explosions of snail populations (incubation time is a couple of weeks), use smaller, immature snails.

Place the microcosms in an area of the laboratory where environmental variation will be minimal. For example, microcosms of a single experiment should receive the same sun exposure (same side of the room). Cross seed samples of water and sediment for the first two weeks to insure replication between the experimental and control microcosms. Allow the microcosms to equilibrate in the laboratory for three weeks. During this time, a relatively stable community and metabolic state should be established. During the following week, make initial measurements on the microcosms and initiate experimental manipulations. Then monitor the effects of the experimental manipulations for a period of two weeks.

POTENTIAL EXPERIMENTAL MANIPULATIONS

The research ideas presented below are straightforward possibilities. However, these ideas should not limit you; they are some of the more practical, but may not be the "best," nor the only, possibilities for microcosm experiments. These ideas include experiments that relate to ecological and environmental problems in limnology.

1. Alteration of decomposition rate due to chemical changes in water. Sample hypotheses:
    a. An increased availability of nitrogen and phosphorus stimulates decomposition of organic matter.
    b. The presence of low pH levels (pH 3 or 4) inhibits bacterial decomposition of dead organic matter (detritus).
2. Changes in primary productivity and/or species composition due to nutrient availability, pH, and temperature. Sample hypotheses:
    a. An increased availability of nitrogen and phosphorus stimulates levels of phytoplankton production.
    b. Increases in temperature cause a change in both species composition (perhaps from diatoms to cyanobacteria) and rates of primary production.
    c. High pH levels due to high carbonate alkalinity cause changes in species composition and rates of production.
    d. Low pH levels cause changes in species composition and rates of production.
3. Variation of N/P ratio. Sample hypothesis: A decrease in the N/P ratio causes the dominant phytoplankton taxa to shift from a diatom/green algal assemblage to a cyanobacteria (nitrogen-fixing) assemblage.
4. Effects of herbivory on aquatic ecosystems. Sample hypotheses:
    a. The presence of herbivores (snails or zooplankton) decreases primary production.
    b. The presence of herbivores alters the size composition of the phytoplankton, such that small or large phytoplankton are excluded selectively.
5. Effects of oil and/or pesticides on aquatic ecosystems. Sample hypotheses:
    a. The presence of oil or pesticides decreases rates of bacterial mineralization.
    b. The presence of oil slicks or pesticides decreases/increases rates of primary production and/or of secondary production.

Chemostat Data Sheet

Control

	pH	Temper-ature (°C)	NH_4-N (mg/l)	NO_3-N (mg/l)	PO_4-P (mg/l)	Chl a (µg/l)	Caro-tenoids (µg/l)	Phytoplankton	
								Species A (cells/ml/ day)	Species B (cells/ml/ day)
Week 1 Section A Section B Section C Section D									
Week 2 Section A Section B Section C Section D									
Week 3 Section A Section B Section C Section D									
Enriched									
Week 1 Section A Section B Section C Section D									
Week 2 Section A Section B Section C Section D									
Week 3 Section A Section B Section C Section D									

Species A =
Species B =

ANALYTICAL PROCEDURES

Because of the relatively small volumes involved, results must be based on in situ measurements or on analyses of very small samples from a microcosm.

To determine the numbers of bacterioplankton, phytoplankton, or zooplankton, collect 5 ml of water from the middle of the water column with a large-bore pipet. Transfer this sample to a small bottle and add 1 drop of Lugol's solution to preserve. Follow the enumeration procedures described in Exercises 10, 11, and 19.

Periphyton may be removed from the surface of the sediment or glass walls with a Pasteur pipet. Take no more than 5 ml/sample. Store and preserve as was done with the plankton samples. Analyze according to the procedures given in Exercise 22.

Follow the procedures given in Exercises 2 (light and temperature), 7 (inorganic nutrients), 8 (pH and alkalinity), 9 (dissolved organic matter), and 10 (plant pigments) for the various physical and chemical analyses. Use the smallest sample possible for chemical analyses of the water column and/or sediment. In establishing the sampling interval, bear in mind the total amount of water or sediment to be removed from the microcosm during the course of the experiment.

In situ procedures have been used successfully in studies of ecosystem metabolism in microcosms [e.g., Beyers (1963, 1965)]. Changes in pH or in dissolved oxygen concentration may be recorded continuously, and without consumptive sampling from the microcosms, with electronic probes (see Exercises 6 and 7). For example, pH has been used to measure the change in total dissolved carbon dioxide concentration in an application of the carbon dioxide diurnal rate of change curve method in microcosms (Beyers, 1963, 1965). The microcosms may be sealed to minimize the diffusion of gases across the air-water interface. Whole microcosms may be darkened to estimate ecosystem respiration in short-term experiments.

These ecosystems are not static. Changes occur rapidly, especially during the initial phases of manipulation.

Questions

1. Compare and contrast your microcosm and the pond(s) from which it was established, in terms of physical, chemical, and biological parameters (e.g., stratification, food webs, and so on). What are the properties peculiar to the microcosm?
2. Can you draw any conclusions from your data about ponds or ecosystems in general? Has the microcosm approach provided "relevant" insight concerning aquatic ecosystems, or is the observed behavior of the microcosm simply an anomaly?
3. What do you view as being the major advantages and disadvantages of working with microcosms? [See Beyers (1964).]
4. What is the difference between a control and a reference jar in microcosm studies? [See Likens (1985).]

References

Barlow, J.P., W.R. Schaffner, F. deNoyelles, Jr., and B. Peterson. 1973. Continuous flow nutrient bioassays with natural phytoplankton populations. pp. 299–319. *In*: Bioassay Techniques and Environmental Chemistry. Ann Arbor Science Publ., Ann Arbor, MI.

Beyers, R.J. 1963. The metabolism of twelve aquatic laboratory microecosystems. Ecol. Monogr. *33*:281–306.

Beyers, R.J. 1964. The microcosm approach to ecosystem biology. Amer. Biol. Teacher *26*:491–498.

Beyers, R.J. 1965. The pattern of photosynthesis and respiration in laboratory microecosystems. pp. 63–74. *In*: C.R. Goldman, Editor. Primary Productivity in Aquatic Environments. Mem. Ist. Ital. Idrobiol. 18 Suppl. Univ. of California Press, Berkeley.

deNoyelles, F. and W.J. O'Brien. 1974. The in situ chemostat—a self-contained continuous culturing and water sampling system. Limnol. Oceanogr. *19*:326–331.

Hasler, A.D. and W.E. Johnson. 1954. Rainbow trout production in dystrophic lakes. J. Wildl. Manage. *18*:113–134.

Herbert, D., P.J. Phipps, and D.W. Tempest. 1965. The chemostat: Design and instrumentation. Lab Prac. *14*:1150–1161.

Likens, G.E. 1985. An experimental approach for the study of ecosystems. J. Ecol. *73*:381–396.

Likens, G.E., F.H. Bormann, N.M. Johnson, D.W. Fisher, and R.S. Pierce. 1970. Effects of forest cutting and herbicide treatment on nutrient budgets in the Hubbard Brook watershed-ecosystems. Ecol. Monogr. *40*:23–47.

Odum, E.P. 1971. Fundamentals of Ecology. 3rd Ed. W.B. Saunders, Philadelphia. 574 pp.

Schindler, D.W. 1974. Eutrophication and recovery in experimental lakes: Implications for lake management. Science *184*:897–899.

Warington, R. 1851. Notice of observation on the adjustment of the relations between animal and vegetable kingdoms. Quart. J. Chem. Soc., London *3*:52–54.

Diurnal* Changes in a Stream Ecosystem: An Energy and Nutrient Budget Approach

Numerous physical, chemical, and biological processes interact to produce a stream ecosystem. Small streams tend to reflect the conditions of the drainage area and usually are chemically and physically dynamic and biologically rich.

In this exercise we will attempt to observe various properties of a small stream ecosystem. By measuring and integrating some of the major physical and biological components of the energy and nutrient budgets (energy and mass balances) during a 24-h period, it will be possible to evaluate aspects of the functional role of streams within the landscape as a whole. We will be interested in the fate of solar radiation as it enters the ecosystem, how much is used to heat the water, how much potential energy is lost as water flows downhill, how much of the solar energy is utilized by plant photosynthesis in the stream and converted to stored chemical energy, and how much of the stored chemical energy is used by various components of the stream ecosystem. Some attention also will be given to the flux of nutrients through the stream ecosystem.

Small steams provide excellent opportunities for such studies. Sampling can be done relatively efficiently and inexpensively. In addition to the techniques and approaches discussed below, see Odum (1957), Teal (1957), Minckley (1963), Hall (1972), Manny and Wetzel (1973), and Fisher and Likens (1973) for further information about such studies of stream and spring ecosystems.

---

* *Diurnal* refers to an event that occurs in a day or recurs each day. *Diel* (= day) is a more recent term that refers to events that recur at intervals of 24 h or less (Odum, 1971). In studies of community periodicity, when whole groups of organisms exhibit synchronous activity patterns in the day-night cycle, *diurnal* is used sometimes in a more restricted sense, referring to animals that are active only during the day versus others that are active only during the period of darkness (= *nocturnal*) and still others only during twilight periods (*crepuscular*).

LOGISTICS

This exercise is designed to be done over a 24-h period, starting at midnight (alternatively, sunrise or sunset). Students may choose to remain at the study site for the entire 24 h or alternate on 4-h shifts. Establishing a field camp with tents facilitates the collections and analysis of samples. If the class were large enough, several camps could be set up at various locations along the stream to facilitate the study of spatial variation. Since this exercise is designed to be an enjoyable outdoor experience as well as a learning experience, it is important for each student to have adequate clothing, boots, rain gear, and shelter.

SAMPLE AREAS

Select study areas along the stream that are relatively homogeneous, physically and biologically. To accentuate response to environmental variables, it would be desirable to select an open segment of stream after it flows out of or into a forested area. Preferably, a biologically rich stream should be studied. It is advisable to avoid areas immediately below waterfalls or very swift rapids, as high O_2-diffusion rates or entrained air bubbles may interfere with some chemical analyses. When possible, select study sites that are about 1-h apart as determined by the flow rate of the stream (see Exercise 5). These sites will be used for studies of changes in heat storage, nutrient flux, and ecosystem metabolism.

GENERAL SAMPLING SCHEDULE

Measurements of solar radiation, temperature (water and air), dissolved oxygen, pH, dissolved inorganic phosphorus, nitrate, total alkalinity, dissolved organic carbon, particulate organic carbon, and flow should be made at 2400, 0300, 0600, 0900, 1200, 1500, 1800, 2100, and 2400 h. Water samples should be taken from well-mixed parts of the stream where stratification is minimal. Intially, samples should be obtained at several locations across the stream channel and at different depths to delineate any spatial patterns.

During this exercise it will become apparent how arbitrary field measurements can be (especially at night!) unless they are done carefully and with replication. Make measurements as precisely as possible and keep notes about any variation in procedure or unusual circumstances. You will be attempting to integrate and evaluate, with a few careful measurements, many of the complex interactions occurring within the entire ecosystem.

If possible divide the laboratory into groups of students, each group having responsibilities as listed below.

	Type of data to be collected	Materials or method required
I	*Physical*	
Leader ———		
Others ———	Solar radiation	See Exercises 2 and 4
	Air temperature	See Exercises 2 and 4
———	Water temperature	See Exercises 2 and 4
———	Channel morphology	See Exercise 5
———	length	
	slope	
———	cross-section	
	Stream water	See Exercise 5
———	velocity	
———	discharge	
II	*Chemical #1*	
Leader ———		
Others ———	Dissolved oxygen	See Exercise 6
———	pH	See Exercise 8
———	PO_4^{3-}	See Exercise 7
———	NO_3^-	See Exercise 7
———	Alkalinity	See Exercise 8
———		
III	*Chemical #2*	
Leader ———		
Others ———	Dissolved organic carbon	See Exercise 9
———	Particulate matter	See below
———		
———		
———		
IV	*Biological: Metabolism*	
Leader ———		
Others ———	Diurnal O_2 curve	See below
———	gross primary production	
———	respiration	
———	net primary production	
———		
———		

EXERCISES

One or more of the following options may be undertaken, depending upon the size of the class and the amount of time to be allotted to this exercise. Nevertheless, the general physical measurements (i.e., such as channel morphology, current velocity, and discharge) are common to all options and must be completed. Thus, it might be advantageous to undertake the various studies outlined in the options below on the same stream that was used in Exercise 5.

OPTION 1. SOME ASPECTS OF AN ENERGY BUDGET

1. Collect and plot the values for net radiation at each of the two study sites. Use a portable net radiometer if available; otherwise, adjust estimated values from a local climatologic data station for the amount of forest canopy if the section of the stream is located in a forested area (see Exercise 4). Alternatively, a photographic light meter, calibrated against measurements at a local climatologic data station, could be used to approximate the input of solar radiation.

2. Collect and plot the air temperature, 10 cm above the water surface, and the surface water temperature at each study site. Plot the direction of sensible heat flux between air and water throughout the diurnal period.

3. Determine and plot the instantaneous amount of heat stored in the water at each sampling time for each study site throughout the diurnal period: $\theta_w = T_w \cdot V \cdot s \cdot \rho$, where θ_w is the heat content of the stream water in calories*, T_w is water temperature in °C, V is volume in cm^3 of discharge, s is specific heat of water in cal/g-°C, and ρ is density in g/cm^3. Superimpose these curves on the same graph, adjusting the real time for the downstream site by the time of flow between the two sites. Compute with an electronic digitizer, by planimetry, or by counting squares the change in heat content between the two sites throughout the diurnal period (see Exercise 3). Construct a graph to show this change in stored heat with time between the two sites [see e.g., Wright and Horrall (1967)].

4. Calculate the difference in potential energy between the two sites: Difference in potential energy (kg-m) = $(W_1 h_1) - (W_2 h_2)$, where W_1 and W_1 are weights of 1 m^3 of water in kilograms at upstream and downstream sites, respectively, and h_1 and h_2 are heights (in m above mean sea level) of the gravitational centers of the channel cross sections at the two sites.

Questions

1. What other components of the heat budget have not been evaluated in this exercise? Do you think these components would be important? Explain.
2. Why does the stream lose heat at night? Would the heat loss be greater or less than that expected from a pond of the same depth? Why?
3. At what times during the diurnal period were the maximum and minimum air and water temperatures observed? How do you explain this?
4. What do you think the biological and chemical responses are to these temperature fluctuations?
5. What would be the effects of a rainstorm on the heat budget of a stream?
6. How would you expect the heat budget of a stream to be affected by the presence or absence of forest vegetation? [See Burton and Likens (1973).]
7. How might the biological community of the stream utilize the potential energy lost as the water flows downstream?

OPTION 2. SOME ASPECTS OF A CHEMICAL MASS BALANCE (INPUT/OUTPUT BUDGET)

1. Determine and plot the concentrations of PO_4^{3-}, NO_3^-, alkalinity, H^+, and dissolved organic carbon for each site throughout the diurnal period. Optionally, particulate matter can be collected with a 1-mm mesh net stretched across the stream. The amounts of finer particulate matter can be determined by filtering a subsample of water, which had passed

---

*1 g calorie (mean) × 4.1862 = 1 joule.

through the 1-mm net, through a tared, 0.45-μm pore size Millipore filter and then measuring its dry weight [see Eaton et al. (1969) for precautions]. Samples also can be filtered through precombusted glass fiber filters (see Exercises 7 and 9), and then the change in weight after combustion at 500°C for 2h can be used to differentiate between the inorganic and organic components. However, studies of particulate matter may not be compatible with the other activities in this exercise, since great care must be taken so that the sediments are not disturbed upstream of the sites where particulate matter is being collected.

2. Based on discharge and concentration measurements, calculate and plot the flux (in g/sec) of PO_4^{3-}, NO_3^-, alkalinity, H^+, and dissolved organic carbon for each site throughout the diurnal study. Determine the drainage area for each site from a topographic map and calculate the input and output flux for each segment of the stream in terms of g/ha-sec or g/ha-day.

3. To obtain some indication of upstream-downstream changes, plot the instantaneous nutrient flux (g/sec) at the upstream site and superimpose on this graph the flux for the downstream site, after adjusting its real time by the time of flow between the two sites. Compute with an electronic digitizer, by planimetry, or by counting squares the change in chemical flux between the two sites throughout the diurnal period. Plot this change in flux with time.

4. Compare the total flux (g/day) for each chemical at the upstream site with the daily flux for the downstream site. Which site had the largest flux? Why?

Questions

1. What kinds of inputs or losses were not measured in this study? How could you measure these? [See Fisher and Likens (1973) and Likens et al. (1977).]
2. How do you explain the differences in concentrations during the diurnal cycle? Between upstream and downstream sites? [See Manny and Wetzel (1973).]
3. What would be the effects of a rainstorm on the results?
4. How might changing concentrations be sampled realistically or optimally in a long-term study (a year or more)? [See Wetzel and Manny (1977) and Likens et al. (1977).]
5. Which was more important to total flux, discharge or concentration? Why?
6. Why do concentrations (mg/l) and fluxes (kg/day) of particulate organic matter fluctuate much more than dissolved organic matter or many nutrients on a diurnal or annual basis? What effects could growth patterns of aquatic and terrestrial vegetation or activity of animals have? [See Manny and Wetzel (1973), Wetzel and Otsuki (1974), and Likens et al. (1977).]

OPTION 3. ECOSYSTEM METABOLISM

An estimation of ecosystem metabolism (production and respiration) can be made from changes in concentration of dissolved oxygen or from changes in pH. We shall describe the two-station method, which is based on changes in dissolved oxygen concentration over a diurnal period (Odum, 1956; Owens, 1969). The procedures for both the O_2 and pH methods are described in detail in Hall and Moll (1975). The dissolved oxygen method is based on the following relationship:

$$q = p - r \pm d + a$$

where q is the rate of change of oxygen in g O_2/m³-h, p is the rate of gross primary productivity, r is the rate of respiration, and d is the rate of oxygen diffusion or reaeration flux across the air-water interface, and a is the rate of drainage accrual. Ideally, the study should be done in a situation where a is negligible. Correction for the reaeration flux d is a very important parameter, and can be estimated by using the mean oxygen deficit over the reach of stream being examined (Young and Huryn, 1998). The mean oxygen deficit is then multiplied by a

Table 24.1. Approximate diffusion constants (k) for gaseous oxygen across the air-water interface in streams.[a]

Areas[b]	$k(\text{g/m}^2\text{-h-atm})$[c]	Remarks
Pools	0.15 (0.05–0.3)	Higher if windy
Riffles	0.8 (0.5–1.5)	Higher for waterfalls

[a] From C.A.S. Hall (personal communication).
[b] Prorate according to area in pools and riffles.
[c] $k(\text{g/m}^3\cdot\text{h}\cdot\text{atm}) = k/\text{mean depth in m}$.

diffusion constant, k_{oxygen}, which may be approximated in this exercise by selecting from the values given in Table 24.1 for shallow streams. Then:

$$\text{Reaeration flux of oxygen (mg/L)} = (\text{DO}_{\text{deficit 1}} + \text{DO}_{\text{deficit 2}})/2 \times k_{oxygen} \times T$$

where the saturation deficit for dissolved oxygen (DO) at station 1 and station 2 is the difference between 100% saturation at the in situ temperature and pressure and the observed saturation percentage (see Exercise 6 for methods of calculating percentage saturation). For example, 80% saturation of dissolved oxygen at a location in a stream at a given time would have a saturation deficit of 0.20. The mean decimal equivalent of the saturation deficit between the two stations is then determined. T is the travel time in minutes along the specific reach of stream. p and r are obtained from a graphical analysis of the data (see Fig. 24.1).

Metabolism is thus estimated with the difference in the upstream-downstream data by the graphical technique in Fig. 24.1:
Net metabolism = $\Delta \text{DO}_{\text{light}}$ − reaeration
Dark metabolism = $\Delta \text{DO}_{\text{dark}}$ − reaeration
Gross community primary productivity (GPP) = net metabolism − respiration$_{\text{light}}$

Community respiration (CR$_{24}$) = Average night$_{\text{resp}}$ scaled for 24h. (see Marzolf, et al., 1994 but corrected by Young and Huryn, 1998).

Procedures

1. Measure the dissolved oxygen and water temperature every 3 h at the upstream and downstream sites throughout a 24-h period (see Exercise 6 for methods of determining dissolved oxygen).
2. Using data in Table 24.1 and your data for the stream, determine the diffusion constant for both the upstream and downstream stations along the stream.
3. Prepare a graphical analysis of the rate of O_2 change between the two stations similar to Fig. 24.1 but make corrections based on the mean values of temperature and dissolved oxygen for the two stations before calculating and plotting the rate-of-change curve for dissolved oxygen.
4. Using planimetry or by counting squares, determine the area of the curve that represents gross primary productivity, net primary productivity, and ecosystem respiration as g O_2/m^3-day. Based on an assumed or calculated respiratory quotient (see Exercise 29), calculate the production and respiration values in terms of carbon.
5. *Note*: There may be some advantages (less data to manipulate, fewer compounded errors, minimized stream heterogeneity) in using the single-curve method in this class exercise for estimating stream metabolism [see Hall and Moll (1975)]. The procedure is the same as above, except that the rate of O_2-change curve is calculated for time intervals from data at only one station. If possible, divide the class in half, obtain data from both stations, and

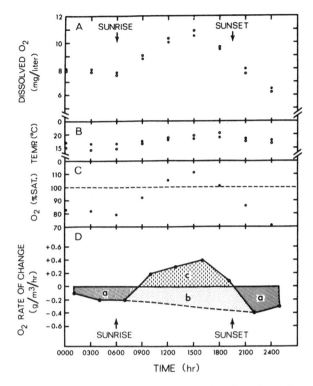

Figure 24.1. Graphical analysis of ecosystem metabolism based on changes in dissolved oxygen concentrations during a diurnal period. Dots indicate upstream stations, and open circles indicate downstream stations. The curve for the downstream station has been shifted by 1 h to the left on the time axis to account for the flow time between the two stations. In D, the sum of the two cross-hatched areas (a) represents an estimate of gross, nighttime ecosystem respiration; the cross-hatched, stippled area (b) is an estimate of gross daytime ecosystem respiration; and the stippled area (b + c) is an estimate of gross primary production in this segment of the stream. The rate of change curve is plotted at the midpoint between sampling times for the two stations but has not been corrected for diffusion. These data might be typical for an unshaded, productive Temperate Zone stream in July.

compare the results based on an average of the two single-curve analyses and the double-curve analysis.

Questions

1. Why is it necessary to make a correction for exchange of oxygen (diffusion) with the atmosphere? [See Owens (1969).] Will errors in this parameter significantly affect the overall estimates of primary productivity and respiration?
2. What is the meaning of the net primary productivity value as determined by this method? [See Wetzel (1975) and Wetzel and Ward (1992).]
3. How might you obtain a better estimate of net productivity? [See Kelly et al. (1974) and Gallegos et al. (1977).]
4. What is the P/R ratio for the whole stream? If the value were less than 1, where would the extra energy come from? If greater, where would the extra energy be going?
5. What are the main factors affecting stream metabolism? How do these differ from one stream segment to another? From day to day or season to season?

6. In some streams, the dissolved oxygen values for the downstream station may be similar to the upstream values. What does this mean? How is the rate-of-change method used in this case?

7. Attempt to construct a possible food web for the stream ecosystem. What are the energy relationships?

OPTION 4. LABORATORY ANALYSES

Using the data in Fig. 24.1B and Table 24.2:

1. Calculate the channel cross-sectional areas for Stations I and II.
2. Calculate and plot the change in heat storage for each station throughout the diurnal period.
3. Calculate the difference in heat storage (budget) between the upstream and downstream stations for the diurnal period (see Option 1).
4. Using data for NO_3^- and Na^+ (Table 24.2), complete Option 2.
5. Calculate the percent saturation values for the downstream station on Fig. 24.1 and plot them.
6. Determine a diffusion constant for each station, and correct the O_2 rate-of-change curves (Fig. 24.1D) for diffusion. Plot the corrected values and complete Parts 3 and 4 of Option 3.
7. Answer the questions for Options 1, 2, and 3.

Table 24.2. Data for discharge and concentrations of nitrate and sodium during a 24-hr period in the stream ecosystem described in Fig. 24.1.[a]

Time (h)	Discharge (l/sec)	NO_3^- ($\mu g/l$)	Na^+ (mg/l)
	Station I		
0000	0.120	140	1.1
0300	0.118	150	1.1
0600	0.125	155	1.1
0900	0.130	154	1.1
1200	0.120	151	1.2
1500	0.110	145	1.1
1800	0.108	140	1.1
2100	0.103	142	1.1
2400	0.099	138	1.2
	Station II		
0100	0.121	138	1.2
0400	0.118	145	1.1
0700	0.126	153	1.1
1000	0.130	151	1.1
1300	0.121	150	1.1
1600	0.111	142	1.2
1900	0.108	135	1.1
2200	0.104	136	1.1
0100	0.100	135	1.1

[a] The mean depth at Station I (upstream) and Station II (downstream) was 1.3 m. The time of flow between these stations was 1 h.

Apparatus and Supplies (in Addition to Those Cited in Other Exercises)

1. Clipboards.
2. Flashlights.
3. Alarm clock.
4. Graph paper (linear and log).
5. Portable net radiometer (e.g., C.S.I.R.O. Net Radiometer, Model CN2, Middleton and Co. Pty., Ltd., Port Melbourne, Victoria, Australia; or a small clock-driven pyrheliometer, e.g., Belfort Instruments Inc., Baltimore, MD).
6. Underwater thermometer.
7. Electronic digitizer or planimeter.

References

Burton, T.M. and G.E. Likens. 1973. The effect of strip-cutting on stream temperatures in the Hubbard Brook Experimental Forest, New Hampshire. BioScience 23(7):433–435.

Eaton, J.S., G.E. Likens, and F.H. Bormann. 1969. Use of membrane filters in gravimetric analyses of particulate matter in natural waters. Wat. Resour. Res. 5(5):1151–1156.

Fisher, S.G. and G.E. Likens. 1973. Energy flow in Bear Brook, New Hampshire: An integrative approach to stream ecosystem metabolism. Ecol. Monogr. 43(4):421–439.

Gallegos, C.L., G.M. Hornberger, and M.G. Kelly. 1977. A model of river benthic algal photosynthesis in response to rapid changes in light. Limnol. Oceanogr. 22:226–233.

Hall, C.A.S. 1972. Migration and metabolism in a temperate stream ecosystem. Ecology 53:585–604.

Hall, C.A.S. and R. Moll. 1975. Methods of assessing aquatic primary productivity. pp. 19–53. In: H. Lieth and R.H. Whittaker, Editors. Primary Productivity of the Biosphere. Springer-Verlag, New York.

Kelly, M.G., G.M. Hornberger, and B.J. Cosby. 1974. Continuous automated measurement of rates of photosynthesis and respiration in an undisturbed river community. Limnol. Oceanogr. 19:305–312.

Likens, G.E., F.H. Bormann, R.S. Pierce, J.S. Eaton, and N.M. Johnson. 1977. Biogeochemistry of a Forested Ecosystem. Springer-Verlag, New York. 146 pp.

Manny, B.A. and R.G. Wetzel, 1973. Diurnal changes in dissolved organic and inorganic carbon and nitrogen in a hardwater stream. Freshwat. Biol. 3:31–43.

Marzolf, E.R., P.J. Mulholland, and A.D. Steinman. 1994. Improvements to the diurnal upstream-downstream dissolved oxygen change technique of determining whole-stream metabolism in small streams. Can. J. Fish. Aquat. Sci. 51:1591–1599.

Minckley, W.L. 1963. The ecology of a spring stream, Doe Run, Meade County, Kentucky. Wildl. Monogr. 11:124 pp.

Odum, E.P. 1971. Fundamentals of Ecology. 3rd Ed. W.B. Saunders, Philadelphia. 574 pp.

Odum, H.T. 1956. Primary production of flowing waters. Limnol. Oceanogr. 2:85–97.

Odum, H.T. 1957. Trophic structure and productivity of Silver Springs, Florida. Ecol. Monogr. 27:55–112.

Owens, M. 1969. Some factors involved in the use of dissolved-oxygen distributions in streams to determine productivity. pp. 209–224. In: C.R. Goldman, Editor. Primary Productivity in Aquatic Environments. Univ. California Press, Berkeley.

Teal, J.M. 1957. Community metabolism in a temperate cold spring. Ecol. Monogr. 27:283–302.

Wetzel, R.G. 1975. Primary production. pp. 230–247. In: B.A. Whitton, Editor. River Ecology. Blackwell, Oxford.

Wetzel, R.G. and B.A. Manny. 1977. Seasonal changes in particulate and dissolved organic carbon and nitrogen in a hardwater stream. Arch. Hydrobiol. 80:20–39.

Wetzel, R.G. and A. Otsuki. 1974. Allochthonous organic carbon of a marl lake. Arch. Hydrobiol. 73:31–56.

Wetzel, R.G. and A.K. Ward. 1992. Primary production. pp. 354–369. *In*: P. Calow and G.E. Petts, Editors. Rivers Handbook. I. Hydrological and Ecological Principles. Blackwell Scientific Publications, Oxford.

Wright, J.C. and R.M. Horrall. 1967. Heat budget studies on the Madison River, Yellowstone National Park. Limnol. Oceanogr. *12*(4):578–583.

Young, R.G. and A.D. Huryn. 1998. Comment: Improvements to the diurnal upstream-downstream dissolved oxygen change technique for determining whole-stream metabolism in small streams. Can. J. Fish Aquat. Sci. *55*:1784–1785.

Diurnal Changes in Lake Systems

Environmental conditions may change markedly within aquatic ecosystems during a 24-h period. Solar radiation varies from high intensity at midday to darkness. Surface temperatures and concentrations of dissolved gases also may fluctuate between extremes of day and night, especially in shallow, littoral areas. Organisms within the lake may move vertically or otherwise relocate themselves in response to these environmental changes. For example, various species of zooplankton undergo marked vertical migrations during a 24-h cycle [cf., Hutchinson (1967) and Wetzel (1983, 1999)]. These movements are keyed to light, food availability, and predation pressures. In this exercise, we shall examine some of the relationships between these factors and the vertical movements of zooplankton in a lake.

LOCATION AND MOVEMENT OF THE AVERAGE INDIVIDUAL

Since variability in response is common among living organisms, it is often convenient to follow the movement of the average individual (Worthington, 1931):

$$\text{Depth of average individual} = \frac{x_1 z_1 + x_2 z_2 + \cdots + x_n z_n}{x_1 + x_2 + \cdots + x_n}$$

where z_1 = a depth of 1 m, z_2 = a depth of 2 m, x_1 = number of individuals/m^3 at a depth of 1 m, x_2 = number of individuals/m^3 at a depth of 2 m, and so on to a maximum depth, n. The velocity and amplitude of vertical migration then can be calculated on the basis of the depth of the average individual.

Obviously, by following only the "average individual" the detailed pattern of movement for the entire population is obscured. Detailed tabular data or histograms can be used but often are difficult to interpret because of the bulk of data involved. Pennak (1943) proposed a method of plotting quartile curves which is useful and particularly well suited for illustrating movements in populations with relatively slow movement. Quartiles can be calculated readily from tabulation of cumulative numbers of organisms with depth (e.g., Table 25.1). As an example, the 25, 50, and 75% quartiles for the data presented in Table 25.1 fall between 2 and 3 m, 4 and 5 m, and 6 and 7 m, respectively. By interpolation, 25% of the individuals in this hypothetical species were found between the surface and 2.25 m, 50% were

Table 25.1. Hypothetical data for calculation of quartiles for a species of zooplankton.

Depth (m)	Number of individuals per liter	Cumulated number of individuals
0	3	3
1	5	8
2	5	13
3	5	18
4	7	25
5	11	36
6	5	41
7	5	46
8	6	52
9	2	54
10	3	57

found between the surface and 4.32 m, and 75% were found between the surface and a depth of 6.35 m. Such points can be calculated and plotted for various vertical profiles throughout a 24-h period.

EXERCISES

OPTION 1

This option is designed as a 24-h field trip to a lake or reservoir. As a class exercise, it may be convenient to have various groups sample at different times throughout the 24-h period. If this were done, it would be extremely important to standardize procedures for all groups.

1. Beginning at, say, 1400 h, collect samples of zooplankton every 3 h at 1-m intervals of depth starting at 0.5 m, from the deepest portion of the lake. If the lake is very deep (>20 m), it may be sufficient to collect samples at 2-m intervals. Randomize the order in which the samples are obtained from the various depths but, in any case, work as rapidly as possible to minimize the length of time for sampling. Overlap the starting time by taking additional samples until 1700 h on the second day. Use either the collection device of choice or the one judged to give the most quantitative results in tests conducted in Exercise 11. Preserve the samples and follow the methods given in Exercise 11.
2. Carefully label the sample bottles and record the following information:
   a. Site.
   b. Depth.
   c. Date and time (a concise and unambiguous way of recording such information is as follows: 6 A.M. on July 6, 1977 recorded as 0600 h EST, 6 July 77).
   d. Volume filtered.
   e. Name of collector.
3. At the same intervals of depth, measure and record:
   a. Water temperature (also air temperature) (see Exercise 2).
   b. Total irradiance (see Exercise 2).
   c. Chlorophyll concentrations (see Exercise 10).
   d. Dissolved oxygen concentrations (see Exercise 6).
4. Enumerate the zooplankton in each sample (see Exercise 11).
5. Construct quartile curves (Pennak, 1943) and plot the depth of the average individual for each of the dominant species over the sampling period.

6. Calculate the amplitude and velocity of vertical migration of the average individual of each of the dominant species of zooplankton.
7. Compare these results with diurnal plots of temperature, irradiance, chlorophyll, and dissolved oxygen concentrations. Three-dimensional graph paper (e.g., National 12-186 Isometric-Orthographic) is useful in illustrating these data with depth and time. Alternatively, utilize computer software programs (e.g., Sigma Plot; Surfer, others) that allow the rapid construction of three-dimensional graphical representations of depth (ordinate), time (abscissa), and zooplankton densities (third dimension).

OPTION 2

On samples and data provided by the instructor, enumerate the zooplankton and make calculations and graphs as called for in parts 5 through 7 of Option 1.

OPTION 3

From data presented in Table 25.2, make calculations and graphs as called for in parts 5 through 7 of Option 1.

Problems and Questions

1. Measure the diurnal changes in spectral fractions of underwater irradiance, using a photometer with narrow-band wavelength filters (see Exercise 2). Plot these light distributions versus depth. Does a correlation exist between light distribution changes and the diurnal migratory patterns of zooplankton?
2. Do relationships exist in your data between the vertical amplitude of migration and the distribution of species of zooplankton and the vertical stratification of temperature and dissolved oxygen concentrations?
3. Were significant diurnal changes in the vertical pigment distribution observed? Were these correlated with the migratory patterns of the zooplankton?
4. Determine the in situ grazing rates of zooplankton (if equipment is available; see Exercise 15). How do the rates of ingestion change diurnally?
5. Collect samples of the phytoplankton at the same strata over the diurnal period (see Exercise 10). Enumerate, identify to genera, and categorize the algae into size classes (e.g., 1 to $30\,\mu m$, 30 to $60\,\mu m$, $>60\,\mu m$ in largest dimension). Compare changes in algal distribution diurnally and zooplankton migratory behavior. Did any of the algal populations, especially the cyanobacteria, appear to migrate diurnally? Why?
6. What are the major disadvantages of the average individual and quartile methods of diagramming zooplankton movement? What are the advantages?
7. Would you expect to observe similar vertical movement of live zooplankton in the laboratory if maintained in darkness for 24 h? Why?
8. How might high-frequency sonar be used to follow the movement of zooplankton in lakes?
9. What type of sampler would be better to use in a study of the vertical migration of zooplankton? Why?
10. What is meant by reverse migration? Nocturnal migration? Which migratory patterns are shown by the organisms in Table 25.2?

Apparatus and Supplies

1. Zooplankton sampling devices (see Exercise 11).
2. Preservatives, bottles, and data records (see Exercise 11).
3. Instruments for measuring temperature and irradiance (Exercise 2).

Table 25.2. Data obtained during a 24-h period in August in a clear-water lake in the North Temperate Zone.

Depth (m)	1600h	1950h	0100h	0600h	1215h	1550h
			Zooplankton			
		Kellicottia longispina (number/m³)				
0	0	34	269	404	269	168
2	101	34	269	943	269	202
4	572	1,246	943	471	1,313	741
6	1,212	1,414	1,178	1,010	1,010	370
8	236	67	707	168	0	34
10	101	67	168	34	0	202
		Polyarthra vulgaris (number/m³)				
0	2,862	5,523	7,037	10,034	7,273	8,519
2	2,391	3,872	8,239	13,333	11,852	14,242
4	9,326	6,767	9,124	14,345	15,992	10,676
6	9,124	13,604	9,764	11,313	9,225	10,572
8	3,704	6,228	11,717	7,071	1,650	9,428
10	303	1,011	2,458	1,010	202	5,758
		Ceratium sp. (number/m³)				
0	3,491	3,097	3,673	2,694	3,030	3,636
2	2,222	2,761	2,727	2,997	3,232	3,300
4	2,930	3,333	2,427	3,602	3,299	2,727
6	4,377	3,572	2,390	2,559	4,444	4,309
8	909	67	1,616	1,380	370	640
10	67	201	505	100	101	572

Relative total solar irradiance (cal × 10⁻³/cm²–min)

Depth (m)	1600h	1850h	1950h	2050h	0100h	0455h	0600h	0740h	1215h	1600h
0	301	20	0.58	<0.01	<0.01	<0.01	3.1	111	440	245
2	107	7.3	0.50	<0.01	<0.01	<0.01	3.1	43	160	85
4	40	3.5	0.15	<0.01	<0.01	<0.01	0.62	9.3	49	31
6	13	0.93	0.03	<0.01	<0.01	<0.01	0.44	3.8	7.3	3.8
8	1.4	0.13	<0.01	<0.01	<0.01	<0.01	0.04	0.57	2.0	1.5
10	0.13	<0.01	<0.01	<0.01	<0.01	<0.01	<0.01	<0.01	<0.01	0.05

Temperature (°C)

Depth (m)	1430h	1830h	2200h	0200h	0600h	1000h	1215h	1600h
0	21.9	22.0	21.6	21.2	20.8	20.9	21.3	22.2
2	21.8	21.3	21.6	21.2	20.8	20.7	21.1	22.0
4	15.5	15.1	15.8	15.5	15.3	15.5	15.4	15.5
6	9.1	9.6	9.1	9.3	9.2	9.2	9.3	9.3
8	6.9	7.1	7.2	7.1	7.2	7.1	7.2	7.2
10	6.4	6.5	6.4	6.5	6.4	6.4	6.4	6.4

Dissolved oxygen (mg/l)

Depth (m)	1430h	1830h	2200h	0200h	0600h	1000h	1215h	1600h
0	8.1	7.8	7.8	7.8	8.2	8.2	8.2	8.1
2	8.2	8.2	8.1	8.3	7.9	8.3	8.4	8.2
4	9.7	9.2	9.4	8.7	8.4	8.4	9.1	9.8
6	9.3	10.4	9.1	9.4	9.4	9.2	9.7	10.1
8	4.4	5.2	6.6	6.6	6.6	6.5	6.6	6.6
10	1.6	1.6	1.5	1.6	1.6	1.6	1.5	1.6

4. Equipment and reagents for measuring dissolved oxygen concentrations (Exercise 6) and chlorophyll concentrations (Exercise 10).
5. Microscopes, counting cells, and other supplies for the enumeration of zooplankton and phytoplankton (Exercises 10 and 11).
6. Graph paper (linear and three-dimensional).

References

Hutchinson, G.E. 1967. A Treatise on Limnology. Vol. 2. Introduction to Lake Biology and the Limnoplankton. Wiley, New York. 1115 pp.

Pennak, R.W. 1943. An effective method of diagramming diurnal movements of zooplankton organisms. Ecology 24:405–407.

Wetzel, R.G. 1983. Liminology. Saunders Coll., Philadelphia. 860 pp.

Wetzel, R.G. 1999. Limnology: Lake and River Ecosystems. 3rd Ed. Academic Press, San Diego (in press).

Worthington, E.B. 1931. Vertical movements of freshwater macroplankton. Int. Rev. ges. Hydrobiol. Hydrogr. 25:394–436.

Special Lake Types

Certain of the vast number of lakes distributed over the surface of the Earth have special limnological characteristics. Bog lakes in the northern latitudes, reservoirs, anthropogenically acidified lakes in Scandinavia and eastern North America and below strip-mining operations, meromictic lakes, playa lakes in arid regions, and alpine lakes provide examples of a diversity of exceptional conditions that add to the fascination and enjoyment of limnological studies. We suggest you explore as many of these diverse habitats in your local area as possible. A discussion of several examples follows.

RESERVOIR ECOSYSTEMS

Reservoirs are created when water runoff from the land, such as a river, is accumulated in a basin by an impounding structure, usually a dam. Reservoirs are created predominantly in regions where large natural lakes are sparse or unsuitable (e.g., too saline) for human exploitation. In these regions the climate tends to be warmer than is the case for many natural lakes and results in somewhat higher average water temperatures, longer growing seasons, and precipitation inputs that are closely balanced to, or less than, evaporative losses (Wetzel, 1990a). Drainage basins of reservoirs are much larger in relation to the lake surface areas than is the case among most natural lakes. Because reservoirs are formed almost always in river valleys and at the base of the drainage basins, the morphometry of reservoir basins is usually dendritic, narrow, and elongated (see Exercises 1 and 5).

These physical characteristics affect biological processes in many complex ways, the most important of which are light and nutrient availability to aquatic organisms (Wetzel, 1990a). Reservoirs receive runoff water mainly via large streams, which have high energy for erosion, large sediment-load carrying capacities, and extensive penetration of dissolved and particulate loads into the recipient lake water. Because the inflows are primarily channelized (along the prior river course before it was inundated) and often not intercepted by energy-dispersive and biologically active wetlands and littoral interface regions, runoff inputs to reservoirs are larger, more directly coupled to precipitation events, and extend much farther into the lake per se than is the case in most natural lakes. All of these properties result in high, but irregularly pulsed, nutrient and sediment loading to the recipient reservoir.

Extreme and irregular water-level fluctuations occur in reservoirs often as a result of flooding, land-use practices not conducive to water retention, channelization of the main inflows, and large, irregular water withdrawals, commonly for hydropower generation. Large areas of sediments are alternately inundated and exposed. These manipulations usually prevent the establishment of productive, stabilizing wetland and littoral flora. Erosion and resuspension of floodplain sediments augment high loadings from elsewhere in the drainage basin. Sediments may be shifted between aerobic and anaerobic conditions, which enhances nutrient release. The reduction or elimination of wetland and littoral communities around many reservoirs minimizes their extensive nutrient and physical sieving capacities that function effectively in most natural lake ecosystems (Wetzel, 1979, 1983, 1990b).

A reservoir can be viewed as a very dynamic lake in which a significant portion of its volume possesses characteristics of and functions biologically as a river (Wetzel, 1990a). Often the riverine portion of a reservoir functions analogously to large, turbid rivers in which turbulence, sediment instability, high turbidity, reduced light availability, and other characteristics limit photosynthesis despite high nutrient availability. Although phytoplanktonic productivity of riverine sections of reservoirs can be high per unit water volume, the limited photic zone reduces areal productivity, as is similarly the case in large rivers [e.g., Wetzel (1975), Minshall (1978), Bott (1983), and Wetzel and Ward (1992)]. As turbidity is reduced and the depth of the photic zone increases in the transition to the lacustrine (near the dam) regions of the reservoir, areal primary productivity increases concomitant with greater light penetration and depth of the trophogenic zone. Likewise, nutrient limitations can vary throughout a reservoir as losses of nutrients exceed renewal rates.

Fish biology and productivity are variable in dynamic, constantly changing reservoir ecosystems. High fish productivity often is observed soon after reservoir formation: this productivity has been related to the high productivity of benthic fauna associated with greater habitat variability and refugia among inundated terrestrial vegetation. In addition, high nutrient and organic matter loadings occur during the early "trophic surge" period following the damming of a river. Many reservoirs are not cleared completely of forest and shrub vegetation prior to inundation, particularly in riverine areas. As these habitats decay and decline, fish must shift to alternative, predominantly pelagic, food sources. High reservoir turbidity can decrease visual predation on pelagic zooplanktonic food sources. Fluctuating water levels, high siltation, and heavy predation often result in high mortality of eggs and larval fish in littoral areas.

The environmental conditions of reservoir ecosystems tend to have large, rapid, and erratic fluctuations. Often insufficient time exists for complete population growth and reproduction to occur before a succeeding major disturbance occurs, e.g., a plume intrusion and disturbance of stratification patterns following a major rainfall. These instabilities result in biota that tend to be few and well adapted with broad physiological tolerances (low diversity, less specialization, rapid growth). As in all restrictive, stressed environments, the productivity of adapted organisms can be high, as high as or greater than in more homeostatic natural lakes.

Determine the thermal characteristics of the water by temperature profiles at different positions in a local reservoir from the primary river input along a gradient to the deeper water near the dam. Collect water from the riverine, transitional, and lacustrine sections of the reservoir. Compare the chemical and biological characteristics at different depths of these areas with methods given in the previous exercises.

Questions

1. How might reservoirs be managed to minimize the extreme variability in physical and chemical parameters?
2. What means could be invoked to increase the exposure time of incoming water to the attached microbiota of wetland and littoral communities?
3. What is meant by underflow, overflow, and interflow of influent water? [See Wetzel (1999).] How could distribution of this inflowing water affect the productivity of phytoplankton? Bacteria? Zooplankton? Turbidity?
4. How would the removal of effluent water from the hypolimnion rather than from the surface affect the physical, chemical, and biological characteristics of a reservoir? Of the river below the dam that is receiving the effluent water?
5. If dissolved organic compounds adsorb to the surfaces of silt and other inorganic particulate matter suspended in water, could these compounds supplement the nutrition of zooplankton if ingested?
6. How could rates of siltation/sedimentation is a reservoir be reduced?

MEROMICTIC LAKES

Meromictic lakes are rather rare but, nevertheless, are distributed throughout the world (Walker and Likens, 1975). Because they lack complete circulation, meromictic lakes provide unique opportunities for interesting limnological studies. Typically, the deepest water contains no dissolved oxygen, maintains a relatively constant temperature, and is relatively poorly illuminated. Sediments below this zone are relatively undisturbed and, therefore, may offer an extraordinarily detailed record of the history of the lake and its surroundings [see Exercise 27; Gorham and Sanger (1972) and Frey (1955)]. Often a "plate," or dense accumulation, of photosynthetic bacteria forms in the chemolimnion of meromictic lakes. In Fayetteville Green Lake in New York State, the density of green-sulfur bacteria (*Chlorobium phaeobacteroides*) at a depth of 18 to 20 m can be sufficiently high to make the water appear light magenta in color (Culver and Brunskill, 1969).

The monimolimnion of a meromictic lake represents a harsh environment for most organisms, because of the absence of dissolved oxygen and the presence of reducing substances such as H_2S and NH_3. However, there are certain organisms that can survive in this environment, at least temporarily. These forms include anaerobic bacteria, certain dipteran larvae (e.g., *Chaoborus*), gastrotrichs, and some oligochaetes.

Collect water from the monimolimnion and measure the dissolved oxygen and hydrogen sulfide content [see Exercise 6 and Cline (1969)]. Be careful that the water sample is not contaminated with atmospheric oxygen during collection and analysis for dissolved oxygen. The human nose is very sensitive to hydrogen sulfide. You should be able to smell it as it vaporizes from the samples. Bubbles may form in the bottles after the water sample has been acidified during the analysis for dissolved oxygen. What do you think these bubbles are composed of?

Observe carefully for the presence of any living organism in the samples of water from the monimolimnion. How could you test for the presence of anaerobic bacteria in these samples? See, for example, Jackson (1967).

Collect some *Chaoborus* larvae and add them to sealed samples of monimolimnetic water. How long do they survive? What adaptations do you think these organisms have for surviving in such a harsh environment? See, for example, Brand (1946).

Questions

1. What are the conditions favorable to the formation of a bacterial plate in the chemolimnion of a meromictic lake?
2. The monimolimnion of a meromictic lake has been referred to as a "biological desert." Is this correct? What is meant by a biological desert?
3. How is decomposition accomplished in a meromictic lake? How does this affect the sediment record?
4. Think about the factors that initiate and maintain meromixis in lakes. Where might you go to find a "new" meromictic lake? [See Hutchinson et al. (1970), Judd (1970), and Walker and Likens (1975).]

BOG LAKES, BOGS, AND QUAKING BOGS

Bog lakes and bogs occur over large areas of the temperate region of the world. The formation of bog (mire) ecosystems depends to a great extent on conditions of high humidity and precipitation, which encourage the colonization of *Sphagnum* and other mosses.

Sphagnum mosses common to bogs and bog lakes can act very effectively as cation exchange columns. Cations such as calcium and magnesium are absorbed by the moss, and, as a consequence, hydrogen ions and organic acids are released into the environment [e.g., Anschütz and Gessner (1954), Clymo (1963, 1967), Burgess (1975), and Glime et al. (1982)]. *Sphagnum* mosses grow in profusion around many bog lakes, often encompassing the entire periphery with dense mats. These mats can have profound effects on the ionic composition of precipitation and drainage water that passes through the plants and, therefore, on the chemistry of the lake water.

The effectiveness of cationic exchange of *Sphagnum* can be demonstrated readily. Collect about 40 living specimens (dead plants work also but not as well) of *Sphagnum*; keep the plants moist with water from the site where the plants were collected. Measure the pH, specific conductance, and calcium concentration of 300 ml of a calcium chloride solution containing 1.0 meq/l (20 mg/l) calcium (see Exercises 7 and 8). Immerse the 40 plants of *Sphagnum* into the solution and swirl intermittently. Measure the pH, specific conductance, and calcium concentration of the solution at 1-h intervals for several hours.

Quaking bogs represent a particular development of bogs in relatively deep lake basins of small area. The prolific development of *Sphagnum* mosses leads to rapid accumulation of partially decomposed organic matter in the littoral areas. The dense *Sphagnum* and accumulation of peat deposits can form a dense, spongy mat, which grows all along the periphery toward the lake center, eventually encompassing the entire surface.

Commonly, a concentric zonation of plant communities occurs landward from the edge of the open water. Within the water there often are found certain submersed angiosperms (certain *Potamogeton* and "carnivorous" *Utricularia* species) and floating-leaved macrophytes, such as the water lilies (*Nuphar* and *Nymphaea*). A variety of interesting and unusual plants are common to the lakeward portion of the *Sphagnum* mat: bog cotton (*Eriophorum* spp.), orchids, and insectivorous plants, such as the pitcher plant (*Sarracenia purpurea* L.), and sundews (*Drosera* spp.). The flora of the mat grades into increasing densities of sedges (especially *Carex* spp.), low shrubs, such as poison sumac and the leatherleaf (*Chamaedaphne*), followed by the landward development of tall shrubs and finally bog trees, e.g., black spruce (*Picea mariana*) and tamarack (*Larix laricina*).

Questions

1. The development of *Sphagnum* mosses occurs over vast areas of the world, but under rather specific conditions of climate and hydrology. What are some of these conditions? [See Wetzel (1999), Moore and Bellamy (1974), Glime et al. (1982), and Gorham (1987).]
2. *Sphagnum* mosses have been found to invade the bottoms of some recently culturally acidified lakes in Sweden (Grahn, 1976). What might be responsible for this? What are some of the possible ecological effects of this type of growth in contrast to encroachment on the surface?
3. From estimates of the biomass of mosses in the bog system you examined and the exchange capacity of that moss used in the simple experiment, estimate the cationic exchange capacity for the bog system. How much external loading of cations would be required from external inflow to exceed this capacity? [See Gorham et al. (1985).]
4. What factors lead to reduced rates of decomposition of organic matter in bog ecosystems?
5. Describe the environmental conditions inside the "pitcher" of a pitcher plant. The larvae of the mosquito, *Wyeomyia smithi*, are commonly found living in the water held by leaves of pitcher plants. How do you explain the presence of the *Wyeomyia* in the pitchers? Do you think that pitcher plants are carnivorous? Explain.
6. Why do you think that an insectivorous (*Sarracenia; Drosera*) or carnivorous (*Utricularia*) habit may have evolved in bog plants?

References

Anschütz, I. and F. Gessner. 1954. Der Ionaustausch bei Torfmossen (*Sphagnum*). Flora *141*:178–236.

Bott, T.L. 1983. Primary productivity in streams. pp. 29–53. *In*: G.W. Minshall and J.R. Barnes, Editors. Stream Ecology: The Testing of General Ecological Theory in Stream Ecosystems. Plenum, New York.

Brand, T. 1946. Anaerobiosis in Invertebrates. Biodynamica, Normandy, MO. 328 pp.

Burgess, J.A. 1975. Organic acid excretion and the impact of *Sphagnum* mosses on their environment. Proc. Birmingham Nat. Hist. Phil. Soc. *23*:21–24.

Cline, J.D. 1969. Spectrophotometric determination of hydrogen sulfide in natural waters. Limnol. Oceanogr. *14*:454–458.

Clymo, R.S. 1963. Ion exchange in *Sphagnum* and its relation to bog ecology. Ann Bot. N.S. *27*:309–324.

Clymo, R.S. 1967. Control of cation concentrations, and in particular of pH, in *Sphagnum* dominated communities. pp. 273–284. *In*: H.L. Golterman and R.S. Clymo, Editors. Chemical Environment in the Aquatic Habitat. N.V. Noord-Hollandsche Uitgevers Maatschappij, Amsterdam.

Culver, D.A. and G.J. Brunskill. 1969. Fayetteville Green Lake, New York. V. Studies of primary production and zooplankton in a meromictic marsh lake. Limnol. Oceanogr. *14*:862–873.

Frey, D.G. 1955. Längsee: A history of meromixis. Mem. Ist. Ital. Idrobiol. Suppl. *8*:141–161.

Glime, J.M., R.G. Wetzel, and B.J. Kennedy. 1982. The effects of bryophytes on succession from alkaline marsh to *Sphagnum* bog. Amer. Midland Nat. *108*:209–223.

Gorham, E. 1987. The ecology and biogeochemistry of *Sphagnum* bogs in central and eastern North America. pp. 3–15. *In*: A.D. Laderman, Editor. Atlantic White Cedar Wetlands. Westview Press. Boulder, CO.

Gorham, E. and H.E. Sanger. 1972. Fossil pigments in the surface sediments of a meromictic lake. Limnol. Oceanogr. *17*:618–622.

Gorham, E., S.J. Eisenreich, J. Ford, and M.V. Santelmann. 1985. The chemistry of bog waters. pp. 339–362. *In*: W. Stumm, Editor. Chemical Processes in Lakes. Wiley, New York.

Grahn, O. 1976. Macrophyte succession in Swedish lakes caused by deposition of airborne acid substances. pp. 519–530. *In*: L.S. Dochinger and T.A. Seliga, Editors. Proceedings of the First International Symposium on Acid Precipitation and Forest Ecosystem. USDA Forest Service Tech. Report NE 23.

Hutchinson, G.E., et al. 1970. Ianula: An account of the history and development of the Lago di Monterosi, Latium, Italy. Trans. Amer. Phil. Soc., N.S. *60*(4). 178 pp.

Jackson, D.F. (ed.). 1967. Some Aspects of Meromixis. Dept. Civil Engr., Syracuse Univ., Syracuse, NY. 243 pp.

Judd, J.H. 1970. Lake stratification caused by runoff from street deicing. Water Res. *4*:521–532.

Minshall, G.W. 1978. Autotrophy in stream ecosystems. BioScience *18*:767–771.

Moore, P.D. and D.J. Bellamy. 1974. Peatlands. Elek Science, London. 221 pp.

Walker, K.F. and G.E. Likens. 1975. Meromixis and a reconsidered typology of lake circulation patterns. Verh. Internat. Verein. Limnol. *19*:442–458.

Wetzel, R.G. 1975. Primary production. pp. 230–247. *In*: B.A. Whitton, Editor. River Ecology. Blackwell, Oxford.

Wetzel, R.G. 1979. The role of the littoral zone and detritus in lake metabolism. Arch. Hydrobiol. Beih. Ergebn. Limnol. *13*:145–161.

Wetzel, R.G. 1990a. Reservoir ecosystems: Conclusions and speculations. pp. 227–238. *In*: K.W. Thornton, B.L. Kimmel, and F.E. Payne, Editors. Reservoir Limnology: Ecological Perspectives. Wiley, New York.

Wetzel, R.G. 1990b. Land-water interfaces: Metabolic and limnological regulators. Baldi Memorial Lecture. Verhand. Internat. Verein. Limnol. *24*:6–24.

Wetzel, R.G. 1999. Limnology: Lake and River Ecosystems. 3rd Edition. Academic Press, San Diego (in press).

Wetzel, R.G. and A.K. Ward. 1992. Primary production. pp. 354–369. *In*: P. Calow and G.E. Petts, Editors. The Rivers Handbook. I. Hydrological and Ecological Principles. Blackwell Science Publishers, Oxford.

Historical Records of Changes in the Productivity of Lakes

Changes in climate and in the geomorphology of drainage basins in the past have altered water and nutrient budgets and, as a result, productivity and rates of eutrophication of lake ecosystems. In many cases, human activities have greatly accelerated these changes, but on a much shorter time scale. A record of the resulting alterations in chemistry, flora, and fauna is left in the sediments as static derivatives of dynamic systems. Paleolimnology assesses the sedimentary record and the diagenetic processes that may alter it. An ultimate goal is to gain insight into the past conditions that caused a lake to enter a different level of productivity.

Interpretations about past levels and conditions of productivity of an aquatic ecosystem can be made from analyses of the: (1) physical structure and mineralogy of sediments, (2) inorganic and organic chemical constituents of the sediments, and (3) preserved morphological remains of organisms. Paleontological records rarely are complete. Moreover, lake sediments contain materials from the atmosphere and the drainage basin. Sediments within the basin can be redistributed by water and wind movements. Some remains of interred organisms preserve at different rates under changing lake conditions; others do not preserve at all. In spite of these difficulties, which demand critical interpretation, much information about changes in lake metabolism can be gleaned from the sedimentary record. Accurate interpretation depends on thorough understanding of ongoing biological and physicochemical processes of lakes.

The amount of permanent sedimentation in a lake is the result of the difference between the various inputs from the drainage basin and from the atmosphere (including solar radiation) and the losses to the atmosphere and to drainage. The form and rate of sedimented materials are regulated largely by biogeochemical transformations within the lake basin. Rates of sedimentation may be estimated from: (1) an analysis of the vertical age distribution in sediments, (2) measurement of seston sedimented, and (3) difference in input/output budgets [e.g., Wetzel et al. (1972), White and Wetzel (1975), Moeller and Likens (1978), and Davis and Ford (1985)].

In this exercise, we shall examine relatively recent sediments in lake or reservoir systems. From analyses of selected chemical and biological characteristics of the sediments, insight into processes that have induced changes in productivity of the systems may emerge.

PROCEDURES

Sediment Sampling

1. Use a coring device that contains a translucent or transparent plastic liner to sample the surficial sediments. A number of simple free-fall corers have been designed, such as those described earlier for the sampling of organisms (see Fig. 12.3). These tubular devices contain an acrylic liner and a threaded retainer ring at the tip to hold the liner within the tube. A one-way valve at the top allows water to exit during descent but creates a vacuum on the sample during retrieval.
2. Lower the corer to approximately 5 m above an area of undisturbed sediments and then allow it to fall freely into the sediments. Carefully retrieve the sampler and place a stopper into the bottom opening *before* removing from the water to prevent loss of the sample.
3. Remove the liner from the corer and stopper the upper end. Store erect. Repeat this procedure to obtain several sediment cores, each at least 0.5 m in length.
4. In the laboratory, *carefully* aspirate off the water overlying the sediments. Do not disturb the sediment-water interface.
5. A stopper of a size sufficient to fit inside the liner tube (but tightly to form a watertight seal) and mounted on a rod is inserted into the lower end. The core is then *gently and slowly* forced upward to the top of the tube.
6. Make careful measurements of the total length of the core and precise points (nearest mm) of any layers of sediment that appear to be different. Note any changes in stratigraphy, such as color and texture.
7. Carefully remove the sediment from the center (why?) of 1-cm-thick slices of the core with *clean* spatulas and place into several sample vials. Seal immediately and label. Carefully extrude the sediment core 1 cm upward, cut off residual sediments, and repeat sampling. A core slicer (Fast and Wetzel, 1974) greatly assists this operation, but good samples can be obtained without this aid when done carefully.

Chemical Analyses

Dry Weight. Weigh the portions of the homogenized wet sediments in flamed, tared crucibles. Dry for 24 h, or until a constant weight is obtained at 105°C. Cool under desiccation and reweigh on an analytical balance. Calculate percentages of water and of dry matter in the sediment samples.

Organic Weight. Heat the same samples used for dry weight to 550°C for at least 2 h after the furnace has reached temperature. Cool under desiccation and reweigh. Calculate the amount of organic matter lost on ignition and determine the percentage of organic matter of the sediment samples.

Total Phosphorus. Pulverize and homogenize dried (105°C) sediment samples with clean glass rods. Transfer tared subsamples of 200 to 250 mg to 100-ml volumetric flasks. Proceed with the analysis for total phosphorus as outlined in Exercise 7 (p. 97). The final procedures are modified slightly as follows. An aliquot of known volume of the final solution after persulfate digestion is transferred to a clean centrifuge tube and centrifuged to obtain a clear supernatant. The supernatant then is diluted with distilled water at a ratio of 1:5 before adding the molybdate solution to develop the blue-colored complex. The optical density then is measured spectrophotometrically within 30 min.

Analyses of Pigment Degradation Products

Relative Pigment Units

1. Homogenize about 4 g of wet sediments with a spatula into a tared crucible; weigh to the nearest 0.1 mg. Remove a 1- to 2-g portion of the homogenate and transfer it to a large test tube. Reweigh the crucible immediately to determine the amount removed. Determine the dry and organic weights of the sediment in the crucible, as discussed above.

2. Disperse the wet sediments in the test tube with a glass rod in about 30 ml of 90% aqueous, basic acetone (see Exercise 10) containing 0.5% dimethylaniline. Cover the test tubes with plastic (e.g., parafilm) and store at 4°C in the dark for 24 h; mix intermittently.

3. In a hood, filter the extracted sediments and supernatant acetone through fine filter paper (e.g., Whatman No. 50) into 100-ml glass-stoppered graduated cylinders. Wash the sediments repeatedly with solvent until exactly 100 ml of filtrate is collected.

4. Determine the optical density of the filtrate without delay in a spectrophotometer with a 1-cm cell at 750 nm for a baseline absorbance reading and, in the chlorophyllous pigment region, read the maximum at about 662 nm (locate the peak between 657 and 666 nm). Longer path length spectrophotometric cells may be needed for low pigment product concentrations.

5. A sedimentary degradation pigment unit (SPDU) is defined as the optical density with a 1-cm light path at the wavelength of maximum absorbance in the red end of the spectrum, from the baseline absorbance at 750 nm, under the extraction and dilution conditions described above (Vallentyne, 1955):

$$\text{SPDU/g dry wt} = (OD_{662} - OD_{750}) \left[\frac{1}{(\text{wet wt}) \left(\dfrac{\% \text{ dry matter in sediment}}{100} \right)} \right]$$

By expressing SPDUs per weight of organic matter, the concentrations essentially are independent of variable inorganic fractions of the sediments:

$$\text{SPDU/g org wt} = (OD_{662} - OD_{750}) \left[\frac{1}{(\text{wet wt}) \left(\dfrac{\% \text{ dry matter wet sediment}}{100} \right) \left(\dfrac{\% \text{ organic matter of dry sediment}}{100} \right)} \right]$$

Chlorophyll Derivatives and Carotenoids

1. Homogenize about 2 g of wet sediments from each depth interval with a spatula into a tared crucible; weigh to the nearest 0.1 mg. Transfer a 1- to 2-g sample of the homogenate to a large test tube. Reweigh the crucible immediately to determine the amount removed. Determine the dry and organic weights of the sediment in the crucible, as discussed above.

2. Disperse the wet sediments in the test tube with a glass rod in aliquots of aqueous 90% acetone; at least 300 ml acetone usually are needed to extract all of the pigments. Record the total amount of extract.

3. Remove a 10-ml portion of the extract for the measurement of chlorophyll derivatives. Measure absorbance with a spectrophotometer at the red maximum, at or

near 667 nm, for the plant chlorophyll derivatives. A correction for background absorbance may be substracted from the chlorophyll peak by drawing a baseline between 520 and 800 nm.

4. For the isolation of carotenoids, add a 25-ml sample of the acetone extract to an equal volume of 20% (w/v) of methanolic potassium hydroxide. Agitate the samples on a shaker for 2 h.

5. Separate the carotenoids into epiphasic pigments (without hydroxyl groups) and hypophasic carotenoids (with two or more hydroxyl groups) by partition in a separatory funnel with petroleum ether (30 to 60°C) and with aqueous 90% methanol, respectively. Measure the absorbance with a spectrophotometer at 450 nm of each of the carotenoid derivatives in their respective solvents.

6. Pigment concentrations are expressed as units per gram of organic matter, one unit being equivalent to an absorbance of 1.0 in a 10-cm spectrophotometer cell when dissolved in 100 ml of the appropriate solvent (Sanger and Gorham, 1972).

Diatom Analyses

1. To a sediment sample of known wet weight (ca. 250 mg), *carefully* add 50 ml of concentrated nitric acid and about 10 g of potassium dichromate. Add several boiling chips and boil the mixture gently *in a hood* for 20 min (Hohn and Hellerman, 1963). *Note*: Use extreme caution and wear protective shielding. Allow the material to cool and settle.

2. Decant and discard the supernatant, add distilled water to the residue, mix, and allow the materials to resettle. Repeat until the pH of the mixture tests neutral with litmus paper.

3. Place microscopic coverslips (ca. 18×18 mm) on an aluminum block or similar hot plate that will maintain a uniform, very low heat (e.g., a slide warmer).

4. Thoroughly mix the diatom-distilled water mixture of known volume in the beaker or flask. Immediately remove an aliquot of 200 to 500 μl with an automatic micropipet. Use a fresh pipet for each sample. Place on the coverslip and allow mixture to dry slowly.

5. After all of the samples are distributed on the coverslips and have dried, increase the temperature of the hot plate for 20 min.

6. Place a drop of the mounting medium Hyrax (refractive index 1.65) onto a microscope slide. Invert the coverslip and place onto the drop of Hyrax. The mount can be made permanent by evaporating most of the Hyrax solvent with moderate heating. Label the slide.

7. Microscopically examine the diatom slides from the several sediment layers. Attempt to identify major genera and species. Determine the proportions (percent) of major groups of diatoms for a given amount of sediment, e.g., centrales; araphidine pennate diatoms. In particular, examine the ratios of araphidine pennate to centrales diatoms in the different strata from the surface to deeper levels within the sediments. An increase in the ratio of araphidinean diatoms to centric diatoms is often associated in sediment cores with increasing lake fertility (Stockner, 1972; Moss et al., 1980).

Pollen Analyses

Pollen grains of higher plants and spores of lower plants typically develop a resistant layer outside the cell wall. This layer, the exine, is waxy or resinous and very resistant chemically. The surface of the exine is variously sculptured and assists, along with general shape and size and the shape and arrangement of wall apertures, in identifying the plant species from which the pollen originated [see, e.g., Faegri and Iversen (1964) and Kapp (1969)].

Chemical treatment of a sediment sample greatly facilitates the analysis of pollen by removing the extraneous sedimentary materials and by concentrating the pollen. With care, such methods are quantitative for a given amount of sediment. Many special techniques have been devised [see Brown (1960), Kummel and Raup (1965), and Faegri and Iversen (1964)]. The brief procedures discussed below are used commonly for sediment samples.

1. The samples are treated with acid to remove salts such as calcium carbonate. Expose the sample to dilute (3N) hydrochloric acid at room temperature for a brief period; then centrifuge at a high speed and decant the acid.
2. Organic colloids are removed by boiling the sample with 10% potassium hydroxide for about 10 min. Then shake the sample vigorously (a few drops of alcohol will dissipate the froth) and strain through a sieve of a mesh opening of ca. 0.2 mm. Wash the residue with a jet of distilled water. Then concentrate the fine suspension that has passed the sieve by centrifugation.
3. The final treatment of the sample is acetolysis, to remove cellulose materials, the cellular contents, and the cellulose wall (intine) of the pollen. Wear gloves and use caution during this procedure.
   a. Dehydrate the sample by adding glacial acetic acid and mix with a stirring rod.
   b. Centrifuge and decant the acetic acid.
   c. Slowly add about 10 ml of a mixture of 9 parts acetic anhydride:1 part of concentrated sulfuric acid. Stir and leave the stirring rod in the tube.
   d. Heat gently to the boiling point by immersion for one minute in a boiling water bath, stirring occasionally.
   e. Centrifuge and decant the acetolysis mixture.
   f. Wash with glacial acetic acid.
   g. Wash twice with distilled water by successive stirring, centrifuging, and decanting.
4. The pollen now may be examined directly with a microscope, but it is better to mount on permanent slides.
   a. Wash the sample (step 3 g above) with a few drops of water and 95% alcohol, centrifuge, decant, and discard the supernatant.
   b. Wash the residue with absolute alcohol, centrifuge; decant.
   c. Wash the residue with benzene, centrifuge, and decant.
   d. Add about 1 ml of benzene to the residue, transfer the mixture to a small vial, add high viscosity silicon oil, and allow to evaporate for 24 h.
   e. Transfer a small amount of the silicon oil to a microscope slide and cover with a coverslip. Slides may be sealed with nail polish along the edges.
   f. One wash with tertiary butyl alcohol can be substituted for the alcohol and benzene washes (steps a, b, and c) (Kapp, 1969).
5. Proceed with identification of major groups of pollen at high magnification (400 × or greater), using reference sources such as Kapp (1969). Attempt to discern the changes in relative abundance of known indicator species. For example, increases in pollen of the ragweed (*Ambrosia*) commonly are associated with landscape disturbance such as deforestation and initiation of agriculture.

EXERCISES

OPTION 1. FIELD AND LABORATORY ANALYSES BY SEVERAL METHODS

1. Obtain several cores of surficial sediments from the central depression of a lake or reservoir, as discussed earlier.
2. Divide the class into teams and perform as many of the analyses on the sediment samples as possible. When samples at certain intervals within the sediments must be omitted, concentrate efforts on the surface sediments and those at 3-cm intervals within the lower portion of the core.
3. Reduce the data to summary form and present graphically.
4. Compare the results with the stratigraphy of organic matter, phosphorus, pigment degradation products, and changes in diatom and pollen distributions. Compare to historical events that are known to have occurred in the immediate area around the lake.
5. Answer the questions following Option 4.

OPTION 2. ANALYSES ALONG GRADIENTS

1. Obtain several cores of surficial sediments from a gradient extending from the littoral zone to the central depression of a lake. Alternatively, sample along the gradient of a reservoir from the central depression near the dam toward the primary inlet stream or river.
2. Perform the analyses on the sediment cores as outlined in Option 1.
3. Answer the questions following Option 4.

OPTION 3. LABORATORY ANALYSES

1. Using the sediment cores provided by your instructor, perform the analyses as outlined in Option 1.
2. Answer the questions following Option 4.

OPTION 4. OLIGOTROPHIC VERSUS EUTROPHIC LAKES

1. Collect replicate surface cores of at least 0.5-m length from the central depression of a lake known to be unproductive and from a lake that is extremely productive.
2. Divide the class into teams and perform as many of the analyses on the sediment samples as possible. It is particularly important to sample at close intervals within the sediments near the surface sediment-water interface and the first 10 cm.
3. Compare the stratigraphy of organic matter, phosphorus, pigment degradation products, and changes in diatom and pollen distributions. Consider historical events that are known to have occurred in the drainage basins of the lakes, e.g., deforestation and shifts to agriculture.
4. Answer the following questions.

Questions

1. What do differences in organic matter content of sediment indicate?
2. Pigment degradation products have been used as an indicator of changes in rates of primary productivity of lakes [e.g., Wetzel (1970)]. What are some of the problems associated with such a direct interpretation?

3. How might one evaluate the proportion of organic matter or pigment degradation products entering the sediments from allochthonous sources of the drainage basin versus those produced autochthonously? What factors of the lake ecosystem lessen or accentuate the proportion from allochthonous sources? [See Davis et al. (1985).]
4. The loading of a lake system with phosphorus has been correlated directly with the amount of phosphorus deposited in the sediments. Is such a general correlation reasonable in view of what is known about the effects of contemporary lake responses to increased or decreased loading of phosphorus? Elaborate.
5. If one knew in sufficient detail the species of diatoms found in sediments, as well as that certain diatoms were usually littoral or nearly always pelagic, how might the stratigraphy of diatoms help to explain past changes in the lake? What events could lead to fluctuations in littoral versus pelagic development of algae? [See, e.g., Manny et al. (1978).]
6. How would one accurately evaluate the rates of sedimentation to lake sediments? How would radiocarbon dating help interpretations? Pollen stratigraphy?
7. Why is it important to know the rates of sedimentation when attempting to interpret the chemical and biological properties of sediments over time?
8. How does differential dissolution of diatom frustules confound use of their remains as a paleolimnological parameter? What factors in the ontogeny of a lake might alter the preservation of diatoms?
9. What factors within the lake influence the deposition and the preservation of pigment degradation products in sediments [see, for example, Wetzel (1970), Daley (1973), and Daley and Brown (1973)].
10. What, and how would, certain morphological remains of animals be useful in the interpretation of past lake conditions? [See e.g., Frey (1974), and Crisman (1978).] Morphological remains of higher aquatic plants? [See Birks and Birks (1980).]
11. Discuss the usefulness of reporting data as percent of total as opposed to deposition per unit area.

Apparatus and Supplies

1. Coring device with several plastic liners, cable, stoppers to seal ends, and storage rack to keep liners vertical.
2. Extruding stopper, mounted on a rod, to fit tightly inside of plastic liners.
3. Meter rulers and marking pens.
4. A number of clean spatulas and knives and *clean* sample vials.
5. 90% aqueous, basic acetone (see Exercise 10), dimethylaniline (*Caution*: toxic), methanolic potassium hydroxide, methanol, petroleum ether, separatory funnels, test tubes (25 × 150 mm), Parafilm, glass funnels, Whatman No. 50 filter paper, 100-ml ground-glass-stoppered graduated cylinders, and spectrophotometer.
6. Crucibles, tongs, 105°C oven, 550°C muffle furnace, analytical balance, and desiccators.
7. 100-ml volumetric flasks, reagents and supplies for total phosphorus analysis (see Exercise 7), and clinical centrifuge.
8. Diatom analyses: Concentrated HNO_3, $K_2Cr_2O_7$, beakers, protective face wear and hood, litmus paper, low temperature hot plate with aluminum block or heavy aluminum sheeting, micropipets, and Hyrax mounting medium.
9. Pollen analyses: 3N HCl, centrifuge, polypropylene centrifuge tubes, 10% KOH, absolute and 95% ethanol, 0.2-mm mesh sieve, glacial acetic acid, acetic anhydride, concentrated H_2SO_4, glass stirring rods, boiling water bath, tertiary butyl alcohol or benzene.

References

Birks, H.J.B. and H.H. Birks. 1980. Plant macrofossils in quaternary lake sediments. Arch. Hydrobiol. Ergebn. Limnol *15*.60 pp.
Brown, C.A. 1960. Palynological Techniques. C.A. Brown, 1180 Stanford Ave., Baton Rouge, LA. 188 pp.

Crisman, T.L. 1978. Reconstruction of past lacustrine environments based on the remains of aquatic invertebrates. *In*: D. Walker, Editor. Biology and Quarternary Environments. Australian Acad. Sciences, Canberra.

Daley, R.J. 1973. Experimental characterization of lacustrine chlorophyll diagenesis. II. Bacterial, viral and herbivore grazing effects. Arch. Hydrobiol. *72*:409–439.

Daley, R.J. and S.R. Brown. 1973. Experimental characterization of lacustrine chlorophyll diagenesis. I. Physiological and environmental effects. Arch. Hydrobiol. *72*:277–304.

Davis, M.B. and M.S. Ford. 1985. Late-glacial and Holocene sedimentation. pp. 346–366. In: G.E. Likens, Editor. An Ecosystem Approach to Aquatic Ecology. Mirror Lake and its Environment. Springer-Verlag, New York.

Davis, M.B., R.E. Moeller, G.E. Likens, J. Ford, J. Sherman, and C. Goulden. 1985. Paleoecology of Mirror Lake and its watershed. pp. 410–429. *In*: G.E. Likens, Editor. An Ecosystem Approach to Aquatic Ecology: Mirror Lake and its Environment. Springer-Verlag, New York.

Faegri, K. and J. Iversen. 1964. Textbook of Pollen Analysis. 2nd Ed. Hafner, New York. 237 pp.

Fast, A. and R.G. Wetzel. 1974. A close-interval fractionator for sediment cores. Ecology *55*:202–204.

Frey, D.G. 1974. Paleolimnology. Mitt. Int. Ver. Limnol. *20*:95–123.

Hohn, M.H. and J. Hellerman. 1963. The taxonomy and structure of diatom populations from three eastern North American rivers using three sampling methods. Trans. Amer. Microsc. Soc. *82*:250–329.

Kapp, R.O. 1969. How to Know the Pollen and Spores. W.C. Brown Co., Dubuque, IA. 249 pp.

Kummel, B. and D. Raup (eds). 1965. Handbook of Paleontological Techniques. Freeman, New York. 852 pp.

Manny, B.A., R.G. Wetzel, and R.E. Bailey. 1978. Paleolimnological sedimentation of organic carbon, nitrogen, phosphorus, fossil pigments, pollen, and diatoms in a hypereutrophic, hardwater lake: A case history of eutrophication. 2nd Int. Symp. on Paleolimnology. Polskie Arch. Hydrobiol. *25*:243–267.

Moeller, R.E. and G.E. Likens. 1978. Seston sedimentation in Mirror Lake, New Hampshire, and its relationship to long-term sediment accumulation. Verh. Int. Ver. Limnol. *20*:525–530.

Moss, B., R.G. Wetzel, and G.H. Lauff. 1980. Annual productivity and phytoplankton changes between 1969 and 1974 in Gull Lake, Michigan. Freshwat. Biol. *10*:113–121.

Sanger, J.E. and E. Gorham. 1972. Stratigraphy of fossil pigments as a guide to the postglacial history of Kirchner Marsh, Minnesota. Limnol. Oceanogr. *17*:840–854.

Stockner, J.G. 1972. Paleolimnology as a means of assessing eutrophication. Verhand. Internat. Verein. Limnol. *18*:1018–1030.

Vallentyne, J.R. 1955. Sedimentary chlorophyll determination as a paleobotanical method. Can. J. Bot. *33*:304–313.

Wetzel, R.G. 1970. Recent and postglacial production rates of a marllake. Limnol. Oceanogr. *15*:419–503.

Wetzel, R.G., P.H. Rich, M.C. Miller, and H.L. Allen. 1972. Metabolism of dissolved and particulate detrital carbon in a temperate hard-water lake. Mem. Ist. Ital. Idrobiol. *29* Suppl.: 185–243.

White, W.S. and R.G. Wetzel. 1975. Nitrogen, phosphorus, particulate and colloidal carbon content of sedimenting seston of a hardwater lake. Verh. Int. Ver. Limnol. *19*:330–339.

Effect of Sewage Outfall on a Stream Ecosystem

Human activity has affected profoundly streams and lakes in all parts of the world. Streams have been subjected to additions of gross amounts of domestic sewage, industrial effluents (e.g., wastes from tanneries, pulp mills, creameries, steel mills, and chemical factories), agricultural wastes, oil spills, mining wastes, urban runoff, radioactive materials, pesticides, waste heat, and numerous other pollutants, often because it was considered expedient and economical to have the unwanted materials carried away ("out of sight") by the flowing water. Likewise, under the guise of "progress," streams have been channelized, stabilized, dewatered (for irrigation), and super-watered (artificially increased flow for drinking and power plant needs). In most cases, the effects on the aquatic biota are insidiously cumulative. In some cases, the effects are readily apparent [e.g., acid mine drainage; see Parsons (1986)], but in others the effects accumulate more slowly [e.g., accumulations of trace metals; see Whitton and Say (1975)]. In all cases, a longitudinal gradient develops below the point of insult and, given enough time (distance) without further insult, the stream ecosystem generally recovers to a state of well-being.

In this exercise, we shall examine the effects of a sewage outfall on a stream ecosystem. Select a site, preferably on a small stream, into which sewage is being discharged. In most areas of the United States, this sewage will have already undergone primary and/or secondary treatment [e.g., Bolton and Klein (1971) and Rohlich and Uttormark (1972)]. Nevertheless, exercise health precautions when collecting and analyzing samples from such areas; for example, keep hands away from face and wash thoroughly after handling samples. Select sample sites immediately above and below the sewage outfall. In addition, a few sites 1 km apart above and 2 km apart below the outfall should delineate the longitudinal changes.

PROCEDURES

1. Describe the physical and topographic characteristics of the sample sites (see Exercise 5). Make a sketch of the area and locate the sample sites.
2. Attempt to determine the amount, type, and frequency of sewage discharge into the stream.
3. At each site measure the following:
   a. Water temperature (see Exercise 2).

b. pH (see Exercise 8).

c. Concentrations of Na^+, NH_4^+, Cl^-, PO_4^{-3}, and NO_3^- (see Exercise 7).

d. Concentration and percent saturation of dissolved oxygen (see Exercise 6).

e. Concentrations of dissolved organic matter (see Exercise 9).

f. Concentrations of suspended solids (see Exercise 7).

g. Numbers and biomass of benthic fauna (see Exercises 12 and 13).

h. Visually, quantities of periphyton and "sewage fungus" [see Exercise 22, and Hynes (1963, p. 96)].

i. Colonies (clones) of coliform bacteria.

The following is a simple procedure for determining the presence and abundance of coliform bacteria in water samples. Certain coliform bacteria, which commonly live in the guts of warm-blooded animals, incorporate dye such as eosin or fuchsin and then deposit the dye on the surface of the colony, leaving a metallic sheen. *Escherichia coli* is a common coliform bacterium of humans. Colonies of *E. coli* grown on a medium containing such dye may be distinguished from the majority of bacteria that either do not take up the dye or simply become colored, without forming the sheen. Further tests may be done to show that the colonies thus isolated are, in fact, *E. coli* [see Rodina (1972) and American Public Health Association (1998)].

4. A sample of the bacteria can be collected conveniently with a special stainless steel syringe equipped with a two-way valve and a plastic filter holder containing a sterile, gridded, white filter (e.g., Millipore; see "Apparatus and Supplies").

a. Obtain 10 ml of water from the stream with a graduated cylinder.

b. Remove the *blue* cap from the filter holder and place the plastic connector on the tube into the opening.

c. Remove the *red* cap from the filter holder and gently insert the side opening of the valve on the syringe into the opening.

d. Place the plastic tube into the water in the graduated cylinder. Holding the filter disc parallel to the ground, draw the water through the filter and into the syringe. If the plunger pulls all the way before all the water has been filtered, press the plunger back. Water will be forced out the end of the valve but should not flow back into the filter holder.

e. When all of the water is drawn out of the graduated cylinder, turn the syringe and filter holder so that the water lies above the filter. Expel the excess water and draw the rest of the water down from the tube and through the filter. Be careful not to get air into the filter before all of the water is expelled or the filter will become clogged. Be sure to hold the filter horizontally so that filtration is even.

f. Pump the syringe until the filter is dry.

g. Replace the plastic stoppers with the *red plug on the grid side* and the blue plug on the pad side. This procedure reverses the color positions and indicates that the filter has been used.

h. Collect three water samples at each station. *Be sure to label each chamber* after it is charged with *station number, time, date,* and *collector's name. Do not reuse the plastic tube.*

Coliform bacteria will be retained on the filter and then can be visualized readily by allowing each microorganism to multiply into a colony (clone). Such colonies are referred to commonly as colony-forming units (CFUs), and each CFU is assumed to have arisen from a single microbial cell in the original sample. There are difficulties with this assumption but counting the number of clones is easy and gives a relative index of the amount of bacterial contamination of water.

At the laboratory, remove the plugs from the filter holder and add 1 ml of M-Endo culture medium to the pad side, replace the plugs, and place the holder (grid side down) in an incubator (35°C, 90% relative humidity) for 24 h.

    i. Remove the filter from the filter holder-incubation holder *with forceps*. Some of the bacteria could be pathogenic. Do not touch the filter either to your hands, to bench tops, or to equipment. Flame the tip of the forceps before and after handling the filter. Place the gridded filter on a 5×7.6 cm glass slide and examine it with a dissection microscope.

    ii. Count the total number of *colored colonies* on the grid. Distinguish between clones that are red, orange, or have a metallic sheen.

    iii. The ratio of colonies with a metallic sheen to total colonies is an index of "coliform contamination" and hence a rough measure of fecal contamination. This index is used commonly as a measure of water pollution.

    iv. Discard the filter into a container of alcohol.

5. Plot the distribution of the various variables along the logitudinal gradient of the stream (e.g., temperature, pH, benthic fauna, sewage fungus).

Questions

1. What are the major ecological effects of sewage discharge on a stream ecosystem?
2. Which biotic components of the ecosystem are most sensitive to pollutants? Why?
3. How far downstream can the effects of a sewage outfall be observed? Are the changes in the various parameters related or do they occur independently of each other?
4. What is meant by the "assimilative capacity" of a stream? Is this a legitimate ecological concept?
5. What is meant by "sewage fungus"?
6. In some areas remote from human habitation, there may be large concentrations of coliform bacteria in streams. How would you explain this? [See Hanes et al. (1965).]
7. Do colonies of bacteria growing on a nutrient medium give a quantitative estimate of the size of natural microbial populations? Why?
8. Is there a difference between eutrophication and pollution? [See Likens (1972).]
9. What additional toxic substances may have been present in the sewage discharge? What about viruses? [See, e.g., Bertucci et al. (1977) and Borchardt et al. (1977).]
10. In some places sewage effluents are discharged onto land rather than into the water courses. What are the advantages/disadvantages of this? What are the advantages/disadvantages of discharges onto wetland, marsh systems? [See Hammer (1997) and Kadlec and Knight (1997).]
11. Protozoan populations often increase in size downstream of sewage outfalls. Why do you think this happens?
12. What might be the effect of discharging heated water (e.g., effluent from a power plant) into a stream containing domestic sewage?

Apparatus and Supplies

1. Incubation chamber: 25°C, ca. 90% relative humidity.
2. Forceps.
3. Bacteriological Analysis Monitors (part MHWG 03700, Millipore Corp., Bedford, MA).

4. Syringe, stainless steel with two-way valve (e.g., part XX 6200035, Millipore Corp., Bedford, MA).
5. M-Endo medium [see American Public Health Association (1998)].
6. See appropriate portions of Exercises 5, 6, 7, 8, 9, 12, and 13.

References

American Public Health Association. 1998. Standard Methods for the Examination of Water and Wastewater. 20th Ed. Water Environment Federation, Alexandria, VA. 1510 pp.

Bertucci, J.J., C. Lue-Hing, D. Zenz, and S.J. Sedita. 1977. Inactivation of viruses during anaerobic sludge digestion. J. Water Poll. Control. Fed. *49*:1624–1651.

Bolton, R.L. and L. Klein. 1971. Sewage Treatment—Basic Principles and Trends. Ann Arbor Publ. Michigan. 256 pp.

Borchardt, J.A., J.K. Cleland, W.J. Redman, and G. Oliver (eds). 1977. Viruses and Trace Contaminants in Water and Wastewater. Ann Arbor Sci. Publ., Michigan. 249 pp.

Hammer, D.A. 1997. Creating Freshwater Wetlands. 2nd Ed. Lewis Publishers, Boca Raton, FL. 406 pp.

Hanes, N.B., G.A. Delaney, and C.J. O'Leary. 1965. Relationship between *Escherichia coli*, Type I, coliform and enterococci in water. J. Boston Soc. Civil Engrs. *52*:129–140.

Hynes, H.B.N. 1963. The Biology of Polluted Waters. Liverpool Univ. Press. 202 pp.

Kadlec, R.H. and R.L. Knight. 1997. Treatment Wetlands. Lewis Publishers, Boca Raton, FL. 893 pp.

Likens, G.E. (ed). 1972. Nutrients and Eutrophication. Special Symposia, Vol. I., Amer. Soc. Limnol. Oceanogr., Allen Press, Lawrence, KS. 328 pp.

Parsons, J.D. 1968. The effects of acid strip-mine effluents on the ecology of a stream. Arch. Hydrobiol. *65*:25–50.

Rodina, A.G. 1972. Methods in Aquatic Microbiology. Translated and revised by R.R. Colwell and M.S. Zambruski. Univ. Park Press, Baltimore. 461 pp.

Rohlich, G.A. and P.D. Uttormark. 1972. Wastewater treatment and eutrophication. pp. 231–245. *In*: G.E. Likens, Editor. Nutrients and Eutrophication. Special Symposia, Vol. I. Amer. Soc. Limnol. Oceanogr. Allen Press, Lawrence, KS.

Whitton, B.A. and P.J. Say. 1975. Heavy metals. pp. 286–311. *In*: B.A. Whitton, Editor. River Ecology. Univ. of California Press, Berkeley.

Estimates of Whole Lake Metabolism: Hypolimnetic Oxygen Deficits and Carbon Dioxide Accumulation

Measurements of the supply of organic matter to aquatic ecosystems are complex and require an elaborate research program continuing over at least a year. The inputs from the products of photosynthesis of autotrophic phytoplankton and of littoral flora must be evaluated, as well as the inputs from allochthonous organic matter entering the aquatic ecosystem from the atmosphere and from the drainage basin.

Assuming that the allochthonous inputs of organic matter are small in relation to those synthesized within a lake, and that the lake is sufficiently large to stratify thermally, the autotrophic productivity can be estimated indirectly from long-term changes in either hypolimnetic oxygen deficits or accumulations of dissolved inorganic carbon (DIC). The assumption is that organic matter, which was synthesized in the trophogenic zone, sinks into the hypolimnetic zone and decomposes there. Changes in oxygen or DIC concentrations of the hypolimnetic water reflect the rates of loading of organic matter and of decomposition. However, the relationships are not so tidy in nature. Several major reactions in the development of an an-aerobic hypolimnion are related to the decomposition of organic matter [see Wetzel (1983, 1999) and Schindler (1985)]. These reactions are based on a simplified molecule for organic matter of planktonic material:*

$$CH_2O + O_2 \rightarrow CO_2 + H_2O \text{ (early stages only)}$$

$$CH_2O + 4Fe(OH)_3 + 8H^+ \rightarrow 4Fe^{2+} + CO_2 + 11H_2O$$

$$2CH_2O + SO_4^{2-} \rightarrow S^{2-} + 2CO_2 + 2H_2O$$

$$2CH_2O \rightarrow CH_4 + CO_2$$

$$CO_2 + NH_3 + H_2O \rightarrow NH_4^+ + HCO_3^-$$

The procedures given in this exercise are appropriate for lakes where methanogenesis is the primary decomposition pathway, and the dominant anion in the hypolimnion is bicarbonate. Other confounding problems include, for example, decomposition in the epilimnion, turbulent exchange between the stratified water layers, and photosynthesis in the hypolimnion. Nonetheless, a general, direct relationship does exist between autotrophic productivity and hypolimnetic changes. The

---

*The complete Redfield molecule is $[(CH_2O)_{106}(NH_3)_{16}(H_3PO_4)]$; see Redfield (1958) and Vollenweider (1985).

advantages of these approaches are that they provide whole ecosystem values and integrate diverse processes of production and decomposition of organic matter. The approaches treated here obviously are not applicable to lakes that do not stratify or to those that are stratified permanently (i.e., meromictic lakes).

COMPUTATION OF OXYGEN DEFICITS

The amount of oxygen lost from the hypolimnion of a thermally stratified dimictic lake during the summer stratification period is a useful relative measure of lake productivity [cf., Wetzel, 1983, 1999]. The difference in amount of oxygen present below a given depth at the beginning and at the end of the stratification period is referred to as the *oxygen deficit*. Although there are several possible procedures for the computation of the oxygen deficit, the basic method is to calculate the total quantity of oxygen in the hypolimnion on two dates. The difference in oxygen content then is expressed as a rate of oxygen change per cm^2 of hypolimnetic surface.

Procedures

1. Collect samples of water at close intervals of depth in a stratified lake soon after the onset of stratification and determine the dissolved oxygen concentrations. Repeat after time intervals of about three to four weeks, preferably throughout the period of stratification. The method cannot be used when a portion of the hypolimnion becomes anaerobic.
2. From the bathymetric map of the lake under study, determine the area and volume of water strata for small depth intervals (cf., Exercise 1).
3. Establish the boundary of the hypolimnion during the period from vertical profiles of temperature [see Fig. 6.3 of Wetzel (1983, p. 75, 1999)].

Calculations

1. Make a table of the concentrations of oxygen and total oxygen content of the hypolimnetic strata for the initial date of sampling, as shown in the following examples.

Layer of hypolimnion	Volume of layer (in km^3)	Mean concentration of O_2 (in mg/l)	Total O_2 of layer (in metric tons)
6–7 m	—	—	—
7–8 m	—	—	—
8–9 m	—	—	—
9–10 m	—	—	—
etc.			
Total			(in hypolimnion)

One mg/l is equivalent to 100 metric tons per km^3, and 1 metric ton = 10^9 mg. Make a similar table for the second and subsequent sampling dates during the period of stratification.

2. Determine the area in hectares at the top of the hypolimnion ($1\,ha = 10^8\,cm^2$).
3. From these tables, determine the decrease in total oxygen in the hypolimnion between the time intervals in days:

$$
\begin{aligned}
\text{Oxygen deficit} &= \frac{O_2 \text{ decrease, in tons, in hypolimnion per } XX \text{ days}}{\text{area of upper surface of hypolimnion in ha}} \\[2mm]
&= \frac{O_2 \text{ difference, tons per } XX \text{ days}}{\text{area upper surface hypolimnion in ha}} \\[2mm]
&= \frac{10^9 \text{ mg } O_2 \text{ per } XX \text{ days}}{10^8 \text{ m}^2} \\[2mm]
&= \text{mg } O_2/\text{cm}^2/XX \text{ days} \\[2mm]
&= (\text{mg } O_2/\text{cm}^2/\text{day})(30) \\[2mm]
&= \text{mg } O_2/\text{cm}^2/\text{month}
\end{aligned}
$$

HYPOLIMNETIC CO₂ ACCUMULATION

The principle of this indirect estimate of organic production for the trophogenic zone is similar to that of the oxygen deficit method. Accumulation of total carbon dioxide (ΣCO_2 = dissolved inorganic carbon, DIC) in the hypolimnion (tropholytic zone) from decomposition is proportional to the production of organic matter in the trophogenic zone that is transported to the hypolimnion. The potential of production estimate from hypolimnetic CO_2 accumulation was first suggested by Ruttner (1931) and applied and compared to other estimations of productivity by Einsele (1941). The most comprehensive development and use of the hypolimnetic CO_2 accumulation principle to lakes, and its comparison to other measures, was done by Ohle (1934, 1952, 1956). The latter treatments particularly are recommended for a detailed discussion.

Carbon compounds are both the initial and final products of organic metabolism and, as such, are among the best of parameters with which to evaluate productivity. The hypolimnetic DIC accumulation method circumvents an important limitation of the oxygen deficit method: Because DIC is the end product of all decomposition processes, it permits the inclusion of anaerobic metabolism in hypolimnetic water. There are, however, two important limitations to the use of DIC as a tracer of decomposition:

1. When methanogenesis or fermentation to fatty acids is occurring under reducing conditions, decomposition will be underestimated as DIC is not the sole carbon product of degradation.
2. In hardwater ecosystems DIC accumulation in hypolimnetic waters can originate from the dissolution of carbonates as well as from the decomposition of organic matter.

There are various means to overcome these difficulties. This exercise reviews an early approach (Ohle, 1956) and allows a relative comparison of lake productivities.

The hypolimnetic CO_2 accumulation method requires evaluation of changes in the various components of the CO_2—HCO_3—CO_3 system and their origins. In bicarbonate-poor softwaters, the amount of free CO_2 is very small at the beginning of hypolimnetic stratification. As stratification continues, free CO_2 and bicarbonate accumulate from respiration (aerobic and anaerobic decomposition). In addition, a portion of the bicarbonate is present as "volatile" ammonium carbonate of metabolic origin and as nonvolatile bicarbonates, half of which are of metabolic origin [cf., discussion in Wetzel (1983, 1999)]. The other half is bound CO_2 as $CaCO_3$, most of which becomes incorporated into the sediments. The sum of these CO_2 inputs is assumed to be an estimate of lake metabolism. When the CO_2 accumulation is calculated for the entire hypolimnetic volume, and this accumulation is divided by the volume of the epilimnion, the resulting quantity is termed the *relative assimilation intensity*. The relative assimilation intensity factor is expressed as mg CO_2/l of epilimnion/month.

The method first corrects DIC production in the hypolimnion by an evaluation of sediment carbonate dissolution to estimate "metabolic DIC." The metabolic DIC then is corrected for methane production. A critical assumption of this method is that methanogenesis is the sole mode of anaerobic metabolism in the hypolimnion, i.e., iron, sulfate, and nitrate reductions and fermentative reactions other than methanogenesis are relatively minor both in contributions to decomposition or to alkalinity accumulation in these water strata. This assumption is not true, particularly in softwater lakes. Nonetheless, the method does provide a proximate estimate of whole lake metabolism.

Procedures

1. Collect water samples at close intervals of depth in a stratified lake soon after the onset of stratification and determine the following parameters (can be incorporated directly with determinations of the oxygen deficit):
   a. Ammonium ion concentrations (see Exercise 7).
   b. Concentrations of CO_2 and HCO_3^- (see Exercise 8).
   c. Dissolved oxygen concentrations (see Exercise 6).
   Repeat after time intervals of about three or four weeks, preferably throughout the period of stratification.
2. Determine the temperature of water at meter intervals.
3. From the bathymetric map of the lake under study, determine the area of water strata of close depth intervals and the volume of water in these strata (cf., Exercise 1).
4. Establish the hypolimnion of the lake during the period from vertical profiles of temperature [see Fig. 6.3 of Wetzel (1983, 1999)].

Calculations

1. Make a table of the concentrations of the parameters needed for each of the hypolimnetic strata for the sampling dates as shown in the following example (Table 29.1).
2. The following assumptions are made in the calculations:
   a. The molecular ratio of ammonium bicarbonate $NH_4^+ : HCO_3^- = 18.04 : 61.02$. Then, HCO_3^- of ammonium bicarbonate is estimated as

$$HCO_3^- = \frac{(NH_4^+)(61.02)}{18.04}$$
$$= (3.38)(NH_4^+)$$
$$= \beta$$

Table 29.1. Data sheet for calculation of hypolimnetic CO_2 accumulation.

Layer of hypolimnion	Volume of layer (km³)	β [(3.38) (NH_4)]	b [(0.721) (β)]	α [HCO_3]	x [α-β]	a [(x) (0.361)]	c [CO_2]
6–7 m	—	—	—	—	—	—	—
7–8 m	—	—	—	—	—	—	—
8–9 m	—	—	—	—	—	—	—
9–10 m	—	—	—	—	—	—	—
etc.							
Total							

Layer of hypolimnion	ΣCO_2 [$a + b + c$]	δO_2 [oxygen deficit]	δCO_2 [(δO_2) (1.375)]	γ [(δCO_2) (0.85)]	δCO_2 [$\Sigma CO_2 - \gamma$]	z [(δCO_2) (2)]	ε [$z + \gamma$]
6–7 m	—	—	—	—	—	—	—
7–8 m	—	—	—	—	—	—	—
8–9 m	—	—	—	—	—	—	—
9–10 m	—	—	—	—	—	—	—
etc.							
Total							

If, for example, 1.85 mg NH_4^+/l were determined, the amount of HCO_3^- present as NH_4HCO_3 would be $(1.85)(3.38) = 6.25 = \beta$.

b. The molecular ratio of $HCO_3 : CO_2 = 61.02 : 44.01$. Then, the CO_2 of bicarbonate of the ammonium bicarbonate is

$$CO_2 = \frac{\left(HCO_3^-\right)(44.01)}{61.02}$$

$$= (0.721)\left(HCO_3^-\right)$$

$$= (0.721)(\beta)$$

$$= b$$

In this example, $\beta = 6.25$ and, therefore, $b = (6.25)(0.721) = 4.51$.

c. a = the measured concentration of bicarbonate [HCO_3^-] in mg/l, for example, 32.4 mg HCO_3^-/l.

d. $x = a - \beta$ to remove the error of bicarbonate from NH_4HCO_3. In this example, $x = 32.4 - 6.25 = 26.15$ mg HCO_3^-/l.

e. Converting the total bicarbonate to CO_2, half of which is bound as CO_2 of $CaCO_3$ (conversion of half of excess HCO_3^- to CO_2):

$$a = \frac{x(0.721)}{2}$$

$$= (x)(0.3605)$$

In this example, $a = (26.15)(0.3605) = 9.43$.

f. c = measured concentration of free carbon dioxide, [CO_2] in mg/l. In this example, $c = 13.1$ mg CO_2/l.

g. $\qquad \Sigma CO_2$ = total carbon dioxide, corrected
$\qquad\qquad = a + b + c$

which, in this example, is $\Sigma CO_2 = 9.43 + 4.52 + 13.10 = 27.04$.

h. The oxygen deficit then is calculated and used in the computations of hypolimnetic CO_2 accumulation. $\Delta O_2 = $ *actual* oxygen deficit, i.e., the change (difference) in O_2 concentration observed at any time at that depth, and the saturation value of that same quantity of water at the temperature and pressure at spring turnover. The *relative* areal O_2 deficit, calculated by the procedure discussed earlier in this exercise, is a better measure than the actual oxygen deficit [cf., Wetzel (1983, 1999)] and can be used here when done over the same interval of time. Using the actual O_2 deficit in this example (Ohle, 1952), the measured $[O_2] = 0.56$ mg/l at 6.7°C. At spring turnover, the $[O_2] = 12.27$ mg/l. Hence, $\Delta O_2 = 12.27 - 0.56 = 11.71$ mgO$_2$/l. Using the molecular ratio of CO_2 to O_2, $CO_2/O_2 = 44/32 = 1.375$, then $\delta CO_2 = (\Delta O_2)(1.375)$, which, in this example, is $\delta CO_2 = (11.71)(1.375) = 16.10$.

i. The respiratory quotient (RQ) is the ratio of the molecules of CO_2 liberated during decomposition (respiration) to the molecules of O_2 consumed, i.e., $+\Delta CO_2/-\Delta O_2$. An RQ value of 0.85 is a mean value based on a number of analytical analyses of plant and animal respiration and has been proposed for use in the hypolimnetic CO_2 accumulation method (Ohle, 1952). This general value is reasonable for the oxidation of proteins and fats where oxygen is available as an electron acceptor. When the hypolimnion becomes anaerobic, however, fermentation by bacteria produces excess CO_2, a positive CO_2 anomaly, and volatile organic compounds, such as methane, that diffuse out of the sediments.

Under anaerobic conditions, alternate electron acceptors (e.g., NO_3^-, SO_4^{2-}), other than molecular O_2, are used [cf., Rich and Wetzel (1978) and Rich (1983)]. Thus, respiratory quotients would be greater than during aerobic metabolism. The in situ hypolimnetic RQ values would be expected to vary seasonally with availability of oxygen and changes in redox gradients. For example, RQ values of the hypolimnion have been found to vary inversely with the availability of oxygen and range from less than 1 after spring circulation to nearly 3 under anoxic conditions of summer stratification (Rich, 1975; 1984).

j. Using an RQ of 0.85 and assuming aerobic hypolimnetic conditions,

$$\gamma = (\delta CO_2)(0.85)$$

In the example being discussed,

$$\gamma = (11.71)(1.375)(0.85)$$
$$= (16.10)(0.85)$$
$$= 13.69$$

k. Then the anaerobic change in CO_2 (total CO_2 at the later date (t_2) minus the sum of CO_2 at t_1 minus that formed aerobically), ΔCO_2, is evaluated by

$$\Delta CO_2 = \Sigma CO_2 - \gamma$$

which, in the example, $\Delta CO_2 = 27.04 - 13.69 = 13.35$.

l. This value is then doubled to estimate the equivalent CO_2 under aerobic decomposition:

$$z = (\Delta CO_2)(2)$$

In our example, $z = (13.35)(2) = 26.70$.

m. Then the total hypolimnetic CO_2 accumulation (ε) is the sum of z and γ:

$$\varepsilon = z + \gamma$$

e.g., $\varepsilon = 26.70 + 13.69 = 40.39\,mgCO_2/l$. Then, assuming 30 days/month,

$$\varepsilon \text{ per month} = \frac{\varepsilon(30)}{\text{no. of days of stratification}}$$

If, for example, the period of stratification were four months (120 days), then the total hypolimnetic CO_2 accumulation for the stratified period would be

$$\varepsilon = \text{per month} = \frac{(40.39)(30)}{120} = 10.1\,mgCO_2/l/month$$

n. When the CO_2 accumulation is calculated for each stratum of the hypolimnion and summed for the entire hypolimnion, and this total value is multiplied by the ratio of the volumes of the hypolimnion to epilimnion, a quantity termed the *relative assimilation intensity* can be calculated:

Relative assimilation intensity

$$= \left(\begin{array}{c} \text{hypolimnetic } CO_2 \text{ accumulation} \\ \text{in } mgCO_2/l/month \end{array} \right) \left[\frac{\text{volume hypolimnion } (m^3)}{\text{volume epilimnion } (m^3)} \right]$$

$$= mgCO_2/l/month$$

If, in our example, the volume of the epilimnion were $7.442 \times 10^6\,m^3$, and that of the hypolimnion $4.620 \times 10^6\,m^3$, then:

$$\text{Relative assimilation intensity} = (10.1\,mgCO_2/month)\left(\frac{4.620 \times 10^6}{7.442 \times 10^6} \right)$$

$$= 6.3\,mgCO_2/l/month \text{ for the lake}$$

EXERCISES

OPTION 1. FIELD ANALYSES

1. At a suitably stratified lake or well-stratified reservoir, determine the depths of the epilimnion, metalimnion, and hypolimnion by temperature profiles. Take water samples at as closely spaced depth intervals as time permits from the upper portion of the hypolimnion to the sediments.
   a. Using appropriate care, collect duplicate dissolved oxygen samples (see Exercise 6). Chemically fix the oxygen immediately while still in the field.
   b. Collect samples of water separately for (i) CO_2, alkalinity, and pH analyses (Exercise 8); and (ii) measurement of ammonium ion concentrations (Exercise 7).
2. Perform the analyses of each of the required parameters as soon as possible in the laboratory.
3. From bathymetric maps, determine the volume of each stratum of the hypolimnion and the total volume of the epilimnion and the hypolimnion, as outlined in Exercise 1.
4. A minimum of two sets of such data are needed, separated by about three to four weeks.

Table 29.2. Exemplary data for the calculation of hypolimnetic CO_2 accumulation.

Layer of hypolimnion	Volume of layer (m³)	NH_4^+ (mg/l)	HCO_3^- (mg/l)	CO_2 (mg/l)	mg O_2/l[a] t_1	t_2
5–6 m	1.362×10^6	0.653	53.2	6.3	8.31	8.06
6–7 m	1.038×10^6	0.671	54.6	8.1	8.25	7.69
7–8 m	0.976×10^6	0.698	57.3	9.3	8.00	7.23
8–9 m	0.831×10^6	0.793	60.0	10.0	7.95	6.21
9–10 m	0.661×10^6	0.912	62.5	10.8	7.72	5.21
10–11 m	0.371×10^6	1.312	67.2	11.6	7.06	3.31
11–11.5 m	0.097×10^6	1.850	69.4	13.1	6.84	1.06

[a] Period between t_1 and t_2 = 96 days.

5. Often the measurements needed for these computations of whole lake metabolism can be obtained while performing the analyses for other exercises. The strength of these analyses is increased when they can be compared to other analyses of productivity or biomass, e.g., in situ rates of productivity of phytoplankton, algal biomass, and other parameters.
6. If possible, split the class into teams, each of which performs the analyses of O_2 deficit and CO_2 accumulation on different lake systems, e.g., a relatively oligotrophic lake versus a eutrophic lake.
7. Compare and evaluate the estimates of metabolism by the relative areal O_2 deficit and hypolimnetic CO_2 accumulation techniques of the lake(s).
8. Answer the questions following Option 2.

OPTION 2. LABORATORY ANALYSES

1. From the data provided by your instructor or that given in Tables 6.2 (p. 83) and 29.2, calculate the following:
   a. The hypolimnetic relative areal oxygen deficit.
   b. The hypolimnetic CO_2 accumulation, in mg CO_2/l/month.
   c. The relative assimilation intensity, in mg CO_2/l/month.
2. Answer the following questions.

Questions

1. Based on your data or those provided, how does your estimate of the hypolimnetic oxygen deficit compare to the general categories of productivity of lakes evaluated by other criteria, as well as in relation to estimates of oxygen deficit given below?

	Suggested arbitrary limits (mg/cm²/month)		Approximate observed
	Hutchinson (1957)	Mortimer (Unpublished)	range (mg/cm²/month)
Oligotrophy	<0.5	<0.75	0.1–1.0
Mesotrophy			1.0–1.5
Eutrophy	>1.0	>1.65	>1.5

2. Can one assume, on the basis of the results of the laboratory experiments with water movements (Exercise 3), that the hypolimnetic concentrations of oxygen or CO_2 are relatively unaffected by the exchange of gases among the water strata? How serious do

you feel this error is in your system? How might the basin morphometry of lakes alter the magnitude of this error? When a lake has a high relative depth, z_r, (see Exercise 1, p. 12), how might this error be influenced?

3. If littoral productivities of macrophytic vegetation and sessile algae were large, how would this organic matter and its decomposition influence the calculated hypolimnetic O_2 deficit or CO_2 accumulation? When does decomposition of the macrophytic flora and attendant algae occur? Is some or much of this organic matter transported to the hypolimnion? [See Godshalk and Wetzel (1977) and Wetzel (1990).]

4. Transport of allochthonous organic matter to lakes occurs both as particulate and as dissolved organic matter. Which form predominates? How are these materials transformed during transport? If relatively refractory dissolved organic matter were brought to the lake, how would its presence and decomposition affect the productivity estimated by the hypolimnetic change methods?

5. Are the hypolimnetic O_2 deficit and CO_2 accumulation techniques applicable to winter periods under ice cover? Support your answer. What would be the influence of short renewal times or water replacement of lakes when these methods are used under winter conditions? Why?

6. When a significant portion of the autochthonous organic matter is sedimented and interred permanently in the sediments of the hypolimnion, how would the computations of productivity by the O_2 deficit and CO_2 accumulation methods be affected?

7. If the algae of one lake were dominated by rapidly sedimenting siliceous diatoms and those of another were dominated by small algae more neutrally buoyant (or possessing buoyancy mechanisms), how would this difference affect productivity estimates by the two hypolimnetic methods? Why?

8. In the CO_2 accumulation method, if some sulfate or ferric hydroxide in the sediments were reduced as the redox potential decreases and some bicarbonate was released to the water, what would be the effect on the productivity estimates?

9. Why are these methods not applicable to meromictic lakes?

10. What is the role of "volatile" ammonium carbonate?

11. Is it better to express the results of the oxygen deficit method on an areal or on a volumetric basis? Why?

Apparatus and Supplies

1. Boats, anchors, and water sampling devices, e.g., Van Dorn samplers.
2. Thermistor thermometer.
3. Dissolved oxygen:
   a. Sample bottles and chemical reagents for oxygen fixation (see Exercise 6, p. 82).
   b. Titration apparatus and reagents (see Exercise 6, p. 82).
4. CO_2-ammonia analyses:
   a. Separate sample bottles for CO_2, alkalinity, pH, and ammonia.
   b. Apparatus and reagents for the determinations of CO_2 and alkalinity (see Exercise 8, pp. 117–134).
   c. pH meter and supplies.
   d. Apparatus and reagents for analyses of ammonia (see Exercise 7, pp. 87–88 and a spectophotometer.
5. Bathymetric maps and planimeter; calculators.

References

Einsele, W. 1941. Die Umsetzung von zugeführtem, anorganischen Phosphat im eutrophen See und ihre Rückwirkungen auf seinen Gesamthaushalt. Zeitsch. f. Fischerei *39*:407–488.

Godshalk, G.L. and R.G. Wetzel. 1977. Decomposition of macrophytes and the metabolism of organic mater in sediments. pp. 258–264. *In*: H.L. Golterman, Editor. Interactions between Sediments and Freshwater. Dr. W. Junk B.V., The Hague.

Hutchinson, G.E. 1957. A Treatise on Limnology. Vol. I. Geography, Physics, and Chemistry, Wiley, New York. 1015 pp.

Ohle, W. 1934. Chemische und physikalische Untersuchungen norddeutscher Seen. Arch. Hydrobiol. *26*:386–464, 584–658.

Ohle, W. 1952. Die hypolimnische Kohlendioxyd-Akkumulation als productions-biologischer Indikator. Arch. Hydrobiol. *46*:153–285.

Ohle, W. 1956. Bioactivity, production, and energy utilization of lakes. Limnol. Oceanogr. *1*:139–149.

Redfield, A.C. 1958. The biological control of chemical factors in the environment. Amer. Sci. *46*:206–226.

Rich. P.H. 1975. Benthic metabolism of a soft water lake. Verh. Int. Ver. Limnol. *19*:1023–1028.

Rich, P.H. 1983. Differential CO_2 and O_2 benthic community metabolism in a softwater lake. J. Fish. Res. Bd. Canada *36*:1377–1389.

Rich, P.H. 1984. Further analysis of respiration in a North American lake ecosystem. Verh. Int. Verein. Limnol. *22*:542–548.

Rich, P.H. and R.G. Wetzel. 1978. Detritus in the lake ecosystem. Amer. Naturalist *112*:57–71.

Ruttner, F. 1931. Hydrographische und hydrochemische Beobachtungen auf Java, Sumatra und Bali. Arch. Hydrobiol. Suppl. *8*:197–454.

Schindler, D.W. 1985. The coupling of elemental cycles by organisms: Evidence from whole-lake chemical perturbations. pp. 225–250. *In*: W. Stumm, Editor. Chemical Processes in Lakes. Wiley, New York.

Vollenweider, R.A. 1985. Elemental and biochemical composition of plankton biomass; some comments and explorations. Arch. Hydrobiol. *105*:11–29.

Wetzel, R.G. 1983. Limnology. 2nd Ed. Saunders Coll., Philadelphia. 860 pp.

Wetzel, R.G. 1990. Land-water interfaces: Metabolic and limnological regulators. Baldi Memorial Lecture. Verh. Int. Verein. Limnol. *24*:6–24.

Wetzel, R.G. 1999. Limnology: Lake and River Ecosystems. 3rd Ed. Academic Press, San Diego (in press).

General Chemical Relationships

ATOMIC AND MOLECULAR WEIGHTS

Atomic and molecular weights are basic to all quantitative chemical reactions and are used in all calculations associated with these reactions. An understanding of the principles of their use is necessary to apply them to the calculations involved in chemical determinations. The molecular weight of H_2SO_4 and percentage composition of the elements in this compound may be obtained by referring to the atomic weights as follows:

$$\text{Hydrogen, 2 at wt} = 2 \times 1.008 = 2.016$$

$$\text{Sulfur, 1 at wt} = 1 \times 32.060 = 32.060$$

$$\text{Oxygen, 4 at wt} = 4 \times 16.000 = \underline{64.000}$$

$$\text{Molecular wt } 98.076$$

The percentage by weight of hydrogen is $2.016/98.076 \times 100 = 2.056$. A complete listing of atomic weights and numbers and other elemental parameters can be found in various handbooks.

In all cases, knowing the equation for a given reaction, the weight ratios can be determined from the atomic weights of the elements and the formulas of the compounds involved. Because the weight ratios can be calculated, when the actual weight of any compound entering into the reaction is known, the weights of all of the other compounds involved can also be determined.

VALENCE

The combining power, or ratio in which atoms of one element combine with atoms of other elements, is called *valence*. The valence of hydrogen is 1; the valence of the other elements is determined by their electron charge and is evaluated by examining the formula of the compound in which the element occurs. The positive and negative valences of elements in different compounds assume different values under various conditions, but the sum of positive and negative valences for any one compound must be equal to 0.

NORMAL SOLUTIONS

Some of the more important determinations in water chemistry are dependent on the use of standard solutions. A standard solution contains a known weight of the active substance dissolved in a definite volume of solution. Methods involving the use of such solutions are known as volumetric procedures, since the quantitative result is obtained by the measurement of volumes.

It is convenient to have solutions of known concentrations for use in various determinations. Such solutions simplify the calculations of results, a decided advantage in routine analysis. They can best be prepared by diluting portions of stock solutions in such a manner as to give solutions of the desired strength.

The strength of a standard solution is usually expressed in terms of its normality. A 1 *normal* solution is one which contains 1 gram-equivalent weight of the active substance in 1 liter of solution. The *gram-equivalent weight* of a substance is the weight of that substance which will react with 1 g of hydrogen.

There are two general types of chemical reactions encountered in the volumetric determinations used for water analysis: (1) simple neutralization or double-decomposition reactions, and (2) reactions involving oxidation and reduction.

The reactions of the first type involve a change in position of the various atoms and groups making up the reacting substances. The following equation illustrates this type of reaction:

$$HCl + NaOH \leftrightarrows NaCl + H_2O$$

In a reaction of this type, the gram-equivalent weight of each compound reacting is calculated by dividing the molecular weight of the compound by the number of replaceable hydrogen atoms or their equivalent in that compound:

$$\frac{\text{Molecular weight of a substance}}{\text{No. of replaceable hydrogen atoms}} = \frac{\text{gram-equivalent}}{\text{weight of the substance}}$$
$$\text{or their equivalent}$$

Thus the gram-equivalent weight of HCl is 36.5, since the molecular weight (36.5) is divided by 1, because there is one replaceable hydrogen. In NaOH, one Na can be replaced by one H and is equivalent to one atom of hydrogen. Therefore, the gram-equivalent weight of NaOH is 40.

The second type of reaction, oxidation-reduction, is illustrated by the following equation:

$$2KMnO_4 + 5H_2C_2O_4 + 3H_2SO_4 \leftrightarrows 2MnSO_4 + K_2SO_4 + 10CO_2 + 8H_2O$$

A study of the equation will show that there is more involved than a simple rearrangement of atoms and groups. Mn, for instance, has a positive valence of 7 in $KMnO_4$, but only 2 in $MnSO_4$. The Mn has lost 5 positive valences (has been reduced); since there are 2 Mn atoms, the total loss is 10 valences. Likewise, each C atom in $H_2C_2O_4$ has a valence of +3, while each C atom in CO_2 has a valence of +4. The 10 C atoms thus gain a total of 10 valences (one for each of the C atoms). The total loss of positive valence by one kind of atom in this type of reaction must equal the total gain of positive valences by another kind of atom.

The gram-equivalent weight of a compound entering into an oxidation-reduction reaction is equal to its molecular weight divided by the change in valence of the element in that compound that is being oxidized or reduced:

$$\frac{\text{Molecular weight}}{\text{Change in valence}} = \text{gram-equivalent weight}$$

The normality of a standard solution is the ratio of the weight in grams of the substance in 1 liter to the gram-equivalent weight:

$$\frac{\text{Weight in grams per liter}}{\text{Gram-equivalent weight}} = \text{normality (N)}$$

Solutions of equal normalities are equal in their reaction potential, volume per volume. For example, a volume of 0.1 N hydrochloric acid will react completely with the same volume of 0.1 N sodium hydroxide. The use of normalities simplifies the calculations necessary in obtaining the results of a volumetric analysis.

In routine analyses, such as those of oxygen, alkalinity, or dissolved CO_2, there is a general formula for the calculation of the appropriate correction factor when a standard solution used in the titration procedure is of a normality other than that specified:

$$\frac{\text{Normality of solution used}}{\text{Normality specified}} = \text{correction factor to be applied}$$

MOLAR SOLUTIONS

Molar concentration may be defined as the number of gram-moles of solute per liter of solution. Then, a 1 molar solution (M) contains 1 g-mole of solute per liter of solution, or 1 liter of a 1 M solution contains 1 g-mole of solute. Also 1 liter of a 2 M solution contains 2 g-moles of solute, 1 liter of a 0.5 M solution contains 0.5 g-mole of solute, 2 liters of 1.5 M solution contains 3 g-moles of solute, and 0.5 liters of a 0.2 M solution contains 0.1 g-mole of solute, or

$$\text{(Number of liters of solution)} \times \text{(molarity of solution)}$$
$$= \text{number of gram-moles of solute}$$

or

$$\text{Number of liters of solution} \times M = \text{number of g-moles of solute}$$

then

$$M = \frac{\text{number of g-moles solute}}{\text{number of liters solution}}$$

The molarity of a solution expresses *the number of gram-moles of solute per liter of solution*.

STANDARD SOLUTIONS

Stock solutions should always be made more highly concentrated than those to be used in the various determinations. Strong solutions usually change concentration less upon storage than do weak ones. In preparing the standard solutions for use in the various tests, the stronger solution may be diluted to give the desired normality or molarity. The portion of a stock solution repuired to make 1 liter of a desired normality may be calculated as follows:

$$\frac{100 \times \text{desired normality}}{\text{normality of stock solution}} = \text{ml of stock solution}$$

Standard solutions of some chemicals may be made directly by weight since the chemical can be obtained in a pure state. An example is anhydrous sodium carbonate. Others, such as sulfuric acid and potassium permanganate, must be standardized by titration against some pure chemical.

The general formula that is applicable to all titrations involving two solutions when using the normality system may be stated as follows:

ml of one solution × its normality = ml of the other solution × its normality

Similar relationships pertain to molar solutions.

SOME UNITS

1. Various units are used to describe the amount of concentration of particulate or dissolved substances in water. In limnology, these units usually are based on

Table A.1.1. Formula weights, equivalent weights, and conversion factors for some common ions found in inland waters.

Ion	Formula weight	Equivalent weight[a]	To convert ppm to epm, or mg/l to meq/l, multiply by:	To convert g/liter to moles/liter divide by:
Cations				
Al^{3+}	26.98	8.99	0.111	26.98
Fe^{3+}	55.85	18.6	0.0537	55.85
Fe^{2+}	55.85	27.9	0.0358	55.85
Pb^{2+}	207.20	104.0	0.00965	207.20
Sr^{2+}	87.62	43.8	0.0228	87.62
Zn^{2+}	65.37	32.7	0.0306	65.37
Cu^{2+}	63.55	31.8	0.0315	63.55
Mn^{2+}	54.94	27.5	0.0364	54.94
Ca^{2+}	40.08	20.0	0.0499	40.08
Mg^{2+}	24.31	12.2	0.0823	24.31
K^+	39.10	39.1	0.0256	39.10
Na^+	22.99	23.0	0.0435	22.99
NH_4^+	18.04	18.0	0.0554	18.04
H^+	1.008	1.01	0.992	1.008
Anions				
PO_4^{3-}	94.97	31.7	0.0316	94.97
SO_4^{2-}	96.06	48.0	0.0208	96.06
CO_3^{2-}	60.01	30.0	0.0333	60.01
HCO_3^-	61.02	61.0	0.0164	61.02
NO_3^-	62.01	62.0	0.0161	62.01
Cl^-	35.45	35.5	0.0282	35.45
OH^-	17.01	17.0	0.0588	17.01

[a] Rounded to three significant figures.

weight, volume, or ionic strength. The metric system of weights and measures is now adopted universally in scientific and technical work. The unit of volume is the liter (l), and that of weight is the gram (g), each of which is divided into smaller divisions, e.g., milligram (mg) and milliliter (ml) [milli = 10^{-3}] or micrograms (μg) and microliters (μl) [micro = 10^{-6}].

2. *Weight per weight* is a unit used frequently in limnology. The unit is expressed commonly as parts per million (ppm) or, in saline waters, as parts per thousand (ppt or ‰), based on weight relationships of, for example, one weight part in a million weight parts. One gram of salt dissolved in a million grams of water would have a concentration of 1 ppm. A 1% solution is equivalent to a concentration of 10,000 ppm. A part per billion (ppb) is, in American usage, one-thousandth of 1 ppm, or one microgram per kilogram. Note that the British system of measurement defines one billion as 1×10^{12}.

3. *Weight per volume* usually is expressed as milligrams per liter (mg/l). Since the density of fresh water is, for most purposes, not significantly different from $1.0 \, g/cm^3$, 1 liter of water may be assumed to weigh 1 kg. Thus, 1 mg/l essentially is equivalent to 1 ppm. Corrections for density differences resulting from dissolved substances and temperature are necessary to make these two units strictly comparable. However, if the dissolved substance concentration is less than about 10,000 ppm, the error introduced by the assumption is less than 1%.

4. *Ionic strength.* The equivalent weight of an ion is equal to its formula weight divided by the ionic charge (Table A.1.1). The common expression based on parts by weight of the solution is equivalents per million (epm):

$$epm = \frac{ppm}{equivalent\ weight}$$

or, on a volume basis, neglecting corrections for density, as milliequivalents per liter (meq/l). The sum of all cation equivalent weights should balance the sum of all anion equivalent weights in a freshwater solution.

Basic Definitions Used in Community Analyses

It would be impossible to treat adequately here all of the fundamental statistical analyses required in chemical and biological analyses of fresh waters. However, a few basic definitions are summarized below. Further reading is strongly encouraged. Particularly lucid and limnologically pertinent accounts are given in Elliott (1977), Sokal and Rohlf (1981), and Zar (1984). The data should be evaluated to verify that a normal probability distribution exists. In a normal distribution the continuous variables are distributed in such a manner about the mean that the density on each side of the mean is approximately equal (Fig. A.2.1). The probability of the variables deviating from the mean (μ), called the standard deviation (σ), follows a pattern of

50% of the items fall between $\mu \pm 0.674\sigma$,
95% of the items fall between $\mu \pm 1.960\sigma$,
99% of the items fall between $\mu \pm 2.576\sigma$.

If the data variables do not follow a normal distribution, i.e., are not parametric and are skewed to the left or right of the mean, application of nonparametric statistical evaluations of significant differences should be performed [see Siegal (1956) and Sokal and Rohlf (1981)].

Frequency. The percentage of quadrats occupied by a given species.
Density. The number of individuals of a given species per quadrat (or total number of individuals divided by the area of the sample).
Cover. The fraction of ground surface (or sediment surface) covered by tissue of a given species or all species combined.
Relative density. The number of individuals of species x divided by the total number of all individuals in the sample, with the quotient expressed as a percentage.
Relative cover. The cover of species x divided by the total cover of all species in the samples, with the quotient expressed as a percentage.
Mean (arithmetic) [$\bar{x}$]. Sum of individual observations (Σx) divided by the number of sampling units (n); hence

$$\bar{x} = \frac{\sum x}{n}$$

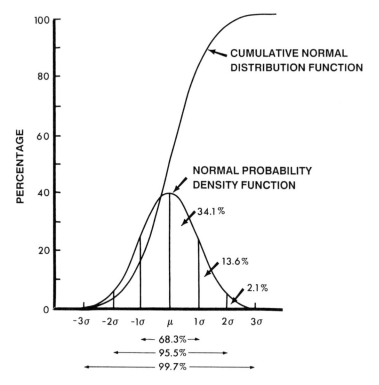

Figure A.2.1. Areas under the normal probability density function and the cumulative normal distribution function. [Modified from Sokal and Rohlf (1981).]

Measurements of Variance

Range. The difference between the highest and lowest counts (values).

Deviation. The quantity by which each individual differs from the arithmetic mean of a sample.

Variance [s^2]. The mean of the squares of the deviations; hence

$$s^2 = \frac{\sum (x - \bar{x})^2}{n - 1}$$

Since the sample variance often is an estimate of the variance of the population, n is usually expressed as the *degree of freedom, n – 1* [this represents a "tax" for using a sample mean $\bar{x}$ (a statistic) instead of the population mean μ in the estimation of population variance; sample variances tend to underestimate population variance].

Standard deviation [s]. A measure of the average departure of individual samples from the mean for all samples. The square root of the variance; hence

$$s = \sqrt{s^2}$$

Standard error (S.E.) of the mean. [$s_{\bar{x}}$ = Standard deviation of the sample means.] An estimate of the average departure of independent means (from a given population) from the mean of means. Said another way, the standard error of the mean indicates the amounts of error in the sample mean ($\bar{x}$) when it is used to estimate the population mean (μ). Usually written as $\bar{x} \pm S.E.$, where $s_{\bar{x}} = s/\sqrt{n}$. As the size of the sample (i.e., n) increases, the standard error decreases, and the estimate of the population mean approaches μ.

Coefficient of variation (C or C.V.). A measure of the relative variability of samples. Ideally one would like to reduce *C.V.* to a minimum.

The term is applied to the standard deviation when it is expressed as a percentage of the sample mean; hence

$$C.V = s\left(\frac{100}{\bar{x}}\right)$$

where s = standard deviation of sample and $\bar{x}$ = sample mean.

MEASUREMENT OF AGGREGATION

In a random distribution, there is a lack of regularity; individuals are aggregated in groups and unequally spaced, some close together and others far apart. Individuals from a randomly distributed population will follow a Poisson distribution. Departures from a $s^2/\bar{x}$ ratio of 1.0 can be tested by the chi-squared (χ^2) goodness-of-fit test (e.g., Fig. A.2.2 and Elliott, 1977:40–43):

$$\chi^2 = \frac{s^2(n-1)}{\bar{x}} \quad \text{or} \quad \frac{\sum(x-\bar{x})^2}{\bar{x}}$$

where: s^2 = sample variance, $\bar{x}$ = arithmetic mean, and n = number of samples.

Although aggregation often has been estimated by the variance/mean ratio (χ^2 or chi-squared test), this estimate of departure of randomness is not acceptable as an estimate of aggregation (Pielou, 1974; Hurlbert, 1990).

Index of Aggregation (Morlalta Index, see Hurlbert, 1990)

When individuals are selected at random from a population of X individuals distributed over an area partitioned into Q quadrats; the probability that both individuals will be in the same quadrat is:

$$\Delta_a = \sum_{i=1}^{Q}\left(\frac{x_1}{X}\right)\left(\frac{x_1-1}{X-1}\right)$$

$$= \sum_{k=2}^{w} q_k\left(\frac{k}{X}\right)\left(\frac{k-1}{X-1}\right)$$

If this sample population of X individuals were distributed at random, so that the number of individuals per quadrat followed a multinominal distribution, Δ_a would be expected to be Δ_p, where

$$\Delta_p = \frac{1}{Q}$$

The ratio of these two probabilities is an *index of aggregation* (I_M) or *patchiness* (Lloyd, 1967), measures of how much more likely it is that two randomly selected individuals will be from the same quadrat than would be the case if the X individuals in the population were distributed at random:

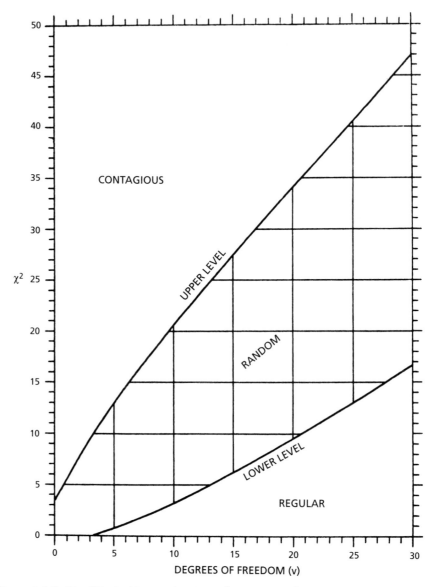

Figure A.2.2. The 5% significance levels of χ^2. When the χ^2 value lies between significance levels, agreement with Poisson series is accepted at 95% probability level ($P > 0.05$). χ^2 below the lower level indicates regular distribution; χ^2 above the upper level indicates contagious distribution. [Reproduced from Elliott (1977) by permission of the Freshwater Biological Association.]

$$I_M = \frac{\Delta_a}{\Delta_p}$$

$$= Q \sum_{i-1}^{Q} \left(\frac{x_i}{X} \right) \left(\frac{x_i - 1}{X - 1} \right)$$

$$= \left(\frac{X}{X-1} \right) \left(\frac{1}{\mu} \right) \left(\frac{\sigma^2}{\mu} + \mu - 1 \right)$$

where:

w = largest number of individuals observed in any quadrat,
k = number of individuals per quadrat,
q_k = number of quadrats containing k individuals,
Q = total number of quadrats,
μ = mean number of individuals per quadrat,
X = total number of individuals present in the Q quadrats,
x_i = number of individuals in the ith quadrat, and
σ^2 = sample of x_i.

If $I_M = 1.5$, for example, the probability that the two randomly selected individuals are from the same quadrat is 50% greater than it would be in the case of a random distribution. Although some specific estimators of I_M have been proposed (see Hurlbert, 1990), usually the parametric form, with μ and σ replaced by $\bar{x}$ and s^2, respectively, is used to calculate patchiness values.

Often an estimated mean value is desired from sampling and the investigator must decide how accurate it must be in relation to how many samples must be taken. Comparison of the sample size with the accuracy of the mean can be done in several ways (Eckblad, 1991). In sampling from a normal distribution, the sample mean can be compared to a theoretical population mean with the t-distribution:

$$t - \text{value} = \frac{\text{sample mean} - \text{population mean}}{\sqrt{\text{sample variance}/\text{sample size}}}$$

The denominator of the above equation is the standard error (SE) of the mean and approaches zero as the sample size increases to infinity. When the SE of a mean is multiplied by the appropriate t-value, a confidence interval for a mean can be estimated. If the t-value is at a 0.05 level of significance, the mean $\pm$ (t-value) (SE) represents a 95% confidence interval for the mean.

If the numerator of the above equation is replaced with *accuracy × sample mean*, the equation can be solved for sample size for a specified accuracy in describing the theoretical population mean:

$$\text{Sample size} \cong \frac{(t - \text{value})^2 (\text{sample variance})}{(\text{accuracy} \times \text{sample mean})^2}$$

If an estimate is available from prestudy sampling about the sample mean and variance desired, this formula can be used to calculate the needed sample size (Eckblad, 1991). Here, the t-value is taken from a t-distribution with a specified level of significance (e.g., 0.05), with degrees of freedom equal to the denominator used in estimating the sample variance. Accuracy would be set at 0.1 to estimate the desired mean within 10%. When the coefficient of variation (standard deviation divided by the sample mean) has been calculated, an equivalent equation would be:

$$\text{Sample size} \cong [(\text{coefficient of variation} \times t\text{-value})/\text{accuracy}]^2$$

The sample-size equations can be rearranged to estimate accuracy for a given sample size as:

$$\text{Accuracy} \cong \frac{(t - \text{value})(\sqrt{\text{sample variance}/\text{sample size}})}{\text{Sample mean}}$$

Similarly, the coefficient of variation can be used in the comparable equation as:

$$\text{Accuracy} \cong \frac{(\text{coefficient of variation})(t\text{-value})}{\sqrt{\text{sample size}}}$$

Thus it is possible to estimate the accuracy of the mean, in terms of percentage deviation from the true mean, as a function of sample size.

DETERMINING ADEQUACY AND EFFICIENCY OF SAMPLING METHODS

Since the techniques of a field scientist must be as efficient as possible, it is desirable to know the minimum number of samples that are required to obtain statistically valid data. It is likewise necessary to determine the relative labor cost of alternative methods. Experimental design is of critical importance to the successful execution of an experiment.

A *minimum adequate sample* factor, $N(\text{min})$, can be determined after a mean ($\bar{x}$) and a standard deviation (s) have been calculated. From the definition of $s_{\bar{x}}$, it follows that:

$$N = \left(\frac{s}{s_{\bar{x}}}\right)^2$$

$N(\text{min})$ is determined by substituting in the equation a value for $s_{\bar{x}}$, which represents the degree of reproducibility desired; this may arbitrarily be designated as a value that is less than or equal to 10% of the mean. In this case,

$$N(\text{min}) = \left(\frac{s}{(\bar{x})(0.1)}\right)^2$$

This number of samples will insure that approximately two-thirds of the sample means, of samples of $N(\text{min})$ size, will fall within $\pm 0.1 \bar{x}$ or 10% of the true mean. When more stringent requirements are necessary (e.g., 95% of the sample means $\pm 0.1 \bar{x}$ of the true means), $N(\text{min})$ must be increased. 95% of the means will fall within $\pm 10\%$ of the true mean if

$$N(\text{min}) = \left(\frac{2s}{(\bar{x})(0.1)}\right)^2$$

Experience shows that this latter level of confidence is costly to achieve in field studies.

One can determine the relative sampling cost (efficiency) by recording the average time required to collect the information at a given sampling point. Do this by recording the minutes (or days) required to obtain a given number of samples. The sample size should be reasonably large and taken under realistic conditions, since fatigue and environmental conditions (such as blistering sun, biting insects, and rough terrain) influence the rapidity with which data can be taken. Divide the sampling time by the number of sampling units needed to provide a sample of specified adequacy. This value represents the relative cost that is necessary to obtain the kind of information sought. Comparisons will demonstrate that all sizes, shapes, and methods of quadrat placement are not equally economical.

References

Eckblad, J.W. 1991. How many samples should be taken? BioScience *41*:346–348.
Elliott, J.M. 1977. Some Methods for the Statistical Analysis of Samples of Benthic Invertebrates, 2nd. Ed. Sci. Publ., Freshwater Biological Association, U.K., No. 25, 156 pp.

Hurlbert, S.H. 1990. Spatial distribution of the montane unicorn. Oikos *58*:257–271.

Lloyd, M. 1967. Mean crowding. J. Animal Ecol. *36*:1–30.

Pielou, E.C. 1974. Population and Community Ecology: Principles and Methods. Gordon and Breach Publs., New York.

Rohlf, F.J. and R.R. Sokal. 1981. Statistical Tables. Freeman, San Francisco.

Siegal, S. 1956. Nonparametric Statistics for the Behavioral Sciences. McGraw-Hill, New York. 312 pp.

Sokal, R.R. and F.J. Rohlf. 1981. Biometry. The Principles and Practice of Statistics in Biological Research. Freeman, San Francisco. 859 pp.

Zar, J.H. 1984. Biostatistical analysis. 2nd Ed. Prentice Hall, Englewood Cliffs, NJ. 718 pp.

Useful Relationships Relative to the Use of Colorimeters and Spectrophotometers

Colorimetric and spectrophotometric analyses are based on several fundamental relationships for the transmission or absorption of light of specific wavelengths through solutions.

1. *Transmittance* (*T*) = transition of light through a solution of known distance or thickness; usually less than 1.0, since it is that fraction of light passed by a solution:

$$T = \frac{I}{I_0}$$

where I_0 = initial or original intensity of incident monochromatic light and I = final intensity of the transmitted radiation after passing through the medium or solution:

$$\%T = \left(\frac{I}{I_0}\right)100$$

2. *Absorption* (*a*) = transformational loss of radiant energy as it passes through a substance, in this case a solution

   a = absorption factor or coefficient of absorption

   a = ratio of the intensity loss by absorption to the total original intensity of radiation

$$a = \frac{I_0 - I}{I_0}$$

$$\%a = \left(\frac{I_0 - I}{I_0}\right)100$$

3. *Absorbance* (*A*) = $\log_{10}(I_0/I)$ or $-\log T$. Also called *optical density* (*OD*).

4. *Lambert's* (*Bouguer's*) *Law of Absorption*: When a ray of incident monochromatic light of intensity I_0 enters an absorbing medium, its intensity, I, decreases exponentially with the thickness, z, of the medium traversed:

$$I = I_0 e^{-kz}$$

where k = a constant depending on the extinction coefficient of the absorbing substance:

$$k = \frac{(4\pi k'n)}{\lambda}$$

k' = index of absorption
n = index of refraction
λ = wavelength (in vacuum)

or

$$2.303 \log_{10}\left(\frac{I_0}{I}\right) = kz$$

5. *Beer's Law*: The intensity of a ray of monochromatic light decreases exponentially with the concentration of the absorbing material (solution):

$$2.303 \log_{10}\left(\frac{I_0}{I}\right) = k'C$$

where C = concentration of the light-absorbing material.

6. *Combining Lambert's and Beer's Laws*:

$$\log_e\left(\frac{I_0}{I}\right) = kCz$$

or

$$2.303 \log_{10}\left(\frac{I_0}{I}\right) = kCz$$

While Lambert's Law is without exception, deviations from Beer's Law can occur when a solute ionizes or dissociates in solution or when the light is not monochromatic. One can test obeyance of the constancy of the transmittance-concentration relationship by plotting $\log(I_0/I)$ or $\log T$ against concentration. Conformity is indicated when the result is a straight line that passes through the origin. When the solutions do not conform, a calibration curve for absorbance versus concentration of standards must be constructed.

Characteristics and Taxonomic Sources of Common Freshwater Organisms

The flora and fauna of fresh waters are enormously diverse, including representatives from nearly all phyla. A detailed summary of even the major organisms and their characteristics is beyond the scope of this book. Nonetheless, the need for a basic taxonomic orientation is common to all biological phases of limnology.

We list common reference works to the taxonomy of North American freshwater plants and animals. The scope and depth of these systematic treatments are highly variable. For certain instructional purposes some of the keys may be sufficient. For many research investigations, however, workers must consult original taxonomic references.

BACTERIA

Most identification of bacteria is based on physiological differentiation by growth on substrates of media of known composition. Few bacteria are identifiable on the basis of morphology alone.

Buchanan, R.E. and N.E. Gibbons (eds). 1974. Bergey's Manual of Determinative Bacteriology. 8th Ed. Williams & Wilkins, Baltimore, 1246 pp.

Holt, J.G. (ed-in-chief). 1982–1988. Bergey's Manual of Systematic Bacteriology. Volume 1. 1982. Gram-Negative Bacteria of Medical or Industrial Importance. Volume 2. 1986. Gram-Positive Bacteria of Medical or Industrial Importance. Volume 3. 1988. Other Gram-Negative Bacteria. Cyanobacteria, Archaebacteria. Volume 4. 1988. Other Gram-Positive Bacteria. This work, under continued development, is the recognized authority on bacterial taxonomy.

Starr, M.P., H. Stolp, H.G. Trüper, A. Balows, and H.G. Schlegel. 1981. The Prokaryotes—A Handbook of Habitats, Isolation, and Identification of Bacteria. Springer-Verlag, New York. 2284 pp.

Van Neil, C.B. and R.Y. Stainer. 1959. Bacteria. pp. 16–46. *In*: W.T. Edmondson, Editor. Freshwater Biology. 2nd Ed. Wiley, New York.

FUNGI

Cooke, W.B. 1963. A Laboratory Guide to Fungi in Polluted Waters, Sewage, and Sewage Treatment Systems. Environmental Health Series 999-WP-1, U.S. Dept. Health, Education, and Welfare. 132 pp.

Ingold, C.T. 1975. An Illustrated Guide to Aquatic and Water-Borne Hyphomycetes (Fungi Imperfecti) with Notes on their Biology. Freshw. Biol. Assoc. U.K. No. *30*. 96 pp.

Nilsson, S. 1964. Freshwater Hyphomycetes. Taxonomy. Morphology, and Ecology. Symboliae Bot. Upsalienses *18*. 130 pp.

Petersen, R.H. 1962. Aquatic hyphomycetes from North America. I. Aleuriosporae (Part I), and key to the genera. Mycologia *54*:117–151.

Petersen, R.H. 1963a. Aquatic hyphomycetes from North America. II. Aleuriosporae (Part II), and blastosporae. Mycologia *55*:18–29.

Petersen, R.H. 1963b. Aquatic hyphomycetes from North America. III. Phialosporae and miscellaneous species. Mycologia *55*:570–581.

Sparrow, F.K. 1959. Fungi. pp. 47–94. *In*: W.T. Edmondson, Editor. Freshwater Biology. 2nd Ed. Wiley, New York.

Sparrow, F.K., Jr. 1960. Aquatic Phycomycetes. 2nd Ed. Univ. Michigan Press, Ann Arbor. 1187 pp.

Webster, J. and E. Descals. 1981. Morphology, distribution, and ecology of conidial fungi in freshwater habitats. pp. 295–355. *In*: G.T. Cole and B. Kendrick, Editors. Biology of Conidial Fungi. Vol. 1. Academic, New York.

ALGAE

The references listed below are largely of a general nature. Many of the detailed taxonomic works are in German and French; see George (1976) for a moderately complete listing.

Bourrelly, P. 1968–1988. Les Algues d'eau Douce. 3 Vol. N. Boubée & Cie, Paris.

Cooke, E.C. 1967. The Myxophyceae of North Carolina. Edwards Bros., Ann Arbor. 206 pp.

Croasdale, H., C.E.M. Bicudo, and G.W. Prescott. 1983. A synopsis of North American desmids. Part II. Desmidiaceae: Placodermae. Section 5. The filamentous genera. Univ. Nebraska Press, Lincoln. 117 pp.

Desikachary, T.V. 1959. Cyanophyta. Indian Council Agricultural Research, New Delhi. 686 pp.

Drouet, F. and W.A. Daily. 1973. Revision of the Coccoid Myxophyceae. Reprint of 1956 Edition, Hafner Press, New York. 222 pp.

Edmondson, W.T. (ed). 1959. Freshwater Biology. 2nd Ed. [Several chapters] Wiley, New York. pp. 95–189.

George, E.A. 1976. A guide to algal keys (excluding seaweeds). Br. Phycol. J. *11*:49–55.

Komárek, J. and B. Fott. 1983. Chlorococcales. pp. 1–1044. *In*: G. Huber-Pestalozzi, Editor. Das Phytoplankton des Süsswassers. *7*(1). Schweizerbart, Stuttgart.

Patrick, R. and C.W. Reimer. 1966. The Diatoms of the United States, Exclusive of Alaska and Hawaii. Vol. I. Monogr. Acad. Nat. Sci. Philadelphia *13*. 688 pp.

Patrick, R. and C.W. Reimer. 1975. The Diatoms of the United States, Exclusive of Alaska and Hawaii. Vol. 2, Part 1. Monogr. Acad. Nat. Sci. Philadelphia *13*. 213 pp.

Prescott, G.W. 1962. Algae of the Western Great Lakes Area. 2nd Ed. Wm.C. Brown, Dubuque, IA. 977 pp.

Prescott, G.W. 1970. How to Know the Freshwater Algae. 2nd Ed. Wm.C. Brown, Dubuque, IA. 348 pp.

Prescott, G.W., H.T. Croasdale, and W.C. Vinyard. 1972. Desmidiales. Part I. Saccordermae, Mesotaeniaceae. North Amer. Flora (N.Y. Bot. Garden), Ser. II(6). 84 pp.

Prescott, G.W., H.T. Croasdale, and W.C. Vinyard. 1975. A Synopsis of North American Desmids. Part II. Desmidiaceae: Placodermae. Section 1. Univ. Nebraska Press, Lincoln. 275 pp.

Prescott, G.W., H.T. Croasdale, and W.C. Vinyard. 1977. A Synopsis of North American Desmids. Part II. Desmidiaceae: Placodermae. Section 2. Univ. Nebraska Press, Lincoln. 413 pp.

Prescott, G.W., H.T. Croasdale, W.C. Vinyard, and C.E.M. Bicudo. 1981. A synopsis of North American Desmids. Part II. Desmidiaceae: Placodermae. Section. 3. Univ. Nebraska Press, Lincoln. 720 pp.

Prescott, G.W., C.E.M. Bicudo, and W.C. Vinyard. 1982. A synopsis of North American Desmids. Part II. Desmidiaceae: Placodermae. Section 4. Univ. Nebraska Press, Lincoln. 700 pp.

Smith, G.M. 1950. The Fresh-water Algae of the United States. 2nd Ed. McGraw-Hill, New York. 719 pp.

Tiffany, L.H. and M.E. Britton. 1952. The Algae of Illinois. Univ. Chicago Press, Chicago. 407 pp.

Vinyard, W.C. 1974. Key to the Genera of Diatoms of the Inland Waters of Temperate North America. Mad River Press, Eureka, CA. 19 pp.

LARGE AQUATIC PLANTS

Beal, E.O. 1977. A Manual of Marsh and Aquatic Vascular Plants of North Carolina with Habitat Data. Tech. Bull. No. Carolina Agric. Exp. Sta. *247.* 298 pp.

Beal, E.O. and J.W. Thieret. 1986. Aquatic and Wetland Plants of Kentucky. Sci. Tech. Ser. *5.* Kentucky Nature Preserves Commission, Frankfort, KY. 314 pp.

Cook, C.D.K. (ed). 1974. Water Plants of the World. Dr. W. Junk B.V., The Hague. 561 pp.

Correll, D.S. and H.B. Correll. 1972. Aquatic and Wetland Plants of Southwestern United States. Wat. Poll. Contr. Res. Ser. 16030 DNL 01/72, Environ. Protection Agency. 1777 pp.

Eyles, D.E. and J.L. Robertson, Jr. 1963. A Guide and Key to the Aquatic Plants of the Southeastern United States. Publ. Health Bull. *286,* U.S. Publ. Health Service. 151 pp.

Fassett, N.C. 1957. A Manual of Aquatic Plants. Univ. Wisconsin Press, Madison. 405 pp.

Godfrey, R.K. and J.W. Wooten. 1979. Aquatic and Wetland Plants of Southeastern United States. Monocotyledons. Univ. Georgia Press, Athens. 712 pp.

Godfrey, R.K. and J.W. Wooten. 1981. Aquatic and Wetland Plants of Southeastern United States. Dicotyledons. Univ. Georgia Press, Athens. 933 pp.

Mason, H.L. 1957. A Flora of the Marshes of California. Univ. California Press, Berkeley. 878 pp.

Muenscher, W.C. 1944. Aquatic Plants of the United States. Comstock Publ., New York. 374 pp.

Prescott, G.W. 1969. How to Know the Aquatic Plants. W.C. Brown, Dubuque, IA. 171 pp.

Steward, A.N., L.R. Dennis, and H.M. Gilkey. 1960. Aquatic Plants of the Pacific Northwest with Vegetative Keys. Oregon State Coll., Corvallis. 184 pp.

Winterringer, G.S. and A.C. Lopinot. 1966. Aquatic Plants of Illinois. Ill. State Museum Popular Sciences Ser. *VI.* 142 pp.

Wood, R.D. 1965. Monograph of the Characeae. *In*: R.D. Wood and K. Imahori, Editors. A Revision of the Characeae. Vol. I. Cramer, Weinheim.

Wood, R.D. 1967. Charophytes of North America. Bookstore, Univ. Rhode Island, Kingston. 72 pp.

Wood, R.D. and K. Imahori. 1964. Iconograph of the Characeae. *In*: R.D. Wood and K. Imahori, Editors. A Revision of the Characeae. Vol. II. Cramer, Weinheim.

FAUNA

Starter Keys for the Identification of Some Common Freshwater Animals

The following keys have been simplified greatly and are intended for use only with common aquatic forms. Most of the microscopic forms, semiaquatic groups, and rare groups have been ignored for the sake of simplicity. Detailed keys to these groups are listed in the Bibliography.

Use the general descriptions found in (1) and (2) below to determine the proper "starting key."

1. Nonarthropods—Use Key A (no jointed legs)
2. Arthropods (jointed legs present)
   a. Crustacea—Use Key B (page 377)
      (possess gills; biramous appendages; 2 pairs of antennae)
   b. Hydracarina (water mites)—further identification difficult
      [see Edmondson (1959)] (4 pairs of legs; no antennae)
   c. Adult insect—Use Key C (page 377)
      [3 pairs of legs; no gills; less than 2 pairs of antennae; usually with obvious flying wings which may be under a hard or leathery cover (Collembola are wingless); body entirely covered with hardened material (chitinous exoskeleton)]
   d. Insect naiads—Use Key D (page 378)
      (appear much like an adult insect but wingless—developing wing buds often apparent; body covering and form like adult; abdomen never soft and membranous)
   e. Insect larvae—Use Key E (page 378)
      (little or no resemblance to an adult insect; jointed legs, temporary non-jointed legs or no legs; if legless, body divided into about 13 segments; much of body soft and with membranous surface)

Key A: Key to some common freshwater animal phyla

1.a. Minute to microscopic in size; bodies not divided into cells; solitary or colonial; sedentary or motile .... Phylum Protozoa
1.b. Bodies divided into cells .... Subkingdom Metazoa 2

    2.a. Body with many segments (always more than 13; insect larvae are worm-like but have only 13 body segments), or unsegmented but with obvious calcareous shell 3
    2.b. Body unsegmented and without a true shell 4

3.a. Body wormlike; segmented, with setae on segments ....
.... Phylum Annelida (worms)
3.b. Body unsegmented, in or under a hard shell ....
.... Phylum Mollusca (mussels and snails)

    4.a. Body of variable shape 5
    4.b. Body wormlike 9

5.a. Sessile, mostly encrusting or mat forming in habit, or gelatinous colonial forms 6
5.b. Sessile or motile; usually not encrusting 7

6.a. Body permeated with pores; no protruding microscopic tentacles; internal spicules; spongy to touch and often green . . . .
. . . . Phylum Porifera (freshwater sponges)

6.b. Body enclosed in gelatinous, calcareous or chitinous covering; anterior end with ciliated tentacles on fold of body wall surrounding mouth . . . . Phylum Bryozoa (moss animals)

7.a. Sessile or planktonic; if sessile, then solitary or colonial; small, stalked form with mouth surrounded by nonciliated tentacles; if planktonic, then solitary, small, bell-shaped medusae with nonciliated tentacles . . . . Phylum Coelenterata

7.b. Sessible or planktonic; solitary or colonial; microscopic; cilia or bristles on anterior end 8

8.a. One or two rings of cilia (or a flat ciliated area) on anterior end; internal jaws (mastax) . . . . Phylum Rotifera

8.b. Large tufts of bristles at anterior end; no internal jaws; no rings of cilia; body mostly covered by spinelike structures . . . . . Phylum Gastrotricha

9.a. Body cylindrical and elongate; no cilia present 10

9.b. Body flattened and elongate; cilia usually present . . . .
. . . . Phylum Platyhelminthes (flatworms)

10.a. Mostly small; moving with whiplike motion; often with bristles; posterior end tapering to a point . . . . Phylum Nematoda (roundworms)

10.b. Long, threadlike; usually dark brown; posterior end not tapered conspicuously . . . . Phylum Nematomorpha (horsehair worms)

Key B: Some important aquatic groups in the class Crustacea

1.a. Bivalve carapace 2

1.b. Carapace not bivalved, or absent 3

2.a. Carapace calcareous; when closed, completely covering head, body, and appendages; two pairs of trunk appendages . . . . Subclass Ostracoda

2.b. Carapace distinctly covers trunk but not head; four to six pairs of trunk appendages; large biramous antennae (except in one form enclosed in a gelatinous case) used for swimming . . . . Order Cladocera

3.a. Trunk appendages flattened and leaflike; carapace absent; eyes stalked; swim "upside down" . . . . Order Anostraca

3.b. Trunk appendages more or less slender and jointed; frequently used for walking or swimming 4

4.a. Variable number of somites; four or five pairs of segmented trunk appendages . . . . Subclass Copepoda
(See Supplementary Key VIII—page 407)

4.b. All body segments except the last one bear appendages, consisting of seven or eight pairs of thoracic and six pairs of abdominal appendages 5

5.a. Carapace absent; all thoracic segments visible; eyes not stalked; body compressed . . . . Order Amphipoda

5.b. Carapace present, eyes stalked . . . . Order Decapoda

Key C: Adult insects

1.a. Wings under a leathery or hard chitinous cover 2
1.b. Wings all membranous, exposed 3

    2.a. Wing cover complete, very hard and rigid . . . . COLEOPTERA
                       (See Supplementary Key I—page 405)
    2.b. Wing cover leathery except at tip where it is membranous and veined
                                      HEMIPTERA
                       (See Supplementary Key II—page 405)

3.a. One pair of wings . . . . DIPTERA (true flies)
3.b. Two pairs of wings 4

    4.a. Wings covered with hairs or scales . . . .
                TRICHOPTERA (hairs); LEPIDOPTERA (scales)
    4.b. Wings without encumbrances . . . . . . HYMENOPTERA (wings hooked
        together into one); EPHEMEROPTERA (rear pair of wings smaller);
        PLECOPTERA (wings all approximately the same size; abdomen not
        especially long); ODONATA (wings all the same size; abdomen long and
        cylindrical).

    Note: COLLEMBOLA have no wings but can always be recognized by the
        long forked appendage at the tip of the abdomen (usually folded under)
        used in springing about on the surface of quiet water.

Key D: Insect naiads

1.a. Fantastically developed lower jaw covering lower half or front of face and
folding back between legs . . . . ODONATA
                       (See Supplementary Key III—page 406)
1.b. Lower jaw more normal 2

    2.a. Platelike gill on rear corners of most abdominal segments; one tarsal
        claw; usually three tails . . . . EPHEMEROPTERA
                       (See Supplementary Key IV—page 406)
    2.b. Gills (if any) as bunches of filaments under body (usually thorax only);
        two tarsal claws; two tails . . . . PLECOPTERA
                       (See Supplementary Key V—page 406)

Key E: Insect larvae

1.a. No jointed legs; body usually worm- or maggotlike . . . . DIPTERA
                       (See Supplementary Key VI—page 406)
1.b. Jointed legs present 2

    2.a. Posterior end of abdomen without hooks or fleshy temporary legs . . . .
                                   COLEOPTERA
                       (See Supplementary Key I—page 405)
    2.b. Hooks at end of abdomen, often on fleshy prolegs 3

3.a. Postabdominal proleg each with two large hooks placed side by side; each
abdominal segment with a pair of stout lateral filaments . . . .
. . . . MEGALOPTERA (Dobson flies)

3.b. Posterior corners of abdomen each dominated by a large hook (often on a proleg) or occasionally by a row of small hooks; larvae usually in a case . . . .
. . . . TRICHOPTERA
(See Supplementary Key VII—page 406)

Supplementary Keys

Key I: Coleoptera

(Beetles; most listed are aquatic through life)

1.a. With fully developed wings and wing covers . . . . adults 2
1.b. No wings . . . . larvae 4

2.a. Crawling, clinging beetles; no swimming hairs. Elmidae (1–3 mm) . . . . .
. . . . Dryopidae (3–8 mm)
2.b. Active swimmers; legs with fringe of swimming hairs 3

3.a. Antennae short, ending in club . . . . Hydrophilidae
(water scavenger beetle)
3.b. Antennae long, slender . . . . Dytiscidae
(predaceous diving beetle)

4.a. Entire body flattened, dishlike . . . . *Psephenus*
(water penny)
4.b. Body more cylindrical 5

5.a. Entire body heavily armored with chitin; very cylindrical; hairless; small jaws . . . . Elmidae
5.b. Abdomen moderately soft with some bristles; large tubular sucking jaws . . . 6

6.a. Legs 4-segmented . . . . Dytiscidae
(predaceous diving beetle)
6.b. Legs 5-segmented . . . . Hydrophilidae
(water scavenger beetle)

Key II: Hemiptera

(True bugs; adults and nymphs aquatic)

1.a. Surface skimmers; antennae longer than head 2
1.b. Fully aquatic; antennae shorter than head 4

2.a. First long segment of hind leg much longer than abdomen . . . . . *Gerris*
(water strider)
2.b. This segment no longer than abdomen . . . . Veliidae 3

3.a. Abdomen about as wide as thorax . . . . *Microvelia*
3.b. Abdomen narrowing sharply posteriorly . . . . *Rhagovelia*

4.a. Wings covering back flatly . . . . Corixidae
(water boatman)
4.b. Wings rooflike over back; swim upside down . . . . Notonectidae
(back swimmers)

Key III: Odonata

(Dragon and damsel flies; naiads aquatic)

1.a. No gills at tip of abdomen . . . . Dragon flies
1.b. Three platelike gills at tip of abdomen . . . . Damsel flies

Key IV: Ephemeroptera

(Mayflies; naiads aquatic)

1.a. Head broad and flat; eyes upward . . . . Heptageniidae 2
1.b. Head rounded; eyes lateral . . . . Baetidae 3

    2.a. Two "tails" . . . . *Iron*
    2.b. Three "tails" . . . . *Stenonema*

3.a. Outer tails long, center one very short; small species . . . . *Baetis*
3.b. All three tails the same length 4

    4.a. Gills on top of center abdominal segments . . . . *Ephemerella*
    4.b. Gills at corners of segments 5

5.a. Gills narrow and forked . . . . *Paraleptophlebia*
5.b. Gills platelike with fingerlike projections . . . . *Blasturus*

Key V: Plecoptera

(Stone flies; naiads aquatic)

1.a. Gills on first two abdominal segments . . . . *Pteronarcys*
1.b. No gills on abdominal segments 2

    2.a. Gills at bases of legs only (check closely) . . . . *Acroneuria*
    2.b. No gills at all . . . . *Alloperla, Isoperla*, and others

Key VI: Diptera

(True flies, some larvae aquatic)

1.a. Prolegs present . . . . Tendipedidae
1.b. Prolegs absent . . . . (Culcidae) 2

    2.a. Antennae prehensile; without mouth brushes [Transparent except for
        pigmented hydrostatic organs (air sacs)] . . . . *Chaoborus*
    2.b. Antennae not prehensile; mouth brushes present (*Aedes, Anopheles*, and
        Culex common) . . . . Culicinae

Key VII: Trichoptera

(Caddis files; larvae aquatic)

1.a. Larva in spiral, snail-like case; anal hooks comblike . . . . *Helicopsyche*
1.b. Case not snail-like (if any); hooks not comblike 2

    2.a. In portable case (may be fastened and closed for pupation) 3
    2.b. In fixed case or none 7

3.a. Tiny purselike case; abdomen greatly enlarged .... *Hydroptila*
3.b. Larger case, not purselike 4

    4.a. Case of pebbles; turtlelike; anal hooks clearly projecting out at rear (but not beyond case) .... *Agapetus*
    4.b. Case turbular; hooks enclosed 5

5.a. Large pebbles like outriggers at side of tube .... *Goera*
5.b. No such pebbles 6

    6.a. Long thin spine between front legs; case sand, leaves, twigs or combination of all .... *Limnephilus*
    6.b. No such spine; case a neat cylinder of sand .... *Psilotreta*

7.a. Three thoracic dorsal plates; bristly body .... *Hydropsyche*
7.b. One thoracic dorsal plate; body smooth 8

    8.a. Obvious dark plate on top of last full adominal segment ....
                            Agapetus (and legs reduced)
                            Rhyacophila (long anal legs)
    8.b. No such plate .... Philopotamidae and Psychomyiidae

Key VIII: Key to the orders of Copepods

1.a. Metasome (cephalothorox) and abdomen (urosome) distinctly different in width and separated by a distinct movable articulation 2
1.b. Metasome not distinctly narrower than abdomen......................
.... Order Harpacticoida
                            common genus *Canthocamptus*

    2.a. Antennae long, usually as long or longer than body, and composed of from 18 to 25 segments; biramous .... Order Calanoida
              common genera *Epischura, Diaptomus*, and *Limnocalanus*
    2.b. Antennae short, usually shorter than the cephalothorax, and composed of from 6 to 17 segments; uniramous .... Order Cyclopoida
                            common genus *Cyclops*

LIFE HISTORY OF COPEPODS*

Because of the unusual life history and morphological differentiation among common planktonic copepods, a brief discussion of their development may be helpful. Freshwater copepods are bisexual. The male attaches a spermatophore to the female. Eggs either are laid into a membranous case and carried until hatching or are cast into the water. The larva that hatches from the egg, known as a Stage I nauplius, has only the three anterior pairs of appendages. There are 11 subsequent moults, each with an increase in size and development of appendages. Thus there are 12 distinct free-swimming stages after the egg, the last of which is the mature adult (Figs. A.4.1 and A.4.2). The first six stages are known as nauplii and the last

---

*Based in part on notes of G.W. Comita (Edmondson and Winberg, 1971).

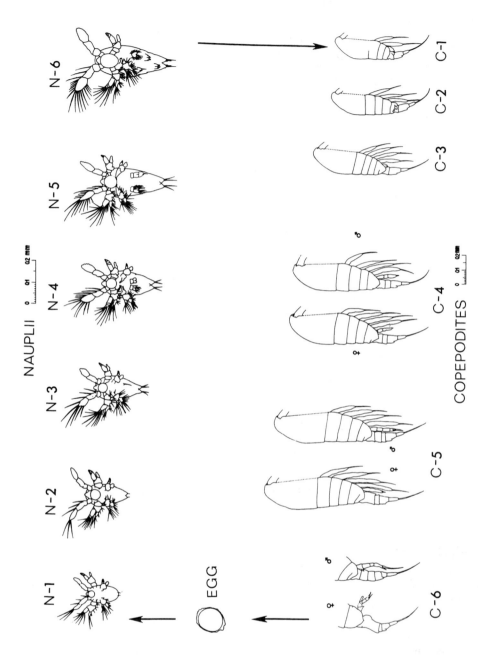

Figure A.4.1. Life history stages of *Diaptomus* (*Eudiaptomus*) *vulgaris*. Nauplii are shown in ventral view. In some, appendages are shown on one side only. Copepodids are shown in lateral view. Only the posterior part of adults (Copepodid 6) is shown. [Modified from Ravera (1953).]

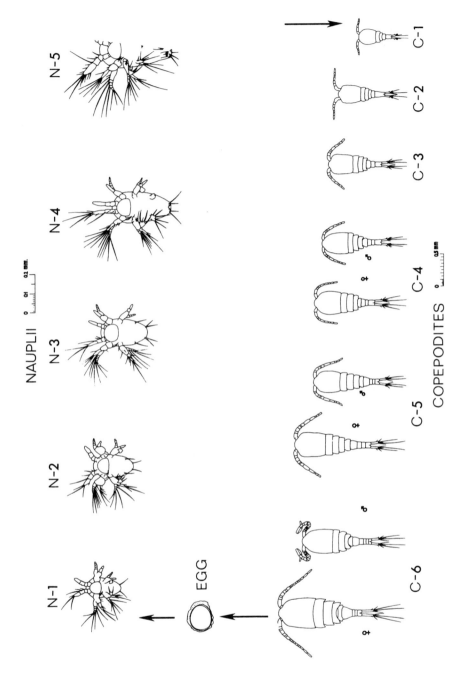

Figure A.4.2. Life history stages of *Cyclops strennus*. Nauplii shown in ventral view (appendages are shown on one side only); copepodids in dorsal view. [Modified from Ravera (1953).]

six as copepodids. When the Stage VI nauplius moults, the changes are greater than the previous ones, and the shape is considerably altered by the elongation of the anterior part and by the development of the abdomen. In some copepods, the first nauplius stage may occur while still in the egg, so that the animal that comes from the egg is actually N-II. In a few cases, the adult may moult.

The course of development varies somewhat among the groups of copepods; even within the Calanoids, details differ among genera. The following table shows, in general, the appendages possessed by each stage of *Diaptomus* and gives a qualitative indication of the degree of development. Although none of the appendages is fully developed until C-VI, the anterior appendages are essentially complete before this stage. Recognition of the copepodid stages is based on segmentation of the body and the details of the appendages.

						Stage						
Appendage	*N-I*	*N-II*	*N-III*	*N-IV*	*N-V*	*N-VI*	*C-I*	*C-II*	*C-III*	*C-IV*	*C-V*	*CVI*
1st antenna	+	+	+	+	+	+	+	+	+	+	+	#
2nd antenna	+	+	+	+	+	+	+	+	+	+	+	#
Mandible	+	+	+	+	+	+	+	+	+	+	+	#
1st maxilla	0	0	×	−	−	−	+	+	+	+	+	#
2nd maxilla	0	0	0	0	−	−	+	+	+	+	+	#
Maxillipeds	0	0	0	0	0	−	+	+	+	+	+	#
1st leg	0	0	0	0	0	−	+	+	+	+	+	#
2nd leg	0	0	0	0	0	−	+	+	+	+	+	#
3rd leg	0	0	0	0	0	0	−	+	+	+	+	#
4th leg	0	0	0	0	0	0	−	−	+	+	+	#
5th leg	0	0	0	0	0	0	0	−	−	+	+	#
Caudal rami	0	×	×	×	×	×	+	+	+	+	+	#

N = nauplius, C = copepodid, # = structure fully formed, + = relatively well developed but not fully formed, − = present as lobed structure or otherwise much undeveloped, × = present as seta or group of setae, 0 = structure absent.

SIMPLIFIED KEY OF COMMON FRESHWATER MACROINVERTEBRATES

The following simple key, courtesy of F.R. Hauer and V.H. Resh, serves only as a visual starting point for identification of common invertebrates. Non-insect taxa are described to the phyla or order level, and insect taxa to the family level. Reprinted by permission from Hauer and Resh (1996).

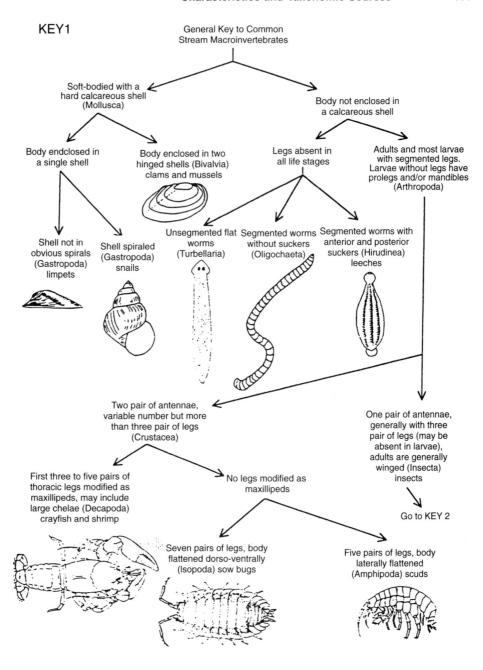

KEY1

General Key to Common
Stream Macroinvertebrates

Soft-bodied with a
hard calcareous shell
(Mollusca)

Body not enclosed in
a calcareous shell

Body endclosed in
a single shell

Body enclosed in two
hinged shells (Bivalvia)
clams and mussels

Legs absent in
all life stages

Adults and most larvae
with segmented legs.
Larvae without legs have
prolegs and/or mandibles
(Arthropoda)

Shell not in
obvious spirals
(Gastropoda)
limpets

Shell spiraled
(Gastropoda)
snails

Unsegmented flat
worms
(Turbellaria)

Segmented worms
without suckers
(Oligochaeta)

Segmented worms with
anterior and posterior
suckers (Hirudinea)
leeches

Two pair of antennae,
variable number but more
than three pair of legs
(Crustacea)

One pair of antennae,
generally with three
pair of legs (may be
absent in larvae),
adults are generally
winged (Insecta)
insects

First three to five pairs of
thoracic legs modified as
maxillipeds, may include
large chelae (Decapoda)
crayfish and shrimp

No legs modified as
maxillipeds

Go to KEY 2

Seven pairs of legs, body
flattened dorso-ventrally
(Isopoda) sow bugs

Five pairs of legs, body
laterally flattened
(Amphipoda) scuds

KEY 2

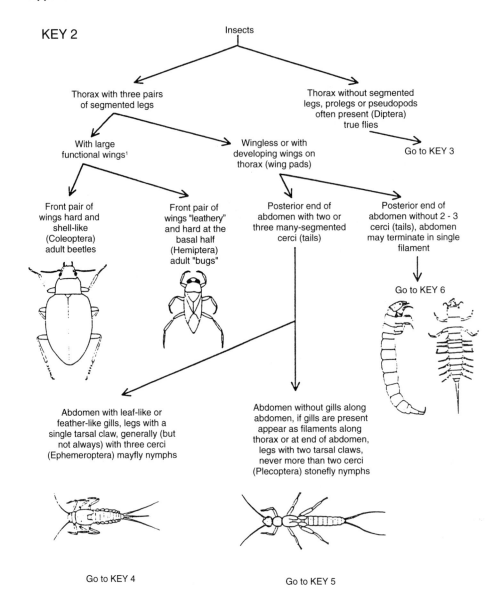

Insects

Thorax with three pairs of segmented legs

Thorax without segmented legs, prolegs or pseudopods often present (Diptera) true flies

Go to KEY 3

With large functional wings[1]

Wingless or with developing wings on thorax (wing pads)

Front pair of wings hard and shell-like (Coleoptera) adult beetles

Front pair of wings "leathery" and hard at the basal half (Hemiptera) adult "bugs"

Posterior end of abdomen with two or three many-segmented cerci (tails)

Posterior end of abdomen without 2 - 3 cerci (tails), abdomen may terminate in single filament

Go to KEY 6

Abdomen with leaf-like or feather-like gills, legs with a single tarsal claw, generally (but not always) with three cerci (Ephemeroptera) mayfly nymphs

Abdomen without gills along abdomen, if gills are present appear as filaments along thorax or at end of abdomen, legs with two tarsal claws, never more than two cerci (Plecoptera) stonefly nymphs

Go to KEY 4

Go to KEY 5

[1]NOTE: Other winged insects may be present from either terrestrial forms or the aerial adults of aquatic larvae

KEY 3

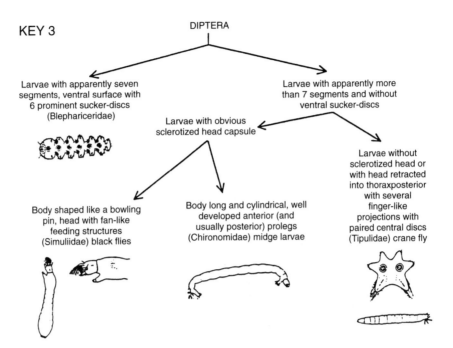

DIPTERA

Larvae with apparently seven segments, ventral surface with 6 prominent sucker-discs (Blephariceridae)

Larvae with apparently more than 7 segments and without ventral sucker-discs

Larvae with obvious sclerotized head capsule

Larvae without sclerotized head or with head retracted into thoraxposterior with several finger-like projections with paired central discs (Tipulidae) crane fly

Body shaped like a bowling pin, head with fan-like feeding structures (Simuliidae) black flies

Body long and cylindrical, well developed anterior (and usually posterior) prolegs (Chironomidae) midge larvae

KEY 4

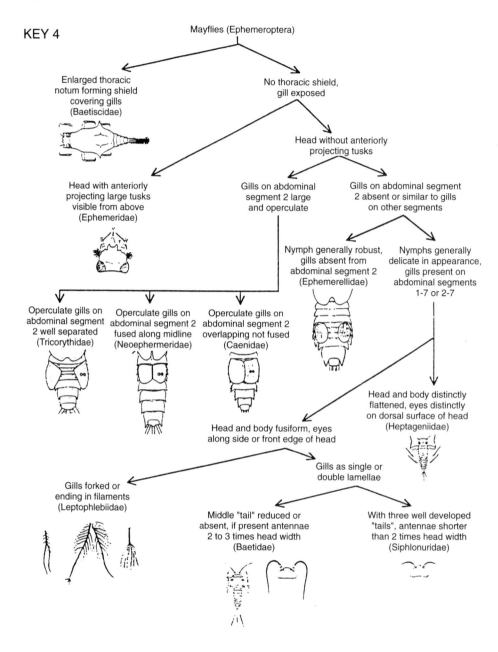

Mayflies (Ephemeroptera)

Enlarged thoracic notum forming shield covering gills (Baetiscidae)

No thoracic shield, gill exposed

Head without anteriorly projecting tusks

Head with anteriorly projecting large tusks visible from above (Ephemeridae)

Gills on abdominal segment 2 large and operculate

Gills on abdominal segment 2 absent or similar to gills on other segments

Nymph generally robust, gills absent from abdominal segment 2 (Ephemerellidae)

Nymphs generally delicate in appearance, gills present on abdominal segments 1-7 or 2-7

Operculate gills on abdominal segment 2 well separated (Tricorythidae)

Operculate gills on abdominal segment 2 fused along midline (Neoephermeridae)

Operculate gills on abdominal segment 2 overlapping not fused (Caenidae)

Head and body distinctly flattened, eyes distinctly on dorsal surface of head (Heptageniidae)

Head and body fusiform, eyes along side or front edge of head

Gills as single or double lamellae

Gills forked or ending in filaments (Leptophlebiidae)

Middle "tail" reduced or absent, if present antennae 2 to 3 times head width (Baetidae)

With three well developed "tails", antennae shorter than 2 times head width (Siphlonuridae)

KEY 5

Stoneflies (Plecoptera)

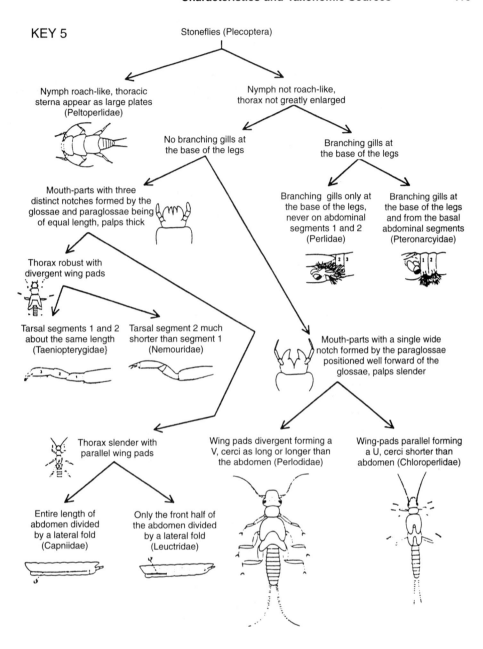

Nymph roach-like, thoracic
sterna appear as large plates
(Peltoperlidae)

Nymph not roach-like,
thorax not greatly enlarged

No branching gills at
the base of the legs

Branching gills at
the base of the legs

Branching gills only at
the base of the legs,
never on abdominal
segments 1 and 2
(Perlidae)

Branching gills at
the base of the legs
and from the basal
abdominal segments
(Pteronarcyidae)

Mouth-parts with three
distinct notches formed by the
glossae and paraglossae being
of equal length, palps thick

Thorax robust with
divergent wing pads

Tarsal segments 1 and 2
about the same length
(Taeniopterygidae}

Tarsal segment 2 much
shorter than segment 1
(Nemouridae)

Mouth-parts with a single wide
notch formed by the paraglossae
positioned well forward of the
glossae, palps slender

Thorax slender with
parallel wing pads

Wing pads divergent forming a
V, cerci as long or longer than
the abdomen (Perlodidae)

Wing-pads parallel forming
a U, cerci shorter than
abdomen (Chloroperlidae)

Entire length of
abdomen divided
by a lateral fold
(Capniidae)

Only the front half of
the abdomen divided
by a lateral fold
(Leuctridae)

KEY 6

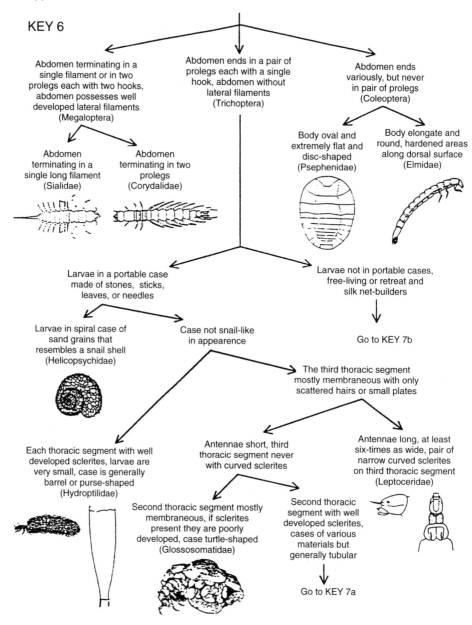

Abdomen terminating in a single filament or in two prolegs each with two hooks, abdomen possesses well developed lateral filaments (Megaloptera)

Abdomen ends in a pair of prolegs each with a single hook, abdomen without lateral filaments (Trichoptera)

Abdomen ends variously, but never in pair of prolegs (Coleoptera)

Abdomen terminating in a single long filament (Sialidae)

Abdomen terminating in two prolegs (Corydalidae)

Body oval and extremely flat and disc-shaped (Psephenidae)

Body elongate and round, hardened areas along dorsal surface (Elmidae)

Larvae in a portable case made of stones, sticks, leaves, or needles

Larvae not in portable cases, free-living or retreat and silk net-builders

Larvae in spiral case of sand grains that resembles a snail shell (Helicopsychidae)

Case not snail-like in appearence

Go to KEY 7b

The third thoracic segment mostly membraneous with only scattered hairs or small plates

Each thoracic segment with well developed sclerites, larvae are very small, case is generally barrel or purse-shaped (Hydroptilidae)

Antennae short, third thoracic segment never with curved sclerites

Antennae long, at least six-times as wide, pair of narrow curved sclerites on third thoracic segment (Leptoceridae)

Second thoracic segment mostly membraneous, if sclerites present they are poorly developed, case turtle-shaped (Glossosomatidae)

Second thoracic segment with well developed sclerites, cases of various materials but generally tubular

Go to KEY 7a

KEY 7

7a

First thoracic sclerites divide by deep furrow, no hump on the first abdominal segment, case often square in cross-section similar to a chimney (Brachycentridae)

First thoracic sclerites not divided by deep furrow, first abdominal segment with lateral humps

Antennae located extremely close to the eye, first abdominal segment without dorsal hump (Lepidostomatidae)

Antennae midway between eye and mandibles, abdominal segment 1 always with dorsal hump, cases variable in materials and design (Limnephilidae)

7b

Each thoracic segment with a single dorsal plate, larvae construct a fixed retreat (Hydropsychidae)

Second and third thoracic segments entirely membraneous

Abdominal segment 9 with a strongly sclerotized plate on the dorsal surface, larvae are free-living. (Rhyacophilidae)

Abdominal segment 9 entirely membraneous

Labrum membraneous and T-shaped, larvae construct sack-shaped nets (Philopotamidae)

Labrum sclerotized and normally shaped

Trochantin of first thoracic leg pointed at apex, larvae construct funnel-shapped capture nets (Polycentropodidae)

Trochantin of first thoracic leg broad at apex appears hatchet shaped, larvae construct tubular-shaped capture nets (Psychomyiidae)

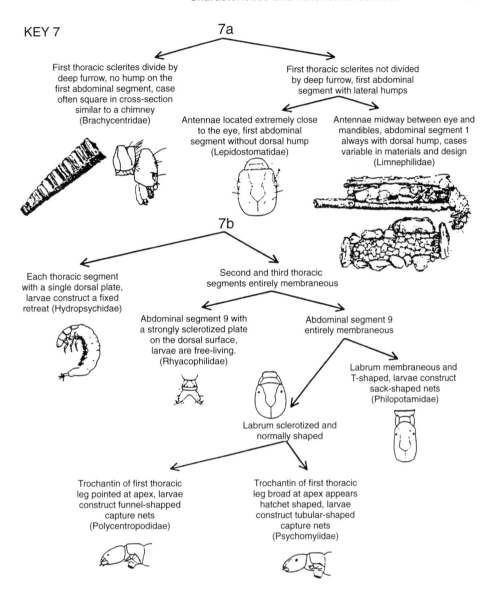

ANIMALS

General Taxonomic Works on Freshwater Animals

Eddy, S. and A.C. Hodson. 1961. Taxonomic Keys to the Common Animals of the North Central States. 3rd Ed. Burgess, Minneapolis. 162 pp.

Edmondson, W.T. (ed.). 1959. Freshwater Biology. 2nd Ed. Wiley, New York. 1248 pp.

Huggins, D.G., P.M. Liechti, and L.C. Ferrington, Jr. 1985. Guide to the Freshwater Invertebrates of the Midwest. Tech. Publ. *11*, Kansas Biological Survey, Univ. Kansas, Lawrence. 221 pp.

Klots, E.B. 1966. The New Field Book of Freshwater Life. Putnam, New York. 398 pp.

Morgan, A.H. 1930. Field Book of Ponds and Streams. Putnam, New York.

Needham, J.C. and P.R. Needham. 1962. A Guide to the Study of Freshwater Biology. 5th Ed. Holden-Day, San Francisco. 107 pp.

Peckarsky, B.L., P.R. Fraissinet, M.A. Penton, D.J. Conklin, 1990. Freshwater Macroinvertebrates of Northeastern North America. Cornell Univ. Press, Ithaca, NY. 442 pp.

Pennak, R.W. 1989. Fresh-water Invertebrates of the United States. Protozoa to Mollusca. 3rd Ed. Wiley, New York. 628 pp.

Thorp, J.H. and A.P. Covich (Editors). 1991. Ecology and Classification of North American Freshwater Invertebrates. Academic Press, San Diego. 911 pp.

Protozoa

Edmondson, W.T. (ed). 1959. Freshwater Biology. [Several chapters, pp. 190–297] Wiley, New York.

Jahn, T.L. 1949. How to Know the Protozoa. Wm. C. Brown, Dubuque, IA 234 pp.

Kudo, R.R. 1966. Protozoology. 5th Ed. Charles C Thomas, Springfield, IL.

Lee, J.J., S.H. Hutner, and E.C. Bovee (ed). 1985. An Illustrated Guide to the Protozoa. Society of Protozoologists, Lawrence, KS. 629 pp.

Smaller Animal Groups

Baker, F.C. 1928. The Freshwater Mollusca of Wisconsin. Part I. Gastropoda. Part II. Pelecypoda. Bull. Wisconsin Geol. Nat. Hist. Surv. *70*. 507 pp; 495 pp.

Brinkhurst, R.O. and B.G.M. Jamieson. 1971. Aquatic Oligochaeta of the World. Univ. Toronto Press, Toronto. 860 pp.

Burch, J.B. 1982. Freshwater Snails (Mollusca: Gastropoda) of North America. EPA—600/3-82-026. Environmental Protection Agency. Cincinnati. 294 pp.

Edmondson, W.T. (ed). 1959. Freshwater Biology. 2nd Ed. [Many chapters] Wiley, New York. 1248 pp.

Goodrich, C. 1932. The Mollusca of Michigan. Univ. Michigan Press, Ann Arbor. 120 pp.

Klemm, D.J. (ed). 1985. A Guide to the Freshwater Annelida (Polychaeta, Naidid and Tubificid Oligochaeta, and Hirudinea) of North America. Kendall/Hunt, Dubuque, IA. 198 pp.

Pennak, R.W. 1989. Fresh-water Invertebrates of the United States. Protozoa to Mollusca. 3rd Ed. Wiley, New York. 628 pp.

Ruttner-Kolisko, A. 1974. Plankton Rotifers: Biology and Taxonomy. Die Binnengewässer *26*(1). 50 pp.

Tarjan, A.C., R.P. Esser, and S.L. Chang. 1977. An illustrated key to nematodes found in fresh water. J. Wat. Poll. Control Fed. *49*:2318–2337.

Williams, W.D. 1972. Freshwater Isopods (Asellidae) of North America. Biota of Freshwater Ecosystems Identification Manual No. 7. Environmental Protection Agency, Washington. 45 pp.

Copepods

Ravera, O. 1953. Gli stadi di sviluppo dei copepodi pelagici del Lago Maggiore. Mem. Ist. Ital. Idrobiol. 7:129–151.

Wilson, M.S. and H.C. Yeatman. 1959. Free Living Copepoda. pp. 735–861. *In*: W.T. Edmondson, Editor. Freshwater Biology. 2nd Ed. Wiley, New York.

Cladocera

Brooks, J.L. 1957. The Systematics of North American Daphnia. Mem. Conn. Acad. Arts Sci. *13*. 180 pp.

Brooks, J.L. 1959. Cladocera. pp. 587–656. *In*: W.T. Edmondson, Editor. Freshwater Biology. 2nd Ed. Wiley, New York.

Goulden, C.E. 1968. The Systematics and Evolution of the Moinidae. Trans. Amer. Phil. Soc., N.S., *58*(6). 101 pp.

Aquatic Insects

Borror, D.J. and R.E. White. 1970. A Field Guide to the Insects of America North of Mexico. Houghton Miffin, Boston. 404 pp.

Brigham, A.R., W.U. Brigham, and A. Gnilka (ed). 1982. Aquatic Insects and Oligochaetes of North and South Carolina. Midwest Aquatic Enterprises, Mohomet, IL.

Chu, P. 1949. How to Know the Immature Insects. Wm. C. Brown, Dubuque, IA 234 pp.

Edmondson, W.T. (ed). 1959. Freshwater Biology. 2nd Ed. [Many chapters] Wiley, New York 1248 pp.

Hauer, F.R. and V.H. Resh. 1996. Benthic macroinvertebrates. Pp. 339–369 In: Methods in Stream Ecology. F.R. Hauer and G.A. Lamberti, Eds. Academic Press, San Diego.

Hilsenhoff, W.L. 1975. Aquatic Insects of Wisconsin, with Generic Keys and Notes on Biology, Ecology and Distribution. Tech. Bull. Wisc. Dept. Nat. Res. *89*. 52 pp.

Lehmkuhl, D.M. 1979. How to Know the Aquatic Insects. Wm. C. Brown, Dubuque, IA.

Mason, W.T., Jr. 1973. An Introduction to the Identification of Chironomid Larvae. U.S. Environ. Protection Agency, Cincinnati. 90 pp.

Merritt, R.W. and K.W. Cummins (Editors). 1996. An Introduction to the Aquatic Insects of North America. 3rd Ed. Kendall/Hunt, Dubuque, Iowa. 862 pp.

Peterson, A. 1951. Larvae of Insects. Part II. Coleoptera, Diptera, Neuroptera, Siphonaptera, Mecoptera, Trichoptera. Edwards Bros., Ann Arbor. 416 pp.

Usinger, R.L. 1956. Aquatic Insects of California. Univ. Calif. Press, Berkeley. 508 pp.

Fishes

A large number of taxonomic works exist for the identification of freshwater fishes. Nearly every state or province has various guides to the identification of fishes. Users should consult local or regional guides. Examples are as follows:

Eddy, S. 1957. How to Know the Freshwater Fishes. Wm. C. Brown, Dubuque, IA 253 pp.

Hubbs, C.L. and K.F. Lagler. 1958. Fishes of the Great Lakes Region. Bull. Cranbrook Inst. Science *26*. 213 pp.

Koster, W.J. 1957. Guide to the Fishes of New Mexico. Univ. New Mexico Press, Albuquerque. 116 pp.

Mettee, M.F., P.F. O'Neil, and J.M. Pierson. 1996. Fishes of Alabama and the Mobile Basin. Oxmoor House, Birmingham, Alabama. 819 pp.

Scott, W.B. and E.J. Crossman. 1973. Freshwater Fishes of Canada. Bull. Fish. Res. Bd. Canada *184*. 966 pp.

Trautman, M.B. 1957. The Fishes of Ohio with Illustrated Keys. Ohio State Univ. Press, Columbus, 683 pp.

SI Conversion Factors

Quantity	Unit		Conversion factor
Length	1 in.	=	25.4 mm
	1 ft	=	0.3048 m
	1 yd	=	0.9144 m
	1 fathom	=	1.8288 m
	1 chain	=	20.1168 m
	1 mile	=	1.60934 km
	1 International nautical mile	=	1.852 km
	1 UK nautical mile	=	1.85318 km
Area	1 in.^{2}	=	6.4516 cm^2
	1 ft^2	=	0.092903 m^2
	1 yd^2	=	0.836127 m^2
	1 acre	=	4046.86 m^2 = 0.404686 ha (hectare)
	1 sq. mile	=	2.58999 km^2 = 258.999 ha
Volume	1 UK minim	=	0.0591938 cm^3
	1 UK fluid drachm	=	3.55163 cm^3
	1 UK fluid ounce	=	28.4131 cm^3
	1 US fluid ounce	=	29.5735 cm^3
	1 US liquid pint	=	473.176 cm^3 = 0.4732 dm^3 (= liter)
	1 US dry pint	=	550.610 cm^3 = 0.5506 dm^3
	1 Imperial pint	=	568.261 cm^3 = 0.5683 dm^3
	1 UK gallon	=	1.201 US gallon
		=	4.54609 dm^3 (liter)
	1 US gallon	=	0.833 UK gallon
		=	3.78541 dm^3 (liter)
	1 UK bu (bushel)	=	0.0363687 m^3 = 36.3687 dm^3
	1 US bushel	=	0.0352391 m^3 = 35.2391 dm^3
	1 in.^{3}	=	16.3871 cm^3
	1 ft^3	=	0.0283168 m^3
	1 yd^3	=	0.764555 m^3
	1 board foot (timber)	=	0.00235974 m^3 = 2.35974 dm^3
	1 cord (timber)	=	3.62456 m^3
Moment of inertia	1 lb ft^2	=	0.0421401 kg m^2
	1 slug ft^2	=	1.35582 kg m^2

Quantity	Unit	Conversion factor
Mass	1 grain	$=$ $0.0647989\,\text{g} = 64.7989\,\text{mg}$
	1 dram (avoir.)	$=$ $1.77185\,\text{g} = 0.00177185\,\text{kg}$
	1 drachm (apoth.)	$=$ $3.88793\,\text{g} = 0.00388793\,\text{kg}$
	1 ounce (troy or apoth.)	$=$ $31.1035\,\text{g} = 0.0311035\,\text{kg}$
	1 oz (avoir.)	$=$ $28.3495\,\text{g}$
	1 lb	$=$ $0.45359237\,\text{kg}$
	1 slug	$=$ $14.5939\,\text{kg}$
	1 sh cwt (US hundredweight)	$=$ $45.3592\,\text{kg}$
	1 cwt (UK hundredweight)	$=$ $50.8023\,\text{kg}$
	1 UK ton	$=$ $1016.05\,\text{kg}$
		$=$ $1.01605\,\text{tonne}$
	1 short ton	$=$ $2000\,\text{lb}$
		$=$ $907.185\,\text{kg}$
		$=$ $0.907\,\text{tonne}$
Mass per unit length	1 lb/yd	$=$ $0.496055\,\text{kg/m}$
	1 UK ton/mile	$=$ $0.631342\,\text{kg/m}$
	1 Uk ton/1000 yd	$=$ $1.11116\,\text{kg/m}$
	1 oz/in.	$=$ $1.11612\,\text{kg/m} = 11.1612\,\text{g/cm}$
	1 lb/ft	$=$ $1.48816\,\text{kg/m}$
	1 lb/in.	$=$ $17.8580\,\text{kg/m}$
Mass per unit area	1 lb/acre	$=$ $0.112085\,\text{g/m}^2 = 1.12085 \times 10^{-4}\,\text{kg/m}^2$
	1 UK cwt/acre	$=$ $0.0125535\,\text{kg/m}^2$
	1 oz/yd^2	$=$ $0.0339057\,\text{kg/m}^2$
	1 UK ton/acre	$=$ $0.251071\,\text{kg/m}^2$
	1 oz/ft^2	$=$ $0.305152\,\text{kg/m}^2$
	1 lb/ft^2	$=$ $4.88243\,\text{kg/m}^2$
	1 lb/in.^{2}	$=$ $703.070\,\text{kg/m}^2$
	1 UK ton/mile^2	$=$ $0.392298\,\text{g/m}^2 = 3.92298 \times 10^{-4}\,\text{kg/m}^2$
Density	1 lb/ft^3	$=$ $16.0185\,\text{kg/m}^3$
	1 lb/UK gal	$=$ $99.7763\,\text{kg/m}^3 = 0.09978\,\text{kg/l}$
	1 lb/US gal	$=$ $119.826\,\text{kg/m}^3 = 0.1198\,\text{kg/l}$
	1 slug/ft^3	$=$ $515.379\,\text{kg/m}^3$
	1 ton/yd^3	$= 1328.94\,\text{kg/m}^3 = 1.32894\,\text{tonne/m}^3$
	1 lb/in.^{3}	$=$ $27.6799\,\text{Mg/m}^3 = 27.6799\,\text{g/cm}^3$
Specific volume	1 in.^{3}/lb	$=$ $36.1273\,\text{cm}^3\text{/kg}$
	1 ft^3/lb	$=$ $0.0624280\,\text{m}^3\text{/kg} = 62.4280\,\text{dm}^3\text{/kg}$
Velocity	1 in./min	$=$ $0.42333\,\text{cm/s}$
	1 ft/min	$=$ $0.00508\,\text{m/s} = 0.3048\,\text{m/min}$
	1 ft/s	$=$ $0.3048\,\text{m/s} = 1.09728\,\text{km/h}$
	1 mile/h	$=$ $1.60934\,\text{km/h} = 0.44704\,\text{m/s}$
	1 UK knot	$=$ $1.85318\,\text{km/h} = 0.514773\,\text{m/s}$
	1 International knot	$=$ $1.852\,\text{km/h} = 0.514444\,\text{m/s}$
Acceleration	1 ft/s^2	$=$ $0.3048\,\text{m/s}^2$
Mass flow rate	1 lb/h	$=$ $0.125998\,\text{g/s} = 1.25998 \times 10^{-4}\,\text{kg/s}$
	1 UK ton/h	$=$ $0.282235\,\text{kg/s}$

Quantity	Unit	Conversion factor
Force or weight	1 dyne	$= \quad 10^{-5}\,\text{N}$
	1 pdl (poundal)	$= \quad 0.138255\,\text{N}$
	1 ozf (ounce)	$= \quad 0.278014\,\text{N}$
	1 lbf	$= \quad 4.44822\,\text{N}$
	1 kgf	$= \quad 9.80665\,\text{N}$
	1 tonf	$= \quad 9.96402\,\text{kN}$
Force or weight per unit length	1 lbf/ft	$= \quad 14.5939\,\text{N/m}$
	1 lbf/in.	$= \quad 175.127\,\text{N/m} = 0.175127\,\text{N/mm}$
	1 tonf/ft	$= \quad 32.6903\,\text{kN/m}$
Force (weight) per unit area or pressure or stress	$1\,\text{pdl/ft}^2$	$= \quad 1.48816\,\text{N/m}^2$ or Pa (pascal)
	$1\,\text{lbf/ft}^2$	$= \quad 47.8803\,\text{N/m}^2$
	1 mm Hg	$= \quad 133.322\,\text{N/m}^2$
	$1\,\text{in. } H_2O$	$= \quad 249.089\,\text{N/m}^2$
	$1\,\text{ft } H_2O$	$= \quad 2989.07\,\text{N/m}^2 = 0.0298907\,\text{bar}$
	1 in. Hg	$= \quad 3386.39\,\text{N/m}^2 = 0.0338639\,\text{bar}$
	$1\,\text{lbf/in.}^2$	$= \quad 6.89476\,\text{kN/m}^2 = 0.0689476\,\text{bar}$
	1 bar	$= \quad 10^5\,\text{N/m}^2$
	1 std. atmos.	$= \quad 101.325\,\text{kN/m}^2 = 1.01325\,\text{bar}$
	$1\,\text{tonf/ft}^2$	$= \quad 107.252\,\text{kN/m}^2$
	$1\,\text{tonf/in.}^2$	$= \quad 15.4443\,\text{MN/m}^2 = 1.54443\,\text{hectobar}$
Specific weight	$1\,\text{lbf/ft}^3$	$= \quad 157.088\,\text{N/m}^3$
	1 lbf/UK gal	$= \quad 978.471\,\text{N/m}^3$
	$1\,\text{tonf/yd}^3$	$= \quad 13.0324\,\text{kN/m}^3$
	$1\,\text{lbf/in.}^3$	$= \quad 271.447\,\text{kN/m}^3$
Moment, torque or couple	1 ozf in. (ounce-force inch)	$= \quad 0.00706155\,\text{N m}$
	1 pdl ft	$= \quad 0.0421401\,\text{N m}$
	1 lbf in.	$= \quad 0.112985\,\text{N m}$
	1 lbf ft	$= \quad 1.35582\,\text{N m}$
	1 tonf ft	$= \quad 3037.03\,\text{N m} = 3.03703\,\text{kN m}$
Energy or Heat or Work	1 erg	$= \quad 10^{-7}\,\text{J}$
	1 hp h (horsepower hour)	$= \quad 2.68452\,\text{MJ}$
	$1\,\text{thermie} = 10^6\,\text{cal}_{IT}$	$= \quad 4.1855\,\text{MJ}$
	1 therm $\quad = 100000\,\text{Btu}$	$= \quad 105.506\,\text{MJ}$
	$1\,\text{cal}_{IT}$	$= \quad 4.1868\,\text{J}$
	1 Btu	$= \quad 1.05506\,\text{kJ}$
	1 kWh	$= \quad 3.6\,\text{MJ}$
Power	1 hp $\quad = 550\,\text{ft lbf/s}$	$= \quad 0.745700\,\text{kW}$
	1 metric horsepower (ch, PS)	$= \quad 735.499\,\text{W}$
Specific heat capacity	1 Btu/lb deg F	
	1 Chu/lb deg C	$= \quad 4.1868\,\text{kJ/kg K}$
	1 cal/g deg C	
Heat flow rate	1 Btu/h	$= \quad 0.293071\,\text{W}$
	1 kcal/h	$= \quad 1.163\,\text{W}$
	1 cal/s	$= \quad 4.1868\,\text{W}$

Quantity	Unit		Conversion factor
Intensity of heat flow rate	$1\,Btu/ft^2\,h$	=	$3.15459\,W/m^2$
Electric stress	$1\,kV/in.$	=	$0.0393701\,kV/mm$
Dynamic viscosity	$1\,lb/ft\,s$	=	14.8816 poise $= 1.48816\,kg/m\,s$
Kinematic viscosity	$1\,ft^2/s$	=	929.03 stokes $= 0.092903\,m^2/s$
Caloric value or	$1\,Btu/ft^3$	=	$0.0372589\,J/cm^3 = 37.2589\,kJ/m^3$
Specific enthalpy	$1\,Btu/lb$	=	$2.326\,kJ/kg$
	$1\,cal/g$	=	$4.1868\,J/g$
	$1\,kcal/m^3$	=	$4.1868\,kJ/m^3$
Specific entropy	$1\,Btu/lb\,°R$	=	$4.1868\,kJ/kg\,K$
Thermal conductivity	$1\,cal\,cm/cm^2\,s\,deg\,C$	=	$41.868\,W/m\,K$
	$1\,Btu\,ft/ft^2\,h\,deg\,F$	=	$1.73073\,W/m\,K$
Gas constant	$1\,ft\,lbf/lb\,°R$	=	$0.00538032\,kJ/kg\,K$
Plane angle	$1\,rad$ (radian)	=	$57.2958°$
	$1\,degree$	=	$0.0174533\,rad = 1.1111$ grade
	$1\,minute$	=	$2.90888 \times 10^{-4}\,rad = 0.0185$ grade
	$1\,second$	=	$4.84814 \times 10^{-6}\,rad = 0.0003$ grade
Radioactivity	$1\,Ci$ (curie)	=	$37\,GBq$ (becquerel)
	$1\,\mu Ci$	=	$37\,kBq$

Index

A

Absorbance, 397
Absorption, 397
Acidity, 113, 121
Acid neutralizing capacity, 119
Aggregation, statistical, 391
Algae
 chemostat, 326
 littoral, 315
 sediments, fossil, 364
 stream productivity, 343
 taxonomy, 400
Alkalinity, 117, 119, 121
Allen curve growth method, 198
Aluminum, 123
Anions, major, 99, 100, 119
Aquatic plants. *See* Macrophytes
Area, 10
Atomic absorption spectroscopy, 103
Azimuths, 2

B

Bacteria
 biomass, 271
 coliform, 370
 decomposition
  dissolved organic matter, 289
  particulate organic matter, 301
 enumeration, 271
 growth, 271
 nucleic acid synthesis, 274
 protein production, 279
 productivity, 277, 280
 respiration, 294
 taxonomy, 399
Base cations, 119
Basin morphometry, 6, 9, 65
Bathythermograph, 24
Bearings, 1, 2
Beer's Law, 398

Benthic fauna, 189, 209
 biomass, 196
 drift, 213
 emergence, 210
 littoral, 317
 P/B ratios, 202
 predation, 257
 productivity, 196
 sampling, 190, 210
 separation from sediments, 195
 taxonomy, 402, 408
Benthos. *See* Benthic fauna
Biomass/biovolume
 bacteria, 271
 benthic fauna, 196
 phytoplankton, 148, 160
 zooplankton, 182
Bogs, 358
Breadth, 10

C

Carbon, inorganic, 113
 acidity, 121
 alkalinity, 117, 119
 bicarbonate, 121
 carbon dioxide, 117, 121
 carbonate, 117
 dissolved inorganic carbon (DIC), 117, 122
 dissolved organic carbon (DOC), 137
 gran titration, 120, 126
 hypolimnetic CO_2 accumulation, 373, 375
 particulate organic carbon (POC), 137, 141
 phytoplanktonic organic carbon, 162
 photosynthetic uptake, 224
Cations, major, 99, 100, 119, 122
Channel characteristics, 66
Chemical mass balance, stream, 342

Chemical relationships, 383
Chemostat, 326
Chloridc, 100
Chromatography
  ion, 104
  high-performance (HPLC), 105
Ciliate feeding rates, 245
Clearance rates, zooplankton, 241, 243
Coefficient of variation, 391
Color, 22
Conductance, specific, 100
Conversion factors, 421
Copepod life histories, 407
Coring samplers, 193, 210
Cover, 389
Crepuscular, 339
Current, 62

D
Decomposition
  dissolved organic matter, 289
  leaf fall in streams, 306
  macrophytes, 305
  particulate organic matter, 301
  phytoplankton, 303
Density, statistics, 389
Density, water, 35, 41
  currents, 39
  salinity, 41
  temperature, 35
Depth determination, 6
  maximum, 12
  mean, 12
  relative, 12
Depth-time diagrams, 26
Depth-volume curve, 13
Deviation, 390
Diatoms, fossil of sediments, 364
Diel, 339
Dissolved inorganic carbon (DIC), 113, 117, 122
Dissolved organic carbon (DOC). *See* Organic carbon, dissolved
Diurnal, 339
  lake changes, 349
  stream changes, 339
  zooplankton migrations, 349
Drainage area, 58, 66
Drift, stream benthos, 213

E
Ekman grab sampler, 190, 210
Electrodes, ion specific, 105
Extinction coefficient, light, 18
Extracellular organic carbon, 232

F
Filtration rates, zooplankton, 241
Fish
  population dynamics, 263
  predation, 257
Flame emission spectroscopy, 103
Flood plain, 60
Floods, 67
Flow cytometry, 159
Flow, discharge, 62
Frequency, 389
Fungi
  biomass, 307
  growth rates, 307
  productivity, 307
  taxonomy, 400

G
Geographic Information System, 5
Geography
  poles, 2
  public lands, 3
Gran titration, 120, 126
Ground water, 57

H
Hardness, 123
Heat
  advective exchange, 45, 48
  budget, 35, 45, 49, 342
  conductive exchange, 45, 48
  content, 45
  ice, 46, 50
  latent exchange, 45, 46
  radiation, 46
  sensible exchange, 45, 46
  snow, 46, 50
  storage, 45, 48, 342
Hess sampler, 210–211
Heteroflagellate feeding rates, 246
Heterotrophy, bacteria. *See* Bacteria
Hydrodynamics, 33
Hydrograph, stream, 63
Hypolimnetic CO_2 accumulation, 375
Hypsographic curve, 12

I
Ice, 50
Inorganic carbon, dissolved, 113
Instantaneous growth method, 197
Ion chromatography, 104
Ion electrodes, 105
Irradiance, 16
Island, 11, 13
Isopleths, 28

L

Lake models, 33
Lake mount, 11, 13
Lambert's Law, 397
Latitude, 1
Length, maximum, 9
Light, 15
   absorption, 18
   attenuation, 18
   color, 22
   extinction coefficient, 18
   irradiance, 16
   measurement, 16
   photosynthetically active radiation,
      16
   radiant flux, 16
   scattered, 18
   Secchi disc, 21
   spectral distributions, 19, 20
   transparency, 21
   turbidity, 21
   units of measurement, 15
Littoral zone, 313
   algae, 315
   benthic fauna, 317
   macrophytes, 314
   phytoplankton, 316
   zooplankton, 316
Longitude, 1

M

Macrophytes
   biomass, 314
   decomposition, 305
   distribution, 314
   emergent plants, 313
   epiphytes, 313
   productivity, 314
   submersed plants, 313
   taxonomy, 401
Maps, 1
   bathymetric, 1, 9
   depth contours, 6
   geographic information system, 5
   land subdivisions, 3
   public lands, 3
   ranges, 4
   townships, 4
Mean, 389
Meromictic lakes,
   characteristics, 357
   hydrodynamics, 39
Microcosms, 333
Microflagellates, 181
Microscopy

   calibration, 152
   epifluorescence, 272
   inverted, 151
Mineralization. *See* Decomposition
Mixing, thermal resistance to, 36
Model ecosystems, 325
Molar solutions, 385
Molecular weights, 383
Morphology, 57, 60, 66
Morphometry, 1, 57

N

Nanoflagellates, 181
Nitrogen compounds, 87
   ammonium nitrogen, 87
   nitrate nitrogen, 88
   nitrite nitrogen, 88
   organic nitrogen, 92
Nitrogen fixation, 93
Normal solutions, 384
Nutrients
   algal, 326
   enrichment, 320, 330
   inorganic, 85
   stream budgets, 342

O

Organic carbon, 137
   algae, 162
   decomposition, 289, 301
   dissolved (DOC), 138
   extracellular, phytoplankton, 232
   particulate (POC), 141
   sediments, 362
Organic matter, 137
Oxbow lakes, 60
Oxygen deficits, 373
Oxygen, dissolved, 73
   deficits, 373
   percentage saturation, 79
   productivity, phytoplankton, 221
   solubility, 80
   spectrophotometric method, 77
   Winkler method, 74

P

Paleolimnology, 361
Palmer-Maloney counting cells, 157
Particulate organic matter
   analysis, 141
   decomposition, 301
P/B ratios, 202
Petersen grab sampler, 192
Petersen population model, 263, 265
pH, 114

pH. (*cont.*)
  measurement, 115
Phosphorus, 93
  organic phosphorus, 96
  sediments, 362
  soluble reactive phosphate, 94
  total phosphorus, 97
Photometers, 19
Phytoplankton
  biomass of species, 148, 160
  biovolume, 149, 160
  counting cells, 157
  decomposition, 303
  displacement method, 159
  enumeration, 148, 153
  filtration techniques, 158
  flow cytometry, 159
  littoral, 316
  mineralization, 303
  organic carbon, 162
  photosynthesis, 219
  picophytoplankton, 156
  pigments, 163
  preservation, 148
  productivity, 219
  sampling, 147
  sedimentation techniques, 149
  volumes, 160
Pigments
  concentrations, 163
  determination, 163, 318
  extraction, 165
  sample preparation, 164
  sedimentary, 363
Planimetry, 10
Plasma emission spectroscopy, 104
Pollen, fossil of sediments, 364
Pollution, streams, 369
Ponar grab sampler, 193
Population models, zooplankton, 252
Predator-prey interactions, 257
Productivity
  bacteria, 277, 280
  benthic fauna, 196
  fungi, 307
  hypolimnetic CO_2 accumulation, 375
  macrophytes, 314
  oxygen deficits, 373
  P/B ratios, 202
  phytoplankton, 219
  whole lake estimates, 373
  zooplankton, 251
Protists, 181, 245
Protistan feeding rates, 245

Protozoans, 181, 245
  taxonomy, 418
Pyrheliometer, 16

R
Randomness, 391
Range, 390
Redfield molecule, 373
Relative depth, 12
Removal-summation productivity method, 197
Reservoirs, 355
Residues, total dissolved, 102
Respiration, bacteria, 294

S
Salinity, major ions, 99
Sampling efficiency, 394
Schnabel population method, 264
Schumacher-Eschmeyer population method, 264
Secchi disc, 21
Sedgwick-Rafter cells, 180
Sediments
  chemical analyses, 362
  decomposition, 296
  pigment degradation products, 363
  stream, 63
Seiches, 37
  internal, 37
Sewage outfall effects, 369
Shore line
  defined, 12
  determination, 6
  development, 12
Silica, 97
Size-frequency productivity method, 200
Slope, 12
Snow, 50
Solar radiation, 15
Solutions
  molar, 385
  normal, 384
  standard, 385
Spectrophotometry, 397
Stability, resistance to mixing, 37
Standard deviation, 390
Standard error, 390
Statistical analyses, 389
Streams
  benthic fauna, 209
  current, 62
  decomposition of leaf fall, 306
  diurnal changes, 339

drainage area, 58
ecosystem metabolism, 343
flooding, 64, 67
flood plain, 60
flow, discharge, 57, 62
hydrograph, 63
morphology, 57, 60, 66
orders, 58
pollution, 369
sediments, 63
sewage outfall effects, 369
Sulfate, 99
Surber sampler, 210

T
Temperate lake, hydrodynamics, 33, 41
Temperature, 15, 23, 435
density, 35
heat budget, 35, 45
heat content, 45
measurement, 23
Thermal resistance to mixing, 36
Transmittance, 397
Transparency, 21
Tropical lake, hydrodynamics, 38
Turbidity, 21

U
Units
conversion factors, 421
measurement, 386

V
Vadose water, 57
Valence, 383
Variance, 390
Volume, 11

W
Water table, 57
Wetlands, 313
Width, 10

Z
Zooplankton
biovolume, 183, 184
clearance rate, 241
copepod life histories, 407
distribution, 175
enumeration, 179
feeding, 241
filtering rate, 241
grazing rate, 241
littoral zone, 316
microflagellates, 181
migrations, 349
preservation, 179
production, 251
protists, 181
protozoa, 181
sampling, 176, 184
taxonomy, 418